Enhanced Heat Transfer Mechanism of Nanofluid MQL Cooling Grinding

Changhe Li
Qingdao University of Technology, China

Hafiz Muhammad Ali
King Fahd University of Petroleum and Minerals, Dhahran, Saudi Arabia

A volume in the Advances in Chemical and Materials Engineering (ACME) Book Series

IGI Global
DISSEMINATOR OF KNOWLEDGE

Published in the United States of America by
IGI Global
Engineering Science Reference (an imprint of IGI Global)
701 E. Chocolate Avenue
Hershey PA, USA 17033
Tel: 717-533-8845
Fax: 717-533-8661
E-mail: cust@igi-global.com
Web site: http://www.igi-global.com

Library of Congress Cataloging-in-Publication Data

Names: Li, Changhe, 1966- author. | Ali, Hafiz Muhammad, 1981- author.
Title: Enhanced heat transfer mechanism of nanofluid MQL cooling grinding / by Changhe Li and Hafiz Muhammad Ali.
Description: Hershey, PA : Engineering Science Reference, 2020. | Includes bibliographical references and index. | Summary: ""This book examines the distribution of airflow field in the grinding zone in nanofluid minimum quantity lubrication cooling grinding; tribological performance in grinding wheel/work piece interface; thermodynamic mechanism in grinding wheel/work piece interface; and surface integrity evaluation"--Provided by publisher"-- Provided by publisher.
Identifiers: LCCN 2019030386 (print) | LCCN 2019030387 (ebook) | ISBN 9781799815464 (h/c) | ISBN 9781799815471 (s/c) | ISBN 9781799815488 (eISBN)
Subjects: LCSH: Grinding wheels. | Lubrication and lubricants. | Nanofluids.
Classification: LCC TJ1293 .L5 2020 (print) | LCC TJ1293 (ebook) | DDC 660/.28422--dc23
LC record available at https://lccn.loc.gov/2019030386
LC ebook record available at https://lccn.loc.gov/2019030387

This book is published in the IGI Global book series Advances in Chemical and Materials Engineering (ACME) (ISSN: 2327-5448; eISSN: 2327-5456)

British Cataloguing in Publication Data
A Cataloguing in Publication record for this book is available from the British Library.

For electronic access to this publication, please contact: eresources@igi-global.com.

Advances in Chemical and Materials Engineering (ACME) Book Series

J. Paulo Davim
University of Aveiro, Portugal

ISSN:2327-5448
EISSN:2327-5456

Mission

The cross disciplinary approach of chemical and materials engineering is rapidly growing as it applies to the study of educational, scientific and industrial research activities by solving complex chemical problems using computational techniques and statistical methods.

The **Advances in Chemical and Materials Engineering (ACME) Book Series** provides research on the recent advances throughout computational and statistical methods of analysis and modeling. This series brings together collaboration between chemists, engineers, statisticians, and computer scientists and offers a wealth of knowledge and useful tools to academics, practitioners, and professionals through high quality publications.

Coverage

- Metallic Alloys
- Coatings and surface treatments
- Thermo-Chemical Treatments
- Artificial Intelligence Methods
- Ceramics and glasses
- Sustainable Materials
- Computational methods
- Electrochemical and Corrosion
- Chemical Engineering
- Numerical Techniques

IGI Global is currently accepting manuscripts for publication within this series. To submit a proposal for a volume in this series, please contact our Acquisition Editors at acquisitions@igi-global.com or visit: https://www.igi-global.com/publish/.

The Advances in Chemical and Materials Engineering (ACME) Book Series (ISSN 2327-5448) is published by IGI Global, 701 E. Chocolate Avenue, Hershey, PA 17033-1240, USA, www.igi-global.com. This series is composed of titles available for purchase individually; each title is edited to be contextually exclusive from any other title within the series. For pricing and ordering information please visit https://www.igi-global.com/book-series/advances-chemical-materials-engineering/73687. Postmaster: Send all address changes to above address. ©

Titles in this Series

For a list of additional titles in this series, please visit: www.igi-global.com/book-series

Applications and Techniques for Experimental Stress Analysis
Karthik Selva Kumar Karuppasamy (Indian Institute of Technology, Guwahati, India) and Balaji P.S. (National Institute of Technology, Rourkela, India)
Engineering Science Reference • © 2020 • 350pp • H/C (ISBN: 9781799816904) • US $245.00

Design of Experiments for Chemical, Pharmaceutical, Food, and Industrial Applications
Eugenia Gabriela Carrillo-Cedillo (Universidad Autónoma de Baja California, Mexico) José Antonio Rodríguez-Avila (Universidad Autónoma del Estado de Hidalgo, Mexico) Karina Cecilia Arredondo-Soto (Universidad Autónoma de Baja California, Mexico) and José Manuel Cornejo-Bravo (Universidad Autónoma de Baja California Mexico)
Engineering Science Reference • © 2020 • 429pp • H/C (ISBN: 9781799815181) • US $255.00

Handbook of Research on Developments and Trends in Industrial and Materials Engineering
Prasanta Sahoo (Jadavpur University, India)
Engineering Science Reference • © 2020 • 370pp • H/C (ISBN: 9781799818311) • US $295.00

Nanocomposites for the Desulfurization of Fuels
Tawfik Abdo Saleh (King Fahd University of Petroleum and Minerals, Saudi Arabia)
Engineering Science Reference • © 2020 • 358pp • H/C (ISBN: 9781799821465) • US $225.00

Advanced Catalysis Processes in Petrochemicals and Petroleum Refining Emerging Research and Opportunities
Mohammed C. Al-Kinany (King Abdulaziz City for Science and Technology, Saudi Arabia) and Saud A. Aldrees (King Abdulaziz City for Science and Technology, Saudi Arabia)
Engineering Science Reference • © 2020 • 257pp • H/C (ISBN: 9781522580331) • US $195.00

Nanotechnology in Aerospace and Structural Mechanics
Noureddine Ramdani (Research and Development Institute of Industry and Defense Technologies, Algeria & Harbin Engineering University, China)
Engineering Science Reference • © 2019 • 356pp • H/C (ISBN: 9781522579212) • US $265.00

Polymer Nanocomposites for Advanced Engineering and Military Applications
Noureddine Ramdani (Research and Development Institute of Industry and Defense Technologies, Algeria & Harbin Engineering University, China)
Engineering Science Reference • © 2019 • 435pp • H/C (ISBN: 9781522578383) • US $225.00

IGI Global
DISSEMINATOR OF KNOWLEDGE

701 East Chocolate Avenue, Hershey, PA 17033, USA
Tel: 717-533-8845 x100 • Fax: 717-533-8661
E-Mail: cust@igi-global.com • www.igi-global.com

Table of Contents

Preface

Green development is a general international trend. Resource and environmental problem being a common challenge faced by human beings, sustainable development has gradually become a global consensus. Especially under the background of international financial crisis and climatic change, promoting green growth and implementing Green New Deal is the common choice for global major economies, and developing green economy and racing to control the commanding height in the future global competition has become an important national strategy. Developed countries have successively implemented the reindustrialization strategy to reshape new competitive advantages of their manufacturing industries, so the influence of green concepts, policies and regulations like clean, high-efficiency, low-carbon and recycling is continuously enhanced, energy and resource utilization efficiency has become an important factor measuring competitiveness of the manufacturing industry in a country and green trade barrier has also become an important means for some countries to seek for competitive advantages (Bock et al., 2001). Green manufacturing is the only route which must be passed by industrial transformation and upgrading. Led by implementation of green manufacturing projects, comprehensively carrying out green manufacturing can not only exert an important realistic effect on relieving the current bottlenecks and constraints of resources and environment and accelerating cultivation of new economic growth points but also it is of far-reaching historical significance to accelerating transformation of the economic development mode, promoting industrial transformation and upgrading and improving international competitiveness of the manufacturing industry (Sharif et al., 2017; Dahmus et al., 2004).

THE CHALLENGES

The flood grinding with mass-flow grinding fluid supply not only increases the treatment cost of grinding fluid but also wastes grinding fluid in quantity (Adler et al. 2006). In full consideration of factors like environmental protection, low carbon, processing quality and economic efficiency, flood grinding process is no longer adaptive to the concepts of green manufacturing and sustainable development (Chetan et al. 2015; Alves et al. 2008). Therefore, it's necessary to develop green, environmentally friendly, high-efficiency and low-consumption grinding fluid or develop new low-consumption, high-efficiency and sustainable green grinding technologies, which can not only reach the technological effect of flood cooling & lubrication but also can ensure machining precision and surface quality of workpiece, lengthen service life of grinding wheel and reduce grinding fluid-induced environmental pollution (Debnath et al., 2014; Baumann et al., 1999).

Minimum quantity lubrication (MQL) grinding is a burgeoning green processing technique (Dogra et al., 2018; Sharma et al., 2016; Zhang et al., 2015). After mixing and atomization, a small quantity of lubricants and gas is jetted at a certain pressure to the grinding zone for cooling lubrication (Babic et al., 2005; Mia et al., 2018). However, the MQL technique has not solved the technical bottleneck of grinding cooling performance according to (Hadad & Sadeghi, 2012; Hadad et al., 2012). This inability to do so greatly limits the application of MQL technique. The high-pressure gas imported to the grinding zone does not have an effective cooling effect may lead to heat accumulation on the workpiece surface, which reduces the surface integrity of the grinding workpiece and shortens the service life of the grinding wheel or even destroys it according to (Ding et al., 2013, 2015).

The key green transformation task for the traditional manufacturing industry is to implement clean transformation of the production process. Taking cutting down pollutant generation from sources as the starting point, the government should reform traditional manufacturing technological equipment and encourage enterprises to adopt advanced and applicable cleaner production technologies to implement upgrading and transformation, and green transformation of fundamental manufacturing technologies and energetically promote green processing technologies without or with few cutting fluids, etc. (Tawakoli et al., 2011).

SEARCHING FOR A SOLUTION

The weak heat transfer effect of MQL encouraged strong research efforts to increase its capacity. Significant improvements can be achieved by introducing nanoparticles into the MQL oil (Sharma et al., 2015; Shen et al., 2008a, 2008b). The burgeoning nanofluids minimum quantity lubrication (NMQL) technique resolves heat transfer in the grinding zone effectively while enhancing the lubricating properties in the zone (Li et al., 2017; Guo et al., 2017).

Nanofluids minimum quantity lubrication (NMQL) is a new-type high-efficiency, low-consumption, clean and low-carbon precision machining mode proposed based on the above background and theory of heat transfer enhancement (Zhang et al., 2015; Wang et al., 2016). The concrete method refers to adding a certain amount of nanoscale solid particles into degradable minimum quantity lubrication (MQL) to prepare nanofluids which is then atomized through high-pressure gas and sprayed into the grinding area in the form of jet flow (Barczak et al., 2010; Tasdelen et al., 2008). High-pressure gas exerts effects of cooling, dust removal and lubricant conveying while MQL exerts lubricating and anti-friction effects; under the same particle volume content, surficial area and heat capacity of nanoscale solid particles are far larger than those of millimeter or micron-order solid particles, so heat conductivity of nanofluids will be substantially enhanced (Rabiei et al., 2015; Amrita et al., 2014). Nanoparticles enhance heat transfer capacity of the fluid in the grinding area and exert effects of enhancing heat transfer and lowering temperature of the grinding area (Sadeghi et al., 2010; Emami et al., 2013). Further study results indicate that nanoparticles have superb anti-wear and anti-friction characteristics and high bearing capacity, they can further improve lubricating and anti-friction tribological properties. NMQL grinding method with nanoparticles participating in the heat transfer enhancement enhances heat transfer capacity and lubricating property of MQL grinding to the greatest extent, thus solving the technical bottleneck namely MQL heat transfer capacity is insufficient and opening a new path for application of MQL to grinding process.

ORGANIZATION OF THE BOOK

The book is organized into 18 chapters. A brief description of each of the chapters follows:

Chapter 1 introduces the application background and characteristics of five kinds of grinding processing methods, briefly describes the enhanced heat transfer mechanism and tribological properties of nanofluids, and points out that nanofluids minimum quantity lubrication (NMQL) solves the technical bottleneck namely minimum quantity lubrication (MQL) heat transfer capacity is insufficient and opening a new path for application of MQL to grinding process. The current status of exploratory research on the mechanism of MQL grinding using nanofluids as cooling lubricants is analyzed. The research characteristics of the new green NMQL grinding technology is described, and puts forward some key problems such as the heat transfer enhancement process of NMQL, the anti-friction and anti-wear tribological mechanism of nanoparticles and the controlled transport strategies of minimal quantity of lubricating droplets. It will be of great scientific significance and pragmatic value to perfecting NMQL grinding technical system.

Chapter 2 provides an insight into the enhancing heat transfer mechanisms of nanofluids. This chapter will discuss in details the various aspects of nanofluids form preparation of nanofluids, characterization and flow and energy transportation mechanisms of nanofluids. Preparation of nanofluids has been performed by using various methods where one step and two step methods are widely known. These methods are discussed in detail for various nanoparticles with stabilizing agents for stable production of nanofluids. Further, the characterization techniques of various aspects of nanofluids including thermal conductivity and factors influencing the thermal conductivity of nanofluids are introduced in depth. Chapter 2 also sheds light on the experimental analysis of flow and heat transfer of nanofluids, natural convection analysis of nanofluids and boiling heat transfer of nanofluids.

Chapter 3 investigates the machining mechanism related to the grinding of NMQL and the three mechanisms of sliding, ploughing and cutting of the abrasive grinding process. A mathematical model of the micro-grinding force is established, and the micro-grinding force can be used to express the force of the grinding zone more accurately. The grinding force component is further divided into the force of cutting and sliding, removing the constant cutting force during the grinding process, retaining the varying sliding friction force, and determining the sliding friction coefficient to characterize the lubrication performance. It reflects the influence of different lubrication conditions on the friction part. The methods for measuring the temperature during grinding are introduced, including direct contact temperature measurement and non-contact temperature measurement. At present, thermocouple temperature measurement is a more commonly used and more accurate temperature measurement method, and the characteristics of each temperature measurement method are analyzed.

Chapter 4 studies the influence degree of temperature on the workpiece, analyzes the heat transfer at the grinding interface, regards the heat effect at the grinding interface as the moving heat source of the workpiece surface, and studies the influence of the heat source on the heat transfer in the form of gradual by point, line and surface on the theoretical foundation in Chapter 3. According to the theory of convective heat transfer on workpiece surface, the influence of heat transfer on workpiece and energy proportion coefficient are studied, which provides theoretical basis for the heat transfer of grinding temperature field model in the finite element simulation software.

Chapter 5 establishes the mathematical model of finite element simulation of temperature field of high efficiency deep grinding under four kinds of cooling lubrication conditions, and the three boundary conditions and the constraints of simulation model are established, and the mesh division and time step algorithm are determined respectively. Using ABAQUS simulation platform and theoretical model to simulate grinding temperature field, the distribution characteristics of grinding temperature field under different working conditions are analyzed from different directions, different grinding depths and different workpiece materials.

Chapter 6 analyzes the cooling and lubricating properties of different cooling and lubricating conditions. The influence mechanism of NMQL on cooling and lubricating effect was revealed with different nanoparticles (MoS_2, CNT, ZrO_2) and different volume concentrations of MoS_2 nanofluids. The experimental results showed that the temperature rise and grinding force obtained under NMQL grinding were the closest to the flood grinding. The specific grinding energy of MoS_2 nanofluids was the lowest. When the volume concentration was 2%, the best cooling and lubricating effect was obtained.

Chapter 7 compares the performances of MQL grinding using castor oil, soybean oil, rapeseed oil, corn oil, sunflower oil, peanut oil, and palm oil as base oils. Results revealed that 1) castor oil-based MQL grinding yields the lowest grinding force but exhibits the highest grinding temperature and energy ratio coefficient; 2) palm oil-based MQL grinding generates the second lowest grinding force but shows the lowest grinding temperature and energy ratio coefficient; 3) viscosity significantly influences grinding force and grinding temperature to a greater extent than fatty acid varieties and contents in vegetable oils.

Chapter 8 studies the influence of NMQL on grinding temperature. Six kinds of nanoparticles were selected to carry out the experimental research, namely MoS_2, ZrO_2, CNT, polycrystalline diamond, Al_2O_3 and SiO_2, which were separately mixed with palm oil to prepare nanofluid. A commonly used nickel base alloy is selected as the workpiece material. The lowest grinding temperature, the related energy proportion coefficient is analyzed that. In addition, the physical properties of nanofluids, such as thermal conductivity, viscosity, surface tension and contact state (contact angle) between droplets and workpiece surface, are discussed to reveal the effect of nanofluids on cooling performance. Based on the boundary layer theory, the mathematical model of convective heat transfer coefficient is established. The model calculation shows that CNT nanofluids have the highest heat transfer coefficient (1.3 × 104w / (MK), which to some extent explains why the grinding temperature of CNT nanofluids MQL is the lowest.

Chapter 9 investigates grinding force, grinding temperature and proportionality coefficient of energy transferred to workpiece of nanofluids with different volume fractions. Based on the analysis of the influence of physical characteristics of nanofluids on experimental results, cooling & heat transfer mechanism of NMQL grinding is studied. The experimental study can provide a certain technical guidance for industrial machining, thus being of great referential significance.

Chapter 10 investigates grinding temperatures obtained through MQL grinding with different concentrations of vegetable oils-based nanofluids as well as their heat transfer mechanism through simulation and experiment.

Chapter 11 studies the interface friction characteristics of grinding wheel / workpiece using seven typical vegetable oils (soybean, peanut, corn, rapeseed, palm oil, castor oil, sunflower oil) as base oils. The results showed that peanut oil, sunflower oil, soybean oil, castor oil and palm oil with higher saturated fatty acid content and palm oil with higher palmitic acid content were all suitable for lubricating oil.

Chapter 12 uses six nanofluids, namely, MoS_2, SiO_2, diamond, CNTs, Al_2O_3, and ZrO_2 as MQL grinding fluids to select the kind of nanoparticles with optimum lubrication performance in grinding nickel alloy GH4169. Experimental results concluded the followings: 1) Nanoparticles with spherical or sphere-like molecular structure and nanofluids with high viscosity demonstrate superior lubrication performances. 2) The polishing effect of nanodiamond particles enhances their surface morphology. 3)

The lubricating property of the six nanofluids is described in the following order: ZrO_2 < CNTs < ND < MoS_2 < SiO_2 < Al_2O_3.

Chapter 13 presents the lubricating properties of different vegetable-oil-based nanofluids through a comparative evaluation between frictional test and grinding experiment. The first experiment aimed to prejudge the lubricating properties of different nanofluids with a frictional test, which simulated the interface state of grinding between the abrasive grains and the workpiece. The second aimed to test and verify the lubricating properties of the same nanofluids through a grinding experiment. The mechanism of oil-film formation of nanofluids in the grinding zone was analyzed by morphology and element analysis of the worn surface. The experimental results show that Al_2O_3 nanofluids have the best tribological properties. Compared with pure base oil, the friction coefficient is reduced by 20%, and the optimal friction surface morphology is obtained. The good anti friction and anti-wear properties of nanofluids are attributed to the formation of the protective oil film formed by chemical reaction on the surface.

Chapter 14 studies the tribological properties of grinding wheel / workpiece interface with different concentration of Al_2O_3 NMQL. The influences of the force ratio, viscosity and contact angle of Al_2O_3 nanofluids with different concentrations on the grinding force and the surface quality of workpieces are discussed. The best concentration of Al_2O_3 nanofluid with good lubrication performance in grinding zone was obtained.

Chapter 15 uses analysis of signal to noise ratio(S/N), variance, microstructure and morphology analysis to study the influence of different typical nanofluid lubricants on the grinding performance of different materials. Results show that, the optimal grinding parameters are using nodular cast iron and MoS_2 nanofluid. Compared with other lubricants, MoS_2 nanofluid can exert a significant effect on reducing wear of grinding wheel. In order to reduce surface roughness, MQL lubricating oils of the following sequence can be used: Al_2O_3 nanofluid, MoS_2 nanofluid and pure palm oil.

Chapter 16 establishes a kinematics model, elastic deformation model, and plastic accumulation model of a single grinding wheel, simulates the grain distribution on the surface of the common grinding wheel by using the grain vibration method and examines the effect of different grinding parameters on the surface topography of the workpiece. The experiments results show that the simulation method can predict the surface roughness of the workpiece, which is a factor in selecting the grinding parameters.

Chapter 17 establishes kinematic model of multiple grains, simulated the grain distribution on the surface of the engineered grinding wheel by using the grain vibration method, and examined the effect of different grinding parameters on the surface topography of the workpiece.

Chapter 18 establishes the mathematical model of the common grinding wheel by using the abrasive vibration method to realize the parameterization of the simulated grinding wheel, to simulate the surface grinding process of the common grinding wheel and evaluate the surface topography of the workpiece. The grinding wheel body with random arrangement of abrasive grains is generated, and the simulation of common grinding wheel morphology is completed. Then, according to the grinding kinematics model, elastic deformation model and plastic accumulation model, the simulation of the process is realized, and the workpiece surface morphology matrix is generated. Finally, the influence of grinding parameters and grinding wheel parameters on the workpiece surface morphology and surface roughness is studied by calculating the surface roughness value.

Changhe Li
Qingdao University of Technology, China

Hafiz Muhammad Ali
King Fahd University of Petroleum and Minerals, Dhahran, Saudi Arabia

REFERENCES

Adler, D. P., Hii, W. S., Michalek, D. J., & Sutherland, J. W. (2006). Examining the role of cutting fluids in machining and efforts to address associated environmental/health concerns. *Machining Science and Technology, 10*(1), 23–58. doi:10.1080/10910340500534282

Alves, S. M., & de Oliveira, J. F. G. (2008). Vegetable based cutting fluid: an environmental alternative to grinding process. In *LCE 2008: 15th CIRP international conference on life cycle engineering: Conference proceedings* (p. 664). CIRP.

Amrita, M., Srikant, R. R., & Sitaramaraju, A. V. (2014). Performance evaluation of nanographite-based cutting fluid in machining process. *Materials and Manufacturing Processes, 29*(5), 600–605. doi:10.1080/10426914.2014.893060

Babic, D., Murray, D. B., & Torrance, A. A. (2005). Mist jet cooling of grinding processes. *International Journal of Machine Tools & Manufacture, 45*(10), 1171–1177. doi:10.1016/j.ijmachtools.2004.12.004

Barczak, L. M., Batako, A. D. L., & Morgan, M. N. (2010). A study of plane surface grinding under minimum quantity lubrication (MQL) conditions. *International Journal of Machine Tools & Manufacture, 50*(11), 977–985. doi:10.1016/j.ijmachtools.2010.07.005

Baumann, W., Grafen, M., & Pollklasner, D. (1999). Assessment of the environmental release of chemicals used in metal-cutting and forming fluids. Emission scenario document–metal extraction industry, refining and processing industry, IC8, Dortmund, Germany.

Bock, W. (2001). Lubricants and the environment: specifications of rapidly biodegradable lubricants. *Proceedings of the International Tribology Conference, Technical Academy Esslingen.*

Chetan, G., Ghosh, S., & Venkateswara Rao, P. (2015). Application of sustainable techniques in metal cutting for enhanced machinability: A review. *Journal of Cleaner Production, 100*, 17–34. doi:10.1016/j.jclepro.2015.03.039

Dahmus, J. B., & Gutowski, T. G. (2004). An environmental analysis of machining. In *ASME 2004 international mechanical engineering congress and exposition* (pp. 643–652). American Society of Mechanical Engineers.

Debnath, S., Reddy, M. M., & Yi, Q. S. (2014). Environmental friendly cutting fluids and cooling techniques in machining: A review. *Journal of Cleaner Production, 83*, 33–47. doi:10.1016/j.jclepro.2014.07.071

Ding, W. F., Xu, J. H., Chen, Z. Z., Yang, C. Y., Song, C. J., & Fu, Y. C. (2013). Fabrication and performance of porous metal-bonded CBN grinding wheels using alumina bubble particles as pore-forming agents. *International Journal of Advanced Manufacturing Technology, 67*(5-8), 1309–1315. doi:10.100700170-012-4567-4

Ding, W. F., Zhu, Y. J., Zhang, L. C., Xu, J. H., Fu, Y. C., Liu, W. D., & Yang, C. Y. (2015). Stress characteristics and fracture wear of brazed CBN grains in monolayer grinding wheels. *Wear*, *332*, 800–809. doi:10.1016/j.wear.2014.12.008

Dogra, M., Sharma, V. S., Dureja, J. S., & Gill, S. S. (2018). Environment-friendly technological advancements to enhance the sustainability in surface grinding-A review. *Journal of Cleaner Production*, *197*, 218–231. doi:10.1016/j.jclepro.2018.05.280

Emami, M., Sadeghi, M. H., & Sarhan, A. A. (2013). Investigating the effects of liquid atomization and delivery parameters of minimum quantity lubrication on the grinding process of Al2O3 engineering ceramics. *Journal of Manufacturing Processes*, *15*(3), 374–388. doi:10.1016/j.jmapro.2013.02.004

Guo, S. M., Li, C. H., Zhang, Y. B., Wang, Y. G., Li, B. K., Yang, M., ... Liu, G. T. (2017). Experimental evaluation of the lubrication performance of mixtures of castor oil with other vegetable oils in MQL grinding of nickel-based alloy. *Journal of Cleaner Production*, *140*(3), 1060–1076. doi:10.1016/j.jclepro.2016.10.073

Hadad, M.J., & Sadeghi, B. (2012). Thermal analysis of minimum quantity lubrication-MQL grinding process. *Int. J. Mach. Tools Manuf.*, *63*, 1-15.

Hadad, M.J., Tawakoli, T., Sadeghi, M.H., & Sadeghi, B. (2012). Temperature and energy partition in minimum quantity lubrication-MQL grinding process. *Int. J. Mach. Tools Manuf.*, *54*, 10-17.

Li, B. K., Li, C. H., Zhang, Y. B., Wang, Y. G., Jia, D. Z., Yang, M., ... Sun, K. (2017). Heat transfer performance of MQL grinding with different nanofluids for Ni-based alloys using vegetable oil. *Journal of Cleaner Production*, *154*, 1–11. doi:10.1016/j.jclepro.2017.03.213

Mia, M., Gupta, M. K., Singh, G., Krolczyk, G., & Pimenov, D. Y. (2018). An approach to cleaner production for machining hardened steel using different cooling-lubrication conditions. *Journal of Cleaner Production*, *187*, 1069–1081. doi:10.1016/j.jclepro.2018.03.279

Rabiei, F., Rahimi, A. R., Hadad, M. J., & Ashrafijou, M. (2015). Performance improvement of minimum quantity lubrication (MQL) technique in surface grinding by modeling and optimization. *Journal of Cleaner Production*, *86*, 447–460. doi:10.1016/j.jclepro.2014.08.045

Sadeghi, M. H., Hadad, M. J., Tawakoli, T., Vesali, A., & Emami, M. (2010). An investigation on surface grinding of AISI 4140 hardened steel using minimum quantity lubrication-MQL technique. *International Journal of Material Forming*, *3*(4), 241–251. doi:10.100712289-009-0678-3

Sharif, M. N., Pervaiz, S., & Deiab, I. (2017). Potential of alternative lubrication strategies for metal cutting processes: A review. *International Journal of Advanced Manufacturing Technology*, *89*(5-8), 2447–2479. doi:10.100700170-016-9298-5

Sharma, A. K., Tiwari, A. K., & Dixit, A. R. (2015). Improved machining performance with nanoparticle enriched cutting fluids under minimum quantity lubrication (MQL) technique: A review. *Materials Today: Proceedings*, *2*(4-5), 3545–3551.

Sharma, A. K., Tiwari, A. K., & Dixit, A. R. (2016). Effects of Minimum Quantity Lubrication (MQL) in machining processes using conventional and nanofluid based cutting fluids: A comprehensive review. *Journal of Cleaner Production*, *127*, 1–18. doi:10.1016/j.jclepro.2016.03.146

Shen, B., Malshe, A. P., Kalita, P., & Shih, A. J. (2008a). Performance of novel MoS2 nanoparticles based grinding fluids in minimum quantity lubrication grinding. *Trans. NAMRI/SME, 36*(357). 364.

Shen, B., Shih, A. J., & Tung, S. C. (2008b). Application of nanofluids in minimum quantity lubrication grinding. *Tribology Transactions, 51*(6), 730–737. doi:10.1080/10402000802071277

Tasdelen, B., Wikblom, T., & Ekered, S. (2008). Studies on minimum quantity lubrication (MQL) and air cooling at drilling. *Journal of Materials Processing Technology, 200*(1-3), 339–346. doi:10.1016/j.jmatprotec.2007.09.064

Tawakoli, T., Hadad, M., Sadeghi, M. H., Daneshi, A., & Sadeghi, B. (2011). Minimum quantity lubrication in grinding: Effects of abrasive and coolant–lubricant types. *Journal of Cleaner Production, 19*(17-18), 2088–2099. doi:10.1016/j.jclepro.2011.06.020

Wang, Y. G., Li, C. H., Zhang, Y. B., Yang, M., Zhang, X. P., Zhang, N. Q., & Dai, J. J. (2016). Experimental evaluation on tribological performance of the wheel/workpiece interface in MQL grinding with different concentrations of Al_2O_3 nanofluids. *Journal of Cleaner Production, 142*, 3571–3583. doi:10.1016/j.jclepro.2016.10.110

Zhang, D. K., Li, C. H., Zhang, Y. B., Jia, D. Z., & Zhang, X. W. (2015). Experimental research on the energy ratio coefficient and specific grinding energy in nanoparticle jet MQL grinding. *International Journal of Advanced Manufacturing Technology, 78*(5-8), 1275–1288. doi:10.100700170-014-6722-6

Zhang, Y. B., Li, C. H., Jia, D. Z., Zhang, D. K., & Zhang, X. W. (2015). Experimental evaluation of MoS_2 nanoparticles in jet MQL grinding with different types of vegetable oil as base oil. *Journal of Cleaner Production, 87*(1), 930–940. doi:10.1016/j.jclepro.2014.10.027

Acknowledgment

Chapter 1-5 of the book is arranged and compiled by Professor Hafiz Muhammad Ali, Chapter 6-9 is arranged and compiled by Min Yang, Chapter 10-12 is arranged and compiled by Wentao Wu, Chapter 13-15 is arranged and compiled by Dongzhou Jia, and Chapters 16-18 are arranged and compiled by Yanbin Zhang. The whole book was compiled and finalized by Professor Changhe Li. Otherwise, this book is financially supported by the following Foundation items: The National Natural Science Foundation of China (51575290 and 51975305), National Natural Science Foundation of China Youth Fund (51905289), Major Research Project of Shandong Province (2017GGX30135, 2018GGX103044, 2019GSF108236 and 2019GGX104040), Shandong Provincial Natural Science Foundation, China (ZR2019PEE008).

This book has received substantial support and assistance from many experts and colleagues in the compilation process with a reference to related literatures by many professors and experts. Hereby sincere gratitude is extended to them.

Restricted to professional level and available time of the editor, there must be omissions and improper parts in this book, which are expected to be criticized and corrected by readers.

Hafiz Muhammad Ali thankfully acknowledges the support of King Fahd University of Petroleum and Minerals, Dhahran, Saudi Arabia.

Introduction

Green development is a general international trend. Resource and environmental problem being a common challenge faced by human beings, sustainable development has gradually become a global consensus. Especially under the background of international financial crisis and climatic change, promoting green growth and implementing Green New Deal is the common choice for global major economies, and developing green economy and racing to control the commanding height in the future global competition has become an important national strategy. Developed countries have successively implemented the reindustrialization strategy to reshape new competitive advantages of their manufacturing industries, so the influence of green concepts, policies and regulations like clean, high-efficiency, low-carbon and recycling is continuously enhanced, energy and resource utilization efficiency has become an important factor measuring competitiveness of the manufacturing industry in a country and green trade barrier has also become an important means for some countries to seek for competitive advantages. Green manufacturing is the only route which must be passed by industrial transformation and upgrading. Led by implementation of green manufacturing projects, comprehensively carrying out green manufacturing can not only exert an important realistic effect on relieving the current bottlenecks and constraints of resources and environment and accelerating cultivation of new economic growth points but also it is of far-reaching historical significance to accelerating transformation of the economic development mode, promoting industrial transformation and upgrading and improving international competitiveness of the manufacturing industry.

The key green transformation task for the traditional manufacturing industry is to implement clean transformation of the production process. Taking cutting down pollutant generation from sources as the starting point, the government should reform traditional manufacturing technological equipment and encourage enterprises to adopt advanced and applicable cleaner production technologies to implement upgrading and transformation, and green transformation of fundamental manufacturing technologies and energetically promote green processing technologies without or with few cutting fluids, etc.

The flood grinding with mass-flow grinding fluid supply not only increases the treatment cost of grinding fluid but also wastes grinding fluid in quantity. In full consideration of factors like environmental protection, low carbon, processing quality and economic efficiency, flood grinding process is no longer adaptive to the concepts of green manufacturing and sustainable development. Therefore, it's necessary to develop green, environmentally friendly, high-efficiency and low-consumption grinding fluid or develop new low-consumption, high-efficiency and sustainable green grinding technologies, which can not only reach the technological effect of flood cooling & lubrication but also can ensure machining precision and surface quality of workpiece, lengthen service life of grinding wheel and reduce grinding fluid-induced environmental pollution.

Nanofluids minimum quantity lubrication (NMQL) is a new-type high-efficiency, low-consumption, clean and low-carbon precision machining mode proposed based on the above background and theory of heat transfer enhancement. The concrete method refers to adding a certain amount of nanoscale solid particles into degradable minimum quantity lubrication (MQL) to prepare nanofluids which is then atomized through high-pressure gas and sprayed into the grinding area in the form of jet flow. High-pressure gas exerts effects of cooling, dust removal and lubricant conveying while MQL exerts lubricating and anti-friction effects; under the same particle volume content, surficial area and heat capacity of nanoscale solid particles are far larger than those of millimeter or micron-order solid particles, so heat conductivity of nanofluids will be substantially enhanced. Nanoparticles enhance heat transfer capacity of the fluid in the grinding area and exert effects of enhancing heat transfer and lowering temperature of the grinding area. Further study results indicate that nanoparticles have superb anti-wear and anti-friction characteristics and high bearing capacity, they can further improve lubricating and anti-friction tribological properties. NMQL grinding method with nanoparticles participating in the heat transfer enhancement enhances heat transfer capacity and lubricating property of MQL grinding to the greatest extent, thus solving the technical bottleneck namely MQL heat transfer capacity is insufficient and opening a new path for application of MQL to grinding process.

Chapter 1
Introduction

ABSTRACT

This chapter introduces the application background and characteristics of five kinds of grinding processing methods, briefly describes the enhanced heat transfer mechanism and tribological properties of nanofluids, and points out that nanofluids minimum quantity lubrication (NMQL) solves the technical bottleneck, namely minimum quantity lubrication (MQL) heat transfer capacity is insufficient and opening a new path for application of MQL to grinding process. The current status of exploratory research on the mechanism of minimum quantity lubrication grinding using nanofluids as cooling lubricants is analyzed. The research characteristics of the new green NMQL grinding technology are described, and the chapter puts forward some key problems such as the heat transfer enhancement process of NMQL, the anti-friction and anti-wear tribological mechanism of nanoparticles, and the controlled transport strategies of minimal quantity of lubricating droplets. It will be of great scientific significance and pragmatic value to perfecting NMQL grinding technical system.

INTRODUCTION

Grinding is one of the most fundamental machining modes. It's especially important that final precision and surface quality of most parts are guaranteed by the grinding technology. The most significant features of grinding are high circular velocity of grinding wheel and high energy consumption (specific grinding energy). Grinding is negative rake cutting of abrasive grain on the grinding wheel surface, and the energy consumed to remove unit volume of materials is far greater than other machining methods (Xu & S, 2001). Most energy generated in the grinding area is converted into heat quantity which is dispersed on abrasive dust, grinding wheel and workpiece (Deng, Liu, & Zhang, 2010). As grinding is different from other machining forms, the contact time between each abrasive grain and workpiece is extremely short under the effect of circular velocity of grinding wheel, and meanwhile, the volume of abrasive dust generated in the grinding process is very small and the ratio of the generated heat quantity which is carried away by abrasive dust is very low. High-energy density in the grinding area has a great bearing on surface quality and usability of the workpiece. Especially when temperature in the grinding area exceeds critical value, this will cause thermal injury (surface oxidation, burning, residual tension

DOI: 10.4018/978-1-7998-1546-4.ch001

and crack) on the workpiece surface so as to degrade anti-fatigue performance and anti-wear performance of parts, and consequently, service life and reliability of parts will be reduced, so will grinding performance and machining precision of grinding wheel (Guo, Yin, Li, & Chen, 2009; Yuan et al., 2010). With heat accumulation on the workpiece surface, its dimensional accuracy and form accuracy will become very poor under the influence of grinding heat (Chen, Shen, Huang, & Xu, 2010). Generally speaking, grinding is the final parts machining procedure and grinding technology and process decide their final precision and surface quality. To effectively control the temperature of the grinding zone and reduce the thermal damage on the surface of the workpiece, large flow grinding fluid is often injected into the grinding zone during processing,which is called cast cooling. Due to the existence of the "gas barrier layer" around the high-speed moving grinding wheel, it is often difficult for the pouring type liquid supply to enter the grinding zone, and the "effective flow rate" (the ratio between the liquid volume that enters the grinding interface actually and the total supply liquid volume) is only 5%-40%. Therefore, most of the grinding fluid can not actually reach the grinding interface, but only acts to cool the workpiece base in the periphery, resulting in insufficient cooling capacity of the conventional liquid supply method and futher leading to the grinding burns and deterioration of the surface integrity of the workpiece. Additionally, the supply of a large amount of grinding fluid will form a fluid introduction force and a fluid dynamic pressure at the wedge-shaped gap of the grinding wheel workpiece, thereby causing deflection of the spindle of the grinding machine and furtherly reducing the actual depth of cut. therefore, the traditional pouring type liquid supply method not only causes large shape and size errors in the processed workpiece, but also wastes a lot of grinding fluid. Green development is an international trend. The current environmental and resource issues have become a common challenge for mankind, and an international consensus has been reached on how to achieve sustainable development. Green manufacturing is undoubtedly the only way for industrial upgrading and transformation. China is a big manufacturing country in the world,there are still "three highs" problems of high emissions, high consumption and high investment. Both energy consumption and pollution emissions are far from the world's advanced manufacturing levels. The Ministry of Industry and Information Technology issued the "Industrial Green Development Plan", and the green manufacturing of basic manufacturing processes to promote the technology of "green manufacturing" with less cutting fluid green processing is the key development direction of the National 13th Five-Year Plan. The development of quasi-dry green manufacturing technology is conducive to alleviating the current environmental resource eastward problem, facilitating the rapid cultivation of new economic growth points, and has a profound historical significance for accelerating the transformation of economic development mode, promoting industrial transformation and upgrading, transforming old and new kinetic energy, and enhancing the international competitiveness of manufacturing industry.

1.1.1 Flood Grinding

In order to reduce temperature in the grinding area and improve machining precision and surface integrity of the grinded workpiece, flood grinding method which supplies mass-flow grinding fluid to the grinding area is extensively used in the production process, and temperature in the grinding area is lowered through heat convection of grinding fluid. Lubrication action is mainly manifested by favorable wetting and anti-friction characteristics of grinding fluid, which can form lubricating oil film on the workpiece surface so as to reduce frictional wear and grinding force of abrasive grain on the grinding wheel surface and improve grinding performance. Cooling effect mainly relies upon heat convection of grinding fluid, and high heat generated in the grinding area is carried away by mass-flow injection of grinding fluid. Grinding fluid is mainly divided into two major types: oil-based grinding fluid and water-based grinding fluid (Gao,Tse, & Mak, 2003). The former has good lubricating property while the latter has superior cooling effect. Besides cooling and lubricating effects, grinding fluid also has rust removal, cleaning and chip removal effects. Heat conductivity coefficient of water is 3-5 times of that of oil. By adding lubricating oil of a certain mass fraction and a small amount of additive into water, low-cost emulsion with superior cooling property can be prepared (Chen, Yin, Ohmori, & Yu, 2013). In view of superior performance of grinding fluid and low cost of water-based grinding fluid, it has been extensively applied to the machining industry.

With enaction and implementation of environmental protection laws in various countries and establishment of ISO14000 environmental management system, new requirements have been proposed for processing technology and level in the machining industry, and meanwhile, this induces industrial circles and scientific researchers to reevaluate the flood cooling grinding. Spanish grinder manufacturer Danobat surveyed costs of grinding fluids (Fig. 1-1), and results showed that (Sanchez et al., 2010): in the total parts machining cost, 75% is machining expense, 7% is tool expense and 18% is expense of cooling liquid, which nearly accounts for 1/5 of total cost and is 2-3 times of that of tool expense. However, in total expense of cooling liquid, cooling liquid supply and filter plants account for 40%, sewage treatment cost for 22%, labor cost for 10%, base fluid of cooling liquid for 14%, energy consumption cost for 7% and others for 7%. It can be seen that base cooling fluid accounts for a very small proportion in the cooling liquid expense, but the proportion occupied by other expenses related to cooling liquid is very high. Due to perfection and requirements of environmental protection laws, treatment cost of cooling liquid will continue to rise. Therefore, the proportion of total cooling liquid expense in the total machining cost will gradually increase.

Scientific researchers (Barczak, Batako, & Morgan, 2010) investigated operating cost of flood grinding and deemed that treatment cost of grinding fluid accounted for 54% of total expense of grinding fluid, energy consumed by revolution of machine tool spindle only occupied 20% of total energy consumed by machining while the proportion of energy consumption related to grinding fluid reached 53%. This indicates that under the major background of environmental protection and low carbon, grinding fluid has already lost its price advantage. Based on analytical investigation of airflow field in the grinding process, related researchers (Ebbrel, Woolley, & Tridimas, 2000; Morgan et al., 2008; Zhou, Zhou, Mao, & Ang, 2009) believed that high-speed revolution of grinding wheel resulted in "air-bond" phenomenon around the grinding wheel, which discouraged grinding fluid from effectively entering the wedge-shaped zone between grinding wheel/workpiece and reduced "effective flow rate" of grinding fluid. Consequently, there was a shortage of grinding fluid in the contact zone between grinding wheel and workpiece with serious heat accumulation, workpiece surface integrity was aggravated after grinding, and even the part was scrapped. In addition, that a large quantity of grinding fluid entered the wedge-shaped zone would generate hydrodynamic pressure under the effect of dynamic pressure, give rise to flexural deflection of grinding wheel spindle (Lee, Nam, Li, & Lee, 2012; Lee, Miyamoto, & Kuriyagawa, 2004), reduce actual grinding depth, degrade grinding precision and machinability of machine tool and parts machining precision, thus failing to satisfy requirements for high parts precision and high surface quality after grinding.

Green development is a general international trend. Resource and environmental problem being a common challenge faced by human beings, sustainable development has gradually become a global consensus. Developed countries have successively implemented the "reindustrialization strategy" to reshape new competitive advantages of their manufacturing industries, so the influence of green concepts,

Figure 1. The proportion of grinding fluid in total processing costs

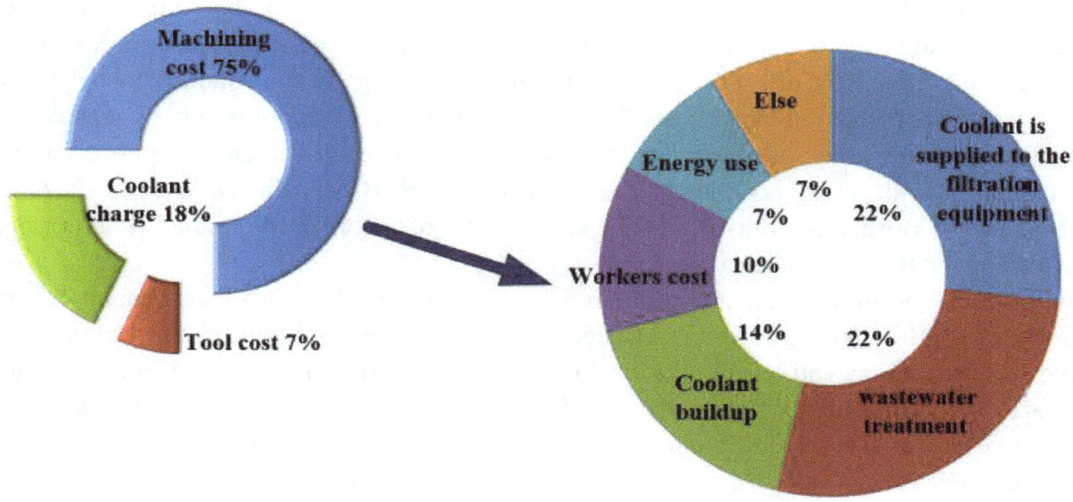

policies and regulations like clean, high-efficiency, low-carbon and recycling is continuously enhanced, energy and resource utilization efficiency has become an important factor measuring competitiveness of the manufacturing industry in a country and green trade barrier has also become an important means for some countries to seek for competitive advantages. Green manufacturing is the only route which must be passed by industrial transformation and upgrading. Led by implementation of green manufacturing projects, comprehensively carrying out green manufacturing can not only exert an important realistic effect on relieving the current bottlenecks and constraints of resources and environment and accelerating cultivation of new economic growth points but also it is of far-reaching historical significance to accelerating transformation of the economic development mode, promoting industrial transformation and upgrading and improving international competitiveness of the manufacturing industry.

The key green transformation task for the traditional manufacturing industry is to implement clean transformation of the production process. Taking cutting down pollutant generation from sources as the starting point, the government should reform traditional manufacturing technological equipment and encourage enterprises to adopt advanced and applicable cleaner production technologies to implement upgrading and transformation, and green transformation of fundamental manufacturing technologies and energetically promote green processing technologies without or with few cutting fluids, etc. Up to 2020, energy will have been saved by over 30% and wastes will have been reduced by over 20%.

Made in China 2025 has clearly proposed that green manufacturing project is one of five major construction projects, which aims at organizing implementation of energy efficiency improvement in the traditional manufacturing industry, clean production, water conservation and pollution control, cyclic utilization and other special technical transformations, and carrying out major demonstration work of energy conservation and environmental protection, comprehensive utilization of resources, remanufacturing and industrialization of low-carbon technology.

The flood grinding which features mass-flow supply of grinding fluid not only increases the treatment cost of grinding fluid but also it is waste of grinding fluid in quantity and poses great harm to the environment. In full consideration of environmental protection, low carbon, machinability and economical efficiency, flood grinding is no longer adaptive to concepts of green manufacturing and sustainable development. Therefore, it's necessary to develop green, environmentally friendly, high-efficiency and low-consumption grinding fluid or develop new clean, low-consumption, high-efficiency and sustainable green grinding technologies, which can not only reach the technological effect of flood cooling & lubrication but also can ensure machining precision and surface quality of workpiece, lengthen service life of grinding wheel and reduce grinding fluid-induced environmental pollution.

1.1.2 Dry Grinding

Dry cutting technology is a kind of green and environmentally friendly processing technology which appeared at the earliest and derived from the automobile industry. It has been successfully applied to machining processes like turning, milling, drilling and boring. It doesn't completely abandon cutting fluid, but instead, it abolishes use of cutting fluid on the precondition that parts machining precision and service life of tools are guaranteed (Li Bin & Deng, 2008). Therefore, dry cutting has proposed higher requirements for cutting tools and machine tools. Cutting tools applied to dry cutting have high hardness, tenacity and abrasive resistance. Meanwhile, the tool surface has low friction coefficient (Jiang, Liu, An, & Chen, 2012). As cutting fluid is not used in the machining process, dry cutting not only satisfies environmental protection requirements but also reduces machining cost and improves working environment.

Dry cutting rapidly removes materials on the workpiece surface by elevating revolving speed of machine tool spindle. It is mainly used for machining of materials with small cutting force and low cutting temperature like aluminum alloy and gray pig iron as chips carry away the heat generated in the machining process. As grinding is different from other machining forms, only 10% of the heat generated in the grinding process will be carried away by abrasive dust, while a large quantity of grinding heat will be transferred to grinding wheel and workpiece (Ding et al., 2015). The heat transferred to workpiece will be aggregated on the workpiece surface layer, and the formed local temperature can reach as high as 1000°C. Studies (Gao & Yuan., 2011; Jin & Stephenson, 2008) show that when dry grinding with grinding fluid used is conducted, the heat aggregated on the workpiece surface will result in serious burning of the workpiece surface, so surface integrity is aggravated, thus lowering workpiece machining precision and surface quality. As a result, grinding wheel loses its grinding performance and its durability and service life will be reduced.

Dry grinding which uses solid lubricating fluid improves lubricating property at the grinding wheel and workpiece interface, but it can't exert a good cooling effect. Moreover, powerful grinding pressure at grinding wheel/workpiece interface will let solid lubricating fluid enter pores on the grinding wheel surface, thus reducing the chip space of grinding wheel so that grinding wheel loses it grinding performance, grinding force is increased, workpiece surface quality is aggravated and meanwhile, service life of grinding wheel is shortened.

1.1.3 Cryogenic Air Grinding

Low-temperature CO_2 and liquid nitrogen are used cooling media of grinding process. Relative to dry grinding, Cryogenic air grinding force, specific grinding energy and temperature in the grinding area are obviously reduced, workpiece surface quality and integrity are significantly improved, and meanwhile, service life of grinding wheel is obviously lengthened and cooling liquid-induced environmental pollution is remitted (An, Fu, Xu, & Zheng, 2006; Bian, Li, He, Qi, & Jia., 2009). Cryogenic air grinding adopted, temperature in the grinding area is effectively controlled, so abrasive grain is always under sharp state, which reduces sliding friction, ploughing and plastic deformation, material is removed in shear way, and then grinding force is reduced, and workpiece surface quality and integrity are improved by a large margin (Zhao, Jiao, & Du, 2009). However, this cooling mode is only applicable to grinding process with small grinding energy, and the fluid supply cost is high, so its application to industrial production is restricted to a certain degree.

1.1.4 Minimum Quantity Lubrication Grinding

Minimum quantity lubrication (MQL) grinding is another green processing technology after Dry grinding and low-temperature grinding technologies (Davim, Sreejith, & Gomes, 2006). It is a grinding technology which mixes minimum quantity lubrication with the gas of a certain pressure for atomization and sprays the mixture into the grinding area to exert cooling and lubricating effects. Cooling and chip removal are realized mainly by relying on high-pressure gas. When high-pressure gas enters the grinding area, it carries away the heat generated by grinding so as to lower temperature in the grinding area and remove abrasive dust generated by grinding. Minimum quantity lubrication conveyed to the grinding area by high-pressure gas is adhered to surfaces of workpiece and abrasive grain, thus forming a lubricating oil film with certain anti-friction and anti-wear characteristics to exert good lubricating effect, and reduce friction coefficient and specific grinding energy as well as grinding force.

Flow quantity of grinding fluid used for MQL grinding of unit grinding wheel width is 30~100ml/h while flow quantity of grinding fluid adopted by flood grinding is 60L/h; however, MQL reaches and even exceeds the flood lubricating effect. Meanwhile, use level of grinding fluid is only parts per thousand and even per ten thousand of traditional machining modes (Sadeghi, Haddad, Tawakoli, & Emami, 2009). Moreover, grinding fluid used in the MQL grinding process is usually degradable synthetic ester or vegetable oil (Davim et al., 2006), which reduces treatment cost of grinding fluid so as to reduce grinding fluid-induced environmental pollution. However, the deficiency of MQL grinding lies in that it's difficult to exert a good cooling effect. Study results indicate that (Sadeghi et al., 2009): high-pressure gas conveyed into the grinding area has not exerted cooling effect as expected, and a very limited quantity of the heat generated by grinding is carried away by high-pressure gas (caused by low heat conductivity coefficient of the gas). Consequently, the heat accumulated on workpiece surface can't be effectively removed, surface integrity of the grinded workpiece is aggravated, workpiece machining precision and surface quality are degraded, service life of grinding wheel is shortened, and even the grinding wheel will be scrapped. Therefore, it's difficult for MQL grinding technology to replace flood grinding and high-pressure airflow fails to solve the technical bottleneck namely heat transfer enhancement under high-energy density in the grinding area, indicating that MQL grinding technology needs further improvement.

1.1.5 Nanofluids Minimum Quantity Lubrication Grinding

In view of the deficiency of MQL grinding namely poor cooling capacity, it's urgent to seek for a new MQL mode to be applied to the grinding process in the academic circles. Study results of heat transfer enhancement (Yao & Wang, 2003) indicate that heat transfer capacity of solid is greater than that of liquid, while that of liquid is greater than that of gas. Table 1 gives heat conductivity coefficients of several common materials (Wen, Lin, Vafaei, & Zhang, 2009). According to the table, under normal

Table 1. Heat conductivity coefficients of several common materials

Material	Engine oil	Water	Copper oxide	Aluminum oxide	Si	Al	Cu	Diamond	Carbon nanotube
Heat conductivity coefficient/W(m·K)$^{-1}$	0.145	0.613	19.6	40	148	237	401	2300	3000

temperature state, heat conductivity coefficient of carbon nanotube is 3000 W/(m·K), that of water is 0.613 W/(m·K), so heat conductivity coefficient of carbon nanotube is several thousand times of that of water. Meanwhile, other solid materials also have such similar relationships with liquid in the aspect of heat conductivity coefficient. Moreover, for similar solid materials with the same quality, the smaller the material volume, the smaller the particle size, the greater the specific surface area of the material, and the higher the heat-conducting property (Choi, Jung, Choi, Lee, & Oh, 2011).

Nanofluids minimum quantity lubrication (NMQL) is a new-type high-efficiency, low-consumption, clean and low-carbon precision machining mode proposed based on the above background and theory. The concrete method refers to adding a certain amount of nanoscale solid particles into degradable minimum quantity lubrication (MQL) to prepare nanofluids which is then atomized through high-pressure gas and sprayed into the grinding area in the form of jet flow. High-pressure gas exerts effects of cooling, dust removal and lubricant conveying while MQL exerts lubricating and anti-friction effects; under the same particle volume content, surficial area and heat capacity of nanoscale solid particles are far larger than those of millimeter or micron-order solid particles, so heat conductivity of nanofluids will be substantially enhanced. Nanoparticles enhance heat transfer capacity of the fluid in the grinding area and exert effects of enhancing heat transfer and lowering temperature of the grinding area. Further study results (Hwang, Park, & Lee, 2006) indicate that nanoparticles have superb anti-wear and anti-friction characteristics and high bearing capacity, they can further improve lubricating and anti-friction tribological properties. NMQL grinding method with nanoparticles participating in the heat transfer enhancement enhances heat transfer capacity and lubricating property of MQL grinding to the greatest extent, thus solving the technical bottleneck namely MQL heat transfer capacity is insufficient and opening a new path for application of MQL to grinding process.

1.1.6 Enhanced Heat Transfer and Tribological Properties of Nanofluids

With rapid development of the nanotechnology, a new cooling liquid-nanofluids is generated in engineering. Nanofluids is sprayed out from the nozzle in the form of fine droplets under the carrying effect of compressed gas. Nanoparticles suspended in the grinding fluid with particle size being 1-100 nm can enter grinding wheel/workpiece interface more easily due to small size effect and high surface energy. By virtue of enormous potentials of nanoparticles which enhance heat conductivity and improve convective heat transfer coefficient and anti-friction and lubricating properties, nanofluids enhances heat transfer capacity and reduces wear of grinding wheel so as to reduce grinding force and temperature in the grinding area and improve grinding performance and workpiece surface quality. Heat transfer characteristics of nanoparticles especially their high heat conductivity and convective heat transfer coefficient can reduce high energy density in the grinding area with minimum quantity lubrication. Study results (Huang, 2005) indicate that relative to conventional grinding fluid, nanofluids has obviously improved heat transfer capacity. Xie (2011) added different nanoparticles in the base fluid to prepare

Table 2. Results of convective heat transfer coefficients for nanofluids

Material	Base fluid	Nanoparticle size	Volume fraction	Increase rate of heat conductivity coefficient	Researcher
Al_2O_3	Water	12 nm	4.3%	30%	Yu et al[35]
Al_2O_3		33 nm	4.3%	15%	Xie et al[34]
Al_2O_3		68 nm	5%	21%	Xie et al[34]
CuO		36 nm	5%	60%	Yu et al[35]
CuO		36 nm	3.4%	12%	Vajjha et al[36]
CuO		50 nm	0.4%	17%	Maddah[38]
SiC		26 nm	4.2%	16%	Xie et al[34]
TiO_2		15 nm	5%	30%	Murshed et al[40]
Cu	Ethanol	10 nm	0.3%	40%	Choi et al[31]
	Water	100 nm	7.5%	78%	Wong et al[37]
MWCNT	Synthetic oil	Particle size 50 nm, length 50 nm	1%	150%	Wong et al[37]
	Glycol	Particle size 15 nm, length 30 nm	1%	13%	Xie et al[34]
	Water	Particle size 100 nm, length 70 nm	0.6%	38%	Hwang et al[32]

a nanofluids and investigated influence laws of nanoparticle size, volume fraction, pretreatment technology as well as physical characteristics of additives (dispersing agent and surfactant) and base fluid, and found that nanofluids stability and homogeneity had a great bearing on heat-conducting property of the nanofluids. Yu (2008) studied the influence of volume fraction of nanoparticles on convective heat transfer coefficient of the nanofluids, which increased with volume fraction of nanoparticles. Even though Al_2O_3 nanoparticles have different particle sizes, heat conductivity coefficients of all Al_2O_3-water based nanofluids have a common change trend namely their heat transfer coefficients increase with volume fraction of added Al_2O_3 nanoparticles. When Al_2O_3 nanoparticles with 1% volume fraction are added, heat conductivity coefficient of Al_2O_3-water based nanofluids increases by about 2% when compared with pure water; when volume fraction of Al_2O_3 elevated to 4%, heat conductivity coefficient of its nanofluids can increase by 10% or so. In addition, Vajjha and Das (2009) measured change laws of heat conductivity coefficient of the nanofluids suspended with CuO nanoparticles. When volume fraction of CuO nanoparticles increases from 1% to 4%, heat conductivity coefficient of nanofluids can increase by 30% when compared with pure water. Wong (2010) studied the influence of nanoparticle size on heat conductivity of the nanofluids. Under the same volume fraction, the smaller the nanoparticle size, the greater the number of nanoparticles contained in the nanofluids, the better the nanofluids stability and the more obviously the heat conductivity is improved. Maddah (2013) studied the influence of temperature on nanofluids heat conductivity coefficient. As the nanofluids temperature rose, heat conductivity coefficient of the nanofluids with 4.0% volume fraction under 80°C was elevated by 10% when compared with that under normal temperature, and heat conductivity coefficients of other nanofluids with different volume fractions all presented increasing trend. The higher the temperature, the greater the heat conductivity coefficient of the nanofluids. Table 1-2 is a summary of study results of convective heat transfer coefficients of nanofluids.

Kalita (2012) conducted an experimental study of NMQL surface grinding of EN24 steel, and supplied flow quantity of MQL was 2.5 ml/min; supplied flow quantity of flood grinding fluid was 8450ml/min. Results indicate that when mass fraction of nanoparticles is 8%, friction coefficients of two oil-based lubricants for MQL grinding are 0.43 and 0.41 respectively which reduce by 31% and 32% respectively when compared with those of paraffin-based MQL and soybean oil-based MQL without addition of nanoparticles; relative to flood lubricant, pure oil-based MQL and MQL added with micron-order particles of 8% mass fraction, specific grinding energy corresponding to MQL added with nanofluids of 8% mass fraction in the paraffin oil reduces by 34%, 29% and 22% respectively. When nanofluids is used for MQL grinding, nanoparticles and base oil form lubricating oil film with anti-friction and anti-wear characteristics at the grinding wheel/workpiece interface, thus reducing friction, sliding friction and ploughing between interfaces (Choi et al., 2011; Hwang et al., 2006), improving lubricating properties of grinding wheel and workpiece surface, reducing grinding force and then reducing friction coefficient and specific grinding energy.

1.2 RESEARCH SIGNIFICANCE

Adding nanoscale solid particles into the degradable lubricating base fluid can prepare the nanofluids. By virtue of enhanced heat transfer capability of nanoparticles and anti-friction and anti-wear tribological characteristics of nanoparticles (Maddah et al., 2013; Wen et al., 2009), the nanofluids is applied to MQL grinding, which not only inherits all advantages of MQL grinding but also solves the critical defect of the traditional MQL grinding namely heat transfer capacity is insufficient, and moreover, superior anti-friction and anti-wear tribological characteristics of nanoparticles are conducive to improving lubricating properties of MQL grinding wheel/workpiece interface and grinding wheel/chip interface so as to greatly improve workpiece machining precision, surface quality and especially surface integrity; meanwhile, service life of grinding wheel is lengthened and working environment is improved. Therefore, it is a new environmentally friendly, resource-saving, sustainable and green grinding method with high utilization efficiency of energy sources.

1.3 REVIEW OF RESEARCH

The research team from Qingdao University of Technology conducted a deep and systematic research on NMQL grinding technology and equipment. Gu L D (2008) implemented an exploratory research on applicability of MQL grinding with the emphasis laid on investigating influence laws of use level of grinding fluid and physical properties of jet flow on surface quality. Han X Y (2010) conducted an experimental study on parameters of high-speed MQL grinding jet flow, and results indicated that nozzle position had a significant influence on MQL high-speed grinding results. Liu Z R (2010) evaluated heat exchange mechanism and surface integrity of surface grinding under NMQL conditions and conducted an experimental study of grinding temperature field. Compared with traditional MQL grinding, surface quality of NMQL grinding, temperature in the grinding area and workpiece surface integrity were obviously improved. Han Z L (2013) conducted theoretical modeling and experimental study of the flow field in the grinding area under NMQL conditions and studied influence laws of jet flow parameters and use level of grinding fluid on velocity field and pressure field in the grinding area. Li J Y conducted a modeling and numerical simulation study of the temperature field in the NMQL grinding process, built heat transfer model of temperature field and proportionality coefficient model for energy conveyed to the workpiece in the NMQL surface grinding process, and then implemented numerical simulation and experimental study of temperature field in the surface grinding process for three workpiece materials: 45 steel, 2Cr13 and nano zirconia ceramics. Wang S (2013) investigated creation mechanism of surface appearance under NMQL grinding and established grinding kinematic model, elastic deformation model and plastic stacking model of a single abrasive grain, engineered grinding wheel and ordinary grinding wheel respectively. MATLAB language was used for programming, grinding processes of a single abrasive grain, engineered grinding wheel and ordinary grinding wheel were simulated through numerical calculation, workpiece surface appearances under different grinding parameters and grinding wheel parameters were generated and evaluated using surface roughness, influence laws of grinding parameters and grinding wheel parameters on surface roughness were obtained, and corresponding experimental verification was conducted. Zhang Q carried out an experimental study of cooling performance of NMQL grinding and experimental comparison of workpiece surface temperatures under flood grinding, Dry grinding, MQL grinding and NMQL grinding. Results showed that NMQL had good lubricating effect, thus creating a new path for application of MQL to grinding process. Zhang carried out (2015) an experimental study of nodular cast iron NMQL grinding performance and evaluated grinding performance from aspects of grinding force, G ratio grinding temperature and surface roughness under four working conditions—dry grinding, flood grinding, MQL grinding and NMQL grinding using K-P36 NC flat grinder. Results showed that NMQL grinding improved heat transfer capacity, the temperature in the grinding area was lowered by 150°C relative to dry grinding, and workpiece surface roughness R_a values obtained through dry grinding and NMQL grinding were 1.2 μm and 0.58 μm respectively, so workpiece surface quality was significantly improved; under lubricating effect of nanoparticles, grinding force was reduced by 15% and 7% respectively relative to those under dry grinding and traditional MQL grinding; G ratio under NMQL grinding was the highest among four working conditions, being 33 while that under dry grinding was only 12, so grinding wheel wear was obviously reduced, and service life of grinding wheel was lengthened. Jia (2014) studied suspended particle distribution characteristics in MQL grinding and relationships of jet flow parameters (nozzle outlet diameter, gas pressure for fluid supply, flow rate of grinding fluid and ratio of gas/liquid flow quantity) with particle size of droplets, analyzed distribution laws of particle sizes of droplets and established mathematical models of particle quantity and volume distribution functions of droplets. Zhang (2015) evaluated lubricating property of MoS_2 NMQL grinding by taking vegetable oil as base oil and used liquid paraffin to conduct a comparative study of friction coefficients and specific grinding energies when soybean oil, palm oil and colleseed oil as MQL base oils respectively. Wang (2015; 2017; 2016;) studied tribological properties at grindign wheel/workpiece interface in NMQL grinding and used a friction tester to verify tribological properties of different nanofluids and compared their grinding conditions. Uner the same experimental conditions, a series of experimental studies was carried out, experimental evaluation of different vegetable base oils, different nanofluids, different nanofluids concentrations and different lubricating fluids which were used for grinding of different workpieces was implemented, and the optimal MQL grinding parameters were finally obtained. Li (2015; 2016; 2017) conducted an analysis and experimental study of temperature and convective heat transfer mechanism in NMQL grinding, and then investigated convective heat transfer mechanism and action laws of MQL grinding of nickel base alloy using different vegetable base oils, different nanofluids and different nanofluids concentrations as well as grinding of different workpiece materials under different nanofluids concentrations.

Professor Mao from Changsha University of Science and Technology (2012; 2013; 2014) conducted a theoretical analysis and experimental study of NMQL grinding and a comparative experimental study of grinding of quenched steel (AISI52100) under four cooling & lubrication modes: water-based Al_2O_3 nanofluids, Dry grinding, flood grinding and pure water MQL, and results indicated that relative to pure water MQL grinding, grinding temperature, grinding force and surface roughness were obviously reduced under water-based Al_2O_3 NMQL grinding. Due to high surface energy, nanoparticles could infiltrate into workpiece/grinding wheel/chip interfaces so as to provide superior lubricating and anti-friction effects and obtain ideal workpiece surface quality. In addition, Professor Mao (2013) also investigated influences of jet flow parameters—jet angle, carrying gas pressure and jet distance (range)—on NMQL grinding performance, and results indicated that when nozzle was above "air-bond" returned flow, ideal grinding performance could be obtained; as carrying gas pressure increased and jet distance reduced, grinding force, grinding temperature and surface roughness reduced.

Professor Kalita and his team (2012) conducted an experimental study of nanofluids surface grinding, and results showed that when nanoparticles with a certain mass fraction were added into the minimal quantity of base oil for MQL surface grinding, nanoparticles and base oil formed a lubricating oil film with anti-friction and anti-wear characteristics at grinding wheel/workpiece interface, thus reducing friction of grinding wheel/workpiece/chip interfaces, improving lubricating property of grinding wheel/workpiece interface, reducing grinding force, reducing friction coefficient and specific grinding energy and elevating G ratio.

Kalita (2012) investigated the influence of nanofluids on tribological properties at grinding wheel/workpiece interface under MQL grinding through an analysis of frictional force and wear. As the nanofluids formed a stable lubricating oil film with low friction coefficient at grinding wheel/workpiece interface, it reduced frictional force and frictional wear, elevated G ratio and improved workpiece surface quality.

Doctor Shen (2010) experimentally studied the influence of flow quantity of supplied nanofluids on nodular cast iron surface grinding temperature. Results indicated that as the supplied flow quantity increased, grinding temperature on the workpiece surface was gradually lowered. When flow quantity of supplied MQL increased from 5ml/min to 30ml/min, the highest grinding temperature on the workpiece surface reduced by 100 °C and it reduced by 250 °C when relative to the highest temperature on the workpiece surface under dry grinding, indicating that supplied flow quantity was an important factor influencing MQL cooling performance. Therefore, properly increasing the flow quantity of supplied MQL could reach ideal cooling effect.

Shen (2008) evaluated MQL grinding performance by adding MoS_2 nanoparticles into different base fluids (paraffin oil, soybean oil and chemical synthetic oil). Relative to flood grinding, under MQL grinding with chemical synthetic oil as the base fluid, grinding force was reduced by 27% and G ratio was elevated by 46%, and the two figures were 21% and 35% respectively under paraffin oil-based MQL grinding and 9% and 15% respectively under soybean oil-based MQL grinding. Supplied fluid quantity under flood grinding was 5400 ml/min while supplied quantity of MoS_2 nanofluids was only 5ml/min, so not only grinding performance was improved but also advantages of production cost and environmental protection were prominent.

Kalita research team experimentally measured grinding temperature of nodular cast iron surface grinding under different cooling & lubrication conditions. Cooling & lubrication modes adopted in this experiment included: flood, MQL, MQL added with micron-order particles and MQL added with nanoscale particles. Added solid particles were MoS_2. Grinding parameters were: revolving speed of grinding wheel 30 m/s, workpiece feed speed 0.06 m/s, grinding depth 10μm, supplied MQL quantity

2.5 ml/min and supplied fluid quantity under flood grinding 8450ml/min. Water-based grinding fluid was used in flood grinding while paraffin-based grinding oil was used in MQL grinding. In MQL grinding, mass fractions of added nanoscale MoS_2 were 2% and 8% respectively, and mass fraction of added micron-order MoS_2 was 8%. Results indicated that cooling performance of MQL medium added with nanoparticles was superior to that added with micron-order particles. Temperature in the grinding area measured in NMQL grinding with mass fraction of nanoparticles being 8% was 160°C, which was lower than grinding temperature under NMQL grinding with mass fraction of nanoparticles being 2%.

Setti (2012) conducted an experimental study of NMQL grinding for Ti-6Al-4V alloy steel and used Al_2O_3 nanofluids with volume fractions of 1% and 4%, and liquid-supply pressure and flow quantity were 1.5bar and 18ml/h. Results indicated that relative to flood grinding and pure water MQL, grinding force was obviously reduced and surface quality was obviously improved.

Lee (2012) experimentally studied NMQL grinding of tool steel, and added diamond and Al_2O_3 nanoparticles with 30 nm particle size into the paraffin oil base fluid to prepare nanofluids with volume fractions of 2%-4% respectively. Experimental results indicated that relative to dry grinding and pure base fluid MQL, grinding force was reduced and surface quality was obviously improved in the NMQL condition. Type, particle size, volume fraction of nanoparticles are of great importance to grinding performance.

1.4 RESEARCH CHARACTERISTICS AND PROBLEM DESCRIPTION

1.4.1 Research Characteristics

In the new green NMQL grinding technology, solid particles participate in heat transfer enhancement of fluid medium, and moreover, MQL media—solid particles—have good flowing property and stability property. Under the same particle volume content, surface area and thermal capacity of nanoparticles are far greater than millimeter or micron-order solid particles, and heat transfer capacity of the nanofluids will be enhanced by a large margin. Due to small size effect of nanomaterial, its behavior is approximate to fluid molecules. Intense Brownian movement of nanoparticles contributes to keeping stable suspension without precipitation, so they can have superior flowing property, stability property and composition homogeneity. Superior tribological characteristics of nanoparticles are conductive to improving lubricating property at grinding wheel/workpiece interface under MQL grinding condition, reducing grinding force and specific grinding energy and further lowering temperature in the grinding area as shown in Fig. 1-2. Application of nanoparticles to MQL grinding will improve cooling and lubricating capabilities to a great extent so that grinding will become a high-efficiency, low-consumption, clean and low-carbon precision machining mode.

According to document retrieval, NMQL grinding is only restricted to local optimization of nanoparticle type, nanoparticle size, base fluid type, volume fraction of nanoparticles in the base fluid, jet flow parameters, use level of grinding fluid, grinding wheel type and workpiece material. Effects of nanoparticles on improving convective heat transfer enhancement in the grinding area and anti-friction and anti-wear characteristics at grinding wheel/workpiece interface are studied from a macroscopic angle, but cooling and lubricating mechanism of nanoparticles in the grinding area is not revealed from microscopic scale—thermodynamics action laws of nanoparticles at abrasive grain/workpiece interface

Figure 2. Anti-friction and anti-wear mechanism of nanoparticles in grinding area

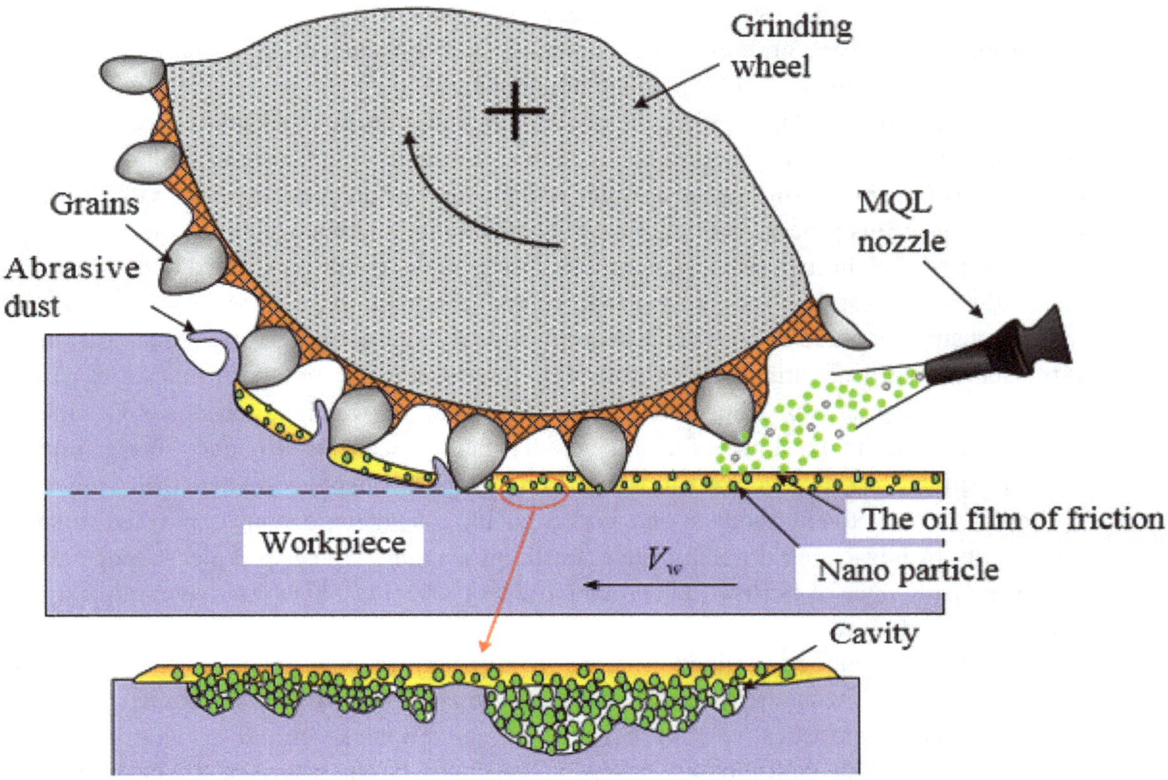

and abrasive grain/chip interface, not to mention proposing key technologies and initiative control strategies influencing the grinding performance from scientific essence.

1.4.2 Description and Explanation of Difficulties

(1) Convective heat transfer enhancement mechanism of NMQL and thermodynamics action laws of grinding area. Under high pressure and three-phase (gas, liquid and solid) jet flow conditions, what is internal relationship between heat conductivity coefficient of nanoparticles and heat transfer capacity of nano jet flow? Under the effects of carrying fluid and Brownian force, how do particles collide and disturb heat transfer? What is scientific essence of collision and convective heat transfer between nanoparticles and carrying fluid medium micro/nano interface? How do size effect, surface effect and interfacial coupling effect of nanoparticles and interaction between Van der Waals force, electrostatic force and Brownian force acting upon nanoparticles promote convective heat transfer enhancement?

(2) Infiltration characteristics of nanoparticles under high pressure of grinding area and formation mechanism of nanoparticle shear oil films between abrasive grain/workpiece, abrasive grain/chip and chip/workpiece. What is action mechanism of contact angle, height and angle of droplets and workpiece interface on cooling in the constrained space of wedge-shaped area and under relatively high motion speed of grinding wheel and workpiece? What is influence mechanism of droplet viscosity, surface tension and surface energy on infiltration into abrasive grain/workpiece interface and abrasive grain/chip

interface? What is action mechanism of impact dynamic characteristics at droplets/workpiece interface on cooling and lubrication? What is scientific essence of infiltration capability and heat exchange of nanoparticle droplets at abrasive grain/workpiece interface and abrasive grain/chip interface? What is anti-friction and anti-wear tribological mechanism of nanoparticles?

(3) Pneumatic atomization MQL grinding refers to mixing minimum quantity lubrication in the high-pressure gas, after which minimum quantity lubrication enters high-temperature grinding area. When aerodynamic force is larger than surface tension and viscous force of droplets, droplets will be crushed and atomized. However, this process is of extremely poor stability and controllability, action result of droplets can't be controlled through jet flow parameters. Only initial energy is applied to droplets at the nozzle outlet and surface energy of droplets is gradually reduced; motion trajectory of jet flow, droplet particle size and distribution and contact state at droplets/workpiece interface can't realize parameter-controlled transportation; penetrating power, absorption force and infiltration performance of jet flow are degraded, and drifting and splashing loss of droplets is serious, which aggravates environmental pollution. In order that minimal quantity of nanometer lubricant can exert its efficiency so as to improve utilization and precipitation efficiency of nanofluids to the greatest extent and reduce the harm caused by drifting droplet particles to environment and worker health, it's urgent to study controlled strategies of minimal quantity of lubricating droplets under constraint grinding wheel/workpiece wedge-shaped interaction space, disturbance of airflow field around grinding wheel and high relative motion speed of grinding wheel and workpiece.

(4) When droplets are under constrained grinding wheel/workpiece wedge-shaped space and workpiece is under rapid motion, what are quantitative relationships of contact form at droplets/workpiece interface with droplet characteristics (velocity, particle size and droplet spectral distribution), lubricant characteristics (viscosity, flash point and surface tension) and interface characteristics? What is the action mechanism of impact dynamic characteristics at droplets/workpiece interface on cooling and lubrication? How contact angle, length and height between droplets and moving workpiece surface and what are their quantitative relationships? What are capillary transportation, infiltration and film formation mechanisms at the constrained grinding wheel/workpiece interface and their scientific essence? Other root problems like anti-friction and anti-wear tribological mechanism of nanoparticles remain to be broken through. Establishing quantitative relation models of workpiece material behaviors, nanofluids physical behaviors and jet flow parameters with capillary infiltration and film formation and exploring into a scientific system for online effect detection and quantitative characterization of shear oil film and multi-phase fluid convection and heat transfer performance analysis will be of great scientific significance and pragmatic value to perfecting NMQL grinding technical system, initiatively controlling film-forming state under different microstructural conditions of grinding wheel/workpiece interface and further improving nanofluids cooling performance.

REFERENCES

An, Q. L., Fu, Y. C., Xu, J. H., & Zheng, H. F. (2006). Application of Cryogenic Pneumatic Mist Jet Impinging in Grinding of Ti-6Al-4V. *Zhongguo Jixie Gongcheng*, *17*(11), 1117–1120.

B., L., & Deng, J. X. (2008). Application of Tribo-Chemical Reaction Technology in Cutting Lubricating. *Tool Engineering, 42*(9), 12-15.

Barczak, L. M., Batako, A. D. L., & Morgan, M. N. (2010). A study of plane surface grinding under minimum quantity lubrication (MQL) conditions. *International Journal of Machine Tools & Manufacture, 50*(11), 977–985. doi:10.1016/j.ijmachtools.2010.07.005

Bian, R., Li, L., He, N., Zhao, W., Qi, B. Y., & Tian, J. (2009). Experimental Research on Tool Wear in High Speed Milling of PH13-8Mo Stainless Steel with Cryogenic Minimum Quantity Lubrication. *Tool Engineering, 43*(7).

Chen, F. J., Yin, S., Ohmori, H., & Yu, J. (2013). Form error compensation in single-point inclined axis nanogrinding for small aspheric insert. *International Journal of Advanced Manufacturing Technology, 65*(1-4), 433–441. doi:10.100700170-012-4182-4

Chen, J. Y., Shen, J. Y., Huang, H., & Xu, X. P. (2010). Grinding characteristics in high speed grinding of engineering ceramics with brazed diamond wheels. *Journal of Materials Processing Technology, 210*(6-7), 899–906. doi:10.1016/j.jmatprotec.2010.02.002

Choi, C., Jung, M., Choi, Y., Lee, J., & Oh, J. (2011). Tribological Properties of Lubricating Oil-Based Nanofluids with Metal/Carbon Nanoparticles. *Journal of Nanoscience and Nanotechnology, 11*(1), 368–371. doi:10.1166/jnn.2011.3227 PMID:21446457

Davim, J. P., Sreejith, P. S., Gomes, R., & Peixoto, C. (2006). Experimental studies on drilling of aluminium (AA1050) under dry, minimum quantity of lubricant, and flood-lubricated conditions. *Proceedings - Institution of Mechanical Engineers, 220*(10), 1605–1611. doi:10.1243/09544054JEM557

Deng, Z. H., Liu, Z. Q., & Zhang, X. H. (2010). Research of the Science and Technology Development in High-speed and Efficient Processing Field. *Journal of Mechanical Engineering, 46*(23), 106-120.

Ding, W. F., Zhu, Y. J., Zhang, L. C., Xu, J. H., Fu, Y. C., Liu, W. D., & Yang, C. Y. (2015). Stress characteristics and fracture wear of brazed CBN grains in monolayer grinding wheels. *Wear, 332*, 800–809. doi:10.1016/j.wear.2014.12.008

Ding, Y., & Wen, D. (2005). Particle migration in a flow of nanoparticle suspension. *Powder Technology, 149*(2), 84–92. doi:10.1016/j.powtec.2004.11.012

Ebbrell, S., Woolley, N. H., Tridimas, Y. D., Allanson, D. R., & Rowe, W. B. (2000). The effects of cutting fluid application methods on the grinding process. *International Journal of Machine Tools & Manufacture, 40*(2), 209–223. doi:10.1016/S0890-6955(99)00060-7

Gao, H., & Yuan, H. P. (2011). Development of a Wheel for Dry Grinding of CFRP Parts and Investigation on Its Machining Performance. *Acta Armamentrii, 32*(2), 186–191. doi:10.1016/j.jocn.2009.11.017

Gao, Y., Tse, S., & Mak, H. (2003). An active coolant cooling system for applications in surface grinding. *Applied Thermal Engineering, 23*(5), 523–537. doi:10.1016/S1359-4311(02)00214-4

Gu, L., Meng, G. Y., Shang, S. S., & Li, C. H. (2008). Selections and Calculations of Fluid Supply Parameters in Near Dry Grinding for Green Manufacturing. *Coal Mine Machinery, 29*(12), 94–96.

Guo, L., Yin, S. H., Li, B., & Chen, X. (2009). Feature Investigation of Acoustic Emission Signals under a Simulative Environment of Grinding Burn. *Zhongguo Jixie Gongcheng, 20*(4), 413–416. doi:10.11092009.5379889

Han, X. Y., Meng, G. Y., & Wang, W. (2010). Analysis of Surface Integrity for MQL in Grinding. *Manufaturing Technology& Machine Tool,* (1), 109-111.

Han, Z. L., & Li, C. H. (2013). Simulation and experimental investigation in pressure field of grinding zone with nanofluid. *Journal of PLA University of Science and Technology, 14*(5), 558–564.

Hwang, Y., Park, H. S., Lee, J. K., & Jung, W. H. (2006). Thermal conductivity and lubrication characteristics of nanofluids. *Current Applied Physics, 6*, 67–71. doi:10.1016/j.cap.2006.01.014

Jia, D., Li, C., Zhang, D., Wang, S., & Hou, Y. (2014). Investigation into the formation mechanism and distribution characteristics of suspended microparticles in MQL grinding. *Recent Patents on Mechanical Engineering, 7*(1), 52–62. doi:10.2174/2212797606666131231000234

Jiang, L., Liu, Z. Q., An, Q. L., & Chen, M. (2012). Mechanism research and experimental investigation of thread turning with dry cutting and minimum quantity lubrication. *Manufacturing Technology & Machine Tool,* (12), 88-92. doi:10.100711783-011-0280-z

Jin, T., & Stephenson, D. J. (2008). A study of the convection heat transfer coefficients of grinding fluids. *Cirp Annals-Manufacturing Technology, 57*(1), 367–370. doi:10.1016/j.cirp.2008.03.074

Kalita, P., Malshe, A. P., & Arun, K. S. (2012). Study of specific energy and friction coefficient in minimum quantity lubrication grinding using oil-based nanolubricants. *Journal of Manufacturing Processes, 14*(2), 160–166. doi:10.1016/j.jmapro.2012.01.001

Kalita, P., Malshe, A. P., & Jiang, W. (2010). Tribological study of nano lubricant integrated soybean oil for minimum quantity lubrication (MQL) grinding. *Transactions of NAMRI/SME, 38*, 137-144.

Kalita, P., Malshe, A. P., & Rajurkar, K. P. (2012). Study of tribo-chemical lubricant film formation during application of nanolubricants in minimum quantity lubrication (MQL) grinding. *Cirp Annals-Manufacturing Technology, 61*(1), 327–330. doi:10.1016/j.cirp.2012.03.031

Lee, P. H., Nam, J. S., Li, C., & Lee, S. W. (2012). An Experimental Study on Micro-Grinding Process with Nanofluid Minimum Quantity Lubrication (MQL). *International Journal of Precision Engineering and Manufacturing, 13*(3), 331–338. doi:10.100712541-012-0042-2

Lee, S. K., Miyamoto, Y., Kuriyagawa, T., & Syoji, K. (2004). Effects of minimizing hydrodynamic pressure in ultra-precision mirror grinding. *International Journal of Machine Tools & Manufacture, 44*(10), 1031–1036. doi:10.1016/j.ijmachtools.2004.03.004

Li, B., Li, C., Wang, Y., Zhang, Y., Yang, M., & Zhang, X. (2015). Technological investigation about minimum quantity lubrication grinding metallic material with nanofluid. *Recent Patents on Materials Science, 8*(3), 208–224. doi:10.2174/1874464808666151030212448

Li, B. K., Li, C. H., Zhang, Y. B., Wang, Y. G., Jia, D. Z., & Min, Y. (2016). Grinding temperature and energy ratio coefficient in MQL grinding of high-temperature nickel-base alloy by using different vegetable oils as base oil. *Chinese Journal of Aeronautics, 29*(4), 1084–1095. doi:10.1016/j.cja.2015.10.012

Li, B. K., Li, C. H., Zhang, Y. B., Wang, Y. G., Yang, M., Jia, D. Z., ... Wu, Q. D. (2017). Effect of the physical properties of different vegetable oil-based nanofluids on MQLC grinding temperature of Ni-based alloy. *International Journal of Advanced Manufacturing Technology*, *89*(9-12), 3459–3474. doi:10.100700170-016-9324-7

Liu, Z. R. (2010). Enhancement of heat transfer mechanism and evaluation of grinding surface integrity by nano-particle jet micro-lubrication. Qingdao University of Technology, (12).

Maddah, H., Rezazadeh, M., Maghsoudi, M., & NasiriKokhdan, S. (2013). The effect of silver and aluminum oxide nanoparticles on thermophysical properties of nanofluids. *Journal of Nanostructure in Chemistry*, *3*(1), 1–6. doi:10.1186/2193-8865-3-28

Mao, C., Tang, X. J., Zou, H. F., Huang, X. M., & Zhou, Z. X. (2012). Investigation of Grinding Characteristic using Nanofluid Minimum Quantity Lubrication. *International Journal of Precision Engineering and Manufacturing*, *13*(10), 1745–1752. doi:10.100712541-012-0229-6

Mao, C., Zhang, J., Huang, Y., Zou, H. F., Huang, X. M., & Zhou, Z. X. (2013). Investigation on the Effect of Nanofluid Parameters on MQL Grinding. *Materials and Manufacturing Processes*, *28*(4), 436–442. doi:10.1080/10426914.2013.763970

Mao, C., Zou, H. F., Huang, Y., Li, Y. F., & Zhou, Z. X. (2013). Analysis of heat transfer coefficient on workpiece surface during minimum quantity lubricant grinding. *International Journal of Advanced Manufacturing Technology*, *66*(1-4), 363–370. doi:10.100700170-012-4330-x

Mao, C., Zou, H. F., Zhou, X., Huang, Y., Gan, H. Y., & Zhou, Z. X. (2014). Analysis of suspension stability for nanofluid applied in minimum quantity lubricant grinding. *International Journal of Advanced Manufacturing Technology*, *71*(9-12), 2073–2081. doi:10.100700170-014-5642-9

Morgan, M. N., Jackson, A. R., Wu, H., Baines-Jones, V., Batako, A., & Rowe, W. B. (2008). Optimisation of fluid application in grinding. *Cirp Annals-Manufacturing Technology*, *57*(1), 363–366. doi:10.1016/j.cirp.2008.03.090

Sadeghi, M. H., Haddad, M. J., Tawakoli, T., & Emami, M. (2009). Minimal quantity lubrication-MQL in grinding of Ti-6Al-4V titanium alloy. *International Journal of Advanced Manufacturing Technology*, *44*(5-6), 487–500. doi:10.100700170-008-1857-y

Sanchez, J. A., Pombo, I., Alberdi, R., Izquierdo, B., Ortega, N., Plaza, S., & Martinez-Toledano, J. (2010). Machining evaluation of a hybrid MQL-CO grinding technology. *Journal of Cleaner Production*, *18*(18), 1840–1849. doi:10.1016/j.jclepro.2010.07.002

Setti, D, S., G., & V, R. P. (2012). Application of nano cutting fluid under minimum quantity lubrication (MQL) technique to improve grinding of Ti–6Al–4V alloy. *International Journal of Mechanical, Aerospace, Industrial. Mechatronic and Manufacturing Engineering*, *6*(10), 2107–2111.

Shen, B., Malshe, A. P., & Kalita, P. (2008). Performance of novel MoS2 nanoparticles based grinding fluids in minimum quantity lubrication grinding. *Transactions of NAMRI/SME, 36*, 357-364.

Vajjha, R. S., & Das, D. K. (2009). Experimental determination of thermal conductivity of three nanofluids and development of new correlations. *International Journal of Heat and Mass Transfer*, *52*(21-22), 4675–4682. doi:10.1016/j.ijheatmasstransfer.2009.06.027

Wang, S. (2013). Mechanism and experimental study of surface morphology formation of nano-particle jet micro-lubrication grinding. Qingdao University of Technology, (12).

Wang, Y. G., Li, C. G., Zhang, Y., Li, B., Yang, M., Zhang, X., ... Zhai, M. (2017). Comparative evaluation of the lubricating properties of vegetable-oil-based nanofluids between frictional test and grinding experiment. *Journal of Manufacturing Processes*, *26*, 94–104. doi:10.1016/j.jmapro.2017.02.001

Wang, Y. G., Li, C. H., Li, B., Yang, M., & Zhang, X. (2015). Advances and Recent Patents in the Field of Grinding Temperature Measurement Methods. *Recent Patents on Materials Science*, *8*(1), 55–68. doi:10.2174/1874464808666150304233653

Wang, Y. G., Li, C. H., Zhang, Y. B., Li, B. K., Yang, M., Zhang, X. P., ... Liu, G. T. (2016). Experimental evaluation of the lubrication properties of the wheel/workpiece interface in MQL grinding with different nanofluids. *Tribology International*, *99*, 198–210. doi:10.1016/j.triboint.2016.03.023

Wang, Y. G., Li, C. H., Zhang, Y. B., Yang, M., Li, B. K., Jia, D. Z., ... Mao, C. (2016). Experimental evaluation of the lubrication properties of the wheel/workpiece interface in minimum quantity lubrication (MQL) grinding using different types of vegetable oils. *Journal of Cleaner Production*, *127*, 487–499. doi:10.1016/j.jclepro.2016.03.121

Wang, Y. G., Li, C. H., Zhang, Y. B., Yang, M., Zhang, X. P., Zhang, N. Q., & Dai, J. J. (2017). Experimental evaluation on tribological performance of the wheel/workpiece interface in minimum quantity lubrication grinding with different concentrations of Al_2O_3 nanofluids. *Journal of Cleaner Production*, *142*, 3571–3583. doi:10.1016/j.jclepro.2016.10.110

Wen, D., Lin, G., Vafaei, S., & Zhang, K. (2009). Review of nanofluids for heat transfer applications. *China Particuology, 7*(2), 141-150. doi:JCSP.0.2009-02-008

Wong, K. V., & De Leon, O. (2010). Applications of Nanofluids: Current and Future. *Advances in Mechanical Engineering*. doi:10.1155/2010/519659

Xie, H. Q., Yu, W., Li, Y., & Chen, L. F. (2011). Discussion on the thermal conductivity enhancement of nanofluids. *Nanoscale Research Letters*, *6*(1), 124. doi:10.1186/1556-276X-6-124 PMID:21711638

Xu, X., & Malkin, S. (2001). Comparison of methods to measure grinding temperatures. *Journal of Manufacturing Science and Engineering*, *123*(2), 191–195. doi:10.1115/1.1369358

Yao, Z. P., & Wang, R. J. (2003). *Heat Transfer Science* (Vol. 9). Beijing: Beijing Institute of Technology Press.

Yu, W. H., France, D. M., Routbort, J. L., & Choi, S. U. S. (2008). Review and comparison of nanofluid thermal conductivity and heat transfer enhancements. *Heat Transfer Engineering*, *29*(5), 432–460. doi:10.1080/01457630701850851

Yuan, J. L., Zhang, F. H., Dai, Y. F., Kang, R. K., & Yang, H. (2010). Development Research of Science and Technologies in Ultra-precision Machining Field. *Journal Of Mechanical Engineering*, *46*(15), 161–177. doi:10.3901/JME.2010.15.161

Zhang, D. K., Li, C. H., Zhang, Y. B., Jia, D. Z., & Zhang, X. W. (2015). Experimental research on the energy ratio coefficient and specific grinding energy in nanoparticle jet MQL grinding. *International Journal of Advanced Manufacturing Technology*, *78*(5-8), 1275–1288. doi:10.100700170-014-6722-6

Zhang, Y. B., Li, C. H., Jia, D. Z., Zhang, D. K., & Zhang, X. W. (2015). Experimental evaluation of MoS_2 nanoparticles in jet MQL grinding with different types of vegetable oil as base oil. *Journal of Cleaner Production*, *87*, 930–940. doi:10.1016/j.jclepro.2014.10.027

Zhao, B., Jiao, F., & Du, B. (2009). A study of the material removal properties in ultrasonic lapping of ZTA ceramics with fixed abrasives by using an online acoustic emission monitoring system. *International Journal of Abrasive Technology*, *2*(4), 345–357. doi:10.1504/IJAT.2009.029093

Zhou, Z. X., Zhou, D. W., Mao, C., & Ang, J. (2009). Experimental Study of the Contact Length in Surface Grinding. *Journal Of Hunan University*, *36*(2), 27–31.

Chapter 2
Enhanced Heat Transfer Mechanism of Nanofluids Minimum Lubrication Grinding

ABSTRACT

This chapter will discuss in detail the various aspects of nanofluids, form preparation of nanofluids, characterization and flow and energy transportation mechanisms of nanofluids. Preparation of nanofluids has been performed using various methods where one step and two step methods are widely known. These methods are discussed in detail for various nanoparticles with stabilizing agents for stable production of nanofluids. Further, the characterization techniques of various aspects of nanofluds including thermal conductivity and factors influencing the thermal conductivity of nanofluids are introduced in depth. The chapter also sheds light on the experimental analysis of flow and heat transfer of naofluids, natural convection analysis of naofluids, and boiling heat transfer of nanofluids.

2.1 INTRODUCTION

Nano fluids are the unique family of fluids which are synthetized by dispersing nano size matter as nanowires, nano-particles and nano-tubes in base fluids. The dispersion of solid particles in base fluids have great potential in heat-management fluids (Keblinski et al., 2002). The advancement in nanotechnology along with applications of nanofluids in heat transfer brought breakthrough in this field (Safaei et al., 2016).

The history of nanofluids started with enhancing single-phase fluids thermal conductivity by adding particles which firstly represented by Maxwell (1873) but the technique failed due to sedimentation and clogging problems. Masuda et al. (1993) proceeded the Maxwell work and achieved increment of thermal conductivity by adding micro-sized nanoparticles but the performance was low. Choi and Eastman (1995) introduced nanofluids by dispersing nanosized particles into base fluid. The resulted substance exhibited better performance having good thermophysical properties with minor stability problems. The various studied on progress on nano fluids (Trisaksri and Wongwises, 2007; Özerinç et al., 2010; Wang

DOI: 10.4018/978-1-7998-1546-4.ch002

and Mujumdar, 2007; Wang and Mujumdar, 2008; Li et al., 2009; Kakaç and Pramuanjaroenkij, 2009) concerned with studies of thermophysical properties and convective heat transfer coefficient of nano fluids. Nanofluids are used in increasing rate of heat transfer by enhancement of "h" due to increment in thermal conductivity. When nanoparticles are present in base fluid then each particle gets electrically charged by absorbing and desorbing an electron (Popa et al., 2010).

The Vander Waals forces dominant to attractive forces in case of stable nanofluids which is vital factor in increasing thermal conductivity (Jang and Choi, 2004; Xuan et al, 2003). Thus, the nanofluids (Mukherjee and Paria, 2013) have potential features which distinguished from other fluids such as:

- Increase of thermal conductivity
- High heat transfer capability
- Stability good as compared to other colloidal
- Reduce power of pumping
- Reduction coefficient of friction

2.2 PREPARATION OF NANOFLUIDS

The preparation of nanofluids is to be done by proper mixing of nano particles by utilization of various methods involved in it. The nanofluids preparation can be categorized to two basic categories based on number of steps which are single and two steps preparation method. The Figure 1 indicates various techniques for preparation of nanofluids.

2.2.1 One-Step Method

Single-step involves the nanofluid preparation by combining the production and nanoparticles dispersion with base fluid, but these two steps occurred at a once time as illustrated in Figure 2. The various methods included in single-step method are laser ablation, vapor deposition and submerged arc (Ali et al., 2018).

The nanoparticles are directly produced by condensation of nanoparticles from vapor phase into low vapor pressure fluid called VEROS (vacuum evaporation on running oil substrate) or Physical deposition technique (Akoh et al., 1978). Choi et al., (2001) investigated direct evaporation condensation method which result in best control size of nanoparticles and stable nanofluid without adding additive. Eastman et al. (1996) utilized VEROS method which involved condensation of vapors of metal to nanoparticles and dispersed into base fluid directly.

2.2.2 Two-Step Method

The binary-step method is mainly used in preparation of nano fluids and preparation is generally completed into two steps. In the first step nanoparticles, nano tubes are produced in form of nano powder by sol-gel technique or hydrothermal synthesis and in second step such nano powder is dispersed into base fluid with the aid of homogenizer, magnetic stirring or ultra-sonic devices (Sharma and Gupta, 2016).

The binary-step technique is used for nanofluids preparation on large scale but treatment of drying, storage and transportation cannot be denied. However, the formation of agglomerates is problem in two

Figure 1. Various methods in nanofluids preparation

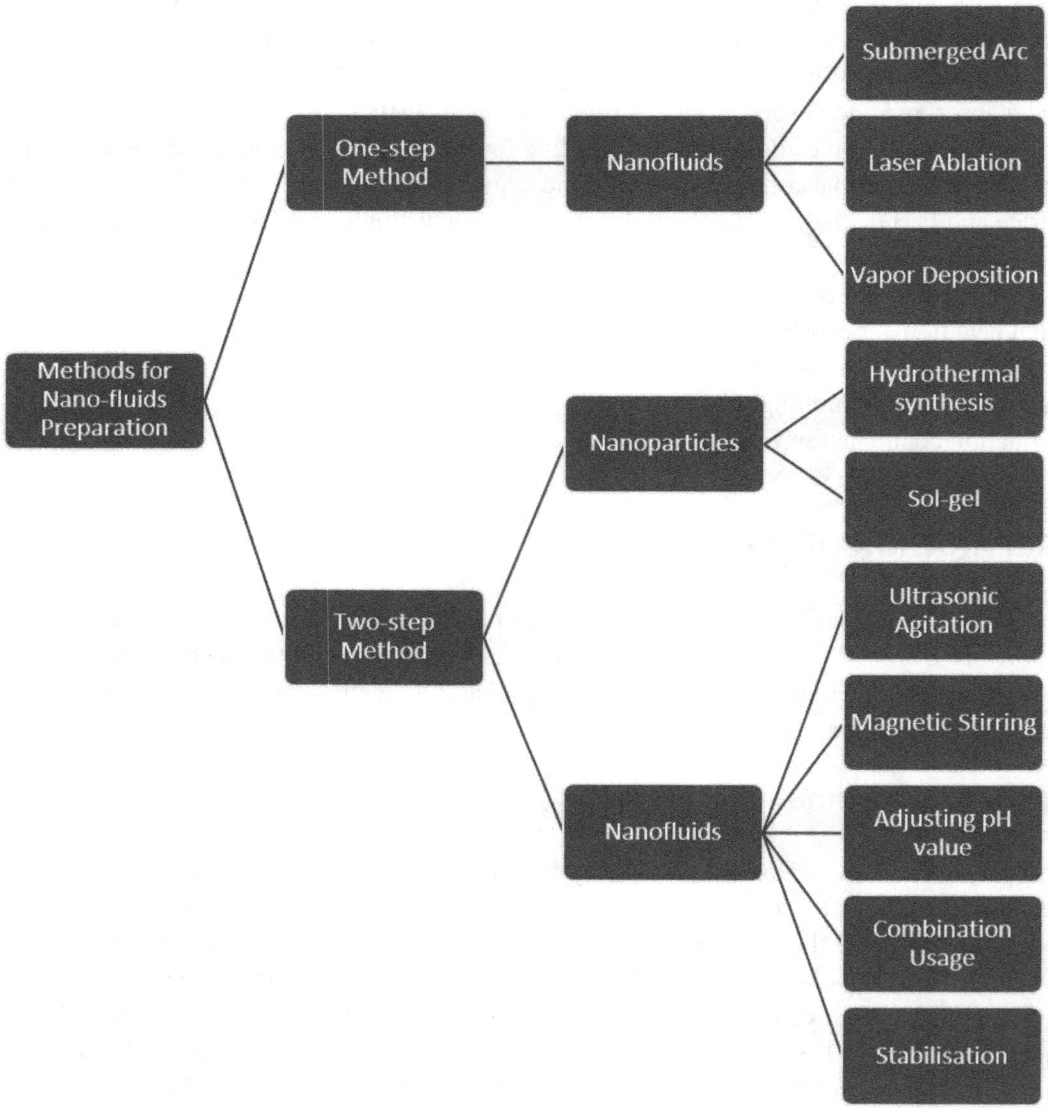

step method which can be minimized by adding surfactants while the one step method is free from this problem (Mohammed et al., 2011; Yu et al., 2012).

2.2.3 Dispersion Technology of Nanofluids

The dispersion behaviour of nanofluids largely influenced by microscopic forces that are exerted on nano particles. The sedimentation of nanoparticle occurred due to gravity effect on single nanoparticle due to density difference (Yu et al., 2017). In the same time the nanoparticle undergoes Brownian force which balance the gravitational sedimentation and result in dispersed in base fluid indicated in Fig. 4.

Figure 2. Preparation of nanofluid using Single Step method (Ali et al., 2018)

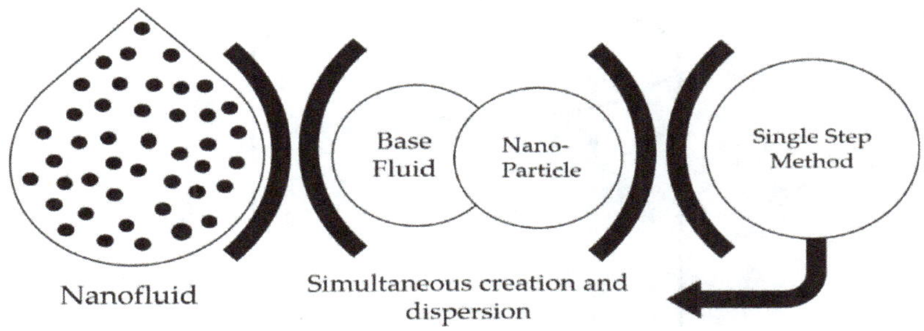

Along these forces, the interparticle forces affect the dispersion behaviour of nanoparticle (Israelachvili, 2011; Russel et al., 1991). As the interaction of nanoparticle take place, neighboring nanoparticles exerts interaction forces due to van der Waals forces. The electrostatic forces between two neighboring particles is given by Pierre (2013);

$$Electrostatciforce = \epsilon\epsilon_o\xi^2 \quad (1)$$

where ϵ the permittivity of base fluid is, ϵ_o is the vacuum permittivity and ζ is the electrostatic potential of nano-particles. The interaction between neighboring nanoparticles depends mainly on distance between nanoparticles and shape of nanoparticles also depend on this factor.

Figure 3. Preparation of nanofluid using Two Step method (Sharma and Gupta, 2016)

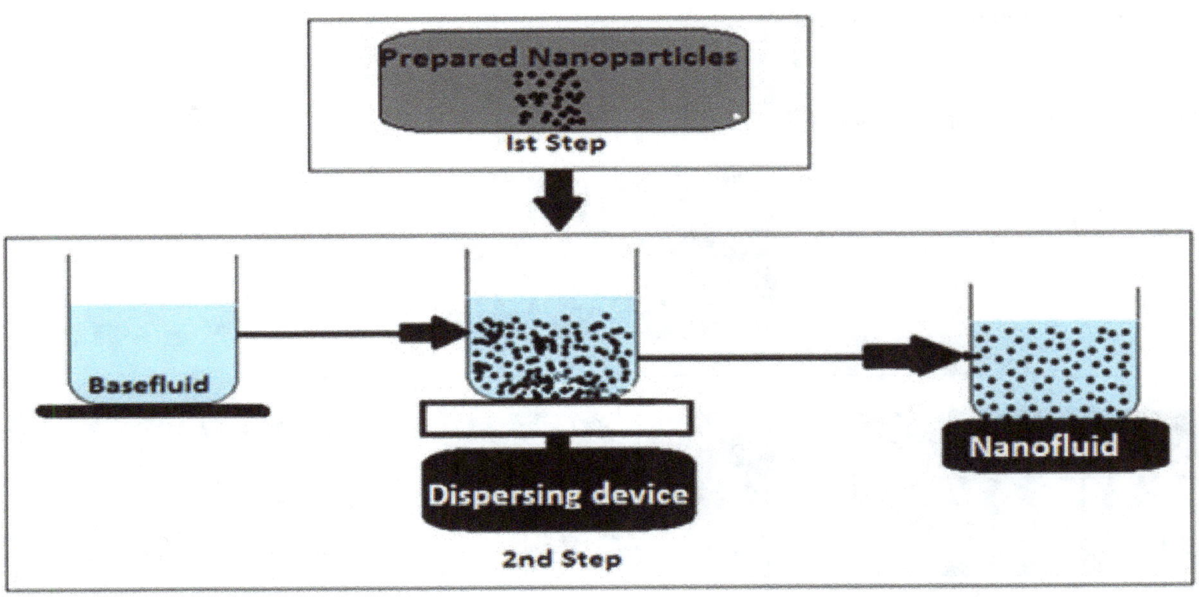

Figure 4. Schematic of nanofluid dispersion show various types of forces (Yu et al., 2017)

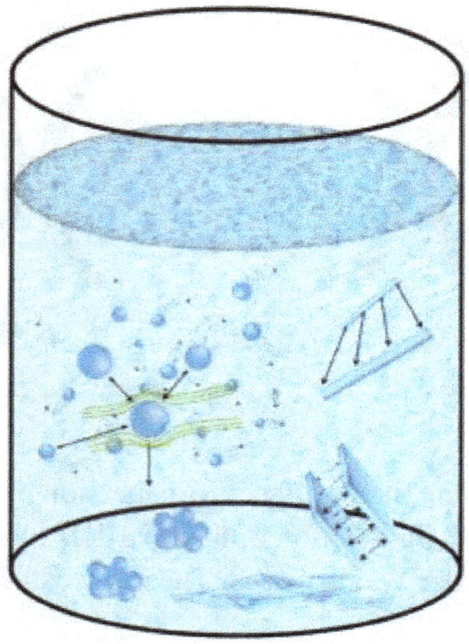

There are two mainly dispersion mechanisms are used namely electrostatic stabilization and steric stabilization that are used to enhance the repulsive forces between nanoparticles for balancing of van der Waals forces. While the above two method combined to give electrosteric dispersion. The different types of dispersion methods are shown in Fig. 5.

In electrostatic stabilization the repulsive forces exists in same surface charges of nanoparticles which can be created by following methods (Brinker and Scherer, 2013);

- Adsorption of ions
- Substitution of ions
- Dissociation of charge specie

Figure 5. Various methods for dispersion

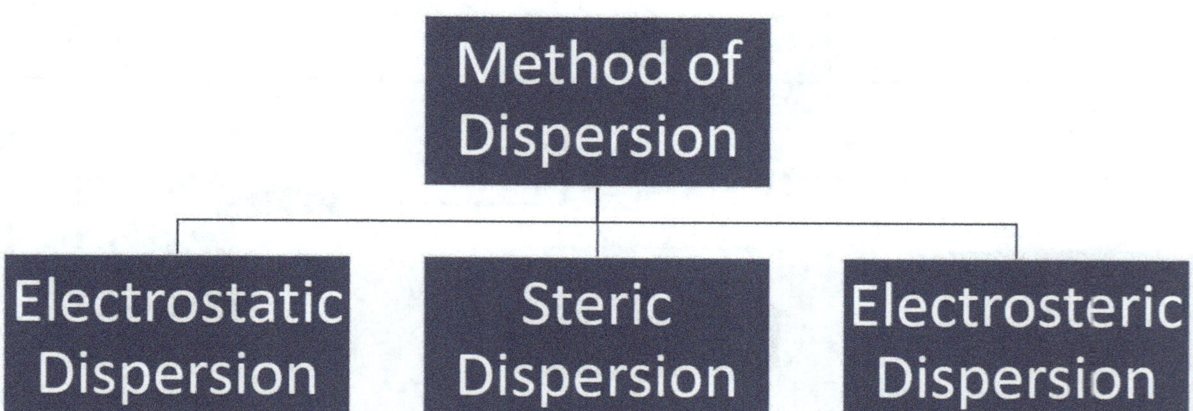

Figure 6. Types of stabilization mechanism (A) electrostatic stabilization (B) steric stabilization

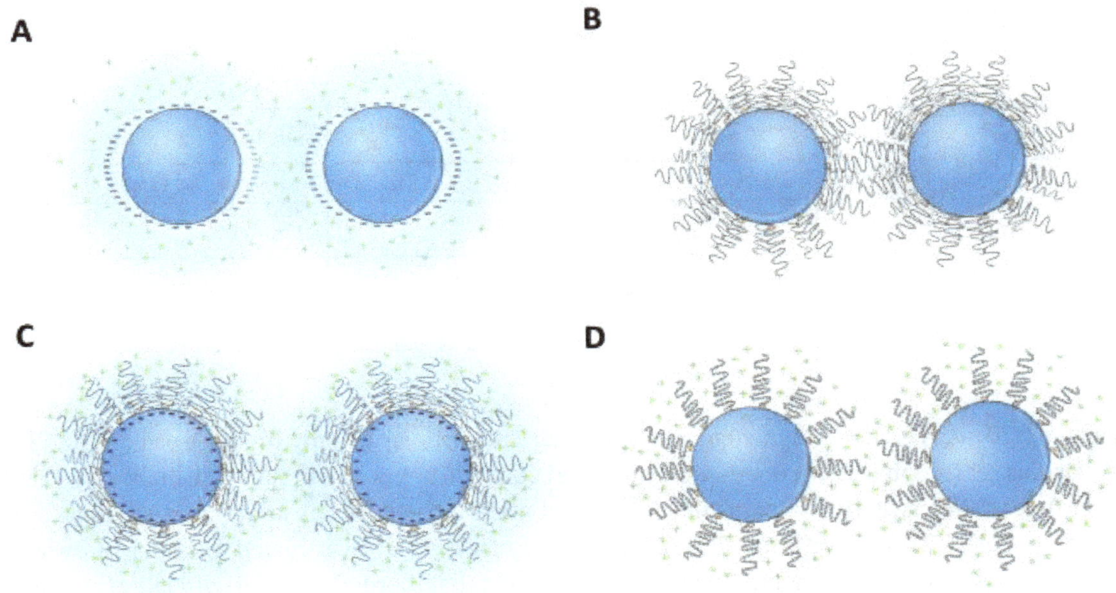

- Physical adsorption of charge specie
- Depletion of electrons at surfaces

In this technique an equalized amount of counter ions will surrounded on charge particles to gain charge neutrality and gain dispersion stability. The electrostatic stability of particles can be interpreting by DLVO theory which described that total electrostatic stabilized particles is sum of van der Waals interaction and electrostatic repulsion as shown in Fig. 6A.

The widely used dispersion technique (Napper, 1983; K. and Birdi; 2015) is steric stabilization for colloidal dispersion as shown in Figure 6B. The macromolecules are attached on surface of nanoparticles as steric barrier. The polymers are attached by chemical, physical adsorption or chemical bonding. The high concentration of polymers materials results in osmotic effect which results in good steric stabilization.

The above two techniques are used in stabilization of nanoparticles by combination of both which is also called electro steric stabilization. In this method the polymer chains are attached to charged surfaces or uncharged surfaces as shown in figure 6C and 2-6D to get better dispersion.

2.3 CHARACTERISTICS OF NANOFLUIDS

2.3.1 Organization Structure of Nanofluids

The type of base fluid and shape of particles are key components for determining structure of nanofluid. The particle shape may be spherical or cylindrical like nano rods, nano wires and nano sheets

Table 1. Particle shape derived from transmission electron microscope (Meriläinen et al., 2013)

Material	Shape	Size (nm)
SiO_2	Irregular	28-110
SiO_2	Irregular	65 ± 34
SiO_2	Spherical	65 ± 1.8
MgO	Rod	15-47
MgO	Rod	21 ± 10
Al_2O_3	Nearly spherical	14-53
Al_2O_3	Irregular	8.2 ± 3.1

(Yang et al., 2017). The lengt/diameter ratio of cylindrical particles usually high (Özerinç et al., 2010). Thus, various thermophysical properties like thermal conductivity, viscosity also influence by shape of nano particles. The viscosity of nanofluids having cylindrical particles has larger viscosity than spherical nano particles (Timofeeva et al. 2009). Timofeeva et al. 2009 investigated alumina-EG/water for thermal conductivity measurement. The authors concluded that thermal conductivity largely depend on shape of nano particle and platelets blade has high thermal conductivity. Jeong et al. (2013) concluded shape of the nano particle on thermal conductivity increment has vital effect and enhanced upto 12 % and 18 % in case of spherical and rectangular nano particles based on ZnO fluids. Murshed et al. (2005) resulted that in TiO_2 nanofluids that rod like nanoparticles have larger thermal conductivity than spherical nano particles. Merilainen et al.(2013) investigated the various shapes of nanoparticles and concluded that three out of seven samples contained irregular shapes. The rod-shape Al_2O_3 particles form short particle chain and MgO have rod shape which contained big agglomerates.

2.3.2 Thermal Conductivity of Nanofluids

The thermal conductivity is the thermophysical property which indicates heat transfer phenomena in nanofluid. The increment in thermal conductivity is associated to convection current between nanoparticles and base fluid and higher thermal conductivity indicates higher performance of nanofluids. The thermal conductivity (Singh, 2008) increment is defined as $\frac{k_{eff}}{k_o}$ ratio which is expressed in percentage and k_0 is thermal conductivity of base fluid and can be written as

$$\frac{k_{eff}}{k_o} = 1 + C_k \varnothing \tag{2}$$

where C_k is thermal conductivity enhancement coefficient.

The Maxwell model (Levin and Miller, 1981) is considered a first model that described thermal conductivity measurement of solid liquid suspension. The model is constructed on the basis of spherical nanoparticles having small concentration which dispersed in base fluid. The model is

$$k_{eff} = k_f \frac{k_{np} + 2k_f + 2\varnothing\left(k_{np} - k_f\right)}{k_{np} + 2k_f - \varnothing\left(k_{np} - k_f\right)} \tag{3}$$

Nan et al. (1997) summarized Maxwell model with addition factor of thermal resistance along particle size, shape and concentration. The model can be expressed as

$$k_{eff} = k_f \left[\frac{2k_f + (1+2\alpha)k_{np} + 2\phi\left[(1-\alpha)k_{np} - k_f\right]}{2k_f + (1+2\alpha)k_{np} - \phi\left[(1-\alpha)k_{np} - k_f\right]} \right] \tag{4}$$

The above expression contained dimensions less parameter α which describes the fluid particle interaction and is expressed as

$$\alpha = \frac{Rk_f}{r_{np}} \tag{5}$$

Where R is thermal boundary resistance coefficient and r is dispersed nanoparticles radius.

Timofeeva et al. (2007) suggested thermal conductivity model based on effective medium theory which can be written as;

$$k_{eff} = k_f\left(1 + 3\varnothing\right) \tag{6}$$

Xue (2005) also presented model for measuring thermal conductivity of carbon nanotube nanofluids given as;

$$k_{eff} = k_f \frac{1 - \varnothing + 2\varnothing \dfrac{k_{np}}{k_{np} - k_f} + ln\dfrac{k_{np} + k_f}{2k_f}}{1 - \varnothing + 2\varnothing \dfrac{k_f}{k_{np} - k_f} + ln\dfrac{k_{np} + k_f}{2k_f}} \tag{7}$$

The thermal conductivity of nanofluid was also proposed by model proposed by Hamilton and Crosser (1962);

$$k_{eff} = k_{bf} \left[\frac{k_{np} + (n-1)k_{bf} - (n-1)\varnothing\left(k_{bf} - k_{np}\right)}{k_{np} + (n-1)k_{bf} + \varnothing\left(k_{bf} - k_{np}\right)} \right] \tag{8}$$

where $n = \dfrac{3}{\omega}$ in which n and ω present the empirical shape factor and sphericity respectively.

Table 2. Thermal conductivity expressions represented by various authors

Authors	Expression of model	Remarks
Maxwell Mode(Levin and Miller, 1981)	$k_{eff} = k_f \frac{k_{np} + 2k_f + 2\varnothing(k_{np} - k_f)}{k_{np} + 2k_f - \varnothing(k_{np} - k_f)}$	For spherical particles
Timofeeva et al.(2007)	$k_{eff} = k_f(1 + 3\varnothing)$	For spherical particles
Xue(2005)	$k_{eff} = k_f \frac{1 - \varnothing + 2\varnothing \frac{k_{np}}{k_{np} - k_f} + ln \frac{k_{np} + k_f}{2k_f}}{1 - \varnothing + 2\varnothing \frac{k_f}{k_{np} - k_f} + ln \frac{k_{np} + k_f}{2k_f}}$	For nano tubes
Wie et al. (2017)	$k_{eff} = k_f \left[1 + \frac{\frac{3fq(p)}{p_o}}{1 - \frac{fq(p)}{p_o}} \right]$	For inclusion nanolayer thickness, temperature and volume fraction
Patel et al. (2010)	$k_n = k_f \left[1 + 0.135 \left(\frac{k_{np}}{k_f} \right)^{0.273} \left(\frac{T}{20} \right)^{0.547} \left(\frac{100}{d_{np}} \right)^{0.234} \phi^{0.467} \right]$	For spherical nanoparticles having 10-150 nm diameter
Patel et al. (2008)	$k_n = k_f \left(1 + \frac{k_p \phi d_f}{k_f (1 - \phi) d_p} \right)$	For carbon nanotubes nanofluids

Pryhnikov et al. (2017) determined nanofluids thermal conductivity by non-stationary hot wire method. The Wheatstone bridge was used to electrical resistance measurement of hot wire. The basic circuit is shown in Figure 7. The bridge has two resistors and a resistor box and around fifty different nanofluids having water, ethylene glycol and engine oil as base fluids were under consideration. The particles concentration was from 0.25-8 % and size range from 10-50 nm.

Murshed et al. (2008) conducted experiments to measure thermal conductivity of nanofluids. The thermal conductivity was measured by hot wire method at 20-60 $^{\circ}C$. The schematic is indicated in Figure 8 and setup was calibrated by measuring thermal conductivity of base fluid. The value of "K" was found to be high as compared to base fluid. The thermal conductivity was also seen to be dependent on temperature strongly.

Figure 7. The electric circuit for experimental thermal conductivity measurement (Pryhnikov et al., 2017)

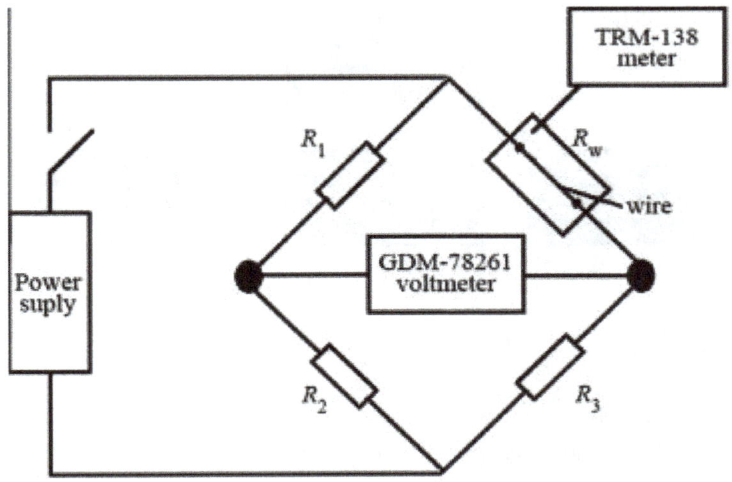

2.3.3 Influencing Factors of Thermal Conductivity of Nanofluids

The various factors influence the thermal conductivity of nano fluids are nano particle size used in nano fluid, particle concentration in nano fluid particle shape used in nano fluid, the type of base fluid used, the effect of temperature and pH. The summary of different factors that affect the thermal conductivity of nano fluids are shown in Fig. 9.

Figure 8. Hot wire method for thermal conductivity measurement (Murshed et al., 2008)

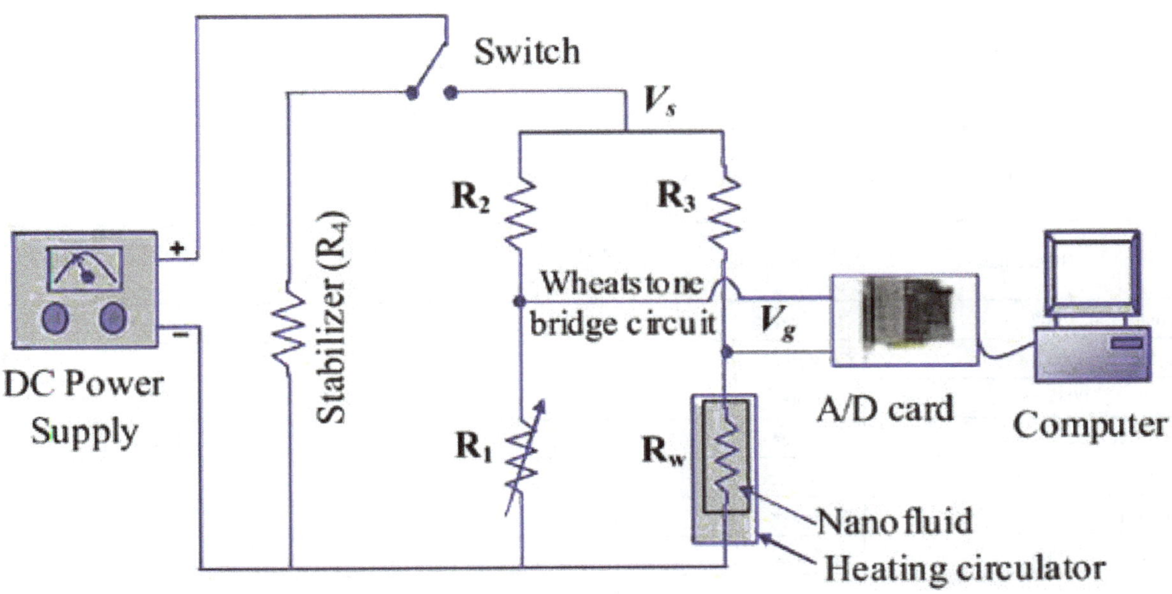

Figure 9. Various factors that influence nanofluids thermal conductivity

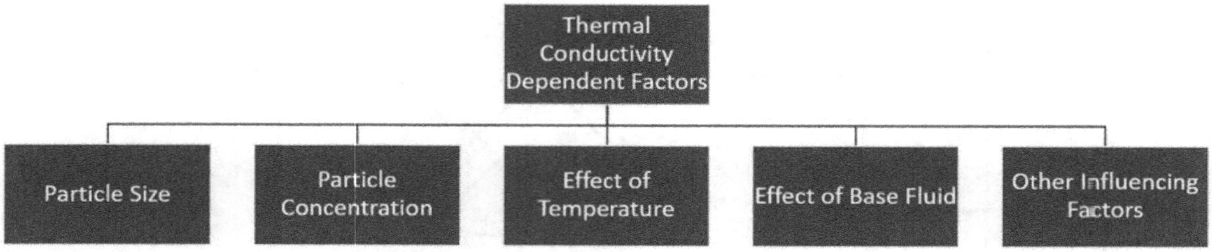

(1) Effect of Particle Size

Due to smaller size, nano-particles have surface area greater resulted in higher heat transfer rate. Nano particles have high thermal conductivity, but heat capacity is low (Ali et al., 2018). The particle size effect nanofluids thermal conductivity and thermal conductivity of nanofluid investigated by various authors with particle size variation. Chon et al. (2005) found Al_2O_3-water nanofluids thermal conductivity having 1 % volume loading. The thermal conductivity was enhanced around two time when size of nano-particle increased from 47-150 nm.

The increased in thermal conductivity of Fe_3O_4 particles also observed by Shima et al. (2009) with 1 % and 5.5 % volume concentration along oil as base fluid at 25 °C. They observed that increment in thermal conductivity occurred with particle size range 2.8-9.5 nm.

The metal nanoparticles were also studied by various researchers (Patel et al., 2003; Pal et al., 2010) which included Au-water and Cu-water nanofluid and conclusion showed that thermal conductivity vary with particle size. Thus, smaller sized of nanoparticles contribute to increment in thermal conductivity. This is the reason that large Brownian motion and surface effects dominant with small particle size. But this is not perfect in each case.

Table 3. Summary of various authors studied nanofluids

References	Nanofluid	Size of Particle diameter nm	Concentration volume %	Temperature °C
Xuan et al., 2002	Cu-water	26	1 & 2	30-60
Liu et al., 2006	Cu-water	50 to100	0.1	Room
Li et al., 2008	Cu-water	25	0.2-0.8 (wt)	Room
Yang et al., 2012	Cu-water	50 to100	0.1	Room
Putnam et al., 2006	Au-toluene	2	0.11-0.36	Room
Kim et al., 2009	Au-water	7.1 to 12.1	0.018	Room
Kang et al., 2006	Ag-water	8 to 15	0.10-0.39	Room
Qiou, 2010	Ag-water	60	0.03-0.121	20-60

(2) Effect of Particle Concentration

The particle concentration influences greatly on thermal conductivity of nanofluids. Many researchers used various type of particles with different base fluids to check thermal conductivity. The particles included metals, oxides, carbon nanotubes to investigate the influence. Generally small size wide spread nanoparticles are helpful in determining the effect of thermal conductivity factor.

Kang et al. (2006) investigated thermal conductivity enhancement while keeping other factors constants. The author observed that increment in thermal conductivity from 3-11 % as increasing concentration of volume from 0.1 % to 0.39 %. Duangthongsuk and Wongwises (2009) concluded that as volume concentration of TiO_2 in water increased from 0.2-2 % the thermal conductivity was enhanced from 7.2-13.2 %.

Kim et al. (2009) concluded that with increment of 0.018 % volume fraction of Au nanoparticles increased thermal conductivity of fluid about 9 % . Murshed et al. (2010) resulted increment in thermal conductivity of TiO_2 was higher at large of concentration of particles as compared to low particle loading.

(3) Temperature Effect

The variation in temperature of nanofluids changes nanoparticles Brownian motion with largely influence the thermal conductivity of nanofluids. The changes of temperature on Al_2O_3, CuO nanofluids was studied by Das et al. (2003). The multiple readings were taken between temperature range of 21-51 °C and enhancement in temperature was 9.4 to 24.3 %.

The effect of temperature with carbon nanofluids also investigated by Ding et al. (2006). The authors measured the thermal conductivity of MWCNT/water nanofluids. The hot wire method was applied to find thermal conductivity. Temperature was measured at various temperature of 20,25,30 °C and weight of particle changed between 0.1-1 %. The maximum increment of 80 % was found at 30 °C having 1 % weight of MWCNT/water nanofluids. Xuan et al. (2002) resulted that thermal conductivity enhanced from 14 to 23 % as temperature increased from 30 to 60 °C in 1 % volume while 24 to 33 °C for 2 % volume loading.

(4) Effect of Base Fluid

By changing base fluid the viscosity vary and result in variation of thermal conductivity. Li and Peterson (2010) determined effect of double layer forming around nanoparticles and exhibit dependence of thermal conductivity on base fluid. Chopkar et al. (2008) investigated water and ethylene glycol which were used as base fluid to check variations effects of both fluids. The nanoparticles of Al_2Cu and Ag_2Al were under examined and results indicated that water nanofluids was higher thermal conductivity. Moreover 100 % thermal conductivity achieved from 2 % volume of Ag_2Al/water. Sonawane et al. (2015) investigated various types of base fluid with nano TiO_2 composition. The ultrasonic sol-gel method was used for nano-fluids preparation. The author report also revealed that lower base fluid viscosities produced higher increment in thermal of nanofluids.

Figure 10. Schematic diagram for experimental setup (Sheikholeslami et al., 2018)

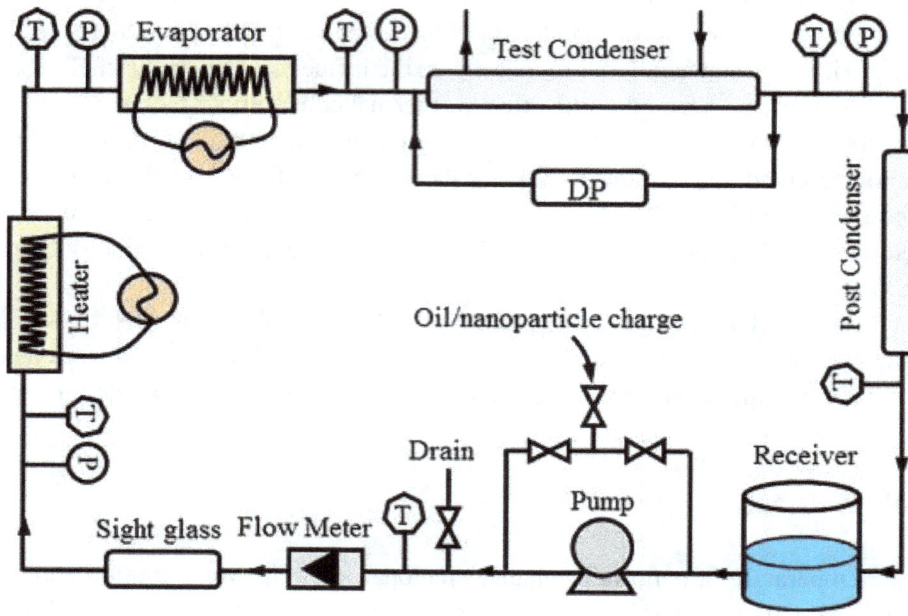

(5) Other Influencing Factors

The various authors also determined various effects like pH, type of particle, type of surfactant that influence thermal conductivity of nanofluids. These factors consider an important part increasing in thermal conductivity. The variation of pH on thermal conductivity has vital factor and it effect is significance (Yang et al., 2017). Similarly, the other factors have their own dependence on thermal conductivity of nanofluids.

2.4 FLOW AND ENERGY TRANSFER MECHANISM OF NANOFLUIDS

The addition of nanofluids enhanced the heat transfer phenomena. The flow of heat transfer indicate by heat transfer coefficient which depends upon thermal conductivity, heat capacity, density (Mohammed et al., 2011).

2.4.1 Experimental Analysis of Flow and Heat Transfer of Nanofluids

Chein and Chuang (2007) studied mixture of CuO-H_2O as cooling agent to investigate variations of heat transfer in microchannel heat sink. The study revealed that heat transfer by nanofluid was higher than water as cooling fluid. The agglomeration was also prevented by enhancing temperature of nanofluids. Chein and Haung (2005) also experimentally studied the effect of heat transfer of nanofluids on microchannel heat sinks. The results indicated that efficiency of heat sinks was greatly enhanced by using nanofluids. The pressure drop was also less due to small size of nanoparticles. Sheikholeslami et al. (2018) analyzed experimentally to estimate the influence of refrigerant base nanofluid inside tube during condensation. The schematic diagram is shown in Fig. 10.

The R600 and CuO were selected as base fluid and nanoparticles and effect of heat transfer was measured. The results indicated that nanoparticles Brownian motion and deposition on tube inner surface enhanced the heat convection in fluid.

Pantzali et al. (2009) conducted experiments to study influence of nanofluids on miniature plate heat exchanger. The nanofluid used was CuO/water and results indicated that at low flow rate the heat transfer was high as compared to high flow rate where main heat transfer is due to convection and nanofluids contribution is limited. Shahrul et al. (2016) studied various types of nanofluids in shell and tube heat exchanger to measure potential of heat transfer. The Al_2O_3, ZnO and SiO_2 nanoparticles were used with water and polyvinylpyrrolidone act as a surfactant. The results showed that increment in heat transfer was about 35 %,26 %, 12 % for Zn-water, Al-water, Si-water respectively.

2.4.2 Natural Convection Analysis of Nanofluids

The various authors investigated the effect of heat transfer by adding nanoparticles in base fluid. Usually nanoparticles addition enhance heat transfer of nanofluid as well as various other factors. Choi et al. (2001) reported that with adding little quantity of nanoparticles traditionally double the thermal conductivity of that fluid.

The heat transfer in nanofluids occurred by Brownian motion which can be defined as particle diffusion motion which is defined by Stokes-Einstein formula

$$D = \frac{k_B T}{3\pi\eta d}$$

(2-9)

Where k_B is Boltzmann constant, d is the fluid diameter and η is fluid viscosity. Keblinki et al. (2002) represented that Brownian motion is less than thermal diffusion due to slow Brownian motion which cannot transport significant amount of energy. The Brownian motion however produced clustering which in turn enhanced value of "k". Xuan and Roetzel (2000) represented that with low particle concentration the value of k can increase up to 20%.

The heat transfer in nanofluids transferred with the help of phonons means by lattice vibration which are scattered by each other. The particles in nanofluids are close to each other and due to Brownian motion, the heat transfer phenomena enhanced.

Kim et al. (2009) carried out tests to measure convective heat transfer of nanofluids for laminar and turbulent conditions. At 3 % volume of alumina nanofluids, the convective heat transfer enhanced by 15 % to 20 %. For amorphous carbonic nanofluid the convective heat transfer increased in laminar flow upto 8 %.

Trisaksri and Wongwises (2007) instigated analytical investigations for calculating convective heat transfer and results indicated that it increased with particle volume fraction but experimentally it was decreased.

2.4.3 Boiling Heat Transfer Characteristics of Nanofluids

The nano-fluids addition in pool boiling heat transfer in increasing critical heat flux has major contribution. The increment in critical heat flux is due nano-particles on surface heating (Kamatchi and Venkatachalapathy, 2015). The boiling phenomena has wide applications related to chemical, desalination, metallurgy etc. The boiling process improved by increasing critical heat flux which can be enhanced by increment in thermal conductivity and turbulence in fluids (Babar and Ali, 2019).

Hashemi and Noie (2017) studied the flow boiling heat transfer characteristics using MWCNT/water in horizontal tube. The surfactant gum Arabic (GA) was used and mix with MWCNT in ratio of 1:1. The effective "h" in case of water and MWCNTsGA/water nanofluids by various mass fluxes is indicated in Figure 11. The experimental results indicated that boiling characteristics improve by using nanofluids and nano fluid heat transfer coefficient was higher than pure water which was 4.3 % at critical heat flux.

Sarfraz and Hormozi (2016) carried out experiments on zirconium oxide nanofluids which were prepared by chemical sol-gel method. The nano powder was dispersed in water-ethylene glycol mixture as base fluid with various particle concentration. The experiments showed that boiling heat transfer coefficient enhanced upto 12 % at volume 0.1 % .The particle deposition on surface result in increasing critical heat flux by 29 % .

Sarfraz and Hormozi (2015) also carried out experiment to study "h" of CuO water based nano fluids having concentration of 0.1-0.4 % . The experimental setup is shown in Figure 12 and have major components as measurement instrument, boiling test section and photographic section.

The results indicated that by adding surfactant the pool boiling "h" was higher than without using it. However, decline in pool boiling "h" was occurred during experiments which was due to reduction of nucleation sites.

Kim et al. (2015) reviewed micro and nanostructured surfaces which enhanced the boiling heat transfer. The nucleation sites increased by creating these surfaces and evaporation mass flux increased. The small size roughness on surface enhance nucleation sites and promote bubble formation and micro sized channel created by surface morphologies give additional passage for critical heat flux achieved.

2.5 Other preparation methods nanofluids

Zhu et al. (2004) proposed single-step chemical method for nano-fluids preparation as single-step physical method is costly and cannot prepared nanofluids on market level. The copper nanofluids were prepared by converting $CuSO_4.5H_2O$ to $NaH_2PO_2.H_2O$ under microwave irradiation and resulted in better dispersed copper nanofluid. The nanofluids based on natural oil having silver nanoparticles were also prepared by this method (Bönnemann at al, 2005). Yu and Xie, 2012 developed continuous flow microfluidic microreactor to prepared Cu nano fluids. By utilizing this technique, Cu nano fluids can be prepared with varying properties as per desired conditions. The copper oxide nanofluids (Zhu et al., 2007) with large solid volume can be synthetized by novel precursor transformation method which used ultrasonic and microwave irradiation. The precursor Copper hydro-oxide converted into copper oxide nanoparticles in water by microwave irradiation. Feng et al. (2006) found aqueous organic phase transfer technique for silver, gold and platinum nanoparticles preparation. This technique is used for kerosene

Figure 11. Experimental "h" for water and MWCNTs-GA/water nanofluid with various mass fluxes: (a) at inlet temperature 70 °C (b) CHF (Hashemi and Noie, 2017)

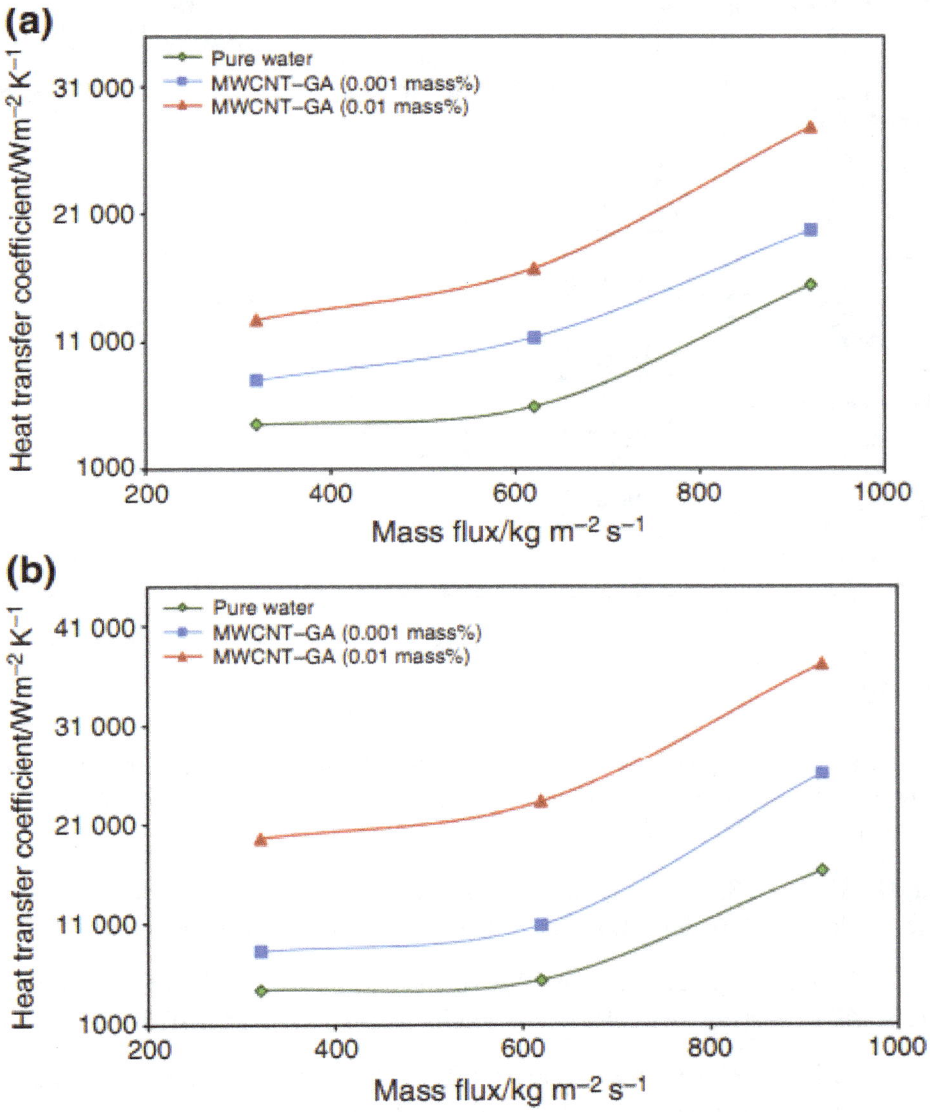

based Fe_3O_4 nanofluids preparation (Yu et al, 2010). The oleic acid is spread on nano particles surface which resulted in better compatibility with kerosene.

Figure 12. Schematic diagram of test sample (Sarfraz and Hormozi, 2015)

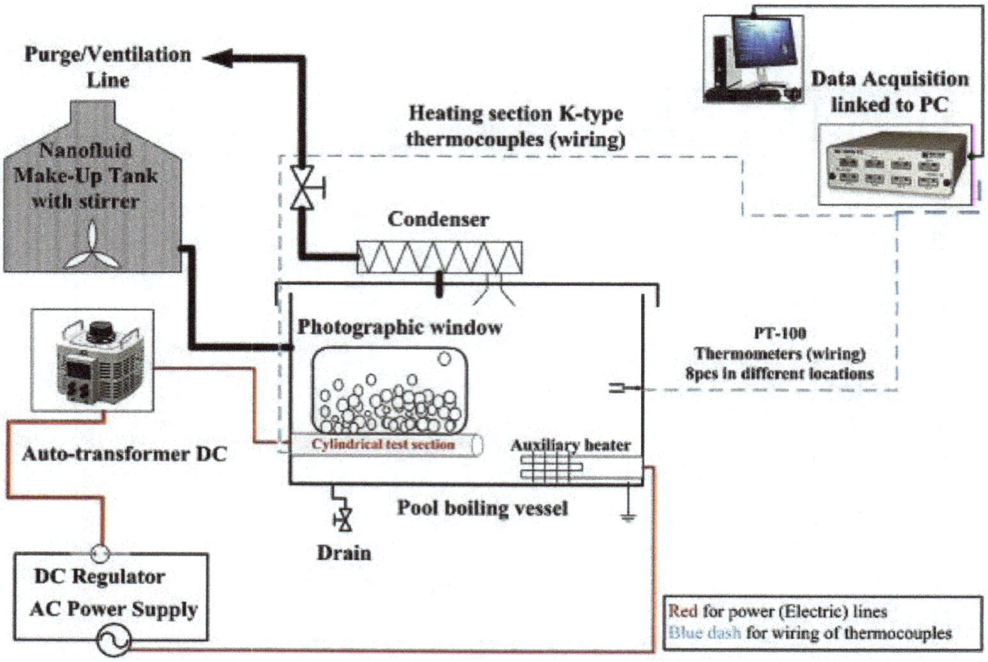

REFERENCES

Akoh, H., Tsukasaki, Y., Yatsuya, S., & Tasaki, A. (1978). Magnetic properties of ferromagnetic ultrafine particles prepared by vacuum evaporation on running oil substrate. *Journal of Crystal Growth*, *45*, 495–500. doi:10.1016/0022-0248(78)90482-7

Ali, H., Babar, H., Shah, T., Sajid, M., Qasim, M., & Javed, S. (2018). Preparation techniques of TiO2 nanofluids and challenges: A review. *Appl. Sci.*, *8*(4), 587. doi:10.3390/app8040587

Ali, H. M., Shah, T. R., Babar, H., & Khan, Z. A. (2018). *Application of nanofluids for thermal management of photovoltaic modules: a review*. Microfluid. Nanofluidics, IntechOpen. doi:10.5772/intechopen.74967

Babar, H., & Ali, H. M. (2019). Towards hybrid nanofluids: Preparation, thermophysical properties, applications, and challenges. *Journal of Molecular Liquids*, *281*, 598–633. doi:10.1016/j.molliq.2019.02.102

Bönnemann, H., Botha, S. S., Bladergroen, B., & Linkov, V. M. (2005). Monodisperse copper-and silver-nanocolloids suitable for heat-conductive fluids. *Applied Organometallic Chemistry*, *19*(6), 768–773. doi:10.1002/aoc.889

Brinker, C. J., & Scherer, G. W. (2013). *Sol-gel science: the physics and chemistry of sol-gel processing*. Academic press.

Chein, R., & Chuang, J. (2007). Experimental microchannel heat sink performance studies using nanofluids. *International Journal of Thermal Sciences*, *46*(1), 57–66. doi:10.1016/j.ijthermalsci.2006.03.009

Chein, R., & Huang, G. (2005). Analysis of microchannel heat sink performance using nanofluids. *Applied Thermal Engineering*, *25*(17-18), 3104–3114. doi:10.1016/j.applthermaleng.2005.03.008

Choi, S. U. S., & Eastman, J. A. (1995). *Enhancing thermal conductivity of fluids with nanoparticles*. Argonne National Lab.

Choi, S. U. S., Yu, W., Hull, J. R., Zhang, Z. G., & Lockwood, F. E. (2001). *Nanofluids for vehicle thermal management*. SAE Technical Paper.

Choi, S. U. S., Zhang, Z. G., Yu, W., Lockwood, F. E., & Grulke, E. A. (2001). Anomalous thermal conductivity enhancement in nanotube suspensions. *Applied Physics Letters*, *79*(14), 2252–2254. doi:10.1063/1.1408272

Chon, C. H., Kihm, K. D., Lee, S. P., & Choi, S. U. S. (2005). Empirical correlation finding the role of temperature and particle size for nanofluid (Al 2 O 3) thermal conductivity enhancement. *Applied Physics Letters*, *87*(15), 153107. doi:10.1063/1.2093936

Chopkar, M., Sudarshan, S., Das, P. K., & Manna, I. (2008). Effect of particle size on thermal conductivity of nanofluid. *Metallurgical and Materials Transactions. A, Physical Metallurgy and Materials Science*, *39*(7), 1535–1542. doi:10.100711661-007-9444-7

Das, S. K., Putra, N., Thiesen, P., & Roetzel, W. (2003). Temperature dependence of thermal conductivity enhancement for nanofluids. *Journal of Heat Transfer*, *125*(4), 567–574. doi:10.1115/1.1571080

Ding, Y., Alias, H., Wen, D., & Williams, R. A. (2006). Heat transfer of aqueous suspensions of carbon nanotubes (CNT nanofluids), Int. *J. Heat Mass Transf.*, *49*(1-2), 240–250. doi:10.1016/j.ijheatmasstransfer.2005.07.009

Duangthongsuk, W., & Wongwises, S. (2009). Measurement of temperature-dependent thermal conductivity and viscosity of TiO2-water nanofluids. *Experimental Thermal and Fluid Science*, *33*(4), 706–714. doi:10.1016/j.expthermflusci.2009.01.005

Eastman, Choi, Li, Thompson, & Lee. (1996). Enhanced thermal conductivity through the development of nanofluids. *MRS Online Proc. Libr. Arch., 457*.

Feng, X., Ma, H., Huang, S., Pan, W., Zhang, X., Tian, F., ... Luo, J. (2006). Aqueous– organic phase-transfer of highly stable gold, silver, and platinum nanoparticles and new route for fabrication of gold Nanofilms at the oil/water Interface and on solid supports. *The Journal of Physical Chemistry B*, *110*(25), 12311–12317. doi:10.1021/jp0609885 PMID:16800553

Hamilton, R. L., & Crosser, O. K. (1962). Thermal conductivity of heterogeneous two-component systems. *Industrial & Engineering Chemistry Fundamentals*, *1*(3), 187–191. doi:10.1021/i160003a005

Hashemi, M., & Noie, S. H. (2017). Study of flow boiling heat transfer characteristics of critical heat flux using carbon nanotubes and water nanofluid. *Journal of Thermal Analysis and Calorimetry*, *130*(3), 2199–2209. doi:10.100710973-017-6661-1

Israelachvili, J. N. (2011). *Intermolecular and surface forces*. Academic press.

Jang, S. P., & Choi, S. U. S. (2004). Role of Brownian motion in the enhanced thermal conductivity of nanofluids. *Applied Physics Letters, 84*(21), 4316–4318. doi:10.1063/1.1756684

Jeong, J., Li, C., Kwon, Y., Lee, J., Kim, S. H., & Yun, R. (2013). Particle shape effect on the viscosity and thermal conductivity of ZnO nanofluids. *International Journal of Refrigeration, 36*(8), 2233–2241. doi:10.1016/j.ijrefrig.2013.07.024

K. and S. Birdi. (2015). *Handbook of surface and colloid chemistry*. CRC Press.

Kakaç, S., & Pramuanjaroenkij, A. (2009). Review of convective heat transfer enhancement with nanofluids, Int. *J. Heat Mass Transf., 52*(13-14), 3187–3196. doi:10.1016/j.ijheatmasstransfer.2009.02.006

Kamatchi, R., & Venkatachalapathy, S. (2015). Parametric study of pool boiling heat transfer with nanofluids for the enhancement of critical heat flux: A review. *International Journal of Thermal Sciences, 87*, 228–240. doi:10.1016/j.ijthermalsci.2014.09.001

Kang, H. U., Kim, S. H., & Oh, J. M. (2006). Estimation of thermal conductivity of nanofluid using experimental effective particle volume. *Experimental Heat Transfer, 19*(3), 181–191. doi:10.1080/08916150600619281

Keblinski, P., Phillpot, S. R., Choi, S. U. S., & Eastman, J. A. (2002). Mechanisms of heat flow in suspensions of nano-sized particles (nanofluids), Int. *J. Heat Mass Transf., 45*(4), 855–863. doi:10.1016/S0017-9310(01)00175-2

Kim, D., Kwon, Y., Cho, Y., Li, C., Cheong, S., Hwang, Y., ... Moon, S. (2009). Convective heat transfer characteristics of nanofluids under laminar and turbulent flow conditions. *Current Applied Physics, 9*(2), 119–123. doi:10.1016/j.cap.2008.12.047

Kim, D. E., Yu, D. I., Jerng, D. W., Kim, M. H., & Ahn, H. S. (2015). Review of boiling heat transfer enhancement on micro/nanostructured surfaces. *Experimental Thermal and Fluid Science, 66*, 173–196. doi:10.1016/j.expthermflusci.2015.03.023

Kim, H. J., Bang, I. C., & Onoe, J. (2009). Characteristic stability of bare Au-water nanofluids fabricated by pulsed laser ablation in liquids. *Optics and Lasers in Engineering, 47*(5), 532–538. doi:10.1016/j.optlaseng.2008.10.011

Levin, M. L., & Miller, M. A. (1981). Maxwell a treatise on electricity and magnetism. *Uspekhi Fiziologicheskikh Nauk, 135*(11), 425–440. doi:10.3367/UFNr.0135.198111d.0425

Li, C. H., & Peterson, G. P. (2007). Mixing effect on the enhancement of the effective thermal conductivity of nanoparticle suspensions (nanofluids), Int. *J. Heat Mass Transf., 50*(23-24), 4668–4677. doi:10.1016/j.ijheatmasstransfer.2007.03.015

Li, X., Zhu, D., & Wang, X. (2008). Study on dispersion behavior and thermal conductivity of Cu-H~2O nanofluids. *J. Funct. Mater., 39*, 162.

Li, Y., Tung, S., Schneider, E., & Xi, S. (2009). A review on development of nanofluid preparation and characterization. *Powder Technology, 196*(2), 89–101. doi:10.1016/j.powtec.2009.07.025

Liu, M.-S., Lin, M. C.-C., Tsai, C. Y., & Wang, C.-C. (2006). Enhancement of thermal conductivity with Cu for nanofluids using chemical reduction method, Int. *J. Heat Mass Transf.*, *49*(17-18), 3028–3033. doi:10.1016/j.ijheatmasstransfer.2006.02.012

Masuda, Ebata, & Teramae. (1993). *Alteration of thermal conductivity and viscosity of liquid by dispersing ultra-fine particles. Dispersion of Al2O3, SiO2 and TiO2 ultra-fine particles.* Academic Press.

Maxwell, J. C. (1873). *A treatise on electricity and magnetism.* Oxford, UK: Clarendon Press.

Merilläinen, A., Seppälä, A., Saari, K., Seitsonen, J., Ruokolainen, J., Puisto, S., ... Ala-Nissila, T. (2013). Influence of particle size and shape on turbulent heat transfer characteristics and pressure losses in water-based nanofluids, Int. *J. Heat Mass Transf.*, *61*, 439–448. doi:10.1016/j.ijheatmasstransfer.2013.02.032

Mohammed, H. A., Al-Aswadi, A. A., Shuaib, N. H., & Saidur, R. (2011). Convective heat transfer and fluid flow study over a step using nanofluids: A review. *Renewable & Sustainable Energy Reviews*, *15*(6), 2921–2939. doi:10.1016/j.rser.2011.02.019

Mohammed, H. A., Bhaskaran, G., Shuaib, N. H., & Saidur, R. (2011). Heat transfer and fluid flow characteristics in microchannels heat exchanger using nanofluids: A review. *Renewable & Sustainable Energy Reviews*, *15*(3), 1502–1512. doi:10.1016/j.rser.2010.11.031

Mukherjee, S., & Paria, S. (2013). Preparation and stability of nanofluids-A Review. *IOSR Journal of Mechanical and Civil Engineering*, *9*(2), 63–69. doi:10.9790/1684-0926369

Murshed, S. M. S., Leong, K. C., & Yang, C. (2005). Enhanced thermal conductivity of TiO2—Water based nanofluids. *International Journal of Thermal Sciences*, *44*(4), 367–373. doi:10.1016/j.ijthermalsci.2004.12.005

Murshed, S. M. S., Leong, K. C., & Yang, C. (2008). Investigations of thermal conductivity and viscosity of nanofluids. *International Journal of Thermal Sciences*, *47*(5), 560–568. doi:10.1016/j.ijthermalsci.2007.05.004

Murshed, S. M. S., Leong, K. C., & Yang, C. (2008). Thermophysical and electrokinetic properties of nanofluids–a critical review. *Applied Thermal Engineering*, *28*(17-18), 2109–2125. doi:10.1016/j.applthermaleng.2008.01.005

Nan, C.-W., Birringer, R., Clarke, D. R., & Gleiter, H. (1997). Effective thermal conductivity of particulate composites with interfacial thermal resistance. *Journal of Applied Physics*, *81*(10), 6692–6699. doi:10.1063/1.365209

Napper, D. H. (1983). *Polymeric stabilization of colloidal dispersions.* Academic Pr.

Özerinç, S., Kakaç, S., & Yazıcıoğlu, A. G. (2010). Enhanced thermal conductivity of nanofluids: A state-of-the-art review. *Microfluidics and Nanofluidics*, *8*(2), 145–170. doi:10.100710404-009-0524-4

Pantzali, M. N., Kanaris, A. G., Antoniadis, K. D., Mouza, A. A., & Paras, S. V. (2009). Effect of nanofluids on the performance of a miniature plate heat exchanger with modulated surface. *International Journal of Heat and Fluid Flow*, *30*(4), 691–699. doi:10.1016/j.ijheatfluidflow.2009.02.005

Patel, H. E., Anoop, K. B., Sundararajan, T., & Das, S. K. (2008). Model for thermal conductivity of CNT-nanofluids. *Bulletin of Materials Science*, *31*(3), 387–390. doi:10.100712034-008-0060-y

Patel, H. E., Das, S. K., Sundararajan, T., Sreekumaran Nair, A., George, B., & Pradeep, T. (2003). Thermal conductivities of naked and monolayer protected metal nanoparticle based nanofluids: Manifestation of anomalous enhancement and chemical effects. *Applied Physics Letters*, *83*(14), 2931–2933. doi:10.1063/1.1602578

Patel, H. E., Sundararajan, T., & Das, S. K. (2010). An experimental investigation into the thermal conductivity enhancement in oxide and metallic nanofluids. *Journal of Nanoparticle Research*, *12*(3), 1015–1031. doi:10.100711051-009-9658-2

Paul, G., Pal, T., & Manna, I. (2010). Thermo-physical property measurement of nano-gold dispersed water based nanofluids prepared by chemical precipitation technique. *Journal of Colloid and Interface Science*, *349*(1), 434–437. doi:10.1016/j.jcis.2010.05.086 PMID:20609848

Pierre, A. C. (2013). *Introduction to sol-gel processing*. Springer Science & Business Media.

Popa, I., Gillies, G., Papastavrou, G., & Borkovec, M. (2010). Attractive and repulsive electrostatic forces between positively charged latex particles in the presence of anionic linear polyelectrolytes. *The Journal of Physical Chemistry B*, *114*(9), 3170–3177. doi:10.1021/jp911482a PMID:20148528

Pryazhnikov, M. I., Minakov, A. V., Rudyak, V. Y., & Guzei, D. V. (2017). Thermal conductivity measurements of nanofluids. *Int. J. Heat Mass Transf.*, *104*, 1275–1282. doi:10.1016/j.ijheatmasstransfer.2016.09.080

Putnam, S. A., Cahill, D. G., Braun, P. V., Ge, Z., & Shimmin, R. G. (2006). Thermal conductivity of nanoparticle suspensions. *Journal of Applied Physics*, *99*(8), 84308. doi:10.1063/1.2189933

Qiao, F. (2010). *Preparation and property Ag, Graphene nanofluids* (Master Diss.). Qingdao Univ. Sci. Technol.

Russel, W. B., Russel, W. B., Saville, D. A., & Schowalter, W. R. (1991). *Colloidal dispersions*. Cambridge university press.

Safaei, M. R., Safdari Shadloo, M., Goodarzi, M. S., Hadjadj, A., Goshayeshi, H. R., Afrand, M., & Kazi, S. N. (2016). A survey on experimental and numerical studies of convection heat transfer of nanofluids inside closed conduits. *Advances in Mechanical Engineering*, *8*(10). doi:10.1177/1687814016673569

Sarafraz, M. M., & Hormozi, F. (2015). Pool boiling heat transfer to dilute copper oxide aqueous nanofluids. *International Journal of Thermal Sciences*, *90*, 224–237. doi:10.1016/j.ijthermalsci.2014.12.014

Sarafraz, M. M., Kiani, T., & Hormozi, F. (2016). Critical heat flux and pool boiling heat transfer analysis of synthesized zirconia aqueous nano-fluids. *International Communications in Heat and Mass Transfer*, *70*, 75–83. doi:10.1016/j.icheatmasstransfer.2015.12.008

Shahrul, I. M., Mahbubul, I. M., Saidur, R., & Sabri, M. F. M. (2016). Experimental investigation on Al2O3–W, SiO2–W and ZnO–W nanofluids and their application in a shell and tube heat exchanger, Int. *J. Heat Mass Transf.*, *97*, 547–558. doi:10.1016/j.ijheatmasstransfer.2016.02.016

Sharma, S. K., & Gupta, S. M. (2016). Preparation and evaluation of stable nanofluids for heat transfer application: A review. *Experimental Thermal and Fluid Science*, *79*, 202–212. doi:10.1016/j.exptherm-flusci.2016.06.029

Sheikholeslami, M., Darzi, M., & Sadoughi, M. K. (2018). Heat transfer improvement and pressure drop during condensation of refrigerant-based nanofluid; an experimental procedure, Int. *J. Heat Mass Transf.*, *122*, 643–650. doi:10.1016/j.ijheatmasstransfer.2018.02.015

Shima, P. D., Philip, J., & Raj, B. (2009). Role of microconvection induced by Brownian motion of nanoparticles in the enhanced thermal conductivity of stable nanofluids. *Applied Physics Letters*, *94*(22), 223101. doi:10.1063/1.3147855

Singh, A. K. (2008). Thermal conductivity of nanofluids. *Defence Science Journal*, *58*(5), 600–607. doi:10.14429/dsj.58.1682

Sonawane, S. S., Khedkar, R. S., & Wasewar, K. L. (2015). Effect of sonication time on enhancement of effective thermal conductivity of nano TiO2–water, ethylene glycol, and paraffin oil nanofluids and models comparisons. *Journal of Experimental Nanoscience*, *10*(4), 310–322. doi:10.1080/17458080.2013.832421

Timofeeva, E. V., Gavrilov, A. N., McCloskey, J. M., Tolmachev, Y. V., Sprunt, S., Lopatina, L. M., & Selinger, J. V. (2007). Thermal conductivity and particle agglomeration in alumina nanofluids: Experiment and theory. *Phys. Rev. E.*, *76*(6), 61203. doi:10.1103/PhysRevE.76.061203 PMID:18233838

Timofeeva, E. V., Routbort, J. L., & Singh, D. (2009). Particle shape effects on thermophysical properties of alumina nanofluids. *Journal of Applied Physics*, *106*(1), 14304. doi:10.1063/1.3155999

Trisaksri, V., & Wongwises, S. (2007). Critical review of heat transfer characteristics of nanofluids. *Renewable & Sustainable Energy Reviews*, *11*(3), 512–523. doi:10.1016/j.rser.2005.01.010

Wang, X.-Q., & Mujumdar, A. S. (2007). Heat transfer characteristics of nanofluids: A review. *International Journal of Thermal Sciences*, *46*(1), 1–19. doi:10.1016/j.ijthermalsci.2006.06.010

Wang, X.-Q., & Mujumdar, A. S. (2008). A review on nanofluids-part I: Theoretical and numerical investigations. *Brazilian Journal of Chemical Engineering*, *25*(4), 613–630. doi:10.1590/S0104-66322008000400001

Wei. (2017). A comprehensive model for the enhanced thermal conductivity of nanofluids. *J. Adv. Res. Phys., 3*.

Xuan, Y., Li, Q., & Hu, W. (2003). Aggregation structure and thermal conductivity of nanofluids. *AIChE Journal. American Institute of Chemical Engineers*, *49*(4), 1038–1043. doi:10.1002/aic.690490420

Xuan, Y., & Roetzel, W. (2000). Conceptions for heat transfer correlation of nanofluids, Int. *J. Heat Mass Transf.*, *43*(19), 3701–3707. doi:10.1016/S0017-9310(99)00369-5

Xuan, Y.-M., Hu, W. F., & Li, Q. (2002). Simulations of structure and thermal conductivity of nanofluids. *Journal of Engineering Thermophysics*, *23*, 206–208.

Xue, Q. Z. (2005). Model for thermal conductivity of carbon nanotube-based composites, Phys. B Condens. *Matter.*, *368*, 302–307.

Yang, J.-C., Li, F.-C., Zhou, W.-W., He, Y.-R., & Jiang, B.-C. (2012). Experimental investigation on the thermal conductivity and shear viscosity of viscoelastic-fluid-based nanofluids, Int. *J. Heat Mass Transf.*, *55*(11-12), 3160–3166. doi:10.1016/j.ijheatmasstransfer.2012.02.052

Yang, L., Xu, J., Du, K., & Zhang, X. (2017). Recent developments on viscosity and thermal conductivity of nanofluids. *Powder Technology*, *317*, 348–369. doi:10.1016/j.powtec.2017.04.061

Yu, F., Chen, Y., Liang, X., Xu, J., Lee, C., Liang, Q., ... Deng, T. (2017). Dispersion stability of thermal nanofluids. *Prog. Nat. Sci. Mater. Int.*, *27*(5), 531–542. doi:10.1016/j.pnsc.2017.08.010

Yu, H., Hermann, S., Schulz, S. E., Gessner, T., Dong, Z., & Li, W. J. (2012). Optimizing sonication parameters for dispersion of single-walled carbon nanotubes. *Chemical Physics*, *408*, 11–16. doi:10.1016/j.chemphys.2012.08.020

Yu, W., & Xie, H. (2012). A review on nanofluids: Preparation, stability mechanisms, and applications. *Journal of Nanomaterials*, *2012*, 1.

Yu, W., Xie, H., Chen, L., & Li, Y. (2010). Enhancement of thermal conductivity of kerosene-based Fe_3O_4 nanofluids prepared via phase-transfer method, Colloids Surfaces A Physicochem. *Eng. Asp.*, *355*(1-3), 109–113. doi:10.1016/j.colsurfa.2009.11.044

Zhu, H., Lin, Y., & Yin, Y. (2004). A novel one-step chemical method for preparation of copper nanofluids. *Journal of Colloid and Interface Science*, *277*(1), 100–103. doi:10.1016/j.jcis.2004.04.026 PMID:15276044

Zhu, H. T., Zhang, C. Y., Tang, Y. M., & Wang, J. X. (2007). Novel synthesis and thermal conductivity of CuO nanofluid. *The Journal of Physical Chemistry C*, *111*(4), 1646–1650. doi:10.1021/jp065926t

Chapter 3
Machining Mechanism of Minimum Quantity Lubrication Grinding

ABSTRACT

To facilitate the analysis of experimental theory, this chapter studied the machining mechanism related to the grinding of NMQL and the three mechanisms of sliding, ploughing, and cutting of the abrasive grinding process. A mathematical model of the micro-grinding force was established, and the micro-grinding force can be used to express the force of the grinding zone. The grinding force component was divided into the force of cutting and sliding, removing the constant cutting force during the grinding process, retaining the varying sliding friction force, and determining the sliding friction coefficient to characterize the lubrication performance. It reflected the influence of different lubrication conditions on the friction part. The methods for measuring the temperature during grinding were introduced, including direct contact temperature measurement and non-contact temperature measurement. At present, thermocouple temperature measurement is a commonly used and more accurate temperature measurement method, and the characteristics of each temperature measurement method were analyzed.

3.1 INTRODUCTION

Grinding machining method, which is of extremely strong technological adaptability, can realize machining of all kinds of material like metal, ceramics, glass, stone material, fireproofing material, concrete, skeleton and composite material, and many difficult-to-machine materials can only be machined through grinding; it can realize all kinds of precision machining like hogging machining, rough machining, precision machining and ultraprecision machining for plane, cylindrical surface, spherical surface, involute tooth, spiral surface and free-form surface. Grinding machining tool, which is a tool with most categories, has been extensively applied to multiple industries like mechanical manufacturing, light industry, building and fireproofing materials, and the number of already known categories has exceeded 3,000. Grinding machining is an important machining method supporting various industries in national economy.

DOI: 10.4018/978-1-7998-1546-4.ch003

As abrasive grain and chip are very fine with small grinding force and numerous abrasive grains participate in grinding, grinding machining can easily obtain a surface with high precision and low roughness and it is the main precision machining means. Especially very fine abrasive grains can be used to obtain nanoscale cutting edge radius so as to realize extremely fine cutting and acquire zero-defect surface with machining precision of 0.1μm and surface roughness of R_a 0.025 ~0.008μm. Therefore, it is the main ultraprecision machining means and has been importantly applied to high-tech fields like modern aerospace, precision machinery and instruments, electronic information, sophisticated weapons and small-scale and micro-machines, etc.

For ordinary materials, grinding machining may be intensely competed by milling, reaming and turning methods (Li et al., 2015; Nieslony et al., 2018). However, for a new generation of materials like ceramics, metallic ceramic composite materials, whisker-reinforcing materials and high-temperature superalloy, hard cutting is never the opponent of grinding machining. For instance, grinding is the only method of machining superalloy workpiece used in jet aircraft engines and also the only method of machining automobile and optical elements made of semiconductor silicon, ceramics and glass.

Grinding machine is habitually regarded as a kind of precision machining method which can acquire bright and clean part surface and accurate tolerance. Indeed, no other methods can compete with grinding machining, but grinding machining is never restricted to this application. Actually, many abrasive materials are exhausted in heavy-load grinding. Grinding machining pursues effectively and rapidly removing materials without consideration of surface quality. Grinding machining can use grinding wheel saw web with 20μm thickness like paper for precision cutting of microelectronic circuit silicon wafers and can clear away billet steel under heavy-load condition at 220kW machine power and 1600cm^3/min removal rate in the casting factory or steel plant.

Another competitive actual application field of grinding machining is machining of ultrahard or crisp materials which can't be effectively machined. In the production of quenched steel parts like cutting tool and bearing ring, grinding machining can easily complete machining of annealed steel or quenched steel, but other methods are restricted. Machining of nonmetallic brittle materials like ceramics, hard alloy and glass can only be completed through grinding machining.

Even though grinding machining is very important in the industry, it has not received deserved attention. Removal of material of the same volume through precision machining is of higher cost than other machining methods, so it is applied as this is unavoidable. Of course, as material removal margin is continuously reduced due to near-net shape precision casting and forging technologies, grinding machining will become more economical as the direct forming method not needing turning and milling.

Among commonly used machining methods, grinding machining is least known and most neglected without any doubt, because grinding machining process is too complicated to figure out. As quantity of cutting edges is large, geometric shape is irregular, grinding velocity is high and grinding depth of each abrasive grain is small and nonidentical, any intention to analyze grinding mechanism is hopeless, and grinding spark flow sprayed from grinding wheel nozzle has added to air of mystery among people.

Grinding machining can acquire high part machining quality and surface precision, so it has been widely applied to machining process. Compared with other cutting methods, grinding force is large and grinding velocity is high. Therefore, its specific grinding energy is greater than other cutting modes by multiple times, a large quantity of heat is generated in the grinding area, and moreover, grinding process can be completed within a very short time, it's difficult for heat to be diffused out, so heat is aggregated in the grinding area in quantity. These heat quantities seriously restrict workpiece surface machining performance, machining precision and efficiency (Li et al., 2016). In addition, both micromorphology and abrasive dust morphology on the workpiece surface will indirectly reflect grinding mechanism under this machining condition.

Grinding machining is one of the most fundamental machining forms. It's especially important that final precision and surface quality of most parts are completed by grinding technology. Grinding machining is negative-rake cutting of abrasive grains on the grinding wheel surface, and energy needed to remove material per unit volume is far greater than other machining methods, especially many difficult-to-machine materials are mostly machined through grinding method (Wang et al., 2017).

3.2 CUTTING MODEL OF ABRASIVE CUTTING EDGE

In the grinding machining process, friction plays a significant role. Both physical cause for friction coefficient and influences of external factors, e.g. nature of workpiece material, temperature, load and lubricant on friction should be taken into consideration in the analysis of frictional action. Abrasive cutting model can be simplified into the process as shown in Fig. 1. spring buffer system represents elastic deformation of the object in the grinding process and system stiffness is k_s. Abrasive grains fixed at one end revolve around the fixation center at the other end.

In the grinding process, as abrasive grains are connected through bonding agent, they have a certain elasticity. Under the action of grinding force, abrasive grains will experience deformation, so actual motion curve of grinding wheel is different from theoretical curve. In addition, workpiece surface will go through elastic recovery after grinding, so the final surface formation curve is also different from actual curve, which will influence elastic-plastic contact area in the abrasive cutting process.

3.2.1 Cutting Action of Abrasive on Workpiece Materials

It's generally believed that cutting process of workpiece by abrasive grains will experience three action mechanisms namely sliding friction, ploughing and cutting as defined in the grinding theory, shown in Fig. 2.

(1) The first phase is sliding friction process as shown in Fig. 2a. Abrasive cutting edge contacts workpiece surface in this process, and only elastic deformation is generated on the workpiece surface. As the cutting edge slides through the surface, elastic deformation and frictional force will gradually rise, but abrasive grains only go through sliding friction on the workpiece surface without cutting phenomenon.

(2) The second phase is ploughing process as shown in Fig. 2b. Based on sliding friction process, a large quantity of heat is converted into heat after continuous grinding so as to reach temperature bearing point of metal itself. When normal force is greater than yield stress of the material, cutting edge of abrasive grains is pressed into the workpiece surface. Metal on the surface is pushed to front end or side face of abrasive grains through plastic deformation, which causes upheaval of the surface so as to form ploughing action.

(3) The third phase is cutting phase also called chip formation process as shown in Fig. 2c. No grinding debris are generated in the abovementioned two process because abrasive grains don't reach critical grinding depth. Abrasive cutting edge results in flow of metal materials so that the front end is upheaved, furrow banks are formed at two sides, grinding debris slide out along cutting edge, and then the cutting process is completed.

Figure 1. Grinding model of abrasive cutting edge

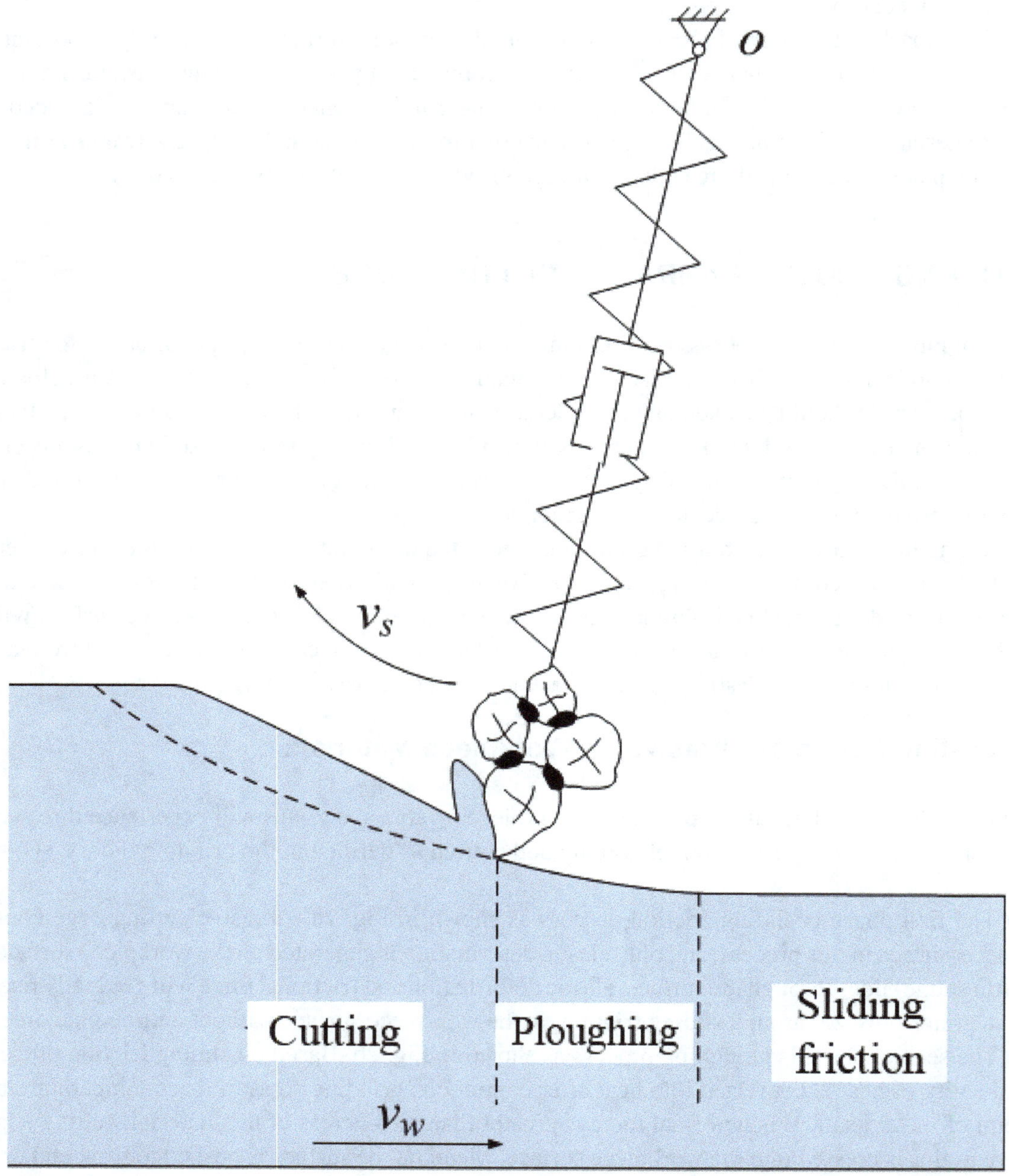

The above three grinding processes are related to cutting thickness in the grinding process. When cutting thickness of abrasive grains is smaller than critical cutting thickness, abrasive grains form sliding friction only on the surface, but no grinding debris are generated. Critical cutting thickness is the minimum penetration value used to generate cutting effect, and its value is related to grinding velocity, workpiece material nature and abrasive grain conditions, etc.

Figure 2. Grinding model of abrasive cutting edge

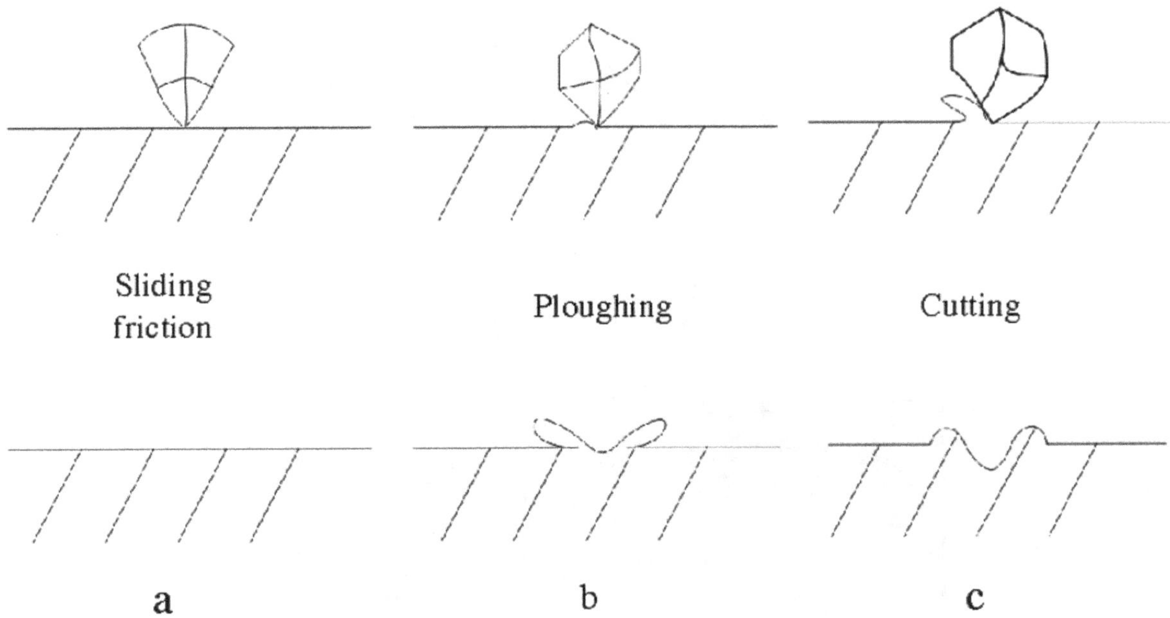

3.3 GRINDING PERFORMANCE EVALUATION PARAMETERS

In the evaluation of grinding performance under one working condition in the grinding machining process, selecting correct and simple evaluation parameters of grinding performance can yield twice the result with half the effort. Commonly used evaluation parameters of grinding performance mainly include: grinding force, grinding heat, G ratio, workpiece surface quality, etc (Zhang, Li, Jia, & Ma, 2014).

3.3.1 Arc Length of Contact in Grinding

In the grinding machining of workpiece surface by grinding wheel, contact length between grinding wheel circumference and workpiece surface is namely grinding contact arc. Grinding contact arc is an important basic parameter in the grinding area in the grinding machining process and an important factor influencing workpiece surface precision and reflecting the grinding process. Elastic-plastic change of grinding wheel and workpiece at contact interface can be obtained through the grinding contact arc. Grinding wheel interacts with workpiece in the contact area, grinding contact arc is namely length of heat source in the grinding area and grinding contact arc influences heat generation and temperature distribution in the grinding area. In addition, it is also closely related to grinding parameters like grinding energy and grinding force as well as integrity of already machined workpiece surface (Wang & Feng, 2013). Grinding contact arc can be derived through geometric calculation using a grinding contact model, both grinding wheel and workpiece are assumed as absolute rigid bodies in this model and grinding contact model is shown in Fig. 3.

Figure 3. The schematic diagram of grinding contact model

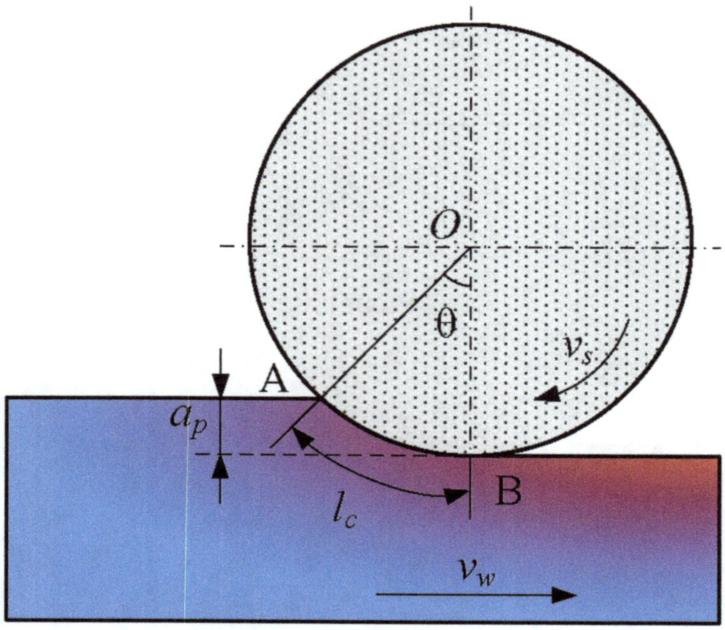

Grinding wheel and workpiece runs in opposite directions under grinding depth of a_p and relative motion velocity of v_w. When grinding wheel contacts workpiece, the arc formed in the contact area under the grinding effect is namely grinding contact arc as expressed by arc AB in Fig. 3:

$$l_c = AB = \frac{d_s\theta}{2} \tag{1}$$

$$\theta = \arccos(1 - \frac{2a_p}{d_s}) \tag{2}$$

As $a_p << d_s/2$, the following can be obtained according to equivalent infinitesimal of trigonometric function:

$$\cos\theta = 1 - \frac{\theta^2}{2} \tag{3}$$

Equation (3-3) is substituted into equations (3-1) and (3-2) to obtain:

$$l_c = \sqrt{a_p \cdot d_s} \tag{4}$$

where: l_c—grinding contact arc (mm);

a_p—grinding depth (mm);
d_s—equivalent diameter of grinding wheel (mm).

In the actual grinding motion process, dynamic contact arc l_k is generated under the influences of motion parameters, elastic-plastic workpiece deformation and thermal deformation, which can be deemed as dynamic correction of static contact arc. In the calculation process, grinding contact arc is regarded as having static similarity with dynamic contact arc l_k, and $l_k \approx l_c = \sqrt{a_p \cdot d_s}$ can be obtained.

3.3.2 Number of Active Cutting Point

Abrasive grain distribution is of great randomness on the grinding wheel surface, so upheaval heights of abrasive grains on the grinding wheel surface are irregular and some abrasive grains can't participate in the grinding effect. Therefore, number of effective grinding edges of abrasive grains which participate in grinding is smaller than total number of grinding edges on the grinding wheel surface, and it is mainly divided into number of static effective grinding edges and number of dynamic effective grinding edges.

(1) Number of static effective grinding edges

Number of static effective grinding edges within unit area of grinding wheel is related to grinding edge density and grinding depth a_p. When the highest upheaval of a_p along the grinding wheel surface gradually increases, number of grinding edges increases due to irregular upheaval heights of abrasive grains. When a_p increases to a certain degree, number of effective grinding edges will no longer increase, because all abrasive grains already participate in grinding, and number N_s of static effective grinding edges within unit area is expressed as:

$$N_s = C_1 a_p^q \tag{5}$$

where: C_1—grinding edge density coefficient;

q—index, $q \approx 1$.

(2) Number of dynamic effective grinding edges

Number N_d of dynamic effective grinding edges is number of effective grinding edges within unit area measured along grinding contact arc, and calculation formula is:

$$N_d = A_g N_s (a_{ds}) = A_g C_1 a_{ds}^q \tag{6}$$

where: A_g—ratio of number of static grinding edges, $A_g \approx 1.2$;

a_{ds}—dynamic grinding depth.

As dynamic grinding depth of each grinding edge is different in the grinding process while dynamic grinding depth is related to grinding motion parameters and geometric parameters, a_{ds} can be expressed as the function related to grinding motion parameters and geometric parameters. Number N_d of dynamic effective grinding edges along actual contact arc between grinding wheel and workpiece can be expressed as:

$$N_d = A_g[c_1]^{\beta}[\frac{v_w}{v_s}]^{\alpha}[\frac{a_p}{d_e}]^{\alpha/2} \quad (7)$$

Where α and β are related to distribution state of abrasive grains on grinding wheel.

3.3.3 Undeformed Chip Thickness

Grinding effect relies on sliding friction, extruding and shear motions between abrasive grains on grinding wheel surface and workpiece surface. Abrasive grains contact workpiece and then leave it so that many fine scratches with different shapes, sizes and complicated positions are formed on the workpiece surface, so shapes of undeformed grinding debris are also different before grinding debris are generated by grinding (Ding, Dai, Yu, Xu, & Fu, 2017).

Undeformed chip thickness influences the whole grinding process. Grinding motion of abrasive grains influences grinding force acting upon abrasive grains so as to impact wear of abrasive grains and service life of grinding wheel. Grinding wheel revolves at high speed, so mechanical power is input. Undeformed chip thickness influences specific grinding energy and decides power consumed by grinding action so as to influence energy input in the grinding contact area. The generated grinding temperature will result in change of workpiece material properties, which influences distribution of residual stress on the already machined surface and then workpiece surface integrity. Undeformed chip shape under the effect of a single abrasive grain during the grinding process is shown in Fig. 4.

Undeformed chip shape can be imaged as curve-edge triangular fish shape as shown in Fig. 5. Curve edge length of undeformed grinding debris is namely grinding contact length l_c between abrasive grains and workpiece, undeformed chip width is b_g and undeformed chip thickness is h.

In actual grinding production, there is no definite calculation formula for undeformed chip thickness mainly because of too strong randomness of abrasive grain position and shape on the workpiece surface. Calculation formulas of undeformed chip thickness proposed by many experts and scholars will be of great importance to study on grinding theories (Agarwal & Rao, 2013).

Expression methods of undeformed chip thickness h generally include maximum undeformed chip thickness h_{max}, average chip thickness and equivalent chip thickness h_{eq}.

Maximum undeformed chip thickness is formed by contact between abrasive grains and workpiece under the maximum grinding depth of abrasive grains and it is an important physical quantity in the grinding process. Grinding machining parameters and abrasive grain characteristics on grinding wheel combined, general expression of maximum undeformed chip thickness h_{max} is obtained as (Malkin & Ritter, 1992):

Figure 4. Undeformed chip shape in the grinding process; 1-abrasive debris, 2-grain, 3-undeformed abrasive debris, 4-workpiece

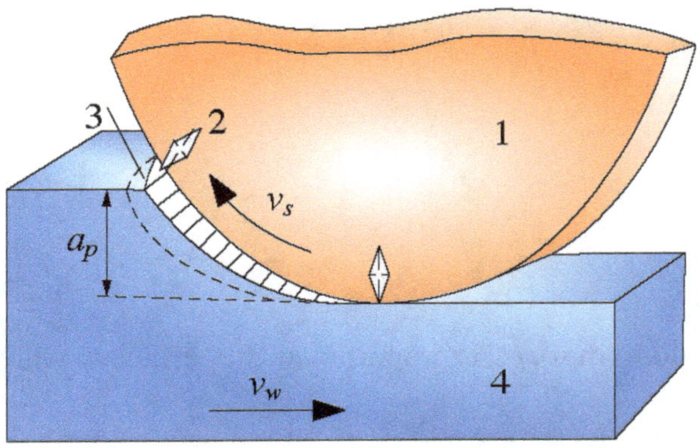

$$h_{\max} = \sqrt{\frac{3}{C_a \cdot \tan\theta} \cdot \frac{v_w}{v_s} \cdot \sqrt{\frac{a_p}{d_s}}} \tag{8}$$

where: C_a—abrasive grain density on grinding wheel surface;

θ—included angle at the bottom of grinding debris;
v_w—workpiece velocity (mm/s);
v_s—linear velocity of grinding wheel (mm/s);
a_p—grinding depth (mm);
d_s—equivalent diameter of grinding wheel (mm).

Shawa and Vyasa (1993) derived chip thickness formula according to the principle of unchanged volume and approximated chip model as a cuboid. Size of chip model is equivalent to:

Figure 5. Undeformed chip model

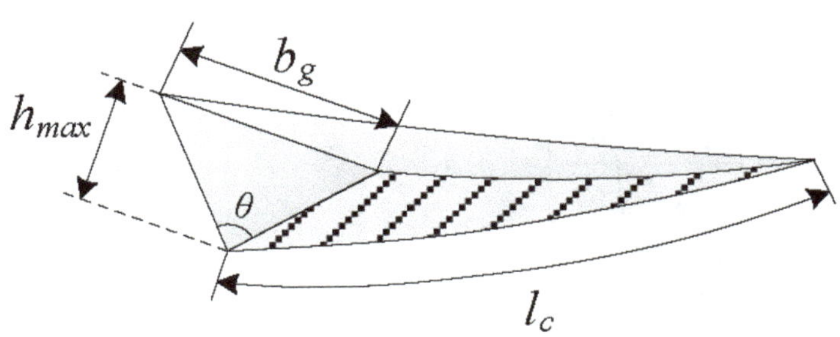

Length: $l_c = \sqrt{a_p \cdot d_s}$ (9)

Width: $\overline{b_g} = C\bar{h}$ (10)

Height: $\bar{h} = \frac{1}{2} h_{\max}$ (11)

where: $\overline{b_g}$ —average chip width (mm), (C is proportionality coefficient, which is related to tip angle of abrasive grains);

$\bar{h}$—average chip width (mm).

Volume of grinding debris generated by single abrasive grain grinding is:

$$V_0 = \frac{V_w}{N_s} = \frac{v_w b a_p}{v_s b N_s} \tag{12}$$

Abraded volume on the workpiece is equal to volume of workpiece material abraded by abrasive grains on grinding wheel, and then:

$$\overline{b_g} \cdot \bar{h} \cdot l_c = \frac{v_w b a_p}{v_s b N_s} \tag{13}$$

Expressions of average chip thickness and maximum undeformed chip thickness can be solved through organization:

$$\bar{h} = \left[\frac{v_w}{v_s} a_p\right] (N_s l_s \bar{b})^{-1} = \left[\frac{1}{N_s l_s C} \frac{v_w}{v_s} a_p\right]^{\frac{1}{2}} \tag{14}$$

$$h_{\max} = \left[\frac{4}{N_s l_s C} \frac{v_w}{v_s} a_p\right]^{\frac{1}{2}} = \left[\frac{4 v_w}{v_s N_s C} \sqrt{\frac{a_p}{d_s}}\right]^{\frac{1}{2}} \tag{15}$$

where: N_s—number of effective grinding edges within unit area of grinding wheel; $(N_s l_s \bar{b})^{-1}$ is dimensionless coefficient, which is related to abrasive grain shape, grinding edge density, chip shape, etc.

As distribution of abrasive grains on grinding wheel surface is of strong randomness, quantity of abrasive grains participating in grinding action can't be determined. Calculation process of undeformed chip thickness is quite complicated, so it's necessary to determine a parameter as a basic parameter used to study grinding process. In 1975, The International Academy for Production Engineering (CIRP) proposed taking $a_p v_w / v_s$ as a basic grinding parameter and named it "equivalent grinding thickness" (Shi, 1998). Equivalent grinding thickness h_{eq} is thickness of chip group formed by grinding edges which simultaneously participate in the grinding process on the unit width of grinding wheel, and it is defined as:

$$h_{eq} = a_p \frac{v_w}{v_s} = \frac{Z'_w}{v_s} \tag{16}$$

where: Z'_w—grinding debris abraded within unit time from unit width of grinding wheel.

Significance of taking equivalent chip thickness h_{eq} as basic grinding parameter: although h_{eq} is a hypothetic thickness of banding grinding debris and influences of workpiece material and grinding performance of grinding wheel are neglected, it can reflect influences of grinding motion parameters—workpiece velocity v_w, linear velocity v_s of grinding wheel and grinding depth a_p—on grinding debris, and it presents linear relations with grinding force, workpiece surface roughness, workpiece abrasion rate and service life of grinding wheel, so it can be a basic parameter reflecting the grinding process (Baeker & Martin, 2006; Dai, Ding, Zhang, Xu, & Su, 2015).

3.3.4 Grinding Force

Grinding force is generated when abrasive grains are grinding the workpiece material, and it mainly derives from elastic-plastic deformation generated by workpiece under the effect of abrasive grains. Moreover, grinding debris and frictional action are formed, and grinding force is related to grinding motion parameters and properties of abrasive grains and workpiece (Chang & Wang, 2008). Grinding force, which is an important parameter in the grinding process, can exert a direct influence on workpiece surface quality, service life of grinding wheel, power consumption and grinding stability, and it is usually used to analyze and diagnose grinding conditions (Hecker et al, 2007).

Total grinding force generated by grinding effect of grinding wheel on workpiece is decomposed into three orthogonal component forces, namely tangential grinding force F_t along tangential direction of grinding wheel surface, normal grinding force F_n along radial direction of grinding wheel and axial grinding force F_a along spin axis direction of grinding wheel when grinding wheel revolves and contacts the workpiece, and F_a is usually neglected as it is small. Due to large negative rake of abrasive grains, normal grinding force F_n is greater than tangential grinding force F_t, grinding force ratio F_n/F_t is usually within 1.5~3, normal grinding force F_n mainly influences workpiece surface quality and tangential grinding force F_t mainly influences grinding power consumption (Guo, Li, Ding, & Liang, 2016; Zhang, Kirpitchenko, & Liu, 2005). Decomposition schematic of grinding force is shown in Fig. 6.

As shown in Fig. 6, during the measurement of grinding force in the grinding process, tangential grinding force F_x and normal grinding force F_y measured through a dynamometer are respectively parallel and

Figure 6. Decomposition schematic of grinding force; 1-wheel, 2-abrasive debris, 3-grinding contact arc, 4-workpiece

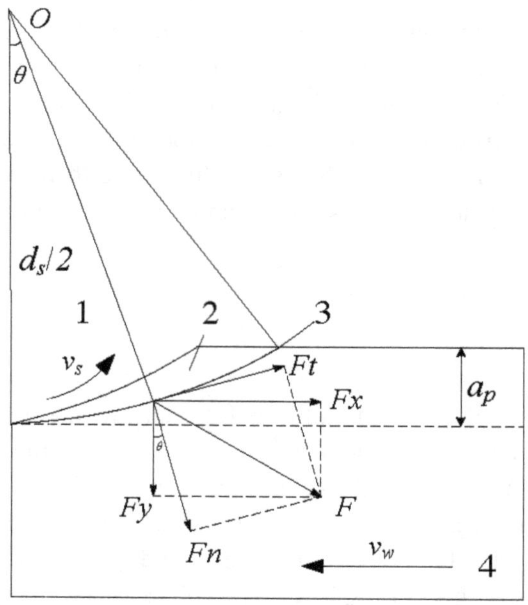

perpendicular to the already machined surface, and their relationships with tangential grinding force F_t and normal grinding force F_n in the actual grinding process are (Sheng, Li, Zhang, Jia, & Zhang, 2014):

$$F_t = F_x \cos\theta - F_y \sin\theta \tag{17}$$

$$F_n = F_x \sin\theta + F_y \cos\theta \tag{18}$$

Tangential grinding force F_t and normal grinding force F_n change with revolving angle θ of grinding wheel (Hecker et al, 2007). When grinding wheel radius $d_s/2$ is far greater than grinding depth a_p, revolving angle of grinding wheel is $\theta \approx 0$, so $F_t \approx F_x$ and $F_n \approx F_y$.

Abrasive grains form grinding debris through sliding friction, ploughing and cutting to complete the grinding process, so tangential grinding force F_t and normal grinding force F_n can be deemed as caused by cutting and sliding friction. Therefore, tangential grinding force F_t and normal grinding force F_n can be decomposed into:

$$F_t = F_{t,c} + F_{t,s} \tag{19}$$

$$F_n = F_{n,c} + F_{n,s} \tag{20}$$

where: $F_{t,c}$ and $F_{n,c}$ are tangential force and normal force caused by cutting;

$F_{t,s}$ and $F_{n,s}$ are tangential force and normal force caused by sliding friction.

Force caused by sliding friction can be expressed as ratio of average pressure p on the contact plane between abrasive grains and workpiece to actual contact area A_a on the grinded plane, where tangential force $F_{t,s}$ caused by sliding friction is in direct proportion to friction coefficient μ at abrasive grains/workpiece interface, and then grinding force is expressed as:

$$F_t = F_{t,c} + \mu p A_a \tag{21}$$

$$F_n = F_{n,c} + p A_a \tag{22}$$

A_a is the product of area of the grinding area and grinded surface area ratio A of grinding wheel, $A_a=bl_cA$, and expression (3-4) of grinding contact arc l_c is substituted to obtain $A_a=b(d_ea_p)^{1/2}A$. After the result is substituted into equations (3-21) and (3-22), the following can be obtained:

$$F_t = F_{t,c} + \mu b(d_e a_p)^{1/2} A \tag{23}$$

$$F_n = F_{n,c} + b(d_e a_p)^{1/2} A \tag{24}$$

According to the above expression of grinding force, tangential grinding force F_t and tangential grinding force F_n can characterize the contact relationship between grinding wheel and workpiece in the grinding process, and friction coefficient can also reflect lubrication conditions on the grinding interface.

Grinding force is the sum of grinding forces acted by single effective abrasive grains in the grinding wheel/workpiece contact area, and tangential grinding force F_t can be expressed as:

$$F_t=N_dF_{gt} \tag{25}$$

where: N_d—number of dynamic effective abrasive grains on grinding wheel, $N_d=l_c/\lambda$, and λ is spacing of dynamic effective abrasive grains;

F_{gt}—tangential force acted by single abrasive grain.

Single abrasive grains generate grinding debris under the grinding effect of grinding edges, so the relation between tangential force F_{gt} of single abrasive grains and maximum undeformed chip thickness h_{max} can be expressed as (Li & Liao, 1997):

$$F_{gt}=k_th_{max}^{\ m} \tag{26}$$

where: k_t—tangential grinding force within unit cross section on the grinding edge;

m—constant, which is decided by grinding parameters and properties of abrasive grains and workpiece;

Maximum undeformed chip thickness h_{max} can be expressed as follow according to geometric grinding relation:

$$h_{\max} = 2\lambda \frac{v_w}{v_s}\sqrt{\frac{a_p}{d_s}} \tag{27}$$

where: λ—spacing of dynamic effective abrasive grains;

d_s—equivalent diameter of grinding wheel (mm).

Equations (3-26) and (3-27) are substituted into (3-25) to obtain formula of tangential grinding force F_t by grinding as:

$$F_t = 2^m \lambda^{m-1} k_t v_w{}^m v_s{}^{-m} a_p{}^{\frac{m+1}{2}} d_s{}^{\frac{1-m}{2}} \tag{28}$$

Grinding force data measured through the grinding experiment are substituted into equation (3-28) to solve unknown parameters of the expression of grinding force. In recent years, foreign and domestic scholars have established related empirical formula (Fuh & Wang, 1997) of grinding force by combining grinding experimental data in consideration of the influence of grinding conditions on grinding force.

3.3.5 Specific Grinding Energy

To remove workpiece materials through grinding, grinding wheel needs to consume power, but grinding forces generated under different grinding conditions are also different, so basic parameter—specific grinding energy E_e—is derived through grinding power and machining conditions, namely the energy consumed to remove unit volume of metal material, and the relational expression is:

$$E_e = \frac{P}{V} = \frac{F_t \cdot (v_s \pm v_w)}{v_w a_p b} \tag{29}$$

where: E_e—specific grinding energy (W/ mm^3);

P—grinding power (W), "+" under up grinding and "-" under down grinding;
V—workpiece volume (mm^3/s);

Significance of specific grinding energy lies in that it can reflect interference mechanism and interference degree between abrasive grains and workpiece (Malkin & Guo, 1991). Specific grinding energy is related to average contact stress acted by abrasive grains on the workpiece surface and is in direct proportion of average contact stress generated by grinded workpiece on the grinding interface (Kurgin, Dasch, Simon, Barber, & Qian, 2012). If workpiece material can be easily grinded or grinding edge of grinding wheel is very sharp, then specific grinding energy is small and will increase due to blunting of grinding edge as the grinding time passes by. If it increases slowly, it means that grinding performance is good. The relationship between specific grinding energy and tangential grinding force reflects lubrication condition on the grinding interface, namely grinding friction coefficient μ. Specific grinding energy E_e is an index used to evaluate lubricating capability of grinding fluid. Sufficient lubrication through grinding fluid in the grinding area can reduce frictional force and grinding force between abrasive grains and workpiece, and at the time, specific grinding energy is low and relatively stable. Under the same grinding parameters, different cooling & lubrication conditions will result in the difference of tangential grinding force F_t in the grinding process, tangential grinding force F_t can be measured using a grinding dynamometer, and grinding parameters are substituted into equation (3-29) to obtain specific grinding energy E_e.

Specific grinding energy can reflect lubrication condition on the grinding interface and estimate grinding force and power consumption in the grinding process (Qian, Xu, Fu, & Lin., 2011). When lubrication condition on the grinding force is poor, tangential grinding force will increase, so will the energy needed to remove unit volume of material. The heat quantity converted through chip formation and frictional effect of workpiece materials will increase, and the finally obtained workpiece surface quality is poor; when lubrication is sufficient on the grinding interface, friction coefficient μ on the grinding wheel/workpiece grinding interface is small, workpiece material can be easily cut tangential grinding force is small, the obtained specific grinding energy is small and stable, and the finally obtained workpiece surface quality is high (Tawakoli, Hadad, Daneshi, & Sadeghi, 2011).

3.3.6 Grinding Heat

Abrasive grains experience three phases—sliding friction, ploughing and cutting—on the workpiece surface, all of which are accompanied by grinding power consumption. Energy consumed by frictional effect between abrasive grains and workpiece and chip formation through grinding of metal materials is converted into heat quantity which is then aggregated in the grinding area to form a heat source. Grinding heat source is constituted by the sum of heat sources of single abrasive grains in the grinding wheel/workpiece contact area. Heat is accumulated in the grinding area, high temperature can be easily formed at workpiece superficial layer, and when a critical temperature value is exceeded, it will cause superficial thermal injuries like burning, residual stress, cracks, etc. Therefore, studying heat generation principles of single abrasive grains in the grinding process will be of great importance to establishment of the grinding heat source model. Schematic diagram of heat source at abrasive grains in the grinding process under ideal conditions is shown in Fig. 7.

1-grain, 2-abrasive debris, 3-shear plane, 4-workpiece

According to Fig. 7, abrasive grains revolve around center of grinding wheel at linear velocity of v_s in the surface grinding process while workpiece does feed motion at velocity of v_w, and there are three heat sources under actions of single abrasive grains and workpiece grinding: 1) located at contact interface between abrasive grains and already machined workpiece surface, AB; 2) located at contact

Figure 7. Schematic diagram of heat source at abrasive grains

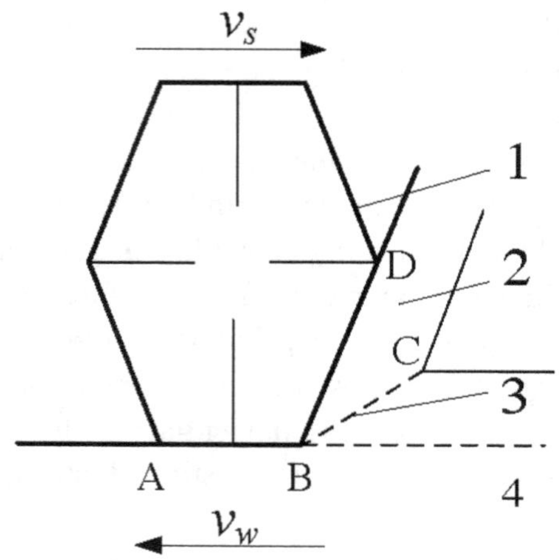

interface between abrasive grains/grinding debris, BD; 3) located at shear plane where grinding debris are generated by abrasive grains, BC.

In 1956, Hahn found that ratio of normal grinding force F_n to tangential grinding force F_t (namely grinding force ratio F_n/F_t) was generally within 1.5~3 in the grinding process, indicating that sliding friction existed at the abrasive grains/workpiece interface, and then proposed "sliding friction hypothesis of abrasive grains". It's deemed that sliding frictional force at abrasive grains/workpiece contact interface is the main cause for grinding heat. As shown in Fig. 7, grinding heat generated at abrasive grains/workpiece contact interface will mainly be transferred onto abrasive grains, grinding debris and workpiece, but only less than 10% of grinding heat will be transferred to grinding debris and carried away, and about 60%-95% of grinding heat is transferred onto workpiece and causes temperature rise on the workpiece surface (Rowe,Black,Mills,Morgan,&Qi,1997). Grinding force and relative motion velocity at abrasive grains/grinding debris contact interface and shear plane of grinding debris are small, so the generated grinding heat is small. Therefore, it's believed that heat source at abrasive grains/grinding debris contact interface is the main heat source for grinding (Mao, Zhou, Zhou, & Jun, 2009).

In the surface grinding process, grinding heat at grinding wheel/workpiece contact interface is the nearly continuous heat input of grinding wheel above the grinding area. Therefore, "banding heat source" model can be used to describe grinding heat source, the whole action process of heat source in surface grinding can be simplified into action of uniformly distributed heat sources on machined workpiece surface, and schematic diagram of heat source model in surface grinding is shown in Fig. 8.

For the convenient of a clear analysis of heat source distribution and transfer, etc., the heat source model is simplified and hypothesized:

(1) Grinding heat source is formed by gathering heat sources of single abrasive grains and grinding heat source is namely ABCD grinding area. Assume that grinding heat source is a plane heat source under continuous distribution and heat production.

Figure 8. Heat source model of surface grinding; 1-wheel, 2-heat flow density, 3-woekpiece

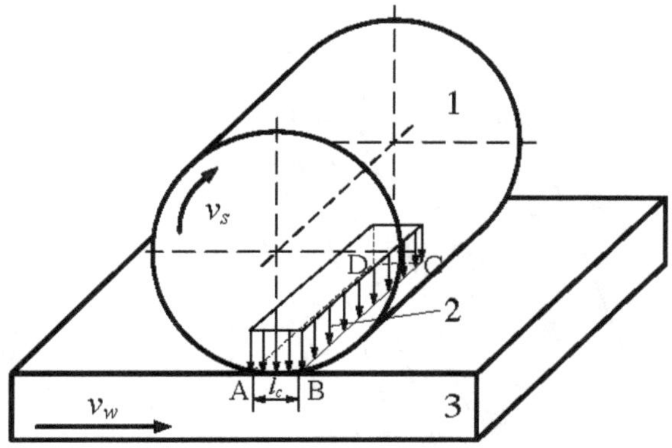

(2) Grinding depth a_p is namely radial feed quantity, generally being micron-order or millimeter order, so grinding heat source can be approximated as plane heat source parallel to motion direction of workpiece, and length of grinding heat source is l_c=AB.
(3) Grinding heat source is generated by grinding effect between grinding wheel and workpiece. Workpiece feed motion results in movement of grinding heat source on workpiece surface in the grinding process, so workpiece feed velocity v_w is equal to movement velocity of grinding heat source (Yang, Lei, Li, & Li, 2011).

Under the above circumstances, heat source model can be approximately regarded as an action problem of heat source of motion plane with infinite width and limited length on semi-infinite heat conductor in the surface grinding (Mao, Zhou, Guo, & Li, 2010). Total heat flux q_{total} generated by action of grinding wheel input energy in the grinding area is:

$$q_{total} = \frac{Q}{S} = \frac{F_t \cdot v_s}{l_c \cdot b} \tag{30}$$

where: q_{total} —total heat flux (J/(m^2·K·s));

Q—grinding power (J/s);
S—contact area between grinding wheel and workpiece (mm^2);
b—grinding width (mm).

3.3.7 Grinding Heat Flex Model

Grinding heat in the grinding area not only influences workpiece material performance but also influences service life of grinding wheel. Therefore, studying distribution types of grinding heat on the workpiece surface will be of great importance to studying grinding mechanism and improving workpiece surface integrity. Under different grinding depths, contact arc formed between grinding wheel and workpiece

Figure 9. Coordinate system of banding heat source motion

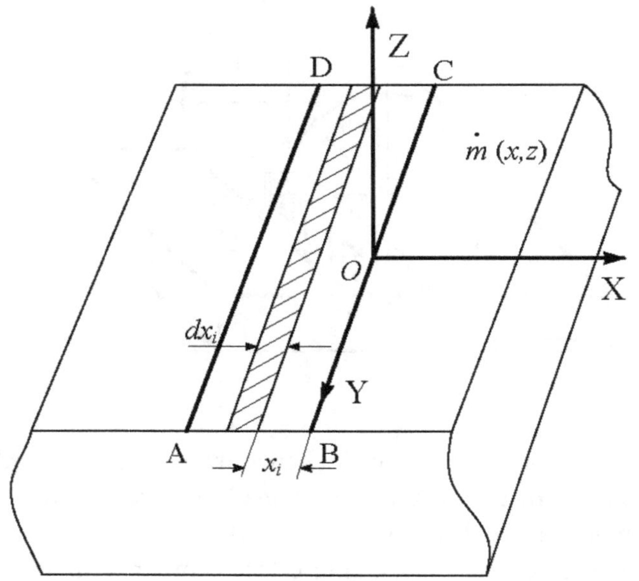

and heat distribution will be changed, so heat source models used to analyze temperature field in the grinding area are also different.

1. Plane Heat Source Model

(1) Heat source Model Under Parallel Movement and Uniform Distribution

In the shallow tangent surface grinding, grinding depth a_p is usually micron-order depth, it's deemed that already machined surface and unmachined surface of the workpiece are located on the same plane, and grinding heat source is parallel to direction of workpiece feed velocity, so heat source model of shallow tangent surface grinding can be evolved into movement of plane banding heat source in the direction parallel to workpiece feed direction on the semi-infinite workpiece surface, and coordinate system of banding heat source is built on the workpiece surface as shown in Fig. 9. ABCD area consists of banding heat sources in the grinding area, grinding contact arc is l_c, and heat source area satisfies $-\infty<y<+\infty$, $0<x<l_c$ and $z=0$ (Komanduri, Hou, & Z, 2000).

A linear heat source dxi with strength of qdx_i among banding heat sources is taken and it moves along X direction at velocity of v_w. Based on heat source theory of Jeager, the influence of linear heat source with width of dxi on temperature rise generated at an arbitrary point m(x,0,z) is expressed as:

$$d\theta_{m(x,0,z)} = \frac{q}{\pi\lambda}\exp[-\frac{(x-x_i)}{2\alpha}]K_0[\frac{v_w}{2\alpha}\sqrt{(x-x_i)^2+z^2}]dx_i \tag{31}$$

where: θ—temperature at an arbitrary point m (x, 0, z), °C;

q—heat source strength, J/(m^{2}K·s);
λ—heat conductivity, W/(m·K);
v_w—motion velocity of linear heat source, m/s;
α—thermal diffusivity, m^2/s; $\alpha=\lambda/c\rho$; c—specific heat, J/(kg·K); ρ—density, g/cm^3;
$K_0(u)$—zero-order second-class corrected Bessel function.

As shown in Fig. 3-9, heat source area ABCD can be regarded as a set of numerous linear heat sources dxi, so through integration of linear heat source dxi from 0 to l_c, temperature rise caused by heat sources in the whole grinding area on an arbitrary point m(x,0,z) can be obtained as below:

$$\theta_{(x,z)} = \frac{q}{\pi\lambda}\int_0^{l_c} e^{-\frac{(x-x_i)}{2\alpha}} K_0[\frac{v_w}{2\alpha}\sqrt{(x-x_i)^2+z^2}]dx_i \tag{32}$$

When z≠0, set $X=\frac{v_w x}{4\alpha}$; $Z=\frac{v_w z}{4\alpha}$; $L=\frac{v_w l_c}{4\alpha}$; $U=\frac{v_w(x-x_i)}{4\alpha}$;

Dimensionless temperature can be obtained by substituting them into equation (3-32):

$$\overline{\theta}=\frac{\pi\lambda v_w}{2\alpha q}\theta = 2\int_{X-L}^{X} e^{-2U} K_0\sqrt{(2Z)^2+(2U)^2}dU \tag{33}$$

When z=0, namely when point m(x,0,0) is located on the workpiece surface, integral substitution is carried out and $u=\frac{v_w(x-x_i)}{2\alpha}$ is set, then $K_0(u)=K_0[\frac{v_w(x-x_i)}{2\alpha}]$,

By substituting it into equation (3-33), expression of temperature distribution of workpiece surface can be obtained as:

$$\theta=\frac{q}{\pi\lambda}\int_{\frac{v_w x}{2\alpha}}^{\frac{(x-l)}{2\alpha}} e^{-u}K_0(u)(-\frac{2\alpha}{v_w})du=\frac{2\alpha q}{\pi\lambda v_w}\int_{\frac{(x-l)}{2\alpha}}^{\frac{v_w x}{2\alpha}} e^{-u}\ K_0(u)du \tag{34}$$

(2) Heat Source Model Under Tilting Movement and Uniform Distribution

For surface grinding with extremely small grinding depth, banding heat source model under parallel movement can be used. As the grinding depth increases, if slow-feed grinding is adopted, grinding depth will reach millimeter order. When grinding contact arc is simplified into a straight line, as there is a certain distance from unmachined surface to already machined surface, an included angle φ is formed between heat source surface and direction of workpiece motion so as to form a plane heat source under tiling movement. If the influence of grinding depth a_p or tilt angle on heat source motion and heat transfer condition is neglected and grinding temperature field is solved according to the heat source model under

Fig. 10. Heat source model under tilting movement

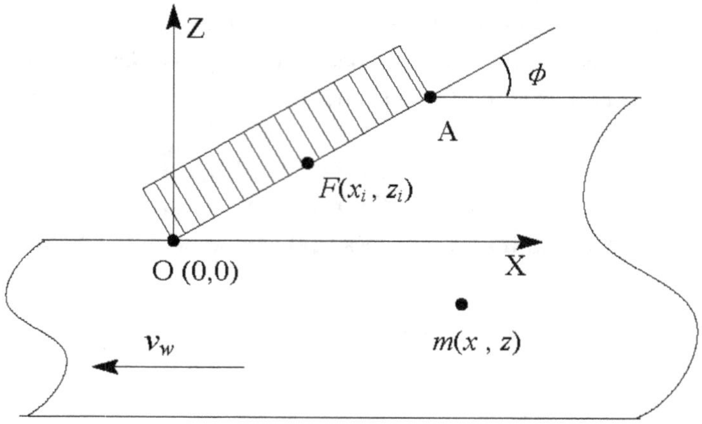

parallel movement, the error will be enlarged, thus reducing calculation accuracy of grinding temperature (Komanduri, Ranga, & Zhen, 2001). Therefore, using plane heat source under tilting movement can predict grinding temperature field more accurately in consideration of the influence of grinding depth ap. Plane heat source model under tilting movement is shown in Fig. 10.

According to Fig. 10, OA is grinding contact arc l_c, point $F(x_i, z_i)$ is linear heat source dli among plane heat sources and it moves along positive direction of axis x at workpiece velocity of v_w. Temperature rise caused by linear heat source dli on an arbitrary point m(x, z) inside the workpiece is:

$$d\theta = \frac{q \cdot dl_i}{\pi\lambda} e^{-\frac{(x - l_i \cdot \cos\varphi) v_w}{2\alpha}} \cdot K_0 \frac{v_w \sqrt{(x - l_i \cos\varphi)^2 + (z - l_i \sin\varphi)^2}}{2\alpha} \tag{35}$$

By taking integral of linear heat source dl_i, temperature rise caused by the whole tilting heat source to point $m(x, z)$ is:

$$\theta = \frac{q}{\pi\lambda} \int_0^{l_c} e^{-\frac{(x - l_i \cdot \cos\varphi) v_w}{2\alpha}} \cdot K_0 \frac{v_w \sqrt{(x - l_i \cos\varphi)^2 + (z - l_i \sin\varphi)^2}}{2\alpha} dl_i \tag{36}$$

Set $X = \frac{v_w x}{4\alpha}; Z = \frac{v_w z}{4\alpha}; L = \frac{v_w l_c}{4\alpha}; U = \frac{v_w (x - l_i)}{4\alpha}$, and they are substituted into equation (36) to obtain dimensionless temperature of the heat source model:

$$\begin{aligned} \bar{\theta} &= \frac{\pi\lambda v_w}{2\alpha q} \theta \\ &= 2\int_{X-L}^{X} e^{-2X + 2X\cos\varphi - 2U\cos\varphi} K_0 \sqrt{4(X - X\cos\varphi + U\cos\varphi)^2 + 4(Z - X\sin\varphi + U\sin\varphi)^2}\, dU \end{aligned} \tag{37}$$

Figure 11. Heat source model under triangular distribution

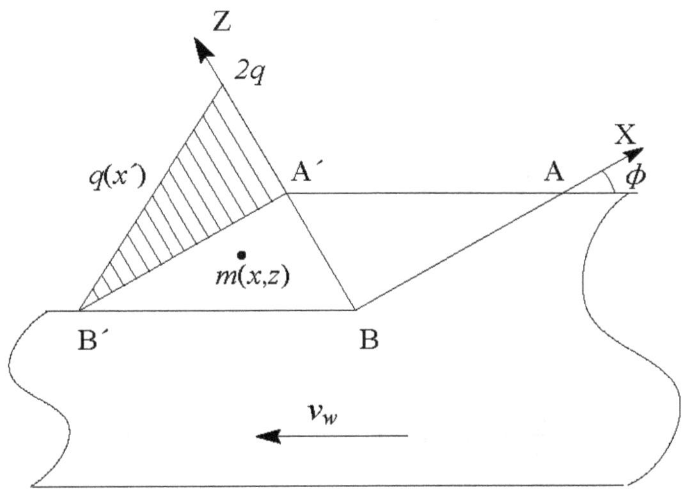

When φ=0, cosφ=1 and sinφ=0, dimensionless temperature of the heat source model under parallel movement and uniform distribution can be obtained, so this model is a special positional condition of heat transfer model under tilting movement.

(3) Heat Source Model Under Triangular Distribution

Abrasive grains grind workpiece material along grinding contact arc to form grinding debris in the grinding process. Grinding thickness of abrasive grains is gradually enlarged along the grinding direction, e.g. under high-efficiency deep-cutting grinding environment, so heat flux generated at the grinding contact interface is not under uniform distribution. Professor Bei J Y proposed a heat source model under triangular distribution as shown in Fig. 11.

Grinding contact length A´B´ is heat source length l_c, average heat flux is l_c, and heat source strength q(x´) at an arbitrary point on the surface is expressed as:

$$q(x') = 2q(1 - \frac{x'}{l_c}) \tag{38}$$

At the time t´(t0≤t´≤te) when the heat source enters grinding surface, transient heat flux of grinding heat source at this point is expressed as q(x´)dt´, and temperature rise caused by the whole heat source to an arbitrary point m(x,z) is:

$$\theta_{(x,z)} = \frac{q\sqrt{t}}{c\rho\sqrt{\alpha\pi}}[4 - \frac{8t}{3t_e} - \frac{2z^2}{3\alpha t_e}]e^{-\frac{z^2}{4\alpha t}} + \frac{qz}{\alpha c\rho}[\frac{z^2}{3\alpha t_e} - 2(1 - \frac{t}{t_e})][1 - erf(\frac{z}{\sqrt{4\alpha t}})] \tag{39}$$

where: $t=x/v_w$, $t'=x'/v_w$ and $t_e=l_c/v_w$.

Figure 12. The arc heat source model

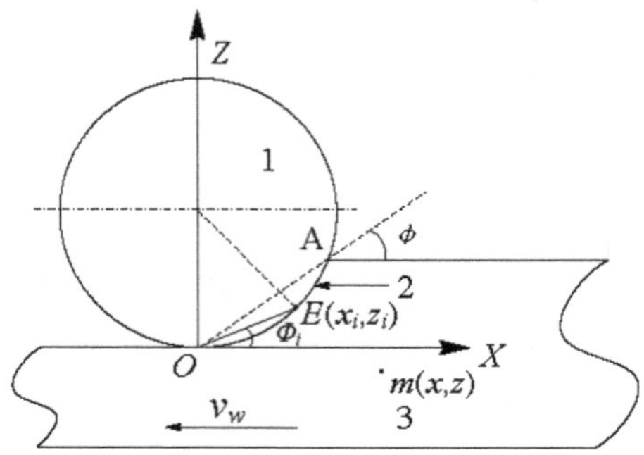

Set $X=\frac{v_w x}{4\alpha}\cos\varphi$ and $L=\frac{v_w l_c}{4\alpha}$ to obtain dimensionless temperature:

$$\begin{aligned}\bar{\theta}&=\frac{\pi\lambda v_w}{2\alpha q}\theta\\&=\frac{1}{L}\frac{\sqrt{\pi}}{\sin\phi}[\frac{\cot\phi}{\sin\phi}+2X]erf(\frac{\sin\phi}{\sqrt{\cos\phi}}\sqrt{L-X}]-\frac{2\sqrt{\pi}}{\sin^2\phi}\sqrt{\cos\phi}\frac{\sqrt{L-X}}{L}\exp[\frac{\sin^2\phi}{\cos\phi}(L-X)]\end{aligned} \quad (40)$$

2. Arc Heat Source Model

Under large grinding depth, if plane heat source is continuously used to analyze the temperature field, it will result in partially large error. Contact arc under deep grinding condition is arc shaped. For the sake of more accurate calculation of temperature in the grinding area, the influence of actual heat source shape on temperature field should be considered. Actual grinding contact arc shape is taken as heat source shape and heat source is regarded as a set of infinite linear heat sources distributed along the grinding contact arc. Arc-shaped heat source model is shown in Fig. 12.

1-wheel, 2-arc heat source, 3-workpiece

As shown in Fig. 12, arc AEO is grinding contact arc l_c, included angle between linear source at an arbitrary point $E(x_i, z_i)$ on the grinding contact angle and already machined surface is φ_i within maximum value being φ, and OE arc length is $l_i=d_e\varphi_i$, where d_e is diameter of grinding wheel, motion direction of heat source is parallel to axis x, velocity is v_w, and temperature rise caused by linear heat source dl_i to an arbitrary point $m(x, z)$ inside the workpiece:

$$d\theta=\frac{q\cdot dl_i}{\pi\cdot\lambda}\cdot e^{-\frac{(x-l_i\cos\phi_i)v_w}{2\alpha}}K_0[\frac{v_w\sqrt{(x-l_i\cos\phi_i)^2+(z-l_i\sin\phi_i)^2}}{2\alpha}] \quad (41)$$

Temperature rise caused to point m (x, z) under the effect of the whole arc heat source is:

$$\theta = \frac{q}{\pi \cdot \lambda} \int_0^{l_c} e^{-\frac{(x-l_i\cos\phi_i)v_w}{2\alpha}} K_0 [\frac{v_w\sqrt{(x-l_i\cos\phi_i)^2+(z-l_i\sin\phi_i)^2}}{2\alpha}] dl_i \tag{42}$$

When z=0, temperature rise on the grinding surface is:

$$\theta = \frac{q}{\pi \cdot \lambda} \int_0^{l_c} e^{-\frac{(x-l_i\cos\varphi_i)v_w}{2\alpha}} \cdot K_0 (\frac{v_w\sqrt{x^2+l_i^2-2xl_i\cos\varphi_i}}{2\alpha}) dl_i \tag{43}$$

Set $X = \frac{xv_w}{4\alpha}; Z = \frac{zv_w}{4\alpha}; L = \frac{l_c v_w}{4\alpha}$ and substitute them into equation (43) to obtain dimensionless temperature as:

$$\begin{aligned} \bar{\theta} &= \frac{\pi\lambda v_w}{2\alpha q}\theta \\ &= 2\int_{X-L}^{X} e^{-2X+2X\cos\phi_i-2U\cos\phi_i} K_0 \sqrt{4(X-X\cos\phi_i+U\cos\phi_i)^2+4(Z-X\sin\phi_i+U\sin\phi_i)^2}\, dU \end{aligned} \tag{44}$$

When φ_i=φ, dimensionless temperature of the heat transfer model of heat source under tilting movement is obtained. When φ_i=0, dimensionless temperature of the heat transfer model of heat source under parallel plane movement and uniform distribution is obtained through substitution (Huang & Liang, 2005). Therefore, heat source model under parallel movement and uniform distribution and heat source model under tilting movement are special circumstances of heat source model under arc distribution, thus verifying correctness of temperature field calculation of each heat source model.

3.3.8 Measuring of Grinding Temperature

In order to obtain relatively accurate grinding temperature, domestic and foreign scholars have carried out a large quantity of experiments and research work, where measuring methods of grinding temperature mainly include two types: direct contact thermometry and indirect contact thermometry. There is no mature measurement technology for grinding temperature so far, but thermocouple thermometry and IR thermometry are quite reliable temperature measuring methods (Wang, Li, Li, Yang, & Zhang, 2015).

1. Direct Contact Thermometry

Direct contact thermometry can be divided into types coating technology and thermocouple {Wang, 2015 #16}thermometry, where among measuring methods of grinding temperature, the most pragmatic one is thermocouple method (Wang et al., 2014).

Figure 13. Schematic diagram of open-top thermometry

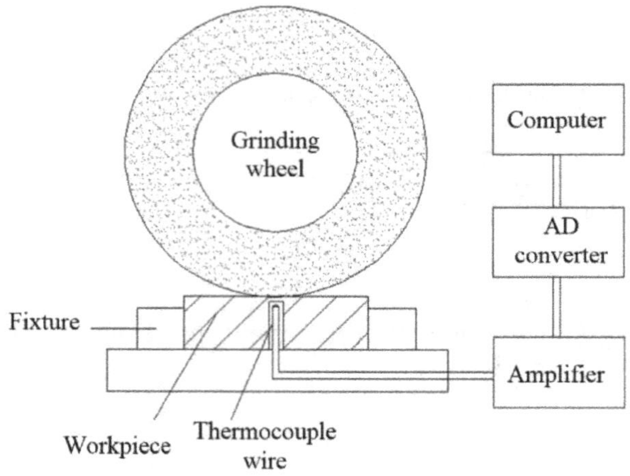

(1) Coating Thermometry

Coating thermometry refers to clamping high-temperature coating materials between two separated workpieces and then determining the temperature by observing change of coating material after grinding (Ding et al., 2019).

(2) Thermocouple Thermometry

Thermocouple thermometry refers to converting grinding temperature signal into electrodynamic potential signal using a thermocouple, amplifying the electrodynamic potential signal and inputting it into the analyzer through digital-to-analog conversion and finally analyzing the measured grinding temperature through a dedicated software (Liu & Zhang, 2006; Shao, 2008). In the high-speed grinding process, as workpiece velocity is very high, action time of heat source on thermocouple is very short, it's impossible for thermocouple to reach the temperature under thermal equilibrium due to its own thermal inertia, and at the time, dynamic calibration of the thermocouple must be carried out.

Basic principle of open-top thermocouple thermometry is shown in Fig. 13, where one or several stepped holes (used for temperature measurement of one workpiece for multiple times) is drilled on the workpiece, prefabricated thermocouple is placed in stepped hole and fixed, and the overall schematic diagram is shown in Fig. 14. Thermocouple is used to convert grinding heat signal into electric signal, and after amplification and digital-to-analog conversion, grinding temperature is obtained through processing using a dedicated software.

As for clamping-type thermometry, grooving or hole opening is carried out firstly on the workpiece, thermocouple foil or wire is clamped or buried into the groove or hole, and then under plastic deformation and high-temperature action in the tangent grinding process, workpiece base is overlapped or welded with top of thermocouple wire to form thermocouple joint when grinding extends to thermocouple wire, and then thermoelectric potential is output, after which is processed through amplification and acquisition systems temperature signal can be obtained (Ji, Ni,& Bai, 2015).

Figure 14. Schematic diagram of artificial thermocouple thermometry

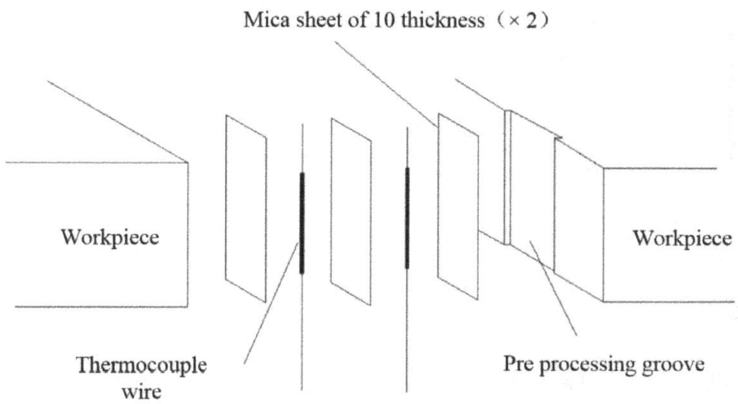

2. Non-Contact Thermometry

As temperature consists of heat flow, heat is a kind of energy and a function constituting fundamental particles of a substance. Temperature expresses disturbance degree of these particles, the greater the distance, the greater the fluctuation of the generated electromagnetic field, the higher the temperature and the greater the energy emitted by radiation, and this constitutes the main basis for IR thermometry. Nowadays four IR thermometric methods are extensively applied namely IR thermal imaging thermometry, IR radiation thermometry, optical fiber IR thermometry and bi-color IR thermometry.

Basic principle of IR thermometry is shown in Fig. 15. Euthermic target generates IR radiation and captures IR radiation energy within its range through the optical system, and the range is decided by optical parts and position of thermometer (Boyd, 1991; Deng, Wang, & Huang, 2001). IR energy focused on the photoelectric detector is converted into corresponding electrical signal which goes through amplification and signal processing. After being calibrated according to internal algorithm of the instrument and target emissivity, the circuit is converted into temperature value of the measured target, which is then displayed and output by a dedicated processing software (Li Sun, & Yuan, 2007; Sun & Li, 2008).

3.3.9 Surface Integrity of Workpiece

Surface integrity is surface texture and superficial state existing on the workpiece surface through machining or processing. The machined workpiece surface is actually not an ideal smooth surface with surface defects such as roughness, furrows and cracks to different degrees. Although it is only a very thin layer, it seriously impacts parts precision, wear resistance, quality of fit and fatigue strength, etc. Workpiece surface integrity mainly includes two parts of content: surface texture and superficial state, where the most representative part of surface texture is surface roughness and superficial state is usually reflected by surface micromorphology (Wang et al., 2017; Yuan et al., 2010).

Figure 15. Schematic diagram of IR thermometry technology

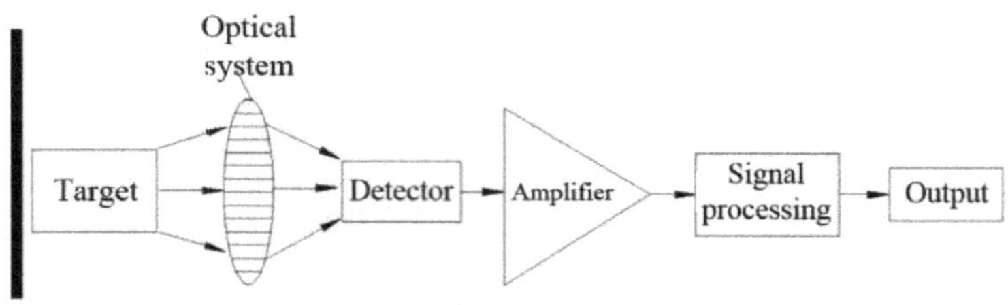

(1) Surface Roughness

After the workpiece goes through grinding machining, mechanical vibration, brittle plastic deformation and scratching of materials in the machining process will unavoidably cause geometric shape errors so that the workpiece will have a certain surface roughness (Aouici, Yallese, Chaoui, Mabrouki, & Rigal, 2012). Surface roughness refers to microscopic geometric shape features consisting of minor peak valleys with small spacing on the machined surface (Lalwani, Mehta, & Jain, 2008). Surface roughness is an important index evaluating surface quality, where there are 5 evaluation parameters, but commonly used are arithmetic average deviation R_a, the maximum height R_z and average width R_{Sm} of surface profile. Surface roughness atlas is shown in Fig. 16.

Arithmetic average deviation Ra of surface profile refers to arithmetic average value of absolute value of the distance from each point on the measured profile to its midcourt line. The greater the Ra value, the rougher the surface.

$$Ra = \frac{1}{l}\int_0^l \left|Z(x)\right| dx \tag{45}$$

where: l is sampling length (mm) and Z(x) is longitudinal coordinate value at coordinate x.

Maximum height Rz of surface profile refers to the sum of peak profile height Zp and maximum valley depth Zv within the sampling length. Rz is very intuitive when used to evaluate height parameters of surface roughness but it is of a certain limitation when reflecting geometric shape characteristics of the measured profile.

Figure 16. Surface roughness atlas

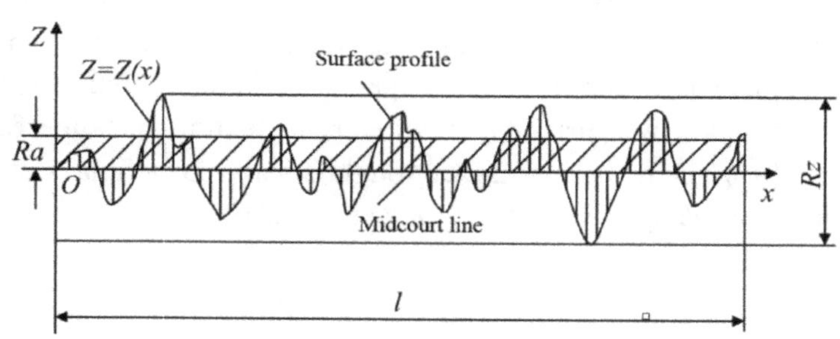

RS_n is average width of surface profile unit, specifically being average value within unit width XS of the profile in a sampling length.

$$RS_m = \frac{1}{m}\sum_{i=1}^{m} XS_i \tag{46}$$

where: m is number of units existing in the sampling length and XSi is width of each unit.

(2) Surface Morphology

Surface roughness and surface morphology in surface features is an important index evaluating surface quality in grinding performance, where surface morphology can qualitatively reflect workpiece surface quality while roughness value can further quantitatively judge workpiece surface quality (Bouacha, Yallese, Mabrouki, & Rigal, 2010). When workpiece surface quality can't be distinguished from surface micromorphology, then surface roughness can be judged from figure so the two have a certain relationship.

After micromorphological graph of the workpiece surface is amplified through SEM and goes through surface analysis, its amplification factor can reach hundreds of and even thousands of times. Therefore, according to surface micromorphological graph and changes of element contents and types on the workpiece surface, geometric microstructure on the surface and workpiece material property can be intuitively observed.

In the grinding machining process, top end of abrasive grains on the grinding wheel surface is embedded into the workpiece sublayer, and cutting will result in furrows on the workpiece surface. Under the influence of other numerous factors, surface morphology becomes more complicated. Workpiece metal material will be re-adhered to the workpiece surface, namely metal material adhered to abrasive grains is re-adhered to the workpiece surface (Bifano, Dow, & Scattergood, 1991; Blackley & Scattergoof, 1991; Venkatachalam, Li, & Liang, 2009). At the time, displacement occurs between workpiece and metal crystal under high temperature and high pressure, followed by plastic ploughing, metal particles will be re-precipitated on two sides of the workpiece furrows by being adhered onto abrasive grains, and then plastic deformation layers are formed. For these plastic deformation layers on the workpiece surface, defects like furrows and material precipitation will be clearly displayed on the micromorphological graph.

3.3.10 Grinding Wheel Wear and Debris Formation

(1) Grinding Wheel Wear

During the grinding process, grinding wheel wear is caused by external mechanical, physical and chemical factors so as to degrade the cutting performance of abrasive grains. When wear due to abrasive grains is serious, it will give rise to grinding vibration, workpiece burning and quality degradation of workpiece surface, etc., and it's especially difficult to guarantee workpiece precision (Liang et al., 2010). Shapes and distribution of commonly used abrasive grains on the grinding wheel surface are random with multiple wear modes. General wear modes include: abrasive wear, crushing wear and blocking adhesive wear. Under abrasive wear mode, abrasive grains gradually become blunt due to sliding friction and chemical reaction between abrasive grains and workpiece surface as well as deformation of abrasive grains, and a

Figure 17. The wear form of grinding wheel

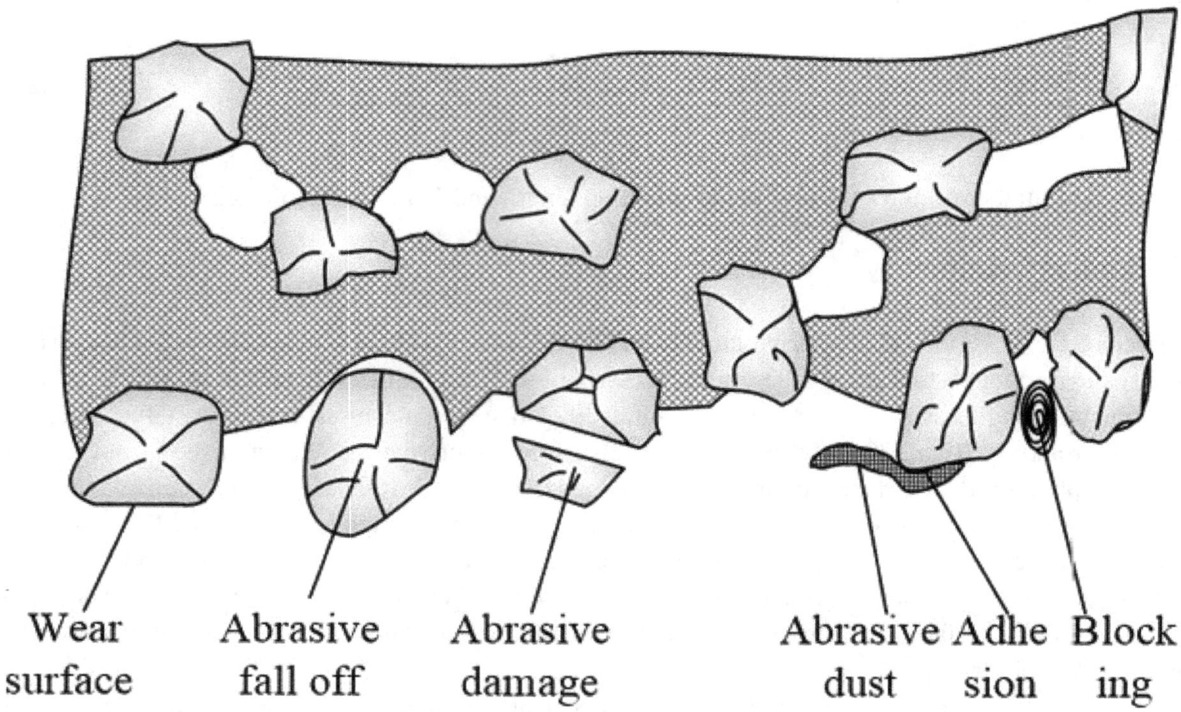

wear plane is formed at top end of abrasive grains. Crushing wear is divided into abrasive crushing and abrasive falling. In the grinding machining process, when force applied onto abrasive grains exceeds their own strength, abrasive grains will fall off in the form of tiny debris. When the bonding agent ruptures, the whole abrasive grain will fall off from the grinding wheel base to form abrasive falling (Zha, Xu, Luo, Xie, & Lun, 2013). When abrasive grain is grinding the material under high temperature and high pressure, the grinded material will be adhered to the surface of abrasive grain. As the grinding process proceeds, material adhesion becomes more serious. In addition, impurities like grinding debris will block grinding wheel pores so as to aggravate grinding wheel wear.

Fig. 17 shows wear form of grinding wheel. Grinding wheel wear type is also related to grinding wheel type, workpiece material, grinding condition and lubricant, etc.

(2)Chip Formation

Lubricating/cooling performance of lubricant in the machining process can be judged by observing morphological features of grinding debris, which is a very simple and effective method, where chip shape can reflect cutting state of abrasive grains and material removal mechanism. Chip morphology especially smoothness of front face of grinding debris namely the contact surface with abrasive grains can reflect lubrication condition under this working condition. Moreover, quantity of grinding debris can approximately reflect the number of effective grinding edges. Width at the root of grinding debris can reflect contact length between grinding wheel and workpiece, and root morphology of grinding debris can indicate formation process of grinding debris (Jiang, Yan, Xu, & Rong, 2014).

Under general conditions, grinding debris obtained through grinding are very tiny. According to different grinding conditions, chip shape can be generally divided into three types: banding grinding debris, mouldering-shaped grinding debris (debris-shaped grinding debris) and molten spherical grinding debris. Grinding, which is typical negative-rake machining, can generate a large quantity of heat within a very short time. In addition, due to high linear velocity of grinding wheel, chip formation is more approximate to adiabatic shear. Local thermal plastic deformation causes reduction of material shear resistance, large negative rake of abrasive grains will make workpiece material be extruded out and then swell, and then mouldering-shaped grinding debris are formed (Liao, Liu, & Tang, 2009; Ma, Du, & Zhang, 2011). Inside spherical grinding debris are empty, their outside structure is dendritic, and they experience melting and solidification process within a very short time. Tiny grinding debris go through oxidizing reaction in the air under high temperature, and spherical hollow structure is formed due to surface tension effect, which is caused by chip fusing and curling. In addition, chip features of different materials are significantly different.

The contact surface where chip is divorced from abrasive grain is front face of chip and the other face is its rear face. Front face is smoother than rear face, and the rear face presents rippled shape due to extruding shear force and thermal stress in the cutting process (Cheng, Xu, Ding, Tian, & Yang, 2011). Morphology of front face of chip can reflect lubricating effect under this working condition to a great degree. When lubrication is insufficient, the surface will become rough, accompanied by obvious sliding friction traces left by abrasive grains.

REFERENCES

Agarwal, S., & Rao, P. V. (2013). Predictive modeling of force and power based on a new analytical undeformed chip thickness model in ceramic grinding. *International Journal of Machine Tools & Manufacture*, *65*, 68–78. doi:10.1016/j.ijmachtools.2012.10.006

Aouici, H., Yallese, M. A., Chaoui, K., Mabrouki, T., & Rigal, J. F. (2012). Analysis of surface roughness and cutting force components in hard turning with CBN tool: Prediction model and cutting conditions optimization. *Measurement*, *45*(3), 344–353. doi:10.1016/j.measurement.2011.11.011

Baeker, M. (2006). Finite element simulation of high-speed cutting forces. *Journal of Materials Processing Technology*, *176*(1-3), 117–126. doi:10.1016/j.jmatprotec.2006.02.019

Bifano, T. G., Dow, T. A., & Scattergood, R. O. (1991). Ducti-Regime Grinding-A- New Technology For machining Brittle Materials. *Journal of Engining For Industry-Transactions of The Asme*, *113*(2), 184–189. doi:10.1115/1.2899676

Blacery, W. S., & Scattergood, R. O. (1991). Ductile-Regime Machining Model for Diamond Turning of Brittle Materials. *Precision Engineering*, *13*(2), 95–103. doi:10.1016/0141-6359(91)90500-I

Bouacha, K., Yallese, M. A., Mabrouki, T., & Rigal, J. F. (2010). Statistical analysis of surface roughness and cutting forces using response surface methodology in hard turning of AISI 52100 bearing steel with CBN tool. *International Journal of Refractory Metals & Hard Materials*, *28*(3), 349–361. doi:10.1016/j.ijrmhm.2009.11.011

Boyd, L. (1991). *Analysis of infrared thermography data for icing applications.* Paper presented at the Aerospace Sciences Meeting. 10.2514/6.1991-666

Chang, H. C., & Wang, J. J. J. (2008). A stochastic grinding force model considering random grit distribution. *International Journal of Machine Tools & Manufacture*, *48*(12-13), 1335–1344. doi:10.1016/j.ijmachtools.2008.05.012

Cheng, Z., Xu, J. H., Ding, W. F., Tian, L., & Yang, S. B. (2011). Simulation of chip formation in grinding titanium alloy TC4 with single abrasive grit. *Diamond & Abrasives Engineering*, *31*(2), 17–21.

Cong, M., Zhou, Z. X., Zhou, D. W., & Yang, J. (2009). Research on Three Dimensional Finite Element Simulation of the Temperature Field in Surface Grinding. *Zhongguo Jixie Gongcheng/china. Mechanical Engineering (New York, N.Y.)*, *20*(5), 589–595. doi:10.1109/MILCOM.2009.5379889

Dai, J. B., Ding, W. F., Zhang, L. C., Xu, J. H., & Su, H. H. (2015). Understanding the effects of grinding speed and undeformed chip thickness on the chip formation in high-speed grinding. *International Journal of Advanced Manufacturing Technology*, *81*(5-8), 995–1005. doi:10.100700170-015-7265-1

Deng, J. P., Wang, G. L., & Huang, P. P. (2001). The IR thermography technique for high temperature measurement. *Experiments & Measurements in Fluid Mechanics*, *15*(1), 43–47. doi:10.3969/j.issn.1672-9897.2001.01.008

Ding, S. H., Ma, G. Z., Chen, S. Y., He, P. F., Wang, Y. W., Wang, H. D., & Xu, B. S. (2019). Research State on Heat Accumulation Behavior and Temperature Control in Thermal Spraying Forming Process. *Materials Review*, *33*(1), 3644–3653.

Ding, W. F., Dai, C. W., Yu, T. Y., Xu, J. H., & Fu, Y. C. (2017). Grinding performance of textured monolayer CBN wheels: Undeformed chip thickness nonuniformity modeling and ground surface topography prediction. *International Journal of Machine Tools & Manufacture*, *122*, 52–66. doi:10.1016/j.ijmachtools.2017.05.006

Fuh, K. H., & Wang, S. B. (1997). Force modeling and forecasting in creep feed grinding using improved bp neural network. *International Journal of Machine Tools & Manufacture, 37*(8), 1167-1178. doi:10.1016/S0890-6955(96)00012-0

Guo, M. X., Li, B. Z., Ding, Z. S., & Liang, S. Y. (2016). Empirical modeling of dynamic grinding force based on process analysis. *International Journal of Advanced Manufacturing Technology*, *86*(9-12), 3395–3405. doi:10.100700170-016-8465-z

Hecker, L. R., Liang, S. Y., Wu, X. J., Xia, P., & Jin, D. G. W. (2007). Grinding force and power modeling based on chip thickness analysis. *International Journal of Advanced Manufacturing Technology*, *33*(5-6), 449–459. doi:10.100700170-006-0473-y

Hecker, R. L., Liang, S. Y., Wu, X. J., Xia, P., & Jin, D. G. W. (2007). Grinding force and power modeling based on chip thickness analysis. *International Journal of Advanced Manufacturing Technology, 33*(5-6), 449–459. doi:10.100700170-006-0473-y

Huang, S. L. (2008). Comparative Study on Thermocouple and Infrared Temperature Measurement Methods. *Electronics Quality, 6*. doi:10.3384/njvet.2242-458X.15v5i1a4

Ji, S., Ni, J. H., & Zhan, B. S. (2015). Research on milling cutting temperature of titanium alloybased on the constantan-workpiece thermal couple method. *Manufacturing Automation, 37*(22), 24-25, 76.

Jiang, F., Yan, L., Xu, X., & Rong, Y. (2014). Mechanical Model of Contact Stress Distribution between Cutting Tool and Chip. *Journal of Mechanical Engineering, 50*(5), 188-193.

Koamndouri, H., & B., Z. (2000). Thermal modeling of the metal cutting process. Part I - Temperature rise distribution due to shear plane heat source. *International Journal of Mechanical Sciences, 42*(9), 1715-1752.

Komandouri, R., & Zhen, B. H. (2001). Thermal modeling of the metal cutting process - Part III: temperature rise distribution due to the combined effects of shear plane heat source and the tool-chip interface frictional heat source. *International Journal of Mechanical Sciences, 43*(1), 89-107.

Kurgin, S., Dasch, J., Simon, D. L., Barber, G. C., & Qian, Z. (2012). Evaluation of the convective heat transfer coefficient for minimum quantity lubrication (MQL). *Industrial Lubrication and Tribology, 64*(6), 376–386. doi:10.1108/00368791211262516

Lalwani, D. I., Mehta, N. K., & Jain, P. K. (2008). Experimental investigations of cutting parameters influence on cutting forces and surface roughness in finish hard turning of MDN250 steel. *Journal of Materials Processing Technology, 206*(1-3), 167–179. doi:10.1016/j.jmatprotec.2007.12.018

Li, B. K., Li, C. H., Zhang, Y. B., Wang, Y. G., Jia, D. Z., & Min, Y. (2016). Grinding temperature and energy ratio coefficient in MQL grinding of high-temperature nickel-base alloy by using different vegetable oils as base oil. *Chinese Journal of Aeronautics, 29*(4), 1084–1095. doi:10.1016/j.cja.2015.10.012

Li, K., & Liao, W. (1997). Modelling of ceramic grinding processes Part I. Number of cutting points and grinding forces per grit. *Journal of Materials Processing Technology, 65*(1–3), 1–10. doi:10.1016/0924-0136(95)02232-5

Li, S. W., Wang, X. B., Xie, L. J., Pang, S. Q., Peng, S., Liang, Z. Q., & Jiao, L. (2015). The milling-milling machining method and its realization. *International Journal of Advanced Manufacturing Technology, 76*(5-8), 1151–1161. doi:10.100700170-014-6345-y

Li, Y. H., Sun, X. G., & Yuan, G. (2007). Accurate measuring temperature with infrared thermal imager. *Optics And Precision Engeering, 15*(9), 1336–1341. doi:10.3321/j.issn:1004-924x.2007.09.005

Liang, S. (1998). A Study on the Grinding Basic Parameter. *Journal of Guangxi University*, 168-171.

Liang, Z. Q., Wang, X. B., Wu, Y. B., Li, Y. Z., Wen, X., & Pang, S. Q. (2010). Status and Progress of Ultrasonic Assisted Grinding Technique. *Acata Armamentrii, 31*(11), 1530–1535. doi:10.1016/S1003-6326(10)60661-9

Liu, Z. X., & Zhang, D. W. (2006). A High-speed Cutting Temperature Measurement Equipment Based on Semi-manual Thermocouple Method. *Modular Machine Tool & Automatic Manufacturing Technique,* (4), 59-61.

Ma, F., Du, S. M., & Zhang, Y. Z. (2011). Review on the Three-dimensional Surface Characteristic Research of Dry Friction Wear Debris. *Lubrication Engineering*, *36*(1), 112–116.

Malkin, S., & Guo, C. S. (1991). Grinding technology: Theory and applications of machining with abrasives. *International Journal of Machine Tools & Manufacture*, *31*(3), 435–436. doi:10.1016/0890-6955(91)90088-K

Malkin, S., & Ritter, J. E. (1992). Grinding Mechanisms and Strength Degradation for Ceramics. *Key Engineering Materials*, *71*(2), 195–190. doi:10.4028/www.scientific.net/KEM.71.195

Mao, C., Zhi, X. Z., Kexi, G., & Xu, Y. L. (2010). Numerical Simulation And Experimental Validation of The Temperture Field In Surface Grinding. *Machining Science and Technology*, *14*(3), 344–364. do i:10.1080/10910344.2010.512198

Nieslony, P., Krolczyk, G. M., Wojciechowski, S., Chudy, R., Zak, K., & Marud, R. W. (2018). Surface quality and topographic inspection of variable compliance part after precise turning. *Applied Surface Science*, *434*, 91–101. doi:10.1016/j.apsusc.2017.10.158

Qian, Y., Xu, J. H., Fu, Y. C., & Tian, L. (2011). Grinding force and specific grinding energy of nickel based superalloy during high speed grinding with cBN wheel. *Dianmod & Abrasives Engineering*, *31*(6), 33–37.

Rowe, W. B., Black, S. C. E., Mills, B., Morgan, M. N., & Qi, H. S. (1997). Grinding Temperatures and Energy Partitioning. *Proceedings Mathematical Physical & Engineering Sciences, 453*(1960), 1083-1104. doi:10.1098/rspa.1997.0061

Shawa, M. C., & Vyasa, A. (1993). Chip Formation in the Machining of Hardened Steel. *CIRP Annals - Manufacturing Technology, 42*(1), 29-33. doi:10.10160007-8506(07)62385-3

Sheng, W., Li, C., Zhang, D., Jia, D., & Zhang, Y. (2014). Modeling the operation of a common grinding wheel with nanoparticle jet flow minimal quantity lubrication. *International Journal of Advanced Manufacturing Technology*, *74*(5-8), 835–850. doi:10.100700170-014-6032-z

Sun, X. G., & Li, Y. H. (2008). Review of the Development of Temperature Measurement Technology with Infrared Thermal Imager. *LASER & INFRARED*, *38*(2), 101–104.

T., L. X., Uu, D. F., & Tang, J. Y. (2009). Finite element analysis of grinding chip formation of single abrasive grain with negative rake angle. *Modern Manufacturing Engineering,* (4), 36-41, 66. doi:10.1109/MILCOM.2009.5379889

Tawakoli, T., Hadad, M. J., Daneshi, A., Sadeghi, M. H., & Sadeghi, B. (2011). Study on the Effects of Abrasive and Coolant-Lubricant Types on Minimum Quantity Lubrication-MQL Grinding. *Advanced Materials Research, 325*(325), 231–237. doi:10.4028/www.scientific.net/AMR.325.231

Venkatachalam, S., Li, X. P., & Liang, S. Y. (2009). Predictive modeling of transition undeformed chip thickness in ductile-regime micro-machining of single crystal brittle materials. *Journal of Materials Processing Technology*, *209*(7), 3306–3319. doi:10.1016/j.jmatprotec.2008.07.036

Wang, S., Hu, Z. Y., Shao, J., Zhang, Z. R., Ye, J. F., & Li, G. H. (2014). Two-color thermally sensitive phosphor coatings for temperature measurement. *Hongwai Yu Jiguang Gongcheng*, (5): 1406–1410.

Wang, W., & Feng, X. Y. (2013). Analysis of Grinding Force and Elastic Deformation in Thread Grinding Process. *Advances in Mechanical Engineering*. doi:10.1155/2013/827831

Wang, Y. G., Li, C. H., Li, B. K., Yang, M., & Zhang, X. W. (2015). Advances and Recent Patents in the Field of Grinding Temperature Measurement Methods. *Recent Patents on Materials Science, 8*(1). doi:10.2174/1874464808666150304233653

Wang, Y. G., Li, C. H., Zhang, Y. B., Yang, M., Zhang, X. P., Zhang, N. Q., & Dai, J. J. (2017). Experimental evaluation on tribological performance of the wheel/workpiece interface in minimum quantity lubrication grinding with different concentrations of Al_2O_3 nanofluids. *Journal of Cleaner Production, 142*, 3571–3583. doi:10.1016/j.jclepro.2016.10.110

Yang, H., Lei, Z., Li, D., & Li, T. (2011). Modeling and analysis of grinding force in surface grinding. *IEEE International Conference on Computer Science & Automation Engineering*.

Yong, H., & Liang, S. Y. (2005). Cutting Temperture Modeling Based On Non-Uniform Heat Intensity And Partition Ratio. *Machining Science and Technology*, *9*(3), 301–323. doi:10.1080/10910340500196421

Yuan, J. L., Zhang, F. H., Dai, Y. F., Kang, R., Yang, H., & Lu, B. H. (2010). Development Research of Science and Technologies in Ultra-precision Machining Field. *Journal of Mechanical Engineering*, *46*(15), 161–177. doi:10.3901/JME.2010.15.161

Zha, T. j., Xu, L., Luo, R., Xie, B., & Shi, L. (2013). Research on wheel wear volume in grinding of high hardness spherical surface. *Manufacturing Technology & Machine Tool,* (4), 96-99.

Zhang, D. K., Li, C. H., Jia, D. Z., & Ma, H., L. (2014). The experimental investigation of mql grinding performance with nanoparticles jet of spheroidal graphite cast iron. *Manufacturing Technology & Machine Tool,* (11), 98-103.

Zhang, N., Kirpitchenko, I., & Liu, D. (2005). Dynamic model of the grinding process. *Journal of Sound and Vibration, 280*(1-2), 425–432. doi:10.1016/j.jsv.2003.12.006

Chapter 4
Heat Transfer Mechanism of Minimum Quantity Lubrication Grinding

ABSTRACT

Grinding temperature rise at the grinding interface has a great bearing on workpiece surface quality and service life of grinding wheel. In order to study influence degree of temperature on workpiece and analyze heat transfer at the grinding interface, as thermal influence at the grinding interface is regarded as a moving heat source on workpiece surface, influence of heat source on heat transfer will be investigated in the gradual expansion form of heat source by point, line, and surface on the theoretical foundation in Chapter 3. The influences of heat quantity transferred to workpiece and energy proportionality coefficient will be studied according to convective heat transfer theory on workpiece surface so as to provide a theoretical foundation for heat transfer of grinding temperature field model in the finite element simulation software.

4.1 INTRODUCTION

The grinding temperature field produced in the grinding process is a very complex process. The influence of grinding heat on the workpiece is a dynamic process (Chen & Wang, 2018). In order to study the law of grinding temperature field conveniently, it is necessary to establish a numerical analysis model of grinding temperature field. The grinding effect between the grinding wheel and the workpiece generates a lot of heat to the workpiece, so that the surface temperature of the workpiece will rise when the temperature reaches the critical value. It is also necessary to analyze the heat in grinding process continuously.

Li et al.(1989) established a mathematical model to simulate the grinding process, and predicted the transient temperature field and stress in the workpiece due to the force and heat generated by the grinding wheel during the grinding process. The residual stress and microstructure were predicted. The model took into account the fully coupled thermo mechanical field equation, which was coupled by the expansion work, plastic work, exothermic heat during phase transformation and the heat generated by

DOI: 10.4018/978-1-7998-1546-4.ch004

the grinding wheel in the energy equation. According to the characteristics of grinding process, Lin and Zhang(2001) established a moving heat source model. Starting from the point of heat source, he analyzed the characteristics of heat source and established the model temperature field. A heat source model with periodic variation was established by Xu et al(1994), and a new theoretical formula for calculating the temperature field of intermittent grinding was proposed. This new formula can not only get the final temperature distribution of the workpiece surface layer, but also find the whole process of the temperature distribution in the grinding process. Using this formula, the temperature distribution characteristics of the workpiece surface in the intermittent grinding process, the influence of the intermittent ratio and period on the temperature distribution, and the cooling effect of the segmented or slotted grinding wheel are studied, González–Santander et al.(2010) and Jiang et al(2016) optimized the heat source model on the basis of his formula. Zhang et al.(2015) studied the influence of temperature field on MEG, analyzed the mechanism of machining process, and established the finite element model of micro end grinding temperature field. On this basis, the finite element simulation experiment was carried out. The results showed that a similar ring temperature field was formed during MEG grinding. The highest surface temperature of the workpiece was located in the center of the wheel face. Koshin and Shipulin(2012) considered the random model of temperature field in cutting area and radial component of grinding force (Koshin & Shipulin, 2012). According to the probability structure of wheel, due to its production process, the interaction between tool and workpiece is random. The temperature prediction model of grinding temperature field was established. Yang et al.(2016) established the largest undeformed chip thickness model and the heat flux density model. The mathematical model was used to establish the convective heat transfer coefficient model of nanofluid under the condition of micro lubrication, and the numerical simulation of the temperature field of bone micro grinding under spray lubrication and nanofluid lubrication conditions was carried out. The results showed that the surface temperature of bone decreased with the increase of the volume fraction of nanoparticles. The micro grinding test was carried out on the fresh bovine femoral dense bone which has the closest mechanical properties to human bone. In terms of atomization mechanism, Mao et al.(2012) analyzed two key factors affecting the heat transfer effect: droplet diameter and droplet velocity. According to the different heat transfer characteristics of the droplets at different wall temperatures, the grinding area was divided into four different heat transfer areas: non boiling heat transfer, nuclear boiling heat transfer, transitional boiling heat transfer and stable film boiling heat transfer. The mathematical model of heat transfer coefficient in the micro lubrication grinding area was established. Zhang et al (2015)established the energy proportion coefficient model of the incoming workpiece in the grinding process, studied the application of nanoparticles in the grinding cooling lubrication, analyzed the influence of cooling lubrication on the grinding surface theoretically through the energy ratio coefficient and specific grinding energy, and continued the verification test. Zhang et al.(2018) aimed at the problems of poor lubrication effect of low temperature air (CA) and poor cooling ability of nanofluid minimum lubrication (NMQL), put forward a new manufacturing process of low temperature air nanofluid minimum lubrication (CNMQL). According to the theory of boiling heat transfer and heat conduction, the heat transfer coefficient and finite difference model under different grinding conditions were established. The temperature field of grinding zone under different cooling conditions was simulated. The results showed that CNMQL has the best cooling effect, followed by CA and NMQL. Based on the model simulation, the temperature field of surface grinding under the cooling conditions of CA, MQL and CNMQL were verified by experiments. Li et al.(2017) established different temperature field finite difference models for different grinding workpieces and different cooling lubricants, and carried out experimental verification to optimize the nanofluid micro lubrication process.

Figure 1. The heat transfer model of transient point heat source

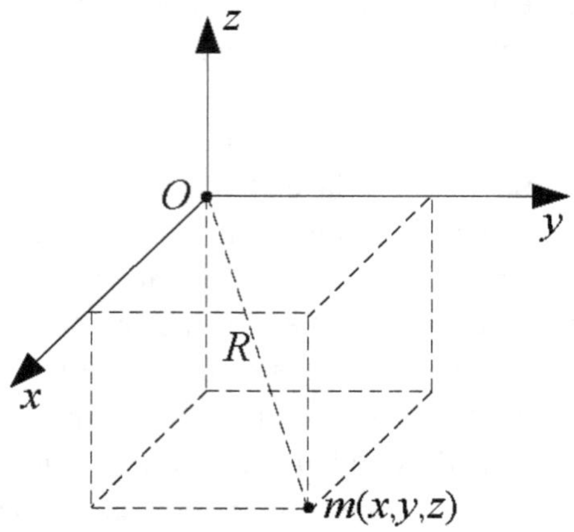

In conclusion, as an important parameter in the grinding process, grinding temperature is closely related to the surface quality, grinding wheel wear, etc., and is also an important basis for grinding parameter correction. Therefore, in the study of the cooling and lubricating effect of nanofluid micro lubrication, it is necessary to conduct a detailed study and analysis of the temperature field, so as to achieve its advantages of environmental protection and ensure the best tool surface quality.

4.2 HEAT TRANSFER MODEL OF GRINDING TEMPERATURE FIELD

Based on moving heat source theory, as the heat source moves on workpiece surface, grinding interface is simplified as a moving surface heat source in the surface grinding, which consists of numerous linear heat sources, and linear heat source consists of numerous point heat sources (Gonzalez-Santander & Monreal, 2019). Therefore, in consideration of the influence of transient exothermic process of heat source on temperature field, heat transfer models of "point-line-surface" heat sources are successively established through the heat source integration method according to Fourier heat transfer law to analyze heat transfer influence.

4.2.1 Heat Transfer Model of Instantaneous Point Heat Source

After grinding wheel slides through one point on workpiece surface, this point becomes a point heat source (Jiang et al., 2016). Assume that heat will be transferred towards three directions while heat dissipation is not considered, and schematic diagram of heat transfer model of transient point heat source is shown in Fig. 1.

According to Fourier heat transfer law, the influence of point heat source with heat quantity of Q on an arbitrary point $m(x,y,z)$ in the temperature field after time dt is expressed by 3D heat transfer differential equation as:

$$\frac{d\theta}{dt}=\alpha[\frac{\partial^2\theta}{\partial x^2}+\frac{\partial^2\theta}{\partial y^2}+\frac{\partial^2\theta}{\partial z^2}] \tag{1}$$

where: α—thermal diffusion coefficient, $\alpha=\lambda/c\rho$;

c—specific heat, J/(kg·K);
ρ—density, g/cm^3;

Point heat source located at origin of coordinates transfers heat to any point $m(x,y,z)$in the workpiece, and after time t, temperature at point m is solved as:

$$\theta=\theta(R,t)=\frac{Q}{c\rho(4\pi\alpha t)^{3/2}}e^{-\frac{R^2}{4\alpha t}} \tag{2}$$

where: R—distance from point heat source to point m, $R^2=x^2+y^2+z^2$;

According to equation (4-2), temperature at this point can be obtained through the distance R from point heat source to this point. Therefore, influence of point heat source at the grinding interface on temperature field is obtained through substitution of coordinates of the point on the workpiece, and temperature at this point can be solved through coordinate values.

4.2.2 Heat Transfer Model of Instantaneous Linear Heat Source

When grinding wheel is grinding the workpiece, length of heat source acting upon workpiece surface is equal to grinding width d_s, so linear heat source is regarded as heat source of transient finite length in the investigation process of heat transfer model of linear heat source. Assume that heat source length is d_s and transient heat quantity is Q, spatial coordinates system O-XYZ is established, and heat transfer model of transient linear heat source with finite length is shown in Fig. 2.

As shown in Fig. 2, axis x is overlapped with heat source, a point $m(x,y,z)$ exists in the workpiece, the influence of linear heat source on temperature of point m is studied, and with the help of the influence of differential line segment dx_i at linear heat source $(x_i,0,0)$ on temperature of point m, temperature field of the whole linear heat source is obtained by taking integral of length of the linear heat source (Krasyuk et al., 2012). Differential line segment dx_i can be regarded as a transient point heat source, and according to equation (2), temperature rise $d\theta$ caused by the influence of differential line segment dx_i on point m is expressed as (Lin et al., 2004):

$$d\theta=\frac{Q}{c\rho(4\pi\alpha t)^{3/2}}e^{-\frac{(x-x_i)^2+y^2+z^2}{4\alpha t}}dx_i \tag{3}$$

Temperature rise caused by the influence of the whole linear heat source on heat transfer at point m can be obtained by taking integral of differential line segment dx_i on the length of linear heat source as follow:

Figure 2. The heat transfer model of transient linear heat source

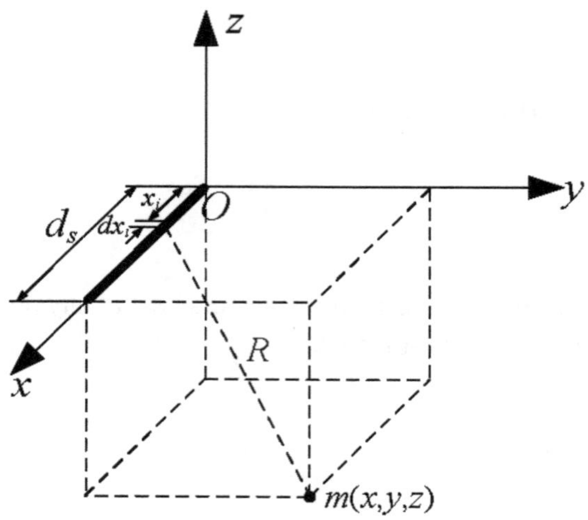

$$\theta = \frac{Q}{c\rho(4\pi\alpha t)^{3/2}} e^{-\frac{y^2+z^2}{4\alpha t}} \int_0^{d_s} e^{-\frac{(x-x_i)^2}{4\alpha t}} dx_i \tag{4}$$

Set $\frac{x-x_i}{\sqrt{4\alpha t}} = u$, then $dx_i = -\sqrt{4\alpha t}du$, and the following can be obtained by substituting it into equation (4-4):

$$\theta = \frac{Q\sqrt{4\alpha t}}{c\rho(4\pi\alpha t)^{3/2}} e^{-\frac{y^2+z^2}{4\alpha t}} \int_{\frac{x-d_s}{\sqrt{4\alpha t}}}^{\frac{x}{\sqrt{4\alpha t}}} e^{-u^2} du \tag{5}$$

After calculation of integral, the following can be obtained:

$$\theta = \frac{Q}{8c\rho(\pi\alpha t)} e^{-\frac{y^2+z^2}{4\alpha t}} [erf(\frac{x}{\sqrt{4\alpha t}}) - erf(\frac{x-d_s}{\sqrt{4\alpha t}})] \tag{6}$$

where: *erf*(*p*)—error function.

According to equation (6), temperature at any point *m* in the temperature field of linear heat source is influenced by coordinate position of this point, indicating that the influence of linear heat source on heat transfer in the temperature field on the workpiece surface is three-dimensional, if influence of error function is neglected, then temperature rise is mainly related to coordinate values of *y* and *z*.

Figure 3. The heat transfer model of transient surface heat source

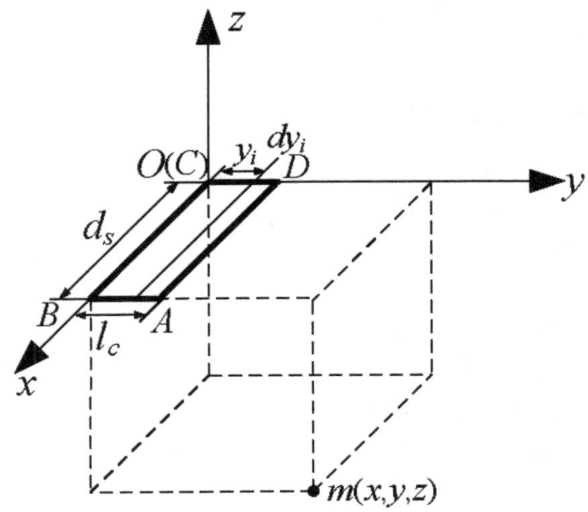

4.2.3 Heat Transfer Model of Instantaneous Surface Heat Source

In consideration of transient influence of actual heat source area at the grinding interface on workpiece temperature, different from the analysis of moving heat source by J. C. Jaeger, influence of motion parameters is not considered in the temperature field of transient surface heat source, but instead, only the influence of heat transfer by heat quantity transiently emitted by one side of heat source on workpiece temperature is considered, and heat transfer model of transient surface heat source is shown in Fig. 3.

As shown in Fig. 3, spatial coordinate system *O*-XYZ is built on workpiece surface, heat source area *ABCD* is above *XY* plane, motion direction of heat source is parallel to axis *Y*, width *BC* of heat source area is equal to grinding wheel width d_s, length of heat source is l_c, average heat source transiently emitted from surface heat source area *ABCD* is *Q*, surface heat source consists of numerous linear heat sources, and there is a linear heat source dy_i at $y=y_i$ in the heat source area, so the influence of the whole surface heat source on temperature of any point $m(x,y,z)$in the workpiece can be regarded as the sum of temperature influence of numerous linear heat sources on any point $m(x,y,z)$.

According to the temperature influence of linear heat source on any point $m(x,y,z)$ in the workpiece in equation (4-6), temperature rise caused by the influence of a linear heat source in the surface heat source on any point $m(x,y,z)$ in the workpiece can be expressed as:

$$d\theta = \frac{Q}{8c\rho(\pi\alpha t)} e^{-\frac{(y-y_i)^2+z^2}{4\alpha t}} [erf(\frac{x}{\sqrt{4\alpha t}}) - erf(\frac{x-d_s}{\sqrt{4\alpha t}})]dy_i \tag{7}$$

If integral of linear heat source dy_i is solved from 0 to l_c in direction *Y*, temperature rise θ caused by the influence of the whole surface heat source on point $m(x,y,z)$ is obtained as:

$$d\theta = \frac{Q}{8c\rho(\pi\alpha t)} e^{-\frac{z^2}{4\alpha t}} [erf(\frac{x}{\sqrt{4\alpha t}}) - erf(\frac{x-d_s}{\sqrt{4\alpha t}})] \int_0^{l_c} e^{-\frac{(y-y_i)^2}{4\alpha t}} dy_i \tag{8}$$

After calculation and organization, the following can be obtained:

$$\theta = \frac{Q}{8c\rho(\pi\alpha t)^{1/2}} e^{-\frac{z^2}{4\alpha t}} [erf(\frac{x}{\sqrt{4\alpha t}}) - erf(\frac{x-d_s}{\sqrt{4\alpha t}})][erf(\frac{y}{\sqrt{4\alpha t}}) - erf(\frac{y-l_c}{\sqrt{4\alpha t}})] \tag{9}$$

According to equation (9), the temperature influence of the whole surface heat source on any point *m(x,y,z)* in the workpiece is mainly related to coordinate position of point *m*. If influence of error function is neglected, the temperature influence of surface heat source on point *m* can be regarded as 1D spatial heat transfer problem and it is only related to the distance from point to surface heat sources.

4.3 GROUNDING BOUNDARY CONDITIONS

In the grinding process, in order to lower workpiece surface temperature and lubricating effect in the grinding area, different cooling and lubrication conditions and cooling media namely boundary conditions at grinding interface are used, which are embodied by convective heat transfer coefficient (Lan & Jiao, 2019). The main cooling and lubrication conditions used in the current grinding machining include dry grinding, flood grinding and MQL, etc. Different boundary conditions are corresponding to different convective heat transfer capabilities, which will result in the difference of effect on lowering workpiece surface temperature.

4.3.1 Boundary Conditions

In the grinding temperature field model, heat quantity continuously input by moving heat source at grinding wheel/workpiece interface, heat transfer in workpiece base and convective heat transfer on workpiece surface follow the first law of thermodynamics and Fourier heat transfer law (Li et al., 2013). Field variable *T(x,y,z,t)* in the transient temperature field satisfies differential equation of heat conduction under thermal balance:

$$\frac{\partial}{\partial x}(k_x \cdot \frac{\partial T}{\partial x}) + \frac{\partial}{\partial y}(k_y \cdot \frac{\partial T}{\partial y}) + \frac{\partial}{\partial z}(k_z \cdot \frac{\partial T}{\partial z}) + \rho Q = \rho c \frac{\partial T}{\partial t} \tag{10}$$

where: k_x, k_y and k_z—heat conductivity coefficients of material along directions x, y and z, W/(m·K);

Q—heat flux, $J/(m^2 \cdot K \cdot s)$;
ρ—density of workpiece material, kg/m^3;
c—specific heat of workpiece material, J/(kg·K);

As shown in equation (10), first three items at the left of the equation express temperature rises caused by heat transfer inside the workpiece, the fourth item is internal heat source in the temperature field, but there is no heat source in the grinding process, so it is non-steady heat transfer model namely Q=0. The right side of the equation is the heat quantity consumed by temperature rise of the workpiece.

Thermal balance analysis of heat conduction problem in the temperature field is to solve differential equation, and the complete heat transfer process description should include differential equation of heat conduction and non-ambiguity conditions, which include geometric conditions, physical conditions, time conditions and boundary conditions (Li et al., 2019). Therefore, initial condition (temperature at t=0) and boundary conditions (given temperature in boundary area or input and output heat fluxes) of the temperature field must be given. According to the heat transfer theory, common boundary conditions can be divided into three types:

The first type is temperature value at the boundary surface of the given object at any time, a boundary condition of forced convection and it is also called Dirichlet condition, and its formula is:

$$T\big|_s = T_w \tag{11}$$

where: T_w—given temperature on boundary surface s. If boundary temperature keeps unchanged, T_w is a constant value. If boundary temperature changes with time, T_w is functional relation of time.

The second type is that normal heat flux at boundary surface of the object at any time is already known, and it is also called Neumann condition (Li et al., 2012). Fourier law gives the relationship between heat flux and temperature gradient, which is equivalent to giving normal temperature change rate on boundary s at any time, and it is expressed as:

$$\left.\frac{\partial T}{\partial n}\right|_s = \frac{Q_w}{\lambda} \tag{12}$$

where: Q_w—heat flux passing through boundary interface s (W/m^2). It is equal to 0 when the boundary is under heat insulation condition. When boundary is under heat conduction, it is a constant value and a time-dependent function.

Differential equation on the second type of boundary can be expressed as:

$$\frac{\partial}{\partial x}(k_x \cdot \frac{\partial T}{\partial x}) + \frac{\partial}{\partial y}(k_y \cdot \frac{\partial T}{\partial y}) + \frac{\partial}{\partial z}(k_z \cdot \frac{\partial T}{\partial z}) = Q_w \tag{13}$$

The third type of boundary condition is convective heat transfer between boundary interface and surrounding medium and it is also called Robin condition. According to Newton's law of cooling, convective heat transfer between workpiece boundary layer and cooling heat transfer medium is expressed as:

$$Q_w = h(T\big|_s - T_a) \tag{14}$$

where: Q_w—convective heat transfer between workpiece boundary interface s and heat transfer medium, W/m^2;

h—convective heat transfer coefficient between boundary and heat transfer medium, W/(m^2·K);
T_a—temperature of cooling heat transfer medium.

Differential equation on the third type of boundary can be expressed as:

$$\frac{\partial}{\partial x}(k_x \cdot \frac{\partial T}{\partial x}) + \frac{\partial}{\partial y}(k_y \cdot \frac{\partial T}{\partial y}) + \frac{\partial}{\partial z}(k_z \cdot \frac{\partial T}{\partial z}) = h(T|_s - T_a) \tag{15}$$

The above gives three types of boundary condition. Heat transfer process in the grinding area can be completely described through differential equation of heat conduction and non-ambiguity conditions (C. Mao, Zhou, Guo, & Li, 2010; C. Mao, Zhou, Ren, & Zhang, 2010).

4.3.2 Convective Heat Transfer

Heat transfer mainly refers to transferring heat from a system to another system, mainly including three forms: heat conduction, convection and radiation. Grinding fluid is used for cooling in the grinding machining process (Nguyen & Zhang, 2010; Parente et al., 2012). Grinding fluid experiences heat transfer on workpiece surface in flowing form on the grinding interface, and the main heat transfer model is convective heat transfer (Parente et al., 2013). Grinding fluid enters grinding area and contacts workpiece surface, one part of heat on workpiece surface is carried away by grinding fluid through convective heat transfer and another part stays at workpiece base and is transferred into the workpiece. Using grinding fluid or enhancing heat transfer performance of grinding fluid can strengthen cooling & lubricating effect at the grinding interface. For the convenient of analysis of heat transfer process at the grinding interface, the area nearby the convective heat transfer layer in the grinding area is gridded, grids in unit volume are replaced by nodes, and schematic diagram of heat conduction relationships among grinding wheel, grinding fluid, workpiece and nodes inside the workpiece is shown in Fig. 4 (Bense et al., 2016).

As shown in Fig. 4, heat at the grinding interface is input under the grinding effect of grinding wheel, while the heat input into the workpiece will be transferred to surrounding material, grinding fluid entering the grinding area contacts workpiece surface, and one part of heat input into the workpiece will be convectively transferred to grinding fluid and then carried away (Peres et al., 2008). Taking one node (m, 1) on the surface as analysis object, grinding wheel directly contacts this node, so transient heat is transferred to this node, and in the meantime, grinding fluid on the surface and adjacent nodes (m-1, 1), (m+1, 1) and (m, 2) all contact node (m, 1), so heat at this node will be convectively transferred to grinding fluid and go through heat conduction with adjacent nodes until temperature at grinding interface becomes stable.

In the grinding area, energy conservation among heat conduction between nodes, heat flux input into grinding wheel/workpiece interface, temperature rise of nodes themselves and convective heat transfer of grinding fluid is expressed as:

Figure 4. The schematic diagram of convection heat transfer in grinding interface

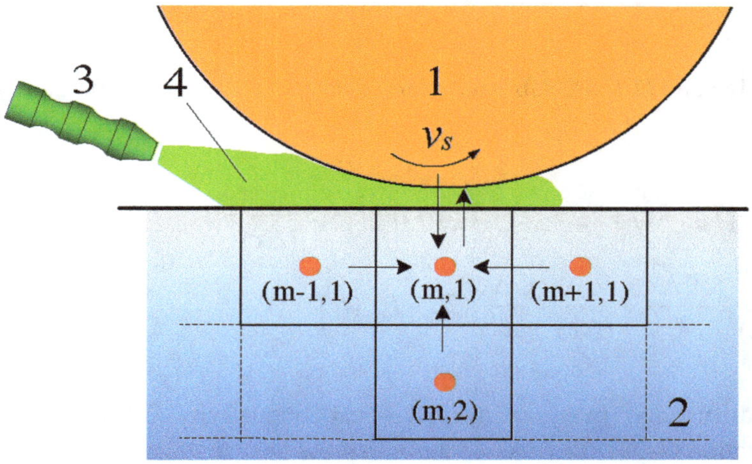

$$q(m-1,1)\rightarrow(m,1)+q(m+1,1)\rightarrow(m,1)+q(m,2)\rightarrow(m,1)+q''A = \rho c V_0\frac{\partial T}{\partial t}+G\cdot A[T_{m,1}(t)-T_a] \quad (16)$$

where: $T_{m,1}$—temperature of node (m,1);

T_a—temperature of grinding fluid;
q´´—heat flux at abrasive grains/workpiece interface;
ρ—density of workpiece material, kg/m3;
c—specific heat of workpiece material, J/(kg·K);
V_0—grid unit volume, V_0=ΔxΔz and Δx=Δz;
A—unit surface area between grids, A=Δx;
G—comprehensive heat transfer coefficient, G=[1/h+Δz/(2k)]-1; h—convective heat transfer coefficient.

1-wheel, 2-workpiece, 3-nozzle, 4-grinding fluid

Heat transfer of node (*m*,1) on workpiece surface with nodes (*m*-1,1), (*m*,2) and (*m*+1,1) is expressed as:

$$q(m-1,1)\rightarrow(m,1)=k\Delta z[\frac{T_{m-1,1}(t)-T_{m,1}(t)}{\Delta x}] \quad (17)$$

$$q(m+1,1)\rightarrow(m,1)=k\Delta z[\frac{T_{m+1,1}(t)-T_{m,1}(t)}{\Delta x}] \quad (18)$$

$$q(m,2) \rightarrow (m,1) = k\Delta x[\frac{T_{m,2}(t) - T_{m,1}(t)}{\Delta z}] \quad (19)$$

Equations (17), (18) and (19) are substituted into equation (16), and temperature at node $(m,1)$ after time Δt is turned into:

$$\begin{aligned} T_{m,1}(t+\Delta t) = & \frac{\alpha \Delta t}{(\Delta x)^2}[T_{m-1,1}(t) + T_{m+1,1}(t) + T_{m,2}(t)] + [1 - 3\frac{\alpha \Delta t}{(\Delta x)^2}]T_{m,1}(t) \\ & + \frac{\alpha \Delta t}{(\Delta x)^2}\frac{q"\Delta x}{k} + \frac{\alpha \Delta t}{(\Delta x)^2}\frac{G\Delta x}{k}(T_a - T_{m,1}(t)) \end{aligned} \quad (20)$$

where: α—thermal diffusivity, (m^2/s),α=λ/(ρc); λ—heat conductivity, W/(m·K).

According to equation (4-20), the first two items express heat conduction between nodes in the workpiece, the third item is the temperature influence of heat flux caused by grinding wheel-input power on workpiece surface, and the final item is convective heat transfer between workpiece surface and grinding fluid. As workpiece surface temperature in the grinding area is temperature of node $(m,1)$, which is always higher than temperature T_a of grinding fluid namely T_a-T_m<0, so grinding fluid exerts the convective heat transfer effect, thus reducing workpiece surface temperature. When a cooling and lubrication condition like flood grinding or MQL is used in the grinding process, workpiece surface temperature is lowered in the form of convective heat transfer, and the only difference lies in comprehensive heat transfer coefficient G or convective heat transfer coefficient h.

Under dry grinding, grinding fluid is not used for convective heat transfer, namely convective heat transfer item in equation (4-20) is equal to 0, and it is expressed as:

$$T_{m,1}(t+\Delta t) = \frac{\alpha \Delta t}{(\Delta x)^2}[T_{m-1,1}(t) + T_{m+1,1}(t) + T_{m,2}(t)] + [1 - 3\frac{\alpha \Delta t}{(\Delta x)^2}]T_{m,1}(t) + \frac{\alpha \Delta t}{(\Delta x)^2}\frac{q"\Delta x}{k} \quad (19)$$

When other grinding parameters remain unchanged, according to the result derived through equation (4-21), temperature of node $(m,1)$ under flood grinding is obviously lower than that under dry grinding. Convective heat transfer capability of grinding fluid is closely related to physical properties of fluid, temperature difference between heat transfer surface and fluid and flow velocity of fluid in the heat transfer process, grinding fluid carries away a part of heat, and as a result, both grinding heat generated at the grinding wheel/workpiece interface and heat transferred to workpiece are obviously reduced when compared with those under flood grinding and dry grinding conditions (Gonzalez-Santander et al., 2011).

4.4 NUMERICAL ANALYSIS MODEL

Grinding temperature field generated in the grinding process is a very complicated process and action of grinding heat upon workpiece is a dynamic process, so in order to conveniently study laws of grinding temperature field, it's necessary to establish a numerical analysis model of grinding temperature field

Figure 5. Grid division of workpiece nodes

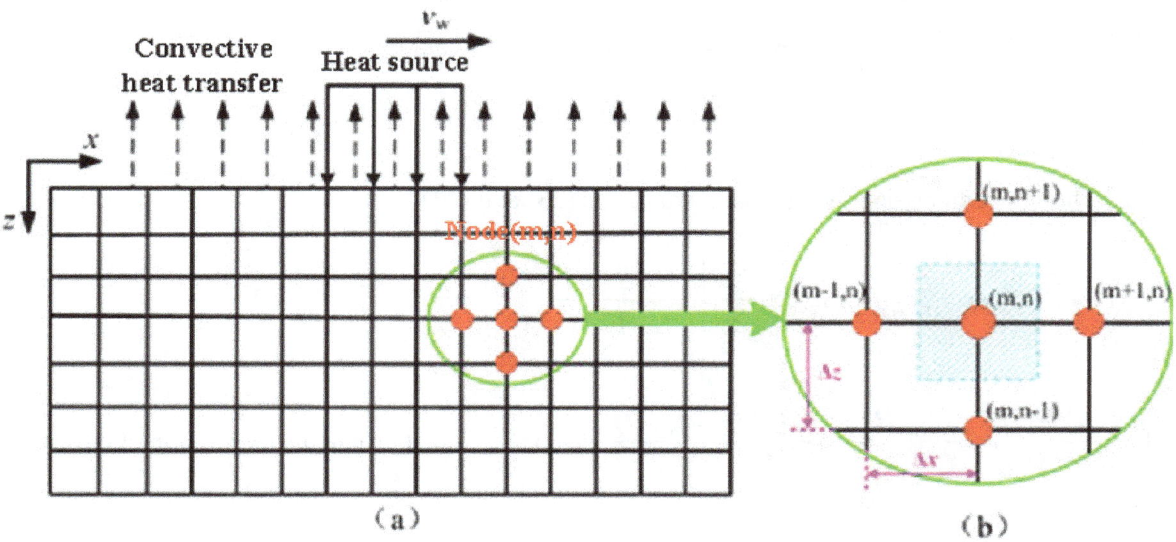

(Sailer et al., 2018; Song et al., 2019). Therefore, a complete numerical analysis and simulation model is built for MQL grinding using finite difference method.

4.4.1 Governing Equation

Grinding temperature field can be regarded as transient heat transfer process without internal heat source, where continuous energy input by moving heat source at the grinding wheel/workpiece interface goes through heat conduction in the workpiece base and convective heat transfer with workpiece surface (Wang et al., 2005). Differential equation of transient heat conduction can be used to describe temperature distribution in the workpiece base, and its 2D differential equation of heat conduction can be expressed as:

$$\frac{\partial \mathrm{T}}{\partial t}=\frac{\lambda_{\mathrm{w}}}{\rho_{\mathrm{w}} c_{\mathrm{w}}}\left(\frac{\partial^2 T}{\partial x^2}+\frac{\partial^2 T}{\partial z^2}\right) \tag{22}$$

Set $a=\lambda/(\rho c)$, where α is thermal diffusion coefficient, m^2/s; c—specific heat of workpiece material, J/(kg·K); ρ is density, kg/m^3.

Figure 5 shows schematic diagram of workpiece node meshing. Shaded area in (b) is the area represented by node (m,n), which consists of heat transfer elements. Heat transfer process is namely heat transfer process between elements.

Taylor's series expansions of function T to node (m,n) are written for nodes $(m+1,n)$ and $(m-1,n)$ respectively:

$$T_{m+1,n}(t)=T_{m,n}(t)+\Delta x\left.\frac{\partial T}{\partial x}\right|_{m,n}+\frac{\Delta x^2}{2}\left.\frac{\partial^2 T}{\partial x^2}\right|_{m,n}+\frac{\Delta x^3}{6}\left.\frac{\partial^3 T}{\partial x^3}\right|_{m,n}+\frac{\Delta x^4}{24}\left.\frac{\partial^4 T}{\partial x^4}\right|_{m,n}+\cdots \quad (23)$$

$$T_{m-1,n}(t)=T_{m,n}(t)-\Delta x\left.\frac{\partial T}{\partial x}\right|_{m,n}+\frac{\Delta x^2}{2}\left.\frac{\partial^2 T}{\partial x^2}\right|_{m,n}-\frac{\Delta x^3}{6}\left.\frac{\partial^3 T}{\partial x^3}\right|_{m,n}+\frac{\Delta x^4}{24}\left.\frac{\partial^4 T}{\partial x^4}\right|_{m,n}+\cdots \quad (24)$$

The following can be obtained through equations (23) and (24):

$$T_{m-1,n}(t)+T_{m+1,n}(t)=2T_{m,n}(t)+\Delta x^2\left.\frac{\partial^2 T}{\partial x^2}\right|_{m,n}+\frac{\Delta x^4}{12}\left.\frac{\partial^4 T}{\partial x^4}\right|_{m,n}+\cdots \quad (25)$$

Then $\left.\frac{\partial^2 T}{\partial x^2}\right|_{m,n}=\frac{T_{m+1,n}(t)+T_{m-1,n}(t)-2T_{m,n}(t)}{\Delta x^2}+o(\Delta x^2)$, and approximate expression of its second derivative is:

$$\left.\frac{\partial^2 T}{\partial x^2}\right|_{m,n}=\frac{T_{m+1,n}(t)+T_{m-1,n}(t)-2T_{m,n}(t)}{\Delta x^2} \quad (26)$$

Similarly:

$$\left.\frac{\partial^2 T}{\partial z^2}\right|_{m,n}=\frac{T_{m,n+1}(t)+T_{m,n-1}(t)-2T_{m,n}(t)}{\Delta z^2} \quad (27)$$

Equations (4-26) and (4-27) are substituted into equation (4-22), and then differential equation of node (m,n) can be obtained:

$$\frac{T_{m,n}(t+\Delta t)-T_{m,n}(t)}{\Delta t}=\frac{\lambda_{\mathrm{w}}}{\rho_{\mathrm{w}}c_{\mathrm{w}}}\left(\frac{T_{m+1,n}(t)+T_{m-1,n}(t)-2T_{m,n}(t)}{\Delta x^2}+\frac{T_{m,n+1}(t)+T_{m,n-1}(t)-2T_{m,n}(t)}{\Delta z^2}\right) \quad (28)$$

As the divided grids are equally divided grids namely $\Delta x=\Delta z$, equation (28) can be expressed as:

$$T_{m,n}(t)=\left[1-4\frac{\lambda_w\Delta t}{\rho_w c_w\Delta x^2}\right]T_{m,n}(t)+\frac{\lambda_w\Delta t}{\rho_w c_w\Delta x^2}\left[T_{m-1,n}(t)+T_{m+1,n}(t)+T_{m,n+1}(t)+T_{m,n-1}(t)\right] \quad (29)$$

4.4.2 Boundary Conditions

In consideration that natural convective heat transfer coefficient is much smaller than convective heat transfer coefficient of nanofluid, other 3 surfaces in Fig. 5 are idealized into heat-insulating surfaces, and the convective heat transfer formula can be obtained through Newton's law of cooling as (Yang et al., 2017):

$$Q_w = h(T_s - T_a) \tag{30}$$

where: Q_w is heat flux of convective heat transfer, W/m^2;

h is convective heat transfer coefficient, W/(m^2·K);
T_s—given temperature on boundary interface s;
T_a is temperature of cooling heat transfer medium.

According to the third-type boundary condition, its differential equation can be expressed as:

$$h(T|_{z=0} - T_0) - \lambda \left.\frac{\partial T}{\partial z}\right|_{z=0} = q_x \tag{31}$$

where initial temperature of workpiece and nanofluid is 20°C.

4.4.3 Convective Heat Transfer Model

Convective heat transfer coefficient of MQL grinding in the grinding area can be divided into two parts: natural convective heat transfer of air and convective heat transfer of grinding fluid sprayed into the grinding area (Alabi et al., 2017).

Natural convective heat transfer of air can be expressed as (Song et al., 2019):

$$h_a = \lambda_a Nu / b \tag{31}$$

where Nu is Nusselt number, $Nu = 0.906\,\mathrm{Re}^{1/2}\,\mathrm{Pr}^{1/3}$, $\mathrm{Re} = v_a \rho_a b / \mu_a$ and $\mathrm{Pr} = \mu_a c_a / \lambda_a$;λ_a is heat conductivity coefficient of air; v_a is air velocity; μ_a is air dynamic viscosity; c_a isobaric specific heat capacity of air.

Yang et al. used distribution method of atomized droplets to calculate convective heat transfer coefficient of grinding fluid sprayed into the grinding area, which accords with actual condition of MQL grinding (Yang et al., 2017).

The area formed when atomized droplets at the nozzle are sprayed onto workpiece surface is idealized into oval shape in the grinding experiment as shown in Fig. 6. According to spray angle β (15°) of nozzle and spray cone angle α=27° and spray height H, boundary equation of oval area is obtained as:

Figure 6. The oval spray boundary

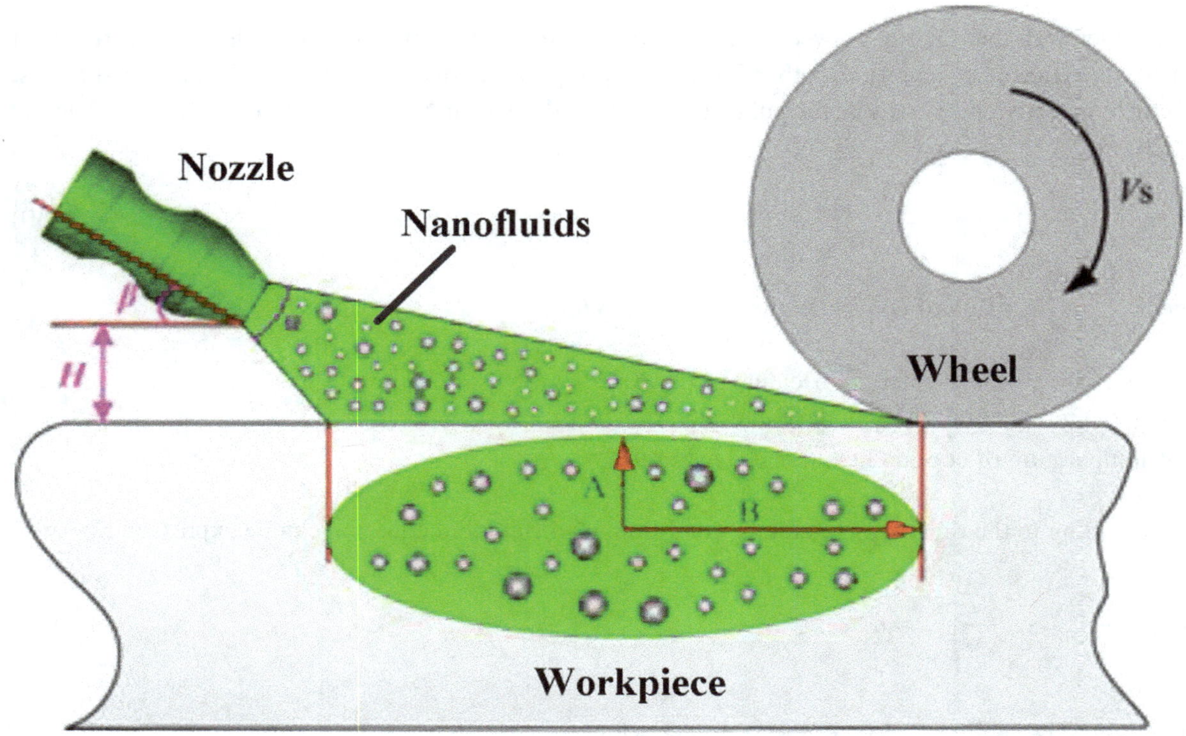

$$\frac{(x-x_0)^2}{A^2}+\frac{(y-y_0)^2}{B^2}=1 \tag{33}$$

where major axis and minor axis of the oval are 2A and 2B respectively and intermediate variable is $C=2\cos(\alpha+\beta/2)\cos(\alpha-\beta/2)$; $A=H\frac{1}{2C}\sin\beta, B=H\sqrt{\frac{1}{C}}\sin\frac{\beta}{2}$.

In order to relatively accurately determine spray size of atomized droplets and their distribution features, actual atomized droplets are idealized into droplet group with diameter of d_{32} (Sauter average diameter). Calculation formula of d_{32} is obtained as (Gonzalez-Santander et al., 2011):

$$\frac{d_{32}}{d_0}=3.07(\frac{\rho_a{}^{0.5}\Delta p d_0}{\sigma^{0.5}\mu})^{-0.259} \tag{34}$$

where: d_0 is nozzle diameter; Δp is internal and external pressure difference of the nozzle; ρ_a is air density; μ is dynamic viscosity of grinding fluid; σ is surface tension of atomized droplets. Drcplets with diameter of d_{32} are idealized into spherical droplets, so their volume is expressed as:

$$V_{d_{32}} = \frac{\pi d_{32}^{\ 3}}{6} \tag{35}$$

Total number N_{total} of atomized droplets from the nozzle is expressed as:

$$N_{total} = \frac{Q}{V_{d_{32}}} \tag{36}$$

where: $Q = Q' \cdot t$;t is total grinding time ($t = a / v_w$, and a is workpiece length); Q' is flow quantity of supplied fluid.

Number Nz of droplets entering the grinding area is:

$$N_Z = N_{tatal} \cdot s/S \tag{37}$$

where s is area of the grinding area, $s = l_c \cdot b$;S is oval area at nozzle boundary, $S = \pi \cdot A \cdot B$.

As the wedge-shaped space formed when grinding is conducted between grinding wheel and workpiece exerts a great constraining effect on lubricant, it can be deemed that droplets entering the grinding area can exert effective cooling & lubricating effect (Tucek et al., 2011; Vieira et al., 2009). When droplets enter the grinding area to exert effective cooling & lubricating effect, spreading phenomenon will occur as shown in Fig. 7.

A is set as spreading area of droplets, and then $\mathrm{A} = \pi \cdot r_{surf}^{\ 2}$, where r_{surf} is spreading radius of droplets and it can be expressed as(Maruda et al., 2016):

$$r_{surf} = \sqrt[3]{\frac{V}{\frac{\pi}{2}\frac{1}{tg\theta}(\frac{1}{\cos\theta}-1)[1+\frac{1}{3}\frac{1}{tg^2\theta}(\frac{1}{\cos\theta}-1)^2]}} \tag{38}$$

where θ is contact angle.

For a single droplet, its heat transfer coefficient h_s is(Gonzalez-Santander, Isidro, & Martin, 2015):

$$\begin{cases} J = cm_0(t_w - t_l) \\ h_s = \dfrac{q}{t_w - t_l} \\ J = qAt_s \end{cases} \tag{39}$$

After organization, $h_s = \dfrac{c \cdot m_0}{A \cdot t}$, where c is specific heat capacity of fluid; m_0 is droplet mass ($m_0 = \rho \cdot V$, V is droplet volume);

Hence, heat transfer coefficient of all droplets under effective heat transfer is:

$$h_n = h_s \cdot N_Z \tag{40}$$

Convective heat transfer coefficient when grinding fluid is sprayed into the grinding area through the atomizing nozzle is obtained as:

$$h = h_n + h_a \tag{41}$$

4.5 GRINDING ENERGY PARTITION

Grinding effect between grinding wheel and workpiece generates a large quantity of heat which is transferred to workpiece so that its surface temperature rises. When the temperature reaches one limiting value, phenomena like burning can easily occur, so it's necessary to continuously analyze heat quantity in the grinding process (Wang et al., 2016; Wang et al., Li et al., 2019). Generation of heat quantity is already analyzed in Chapter Three, but heat distribution needs further analysis, followed by derivation of proportionality coefficient of energy transferred to workpiece and establishment of the relationship between energy proportionality coefficient and grinding temperature. Related cooling & lubrication measures should be taken to improve the grinding area.

4.5.1 Energy Partition

Heat source in the grinding process came from three interfaces: abrasive grain/workpiece interface, abrasive grain/chip interface and workpiece/chip shear surface. Frictional heat is generated at abrasive grain/workpiece interface and abrasive grain/chip interface. Plastic deformation is generated at workpiece/chip shear interface and abrasive grain/workpiece interface, which then generates a large quantity of heat. Heat is transferred to workpiece base, abrasive grain, chip and grinding fluid respectively according to different heat conductivity coefficients of workpiece material, abrasive grain and grinding fluid (Zhang et al., 2018; Zhang et al., 2016). Under flood grinding environment, schematic diagram of heat transfer at the grinding interface is shown in Fig. 8 (Zhang et al., 2015).

In this model, it's deemed that heat is mainly generated at workpiece/chip shear interface and abrasive grain/workpiece interface. Frictional heat generated at abrasive grain/chip interface will be transferred to abrasive grain and chip respectively, this heat is regarded as one part of workpiece/chip, and generated total heat flux is the sum of heat quantity q''_{sh} at workpiece/chip interface and heat quantity q''_{wf} at abrasive grain/workpiece interface and expressed as $q''_{total} = q''_{sh} + q''_{wf}$.

When heat quantity is generated at two interfaces, heat will be transfer to two objects contacting interfaces in way of heat conduction (Zhou et al., 2013). Heat quantity q''_{sh} generated at workpiece/chip interface is transiently transferred to chip, workpiece and abrasive grain, heat quantity q''_c transferred to chip is carried away by chip falling off, the heat transferred from workpiece/chip interface to workpiece and abrasive grain is regarded as one part of heat source at abrasive grain/workpiece interface, heat flux generated at abrasive grain/workpiece interface is q''_{wf}, and total heat flux can be divided into heat

Figure 7. The droplet spreading sketch

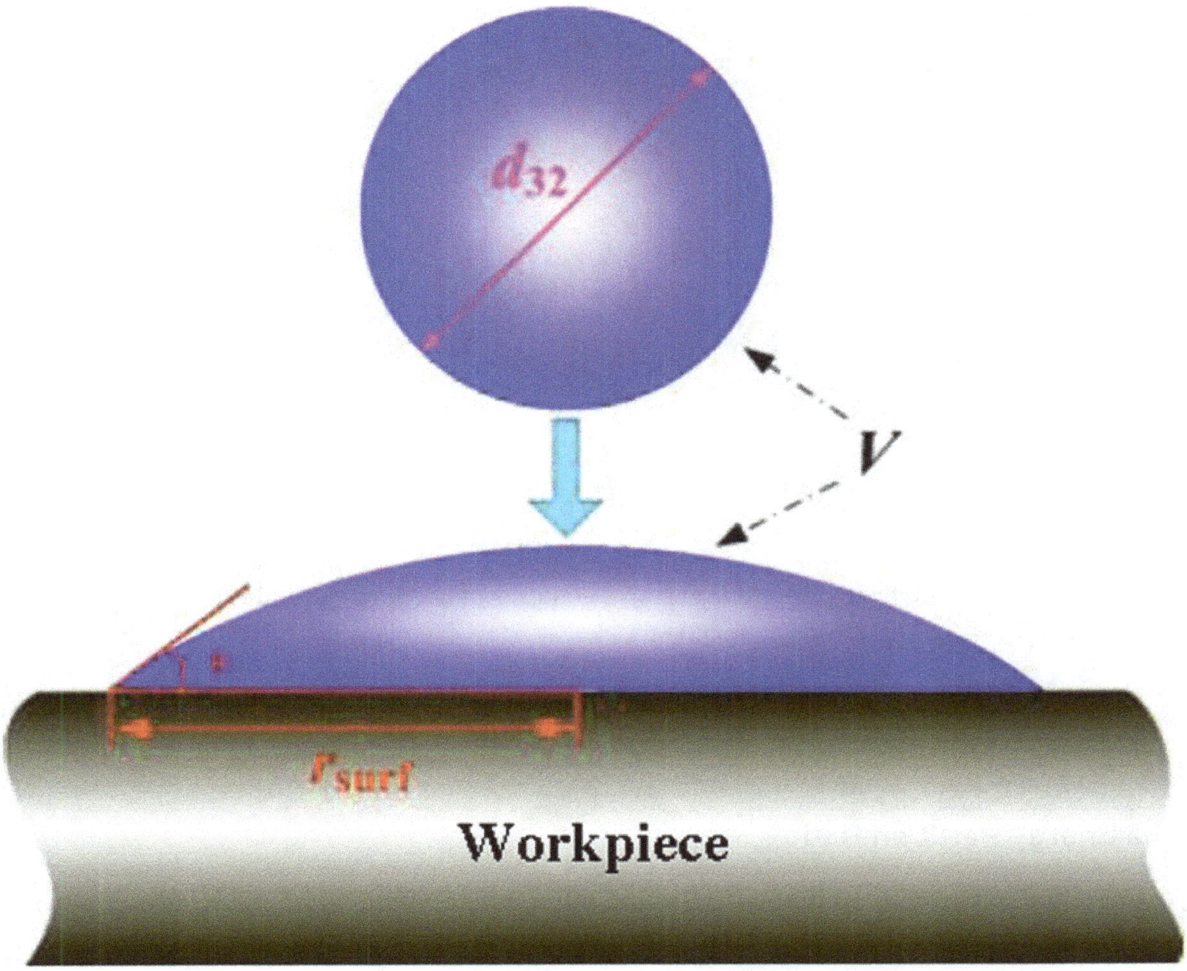

quantity generated at abrasive grain/workpiece interface and heat quantity transferred to chip and it is expressed as $q''_{total} = q''_{wf} + q''_{c}$.

Under dry grinding, heat quantity at abrasive grain/workpiece interface will be transferred to grinding wheel and workpiece and can be expressed as $q''_{wf} = q''_{g} + q''_{w}$. q''_{g} is heat transferred to abrasive grain and q''_{w} is heat staying on workpiece. Therefore, heat distribution relation under dry grinding is:

$$q''_{total} = q''_{w} + q''_{g} + q''_{c} \tag{42}$$

After grinding fluid is used, heat distribution of temperature field in the grinding area will change (Rowe et al., 1997). Due to convective heat transfer effect of grinding fluid, one part q''_{f} of heat will be distributed to grinding fluid and carried away by grinding fluid, one part q''_{wb} will stay on workpiece

Figure 8. Heat transfer in flood grinding; 1-abrasive grain, 2-chip, 3-workpiece, 4-grinding fluid

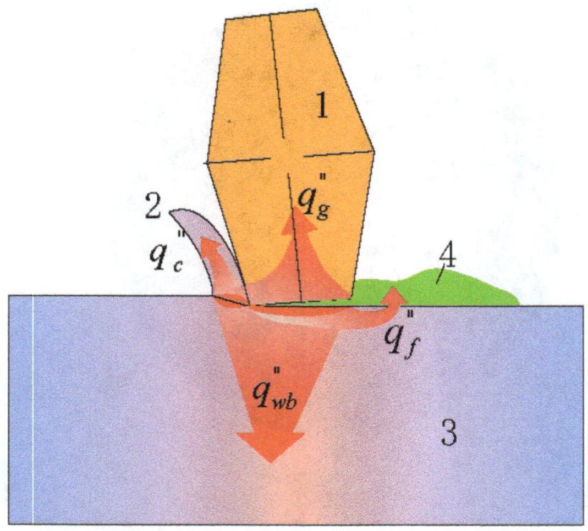

base and one part q_g'' will be transferred to abrasive grain, so when grinding fluid is used, heat distribution relation is expressed as (Jiang et al., 2016):

$$q_{total}'' = q_{wb}'' + q_g'' + q_c'' + q_f'' \tag{43}$$

4.5.2 Energy Partition Ratio

After heat distribution at the grinding interface is defined, it's still necessary to determine concrete energy distribution (Wang et al., 2019). Workpiece surface integrity after machining is considered in the grinding machining process, so calculating energy proportionality coefficient of workpiece and heat transferred to workpiece will be of great importance to taking proper cooling & lubrication measures to prevent phenomena like grinding burning.

Single abrasive grain model proposed by Hahn is used to calculate energy proportionality coefficient from the angle of heat distribution between single abrasive grain and workpiece (Krasyuk et al., 2012). It's believed that heat quantity is generated at abrasive grain/workpiece interface. As the heat carried away by grinding debris is little, heat carried away by grinding debris and dissipated heat due to convection are neglected (Wang et al., 2019). Heat is transferred onto workpiece and abrasive grain. In view that heat conductivity coefficient of abrasive grain is greater than that of grinding fluid under flood grinding, assume that the heat is not transferred to grinding fluid. According to heat conductivity coefficient between abrasive grain and workpiece material, proportionality coefficient of energy transferred to workpiece is expressed as:

$$R = [1 + \frac{k_g}{\sqrt{r_0 \cdot v_s \cdot (k\rho c)_w}}]^{-1} \tag{44}$$

Rowe made correction on this basis(J. L. Jiang et al., 2016):

$$R=[1+\frac{0.974k_g}{\sqrt{r_0\cdot v_s\cdot(k\rho c)_w}}]^{-1} \quad (45)$$

where: R—energy proportionality coefficient;

k_g—heat conductivity coefficient of abrasive grain, W/(m·K);
r_0—contact radius, mm;

This mathematical model is very applicable to analysis of dry grinding and it is not influenced by grinding fluid (Yi et al., 2019). However, under flood grinding, neglecting the influence of grinding fluid on heat transfer will impact precision of energy proportionality coefficient.

As grinding fluid enters the grinding area under flood grinding, a liquid film s formed at grinding wheel/workpiece interface (Hwang et al., 2006). At the time, heat generated at the grinding interface will be distributed onto workpiece, grinding wheel and liquid film, so energy proportionality coefficient R will also change. Heat energy transferred to workpiece within unit time is calculated using a linearized model as (Zhang et al., 2015):

$$E_{wb}=\frac{1}{2}\theta_m b[2(\lambda\rho c)_w v_w l_g]^{\frac{1}{2}} \quad (46)$$

Other heat quantity is transferred to grinding wheel and liquid film. Assume that liquid film is adhered to grinding wheel surface, ratio of actual contact area to nominal contact area between grinding wheel and workpiece is $A_r/A<<1$ for abrasive grain and it is $A_r/A=1$ for liquid film, so energy transferred to grinding wheel is regarded as the sum of energy transferred to abrasive grain and energy transferred to liquid film on grinding wheel surface, and it is expressed as:

$$E_s=E_g+E_f=\frac{1}{2}\theta_m b[2(\lambda\rho c)_s(\frac{A_r}{A})_s v_s l_g]^{\frac{1}{2}}+\frac{1}{2}\theta_m b[2(\lambda\rho c)_f v_s l_g]^{\frac{1}{2}} \quad (47)$$

Hence, proportionality coefficient R of energy transferred to workpiece is calculated as:

$$R=\frac{E_{wb}}{E_{wb}+E_g+E_f}=\frac{1}{1+\sqrt{\frac{(\lambda\rho c)_s}{(\lambda\rho c)_w}\frac{v_s}{v_w}(\frac{A_r}{A})}+\sqrt{\frac{(\lambda\rho c)_f}{(\lambda\rho c)_w}\frac{v_s}{v_w}}} \quad (48)$$

Proportionality coefficient of energy transferred to workpiece can be solved under flood grinding and MQL conditions, but when NMQL cooling condition is used, solid nanoparticles are added in the grinding fluid, thus enlarging heat conductivity coefficient of grinding fluid and improving its heat transfer capacity, and heat quantity transferred to grinding fluid will also increase. Heat transferred to grinding fluid includes heat transferred to base grinding fluid and heat transferred to nanoparticles, and it is expressed as:

$$E_f = \frac{1}{2}\theta_m b[2(\lambda\rho c)_f v_s l_g]^{\frac{1}{2}} = \frac{1}{2}\theta_m b\left\{2[\phi(\lambda\rho c)_p + (1-\phi)(\lambda\rho c)_f]v_s l_g\right\}^{\frac{1}{2}} \tag{49}$$

Grinding fluid added with nanoparticles will impact change of distribution proportionality coefficient R of energy transferred to workpiece. Proportionality coefficient R of energy transferred to workpiece through NMQL grinding is expressed as:

$$R = \frac{E_{wb}}{E_{wb}+E_g+E_f} = \frac{1}{1+\sqrt{\frac{(\lambda\rho c)_s}{(\lambda\rho c)_w}\frac{v_s}{v_w}}(\frac{A_r}{A})+\sqrt{\frac{\phi(\lambda\rho c)_p+(1-\phi)(\lambda\rho c)_f}{(\lambda\rho c)_w}\frac{v_s}{v_w}}} \tag{50}$$

According to equations (30) and (50), heat transferred to workpiece is expressed as:

$$q = Rq_{total} = R\frac{F_t v_s}{l_c b} \tag{51}$$

According to equations (28) and (50), grinding fluid doesn't participate in energy distribution under dry grinding, E_f =0, and the obtained energy proportionality coefficient R is the maximum. Large-flow grinding fluid is used in the flood grinding process, heat transfer capacity is strong, heat quantity E_f flowing into the grinding fluid increases, and proportionality coefficient R transferred to workpiece is small. After nanoparticles with volume concentration of φ are added in the grinding fluid, as heat conductivity λ, density ρ and constant-pressure specific heat cp of nanoparticles are greater than parameter values of base grinding fluid, heat conductivity coefficient of grinding fluid will increase and heat transfer capacity is enhanced. According to the heat transfer theory of solid, heat quantity E_f transferred to grinding fluid under NMQL grinding will also increase when compared with that under MQL grinding, so proportionality coefficient R of energy transferred to workpiece correspondingly reduces when compared with those under dry grinding and MQL grinding conditions. Cooling effect of NMQL is verified through theoretical derivation on the mathematical model.

To sum up the above analysis, average temperature on workpiece surface can be approximately expressed as:

$$\bar{\theta} = \frac{Rq_0 l}{\lambda L^{1/2}} = \frac{RE_e v_w a_p}{[v_w l_g (\lambda\rho c)_w]^{1/2}} \tag{52}$$

where: $\bar{\theta}$—average temperature on workpiece surface, °C;

q_0—heat source intensity, J/(m²K·s);
L—Peclet number, $L=vw/\alpha$ and thermal diffusivity $\alpha=\lambda/(\rho c)$;
E_e—specific grinding energy, W/ mm³;

Through equation (52), proportionality coefficient R of energy transferred to workpiece is obtained so as to derive average temperature on workpiece surface. Based on the relationship between energy proportionality coefficient R and average temperature on workpiece surface, proportionality coefficient R transferred to workpiece is theoretically derived, average temperature on workpiece surface is obtained through conversion according to equation(52), so corresponding cooling and lubrication measures can be taken through average temperature on workpiece surface to lower grinding temperature, prevent grinding burning on workpiece surface and lengthen service life of grinding wheel.

REFERENCES

Alabi, A. A., Akinyemi, O. D., Olowofela, J. A., Salako, F. K., Ajiboye, G. A., & Olurin, O. T. (2017). Temporal variation of ground temperature at depths 2cm to 200cm in an experimental field in Abeokuta, South-Western, Nigeria. *Arabian Journal of Geosciences*, *10*(19). doi:10.100712517-017-3201-z

Bense, V. F., Read, T., & Verhoef, A. (2016). Using distributed temperature sensing to monitor field scale dynamics of ground surface temperature and related substrate heat flux. *Agricultural and Forest Meteorology, 220*, 207-215. doi:.01.138 doi:10.1016/j.agrformet.2016

Chen, Y. J., & Wang, X. (2018). Influences of the Temperature on the Ground State of the Coupling Polaron in an Asymmetric Quantum Pseudodot Under the External Magnetic Field. *International Journal of Theoretical Physics*, *57*(11), 3540–3549. doi:10.100710773-018-3868-2

Dongkun, Z., Changhe, L., Dongzhou, J., Yanbin, Z., & Xiaowei, Z. (2015). *Specific grinding energy and surface roughness of nanoparticle jet minimum quantity lubrication in grinding*. Academic Press.

Gonzalez-Santander, J. L., Isidro, J. M., & Martin, G. (2015). An analysis of the transient regime temperature field in wet grinding. *Journal of Engineering Mathematics*, *90*(1), 141–171. doi:10.100710665-014-9713-6

Gonzalez-Santander, J. L., & Monreal, L. (2019). Efficient temperature field evaluation in wet surface grinding for arbitrary heat flux profile. *Journal of Engineering Mathematics*, *116*(1), 101–122. doi:10.100710665-019-10004-y

González–Santander, J. L., Pérez, J., Córdoba, P. F. D., & Isidro, J. M. (2010). An analysis of the temperature field of the workpiece in dry continuous grinding. *Journal of Engineering Mathematics*, *67*(3), 165–174. doi:10.100710665-009-9335-6

Gonzalez-Santander, J. L., Placeres, J. M. V., & Isidro, J. M. (2011). Exact Solution for the Time-Dependent Temperature Field in Dry Grinding: Application to Segmental Wheels. *Mathematical Problems in Engineering*, *2011*, 1–28. doi:10.1155/2011/927876

Hwang, Y., Park, H. S., Lee, J. K., & Jung, W. H. (2006). Thermal conductivity and lubrication characteristics of nanofluids. *Current Applied Physics*, *6*(1), e67–e71. doi:10.1016/j.cap.2006.01.014

Jiang, J., Ge, P., Sun, S., Wang, D., Wang, Y., & Yong, Y. (2016). From the microscopic interaction mechanism to the grinding temperature field: An integrated modelling on the grinding process. *International Journal of Machine Tools & Manufacture*, *110*, 27–42. doi:10.1016/j.ijmachtools.2016.08.004

Jiang, J. L., Ge, P. Q., Sun, S. F., Wang, D. X., Wang, Y. L., & Yang, Y. (2016). From the microscopic interaction mechanism to the grinding temperature field: An integrated modelling on the grinding process. *International Journal of Machine Tools & Manufacture*, *110*, 27–42. doi:10.1016/j.ijmachtools.2016.08.004

Koshin, A. A., & Shipulin, L. V. (2012). Temperature and Force Stochastic Models in Grinding Processes and Implementation of Them by Parallel Computing. South Ural State University, 31.

Krasyuk, A. M., Lugin, I. V., & P'yankova, A. Y. (2012). Temperature fields in surrounding ground of shallow tube stations. *Journal of Mining Science*, *48*(3), 465–473. doi:10.1134/S1062739148030094

Lan, S. L., & Jiao, F. (2019). Modeling of heat source in grinding zone and numerical simulation for grinding temperature field. *International Journal of Advanced Manufacturing Technology*, *103*(5-8), 3077–3086. doi:10.100700170-019-03662-w

Li, B. K., Li, C. H., Zhang, Y. B., Wang, Y. G., Min, Y., Jia, D. Z., ... Ding, W. F. (2017). Numerical and experimental research on the grinding temperature of minimum quantity lubrication cooling of different workpiece materials using vegetable oil-based nanofluids. *International Journal of Advanced Manufacturing Technology*, *93*(5-8), 1971–1988. doi:10.100700170-017-0643-0

Li, C. F., Mao, J. F., Peng, X., Mao, W., Xing, Z. L., & Zhu, G. D. (2019). Analysis of intermittent operation on the ground temperature recovery of borehole heat exchanger fields. *International Journal of Green Energy*. doi:10.1080/15435075.2019.1671416

Li, C. H., Li, J. Y., Wang, S., & Zhang, Q. (2013). Modeling and Numerical Simulation of the Grinding Temperature Field with Nanoparticle Jet of MQL. *Advances in Mechanical Engineering*. doi:10.1155/2013/986984

Li, Q. J., Zhao, Z. Y., Zhou, H. D., Ke, W. P., Wang, X. M., Fan, C., ... Sun, X. F. (2012). Paramagnetic ground state with field-induced partial order in Nd3Ga5SiO14 probed by low-temperature heat transport. *Physical Review B: Condensed Matter and Materials Physics*, *85*(17), 174438. doi:10.1103/PhysRevB.85.174438

Li, Y. Y., & Chen, Y. (1989). Simulation of Surface Grinding. *Journal of Engineering Materials and Technology*, *111*(1), 46–53. doi:10.1115/1.3226432

Lin, B., Yuan, Q. H., Zhang, H. L., Zhang, J. G., Chen, S. H., & Guo, W. J. (2004). Theoretical Analysis of Surface Grinding Temperature Field by Cup Wheel. *Key Engineering Materials*, *259-260*, 254–258. doi:10.4028/www.scientific.net/KEM.259-260.254

Lin, B., & Zhang, H. L. (2001). Theoretical Analysis of Temperature Field in Surface Grinding with Cup Wheel. *Key Engineering Materials*, *202-203*(202), 93–98. doi:10.4028/www.scientific.net/KEM.202-203.93

Mao, C., Zhou, Z. X., Guo, K. X., & Li, X. Y. (2010). Numerical Simulation and Experimental Validation of the Temperature Field in Surface Grinding. *Machining Science and Technology*, *14*(3), 344–364. doi:10.1080/10910344.2010.512198

Mao, C., Zhou, Z. X., Ren, Y. H., & Zhang, B. (2010). Analysis and FEM Simulation of Temperature Field in Wet Surface Grinding. *Materials and Manufacturing Processes*, *25*(6), 399–406. doi:10.1080/10426910903124811

Mao, C., Zou, H. F., Huang, X. M., Zhang, J., & Zhou, Z. X. (2012). The influence of spraying parameters on grinding performance for;nanofluid minimum quantity lubrication. *International Journal of Advanced Manufacturing Technology*, *64*(9-12), 1791–1799. doi:10.100700170-012-4143-y

Maruda, R. W., Krolczyk, G. M., Feldshtein, E., Pusavec, F., Szydlowski, M., Legutko, S., & Sobczak-Kupiec, A. (2016). A study on droplets sizes, their distribution and heat exchange for minimum quantity cooling lubrication (MQCL). *International Journal of Machine Tools & Manufacture*, *100*, 81–92. doi:10.1016/j.ijmachtools.2015.10.008

Min, Y., Li, C., Zhang, Y., Wang, Y., Li, B., & Hou, Y. (2016). Experimental research on microscale grinding temperature under different nanoparticle jet minimum quantity cooling. *Advanced Manufacturing Processes*, *32*(6), 589–597.

Nguyen, T., & Zhang, L. C. (2010). Grinding-hardening using dry air and liquid nitrogen: Prediction and verification of temperature fields and hardened layer thickness. *International Journal of Machine Tools & Manufacture*, *50*(10), 901–910. doi:10.1016/j.ijmachtools.2010.06.002

Parente, M. P. L., Jorge, R. M. N., Vieira, A. A., & Baptista, A. M. (2012). Experimental and numerical study of the temperature field during creep feed grinding. *International Journal of Advanced Manufacturing Technology*, *61*(1-4), 127–134. doi:10.100700170-011-3711-x

Parente, M. P. L., Jorge, R. M. N., Vieira, A. A., Santos, S., & Baptista, A. M. (2013). Experimental and Numerical Study on the Temperature Field during Surface Grinding of a Ti-6Al-4V Titanium Alloy. *Mechanics of Advanced Materials and Structures*, *20*(5), 397–404. doi:10.1080/15376494.2011.627641

Peres, L. F., Sobrino, J. A., Libonati, R., Jimenez-Munoz, J. C., Dacamara, C. C., & Romaguera, M. (2008). Validation of a temperature emissivity separation hybrid method from airborne hyperspectral scanner data and ground measurements in the SEN2FLEX field campaign. *International Journal of Remote Sensing*, *29*(24), 7251–7268. doi:10.1080/01431160802036532

Rowe, W. B., Black, S. C. E., Mills, B., Morgan, M. N., & Qi, H. S. (1997). Grinding Temperatures and Energy Partitioning. *Proceedings Mathematical Physical & Engineering Sciences*, *453*(1960), 1083-1104.

Sailer, E., Taborda, D. M. G., & Zdravkovic, L. (2018). A new approach to estimating temperature fields around a group of vertical ground heat exchangers in two-dimensional analyses. *Renewable Energy*, *118*, 579–590. doi:10.1016/j.renene.2017.11.035

Song, M. C., Pu, Y. S., Liu, Q., Wang, R., & Wang, B. Q. (2019). Numerical Simulation and Experimental Verification of the Quenching Temperature Field of Grinding Balls. *Materiali in Tehnologije*, *53*(4), 515–520. doi:10.17222/mit.2018.226

Tucek, J., Ohkoshi, S., & Zboril, R. (2011). Room-temperature ground magnetic state of epsilon-Fe2O3: In-field Mossbauer spectroscopy evidence for collinear ferrimagnet. *Applied Physics Letters*, *99*(25), 253108. doi:10.1063/1.3671114

Vieira, A. A., Baptista, A. M., Parente, M. P. L., & Jorge, R. M. N. (2009). On the temperature field during superficial grinding: An experimental study. *International Journal of Advanced Manufacturing Technology*, *40*(11-12), 1084–1092. doi:10.100700170-008-1442-4

Wang, X. Q., Chen, Y. J., & Xiao, J. L. (2019). Effects of temperature and magnetic field on the ground state binding energy of the strong coupling magneto-polaron in an RbC1 asymmetrical semi-exponential quantum well. *International Journal of Modern Physics B*, *33*(21), 1950239. doi:10.1142/S0217979219502394

Wang, X. Z., Yu, T. B., Sun, X., Shi, Y., & Wang, W. S. (2016). Study of 3D grinding temperature field based on finite difference method: Considering machining parameters and energy partition. *International Journal of Advanced Manufacturing Technology*, *84*(5-8), 915–927. doi:10.100700170-015-7757-z

Wang, Y., De-Jin, H. U., Deng, Q. L., & Qing-Xin, X. U. (2005). Analysis of Temperature Field for Laser Assisted Mechanical Dressing and Truing of Diamond Grinding Wheel. *Chinese Journal of Lasers*.

Wang, Z. X., Li, Y., Yu, T. B., Zhao, J., & Wen, P. H. (2019). Prediction of 3D grinding temperature field based on meshless method considering infinite element. *International Journal of Advanced Manufacturing Technology*, *100*(9-12), 3067–3084. doi:10.100700170-018-2801-4

Wang, Z. X., Yu, T. B., Wang, X. Z., Zhang, T. Q., Zhao, J., & Wen, P. H. (2019). Grinding temperature field prediction by meshless finite block method with double infinite element. *International Journal of Mechanical Sciences*, *153*, 131–142. doi:10.1016/j.ijmecsci.2019.01.037

Xu, H., Xu, X., Tao, L., & Zhang, Y. (1994). Analysis of the temperature field of workpiece surface layer during intermittent grinding. *Chinese Journal of Mechanical Engineering*.

Yang, M., Li, C. H., Zhang, Y. B., Wang, Y. G., Li, B. K., Jia, D. Z., ... Li, R. Z. (2017). Research on microscale skull grinding temperature field under different cooling conditions. *Applied Thermal Engineering*, *126*, 525–537. doi:10.1016/j.applthermaleng.2017.07.183

Yi, J., Jin, T., & Deng, Z. H. (2019). The temperature field study on the three-dimensional surface moving heat source model in involute gear form grinding. *International Journal of Advanced Manufacturing Technology*, *103*(5-8), 3097–3108. doi:10.100700170-019-03752-9

Zhang, D., Li, C., Zhang, Y., Jia, D., & Zhang, X. (2015). Experimental research on the energy ratio coefficient and specific grinding energy in nanoparticle jet MQL grinding. *International Journal of Advanced Manufacturing Technology*, *78*(5-8), 1275–1288. doi:10.100700170-014-6722-6

Zhang, J. C., Li, C. H., Zhang, Y. B., Min, Y. M., Jia, D. Z., Hou, Y. L., & Li, R. Z. (2018). Temperature field model and experimental verification on cryogenic air nanofluid minimum quantity lubrication grinding. *International Journal of Advanced Manufacturing Technology*, *97*(1-4), 209–228. doi:10.100700170-018-1936-7

Zhang, J. C., Li, C. H., Zhang, Y. B., Yang, M., Jia, D. Z., Hou, Y. L., & Li, R. Z. (2018). Temperature field model and experimental verification on cryogenic air nanofluid minimum quantity lubrication grinding. *International Journal of Advanced Manufacturing Technology*, *97*(1-4), 209–228. doi:10.100700170-018-1936-7

Zhang, J. H., Yan, Z., Tian, F. Q., Zhang, S., & Guo, L. S. (2015). Kinematics and experimental study on ultrasonic vibration-assisted micro end grinding of silica glass. *International Journal of Advanced Manufacturing Technology*, *78*(9-12), 1893–1904. doi:10.100700170-014-6761-z

Zhang, Z. Y., Shang, W., Ding, H. H., Guo, J., Wang, H. Y., Liu, Q. Y., & Wang, W. J. (2016). Thermal model and temperature field in rail grinding process based on a moving heat source. *Applied Thermal Engineering*, *106*, 855–864. doi:10.1016/j.applthermaleng.2016.06.071

Zhou, L., Huang, S. T., & Zhang, C. Y. (2013). Numerical and experimental studies on the temperature field in precision grinding of SiCp/Al composites. *International Journal of Advanced Manufacturing Technology*, *67*(5-8), 1007–1014. doi:10.100700170-012-4543-z

Chapter 5
Finite Element Analysis of Grinding Temperature Field for NMQL in Nickel-Base Alloy Grinding

ABSTRACT

Temperature is not only an important parameter in machining, but also an important basis for process optimization. Accurate prediction and reasonable analysis of grinding temperature is of great and far-reaching significance to the development and promotion of nanofluid micro-lubrication. In this chapter, the mathematical model of finite element simulation of temperature field of high efficiency deep grinding under four kinds of cooling lubrication conditions is established, and the three boundary conditions and the constraints of simulation model are established, and the mesh division and time step algorithm are determined respectively. Using ABAQUS simulation platform and theoretical model to simulate grinding temperature field, the distribution characteristics of grinding temperature field under different working conditions are analyzed from different directions, different grinding depths, and different workpiece materials.

5.1 INTRODUCTION

In the international research, simulation process has been recognized as an excellent tool to evaluate and optimize the cutting and grinding process. Using the different interaction between tool and workpiece under different factors combination in grinding process, researcher can work out the processing plan which can obtain the optimal workpiece processing quality, the shortest processing time and higher economic benefits at the same time (Li et al., 2010).

Based on the complex relationship among system parameters, machining parameters, process parameters and workpiece quality in grinding process, as well as the high correlation in industrial production process, modeling and simulation in grinding process has become the focus of mechanical engineering

DOI: 10.4018/978-1-7998-1546-4.ch005

research and industrial production research in universities (Hou et al., 2011). This can be demonstrated by the proliferation of relevant literature in the past decade. With the development of computer, high performance computer can provide many new modeling and simulation methods, such as FEA, kinematic geometry model, MD, etc. In addition, each simulation needs to determine the input parameters of the simulation through experiments. The simulation results was in good agreement with the experimental data. Therefore, it was very suitable to use the finite element analysis method to simulate the grinding process. The advantage of using macro concept was that the grinding process can be improved by considering more parameters, such as the friction coefficient between grinding wheel and workpiece. In the future, the heat will affect the phase transformation under the surface of the workpiece and the quality of the workpiece.

Another trend of finite element simulation was to apply software based tools to the production process, provided that input parameters such as temperature and grinding force were known. These tools enable users who do not have a lot of software knowledge to build finite element models. They may be used in industrial production to reduce the cost of modeling and simulation. The goal of modeling was to predict the thermal load, phase transformation and shape error. With the combination of finite element analysis software and CAM software, the calculated error can be considered in the computer program. In this way, the processed workpiece can achieve high precision, and the cost of the post-processing process will also be reduced (Li et al., 2019).

In the simulation of grinding process, the finite element method was widely used. When the finite element method was used to simulate the whole working condition, it can simulate various factors that can affect the grinding temperature field. At present, many scholars have done a lot of research work on the finite element analysis software, and achieved some results. Gu et al.(2004) Combined the three factors of workpiece, grinding wheel and coolant, proposed a two-dimensional grinding thermal finite element model, which fully considered the heat distribution in the grinding process. Jin et al. (2004) had carried on the finite element simulation analysis to the high efficiency deep grinding working condition, which mainly studies the transient heat conduction process. They put forward the idea that when the grinding contact arc length reaches the longest state, the heat conduction can be regarded as the steady state, greatly simplifying the difficulty in the finite element modeling process. Li et al. (2010) had also carried out more in-depth research in the field of grinding finite element simulation, developed the engineering software (GRINDSIM) applied to grinding technology, and has been applied to automobile, bearing and other industries.

Temperature field generated in the grinding process is very complicated and its influence on workpiece temperature is a dynamic process. Heat generation and heat transfer mechanism in the grinding temperature field were explained according to the grinding mechanism in Chapter 3 and Chapter 4, and a numerical model was established for numerical analysis. The established numerical model is combined with finite element software. Grinding process is simulated using a computer software. According to the comparison of simulation software like ABAQUS and ANSYS, ABAQUS is a CAE analysis software based on finite element method with strong heat conduction analysis capability, and moreover, it can solve transient dynamic event, so ABAQUS is taken as the simulation platform. Nickel base alloy material property is given to workpiece model, different boundary conditions are applied for temperature field simulation under dry grinding and flood grinding, and convective grinding heat transfer and influence laws on grinding temperature rise are analyzed through simulation results.

5.2 ESTABLISHMENT OF FINITE ELEMENT SIMULATION MODEL OF GRINDING TEMPERATURE FIELD

ABAQUS finite element software is used as the simulation platform to establish a workpiece model by combining grinding principle and heat transfer law. Theoretical mathematical model is loaded to the simulation model and grinding temperature is calculated using finite element method so as to simulate the actual grinding machining process. According to simulation results, grinding temperature field can be clearly explained with the help of the simulation software, which provides a guidance for actual grinding production (Huang, 2011; Li et al., 2019; Mao et al., 2010). Finite element simulation flowchart of grinding temperature field is shown in Fig. 1.

Figure 1. The simulation flow diagram of grinding temperature field

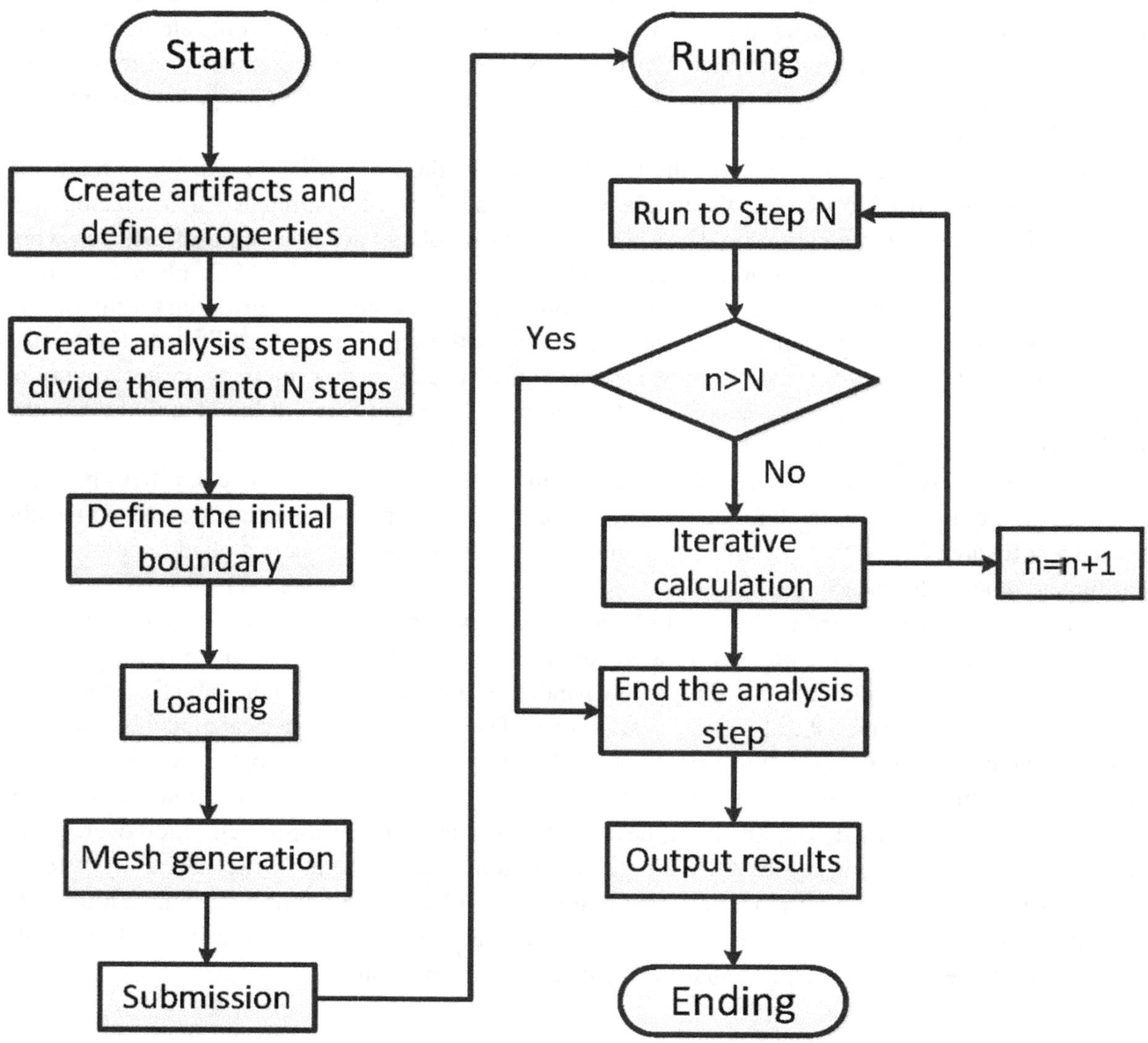

5.2.1 Material Properties of Workpieces

According to the theory of moving heat source of Jaeger (Gonzalez-Santander et al., 2015; Gonzalez-Santander et al., 2010), grinding motion of grinding wheel on workpiece surface is regarded as movement of heat source on workpiece surface, so workpiece model is built, material property is defined and moving heat source is applied to simulate actual grinding process. Dimensions of the built workpiece model are 36mm×20mm×10mm and initial temperature is room temperature 25°C. Study object of grinding machining is GH4169 material, material property parameters defined for workpiece model are shown in Table 1 and technological parameters used for grinding machining are shown in Table 2.

5.2.2 Determination of Analysis Steps

In order to simulate time-dependent movement of heat source on workpiece surface, analysis steps are divided on workpiece surface. Load value and time value of heat source are defined on each analysis step to embody change of heat source (Gavriliev et al., 2012; Gonzalez-Santander et al., 2017; Wang et al., 2019).

During the loading process of heat source to workpiece model, heat sources will be successively loaded onto analysis steps of workpiece surface according a certain time step. Analysis step lengths are divided according to workpiece length L and contact arc length l_c between grinding wheel and workpiece. Grinding contact arc length is calculated according to technological parameters of grinding, workpiece length is L=36mm, grinding depth is a_p=20μm and diameter of grinding wheel is d_s=300mm, so contact arc length l_c=2.45mm namely analysis step length is calculated as through equation (4). Therefore, total number of analysis steps divided on workpiece surface namely heat source loading times is expressed as:

$$N = \frac{L}{l_c} = \frac{36}{2.45} \approx 15 \tag{1}$$

Time step length of analysis step is expressed as:

$$T = \frac{L}{Nv_w} = \frac{36}{15 \times 50} = 0.05s \tag{2}$$

Schematic diagram of analysis steps divided on workpiece model is shown in Fig. 2.

5.2.3 Boundary Conditions

Violent interaction exists between grinding wheel and workpiece surface in the grinding process (Gonzalez-Santander & Monreal, 2019; Gonzalez-Santander et al., 2011; Parente et al., 2012). Heat generated at the grinding interface is simplified into uniformly distributed surface heat source, and continuous heat input into the workpiece will elevate the workpiece surface temperature (Parente et al., 2013). Under dry grinding, high temperature can be easily generated as the heat is aggregated on workpiece surface (Jia et al., 2019; Jiang et al., 2016). Without cooling heat transfer effect of grinding fluid, it will cause

Table 1. The material performance parameter of GH4169

Temperature (°C)	25	100	200	300	400	500	600	700
Heat conductivity λ (W/m·K)	13.4	14.7	15.7	17.8	18.3	19.6	21.2	22.8
Specific heat c(J/kg·K)	435	458	467	481	494	515	539	573
Thermal expansion coefficientα (10^{-5}/K)	—	1.18	1.30	1.35	1.41	1.44	1.48	1.54
Densityρ (kg/m^3)	8240							
Poisson's ratio	0.3							
Melting point (°C)	1260							
Elasticity modulus (GPa)	200							
Yield strength (MPa)	550							
Tensile strength (MPa)	965							

Table 2. The parameters of grinding process

Grinding parameters	Parameter setting
Linear velocity of grinding wheel v_s	30m/s
Workpiece velocity v_w	0.05m/s
Grinding depth a_p	20μm
Grinding width b	20mm

Figure 2. Analysis steps on the workpiece model

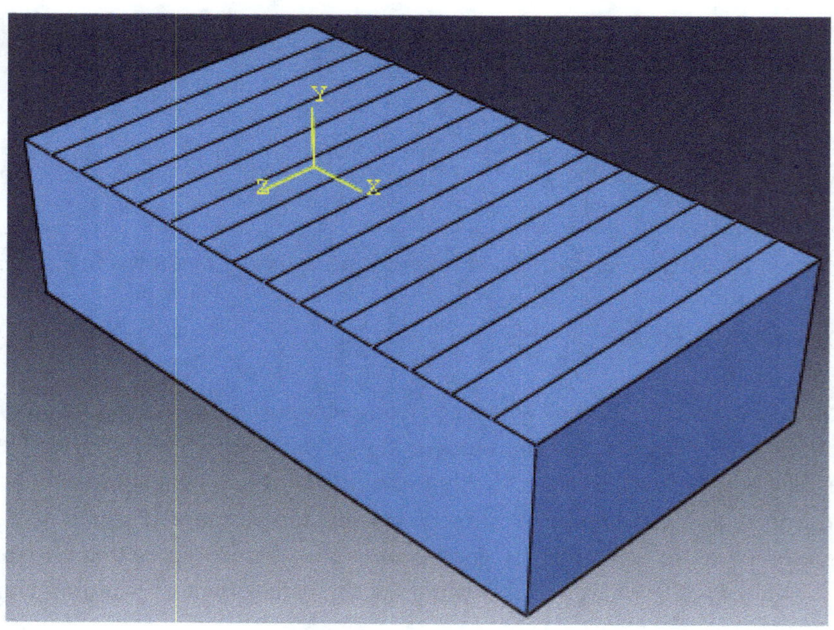

Figure 3. The schematic diagram of the boundary in workpiece surface: (a) dry grinding (b) flood grinding

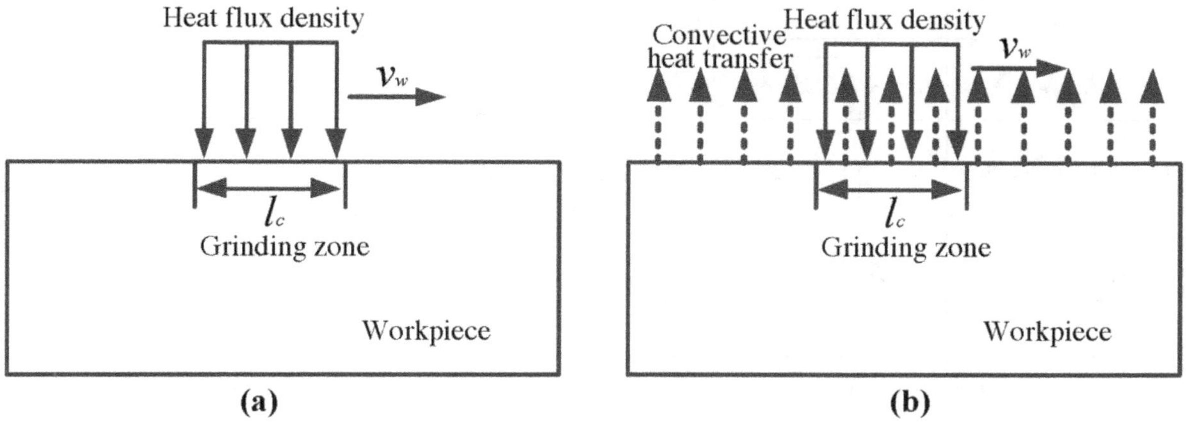

phenomena like grinding burning. Under flood grinding, one part of heat on workpiece surface is carried away by grinding fluid due to its convective heat transfer effect, thus reducing heat transferred to workpiece and reaching the effect of lowering workpiece surface temperature. Schematic diagram of heat input and output under dry grinding and flood grinding namely boundary on workpiece model surface is shown in Fig. 3.

Difference of boundary conditions is embodied by difference of heat transfer coefficient on workpiece surface (Nguyen & Zhang, 2010; Wang et al., 2019). As different grinding fluids have different heat conductivity coefficients, cooling heat transfer coefficients will also become different (Chen & Wang, 2018). As no grinding fluid is used in dry grinding, dry grinding boundary is defined as natural convective heat transfer cooling of air in consideration of actual grinding conditions. As forced convective heat transfer is used in flood grinding, heat transfer capacity is very strong, heat transfer coefficient is about 10^2~10^5W/m^{2}K, difference between water-based grinding fluid and oil-based grinding fluid in the aspect of heat transfer coefficient is mainly due to the difference of heat-conducting properties of their base fluids, and heat transfer coefficient of grinding fluid is also influenced by use level and jet flow velocity of grinding fluid, etc. Heat transfer coefficients under different cooling and lubrication conditions are seen in Table 3 (Lan & Jiao, 2019; Li et al., 2013).

5.2.4 Loading

As grinding depth is shallow, heat source model used in the simulation is a plane heat source model under rectangular uniform distribution according to the analysis given in section 3.3.7. ABAQUS simu-

Table 3. The coefficient of heat transfer under different cooling conditions

Cooling mode	Heat transfer coefficient (W/m^{2}K)
Dry grinding	10
Water-based grinding fluid	2.9×105
Oil-based grinding fluid	2.3×104

Table 4. The load parameter value of the simulation model

Parameter	Numerical calculation model	Parameter value
Grinding contact arc length l_c	$l_c = \sqrt{a_p \cdot d_s}$,equation (3-4)	2.5mm
Tangential grinding force	$F_t = 2^m \lambda^{m-1} k_t v_w{}^m v_s{}^{-m} a_p{}^{\frac{m+1}{2}} d_s{}^{\frac{1-m}{2}}$,equation (3-28)	63.2N
Energy proportionality coefficient	$R = \{1 + [\frac{(\lambda\rho c)_s}{(\lambda\rho c)_w} \frac{v_s}{v_w} (\frac{A_r}{A})]^{1/2} + [\frac{(\lambda\rho c)_f}{(\lambda\rho c)_w} \frac{v_s}{v_w}]^{1/2}\}^{-1}$, equation (4-50)	60.9% (dry grinding) 27.1% (flooding grinding)
Heat flux q	$q = F_t v_s / l_c b$,equation (3-30)	4.51×10^7W/m²s

lation software can't directly load moving heat source to workpiece surface to realize grinding simulation, so heat source movement is indirectly realized through analysis steps and time step length of each analysis step in the simulation. Within the time step length of one analysis step, heat source load under rectangular uniform distribution is loaded to an analysis step area on the workpiece model surface, heat is transferred to workpiece base in this analysis step area to form temperature field, and up to the next time step length, heat source load moves to the next adjacent analysis step area, result of the last analysis step is taken as initial condition of this analysis step, and loading is continuously conducted to realize heat source movement.

Loaded heat flux is heat source intensity generated on the grinding interface, which is decided by grinding parameters like grinding contact arc length l_c and tangential grinding force F_t, and its calculation formula is shown in equation (4-51). Under the influence of convective heat transfer, heat which is actually transferred to workpiece is affected by energy proportionality coefficient R. The calculated load parameter value is assigned to the mathematical model, and load parameters input by the model are shown in Table 4.

5.2.5 Finite Element Meshing

For the convenience of calculation of temperature field, the workpiece model is gridded. Divided grids shall not be too dense or too sparse for workpiece model in order to avoid problems like too long simulating calculation time and poor calculation accuracy, so grid cell type is set as C3D8T after grid division of workpiece model: eight-node thermal coupling hexahedral cells, three-way linear displacement and three-way linear temperature, and the workpiece model after gridding is shown in Figure 4. After job submission, the simulation model establishment is completed.

Figure 4. Grid division of simulation model

5.3 ANALYSIS OF TEMPERATURE FIELD SIMULATION RESULTS IN SURFACE GRINDING

The simulation model of grinding temperature field is completed by establishing workpiece model, defining boundary conditions and loading heat sources. Grinding wheel wear, deformation of workpiece and abrasive grains in the grinding process and other problems are neglected in the simulation model, and the influences of temperature rise caused by grinding heat, which is transferred to workpiece, and convective heat transfer of grinding fluid on grinding temperature rise of workpiece are mainly analyzed.

In order to explain workpiece grinding temperature field in details, simulation program is operated under dry grinding condition, moving heat source transfers grinding heat to workpiece, which causes temperature rise, and grinding temperature fields along grinding motion direction, direction of grinding depth and direction of grinding width are analyzed according to obtained workpiece temperature result.

5.3.1 Temperature Analysis of Grinding Motion Direction

Heat sources are continuously loaded to workpiece surface to simulate grinding feed motion of trending wheel. In each grinding travel, heat source repeatedly acts upon workpiece surface, thus repeatedly influencing the temperature field, so temperature field obtained under heat source effect in each grinding travel is studied, and analysis is made using temperature field cloud atlas and temperature curves obtained through simulation.

Figure 5. Temperature field cloud atlas of cut-in area

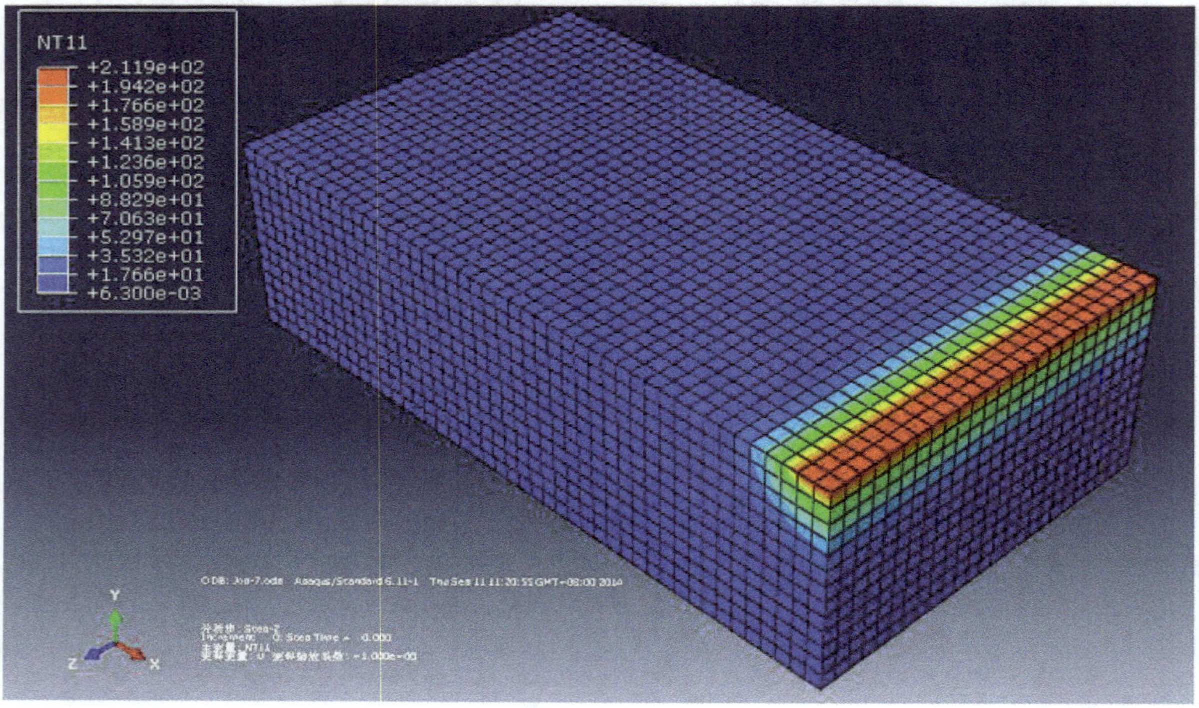

(1) Temperature of Cut-In Area

When grinding motion is started, grinding wheel cuts into workpiece, the heat source is loaded to the first analysis step namely at the first time step ((t=0.05s), temperature field cloud atlas of cut-in area is obtained according to simulation results as shown in Fig. 5.

When the grinding travel starts, grinding wheel contacts and cuts into the workpiece to exert grinding effect, heat generated at the grinding interface is input into the workpiece surface and then transferred from grinding contact area to workpiece base. Grinding temperature generated in the grinding area is low, mainly because workpiece material participating in grinding in the grinding cut-in area is small in volume, so is the consumed energy. The heat starts being transferred towards various directions of the workpiece according to its heat-conducting properties, the formed high-temperature field is small and the workpiece surface temperature gradually rises from the first analysis step to the third analysis step as the grinding motion proceeds.

(2) Temperature in Steady Grinding Area

From the fourth analysis step to the fourteenth analysis step, maximum temperature rise on workpiece surface will not rise any longer, namely the formed temperature field reaches a steady state. In this phase, heat input into the workpiece surface reaches a balance state with convective heat transfer, and maximum temperature rise on workpiece surface tends to be steady, so it becomes a steady grinding

Figure 6. Temperature field cloud atlas in steady grinding area

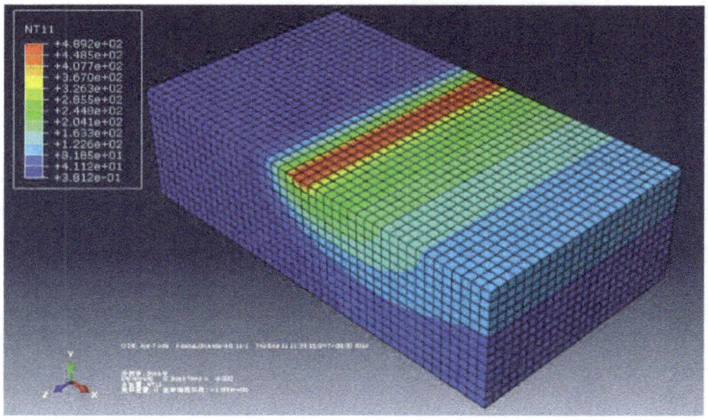

area. The temperature field cloud atlas obtained when heat source is located at the eighth analysis step (t=0.40s) in the middle of steady grinding area is shown in Fig. 6.

As shown in Fig. 6, maximum temperature rise is distributed in the grinding contact area, temperature rises in other areas are all smaller that in the grinding contact area. Temperature fields when heat source moves to all analysis steps in the steady grinding area are similar, maximum temperature rise is kept at 490°C or so. When heat source is located in the eighth analysis step area in the steady grinding area, it reaches the eighth time step, and temperature rising trend on workpiece surface when the eighth analysis step area is in the whole time step is shown in Fig. 7.

According to Fig. 6 and Fig. 7, temperature rise is already caused before the eighth analysis step area reaches the eighth time step at front edge of grinding area due to heat conduction, so a small temperature rise area is formed, which is caused by heat conduction when heat source is at the seventh analysis step. When heat source moves to the eighth analysis step, temperature rise is rapid and reaches the highest, heat input into the workpiece is transferred to the material around heat source in the form of heat conduction to lower the temperature on workpiece surface in the form of convective heat transfer, and after heat source moves, temperature rapidly declines at rear edge of grinding area under the influences of heat transfer velocity and velocity of convective heat transfer-induced cooling, a large temperature field area is formed at rear edge, and the temperature gradually declines with time.

(3) Temperature of Grinding Cut-Out Area

When heat source moves to the fifteenth analysis step, grinding wheel will cut out and leave the workpiece surface namely grinding cut-out area, and temperature cloud atlas in this area is shown in Fig. 5-8. Temperature rise on workpiece surface in grinding contact area is suddenly elevated. According to the heat transfer theory, workpiece material participating in grinding process is gradually reduced in volume when grinding wheel is in cut-out area. When heat quantity generated at the grinding interface remains unchanged, material volume causing heat dissipation in cut-out area is gradually reduced. Under limited convective heat transfer conditions, heat quantity transferred by unit volume of workpiece material increases and grinding wheel generates high surface temperature rise in cut-out area, so when cooling heat transfer is insufficient in cut-out area, it can easily result in phenomena like grinding burn-

Figure 7. The temperature rise in 8th analysis step

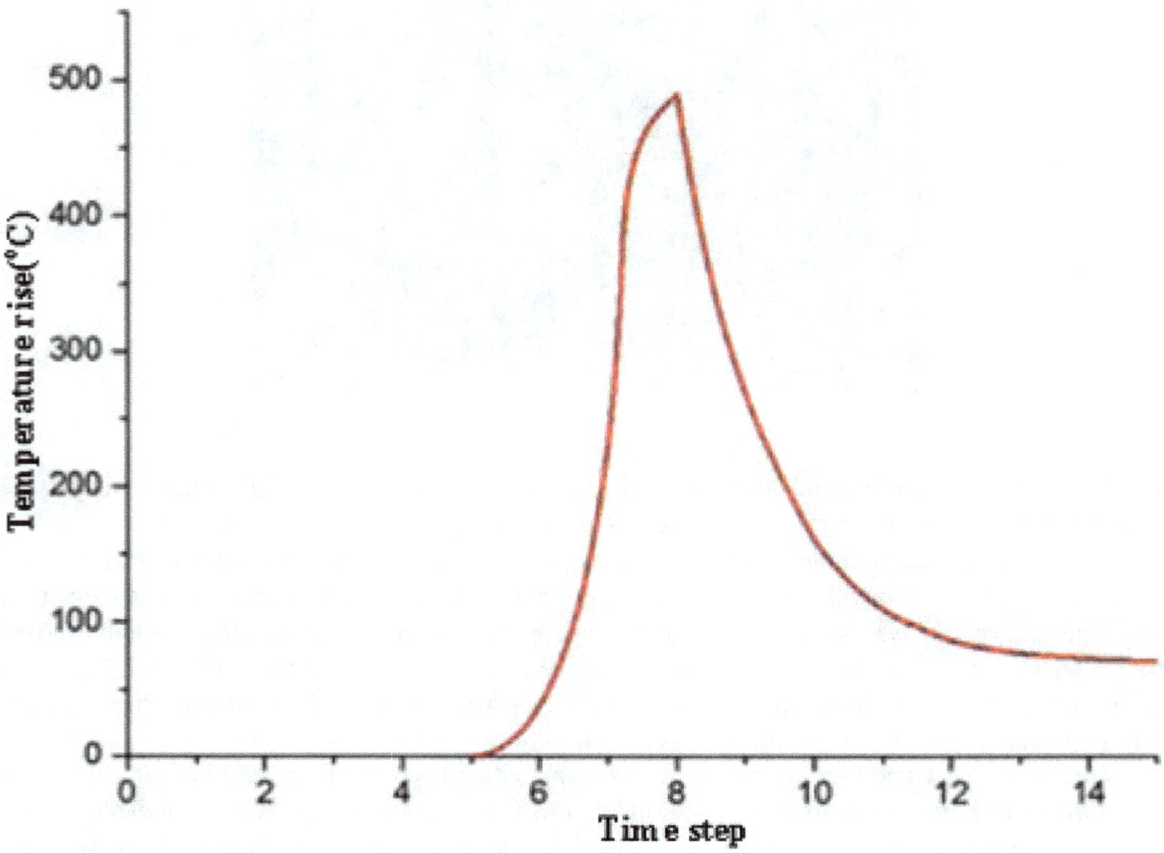

ing, thus certifying that only effective cooling heat transfer can lower the surface temperature (Sailer et al., 2018; Zhou et al., 2013).

Through a combination of grinding temperature rises in grinding cut-in area, steady grinding area and grinding cut-out area, time-dependent temperature rises at analysis steps on workpiece surface along grinding motion direction are shown in Fig. 5-9. Along the grinding motion direction, it can be found that maximum temperature rise on workpiece surface experiences "developing-steady-sudden" processes, which are corresponding to cut-in area, steady area and cut-out area on workpiece surface. Temperature rise gradually increases from the first analysis step to the third analysis step, because heat input into workpiece gradually increases since grinding wheel contacts workpiece until completely participating in grinding process; from the fourth analysis step to the fourteenth analysis step, heat quantity input into workpiece within unit time is constant when grinding power is stable, and energy conservation exists between heat transferred to workpiece base and heat dissipated due to convective heat transfer, so maximum temperature rise at each analysis step is stable; grinding wheel is about to leave workpiece surface at the fifteenth analysis step, workpiece volume participating in the grinding process is reduced, heat transferred to unit volume of workpiece material increases, and temperature rise is rapidly elevated.

Figure 8. Temperature field cloud atlas in cut-out area

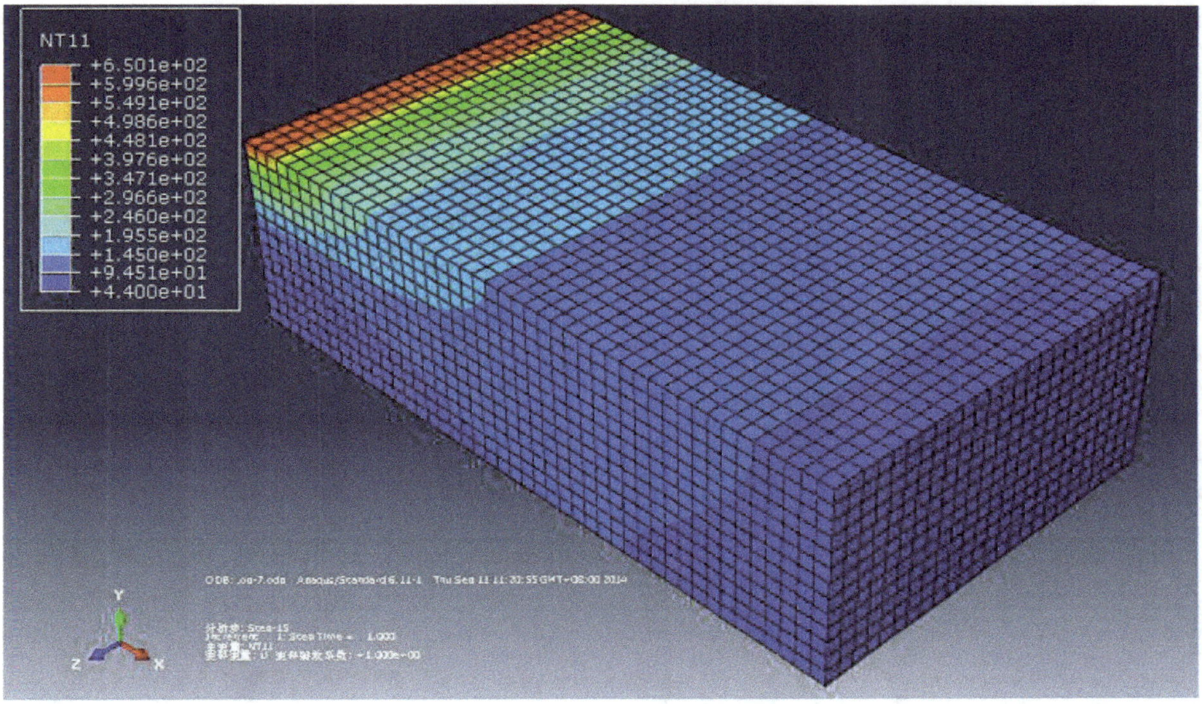

Before heat source reaches each analysis step, temperature field is generated at front edge of heat source motion due to heat transfer effect, and as a result, temperature gradually rises. When heat source moves to the analysis step, maximum temperature rise is rapidly reached. After heat source leaves, temperature is rapidly lowered due to convective heat transfer effect. Temperature slowly declines with time, and adjacent analysis steps interact with each other. Under the influence of temperature field of

Figure 9. Temperature rising trend of analysis steps with time steps on workpiece surface

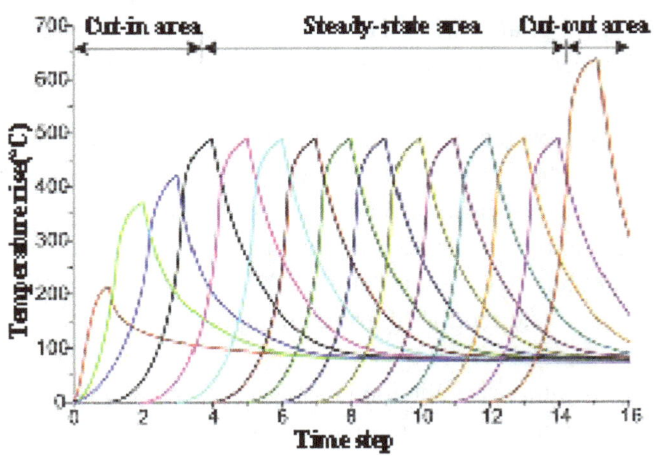

Figure 10. Temperature field cloud atlas of the eighth analysis step along the direction of grinding depth

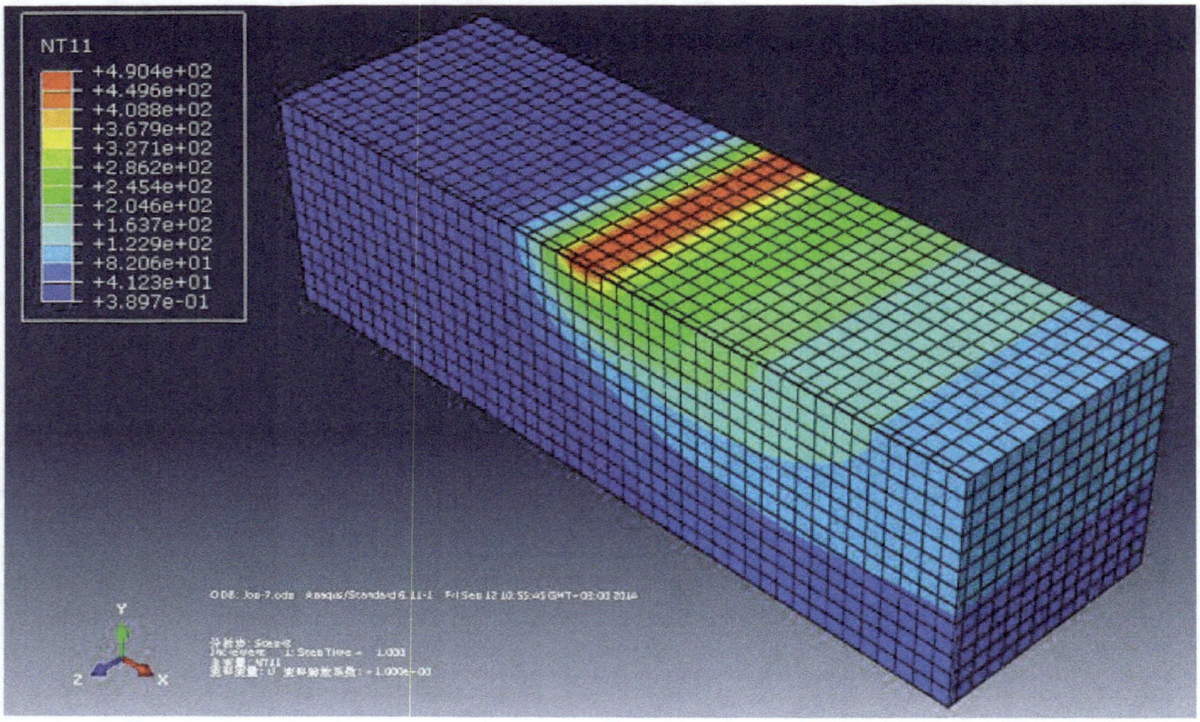

the current analysis step, temperature which causes temperature rise of the next analysis step is taken as initial temperature, and then temperature influence of continuous heat sources on workpiece surface in the grinding process can be completed.

5.3.2 Temperature Analysis of Grinding Depth Direction

One part of heat generated at the grinding wheel/workpiece interface is distributed onto grinding wheel base under the effect of heat conduction and one part is distributed onto workpiece surface (Mao et al., 2010; Song et al., 2019; Su et al., 2018). Heat on the workpiece surface is dissipated into air or grinding fluid due to free convective heat transfer or forced convective heat transfer, and residual heat is transferred from workpiece surface downward inside workpiece base along the grinding depth. Heat forms temperature field inside workpiece base, and heat transferred to different grinding depths of the workpiece is related to heat-conducting property of workpiece material, where heat conductivity coefficient of workpiece has the greatest influence on heat transfer, and the greater the heat conductivity coefficient, the more the heat transferred within unit area, and this influences temperature distribution in the temperature field (Li et al., 2012; Vieira et al., 2009). Workpiece model is split, cloud atlas of temperature field formed along the direction of grinding depth in the workpiece base when heat source acts at the eighth analysis step is shown in Fig. 10.

According to Fig. 10, temperature rise obtained through heat transfer from workpiece surface downward along the direction of grinding depth is gradually lowered, the highest temperature rise is located

Figure 11. The temperature rise trend in 8th analysis step along the direction of grinding depth

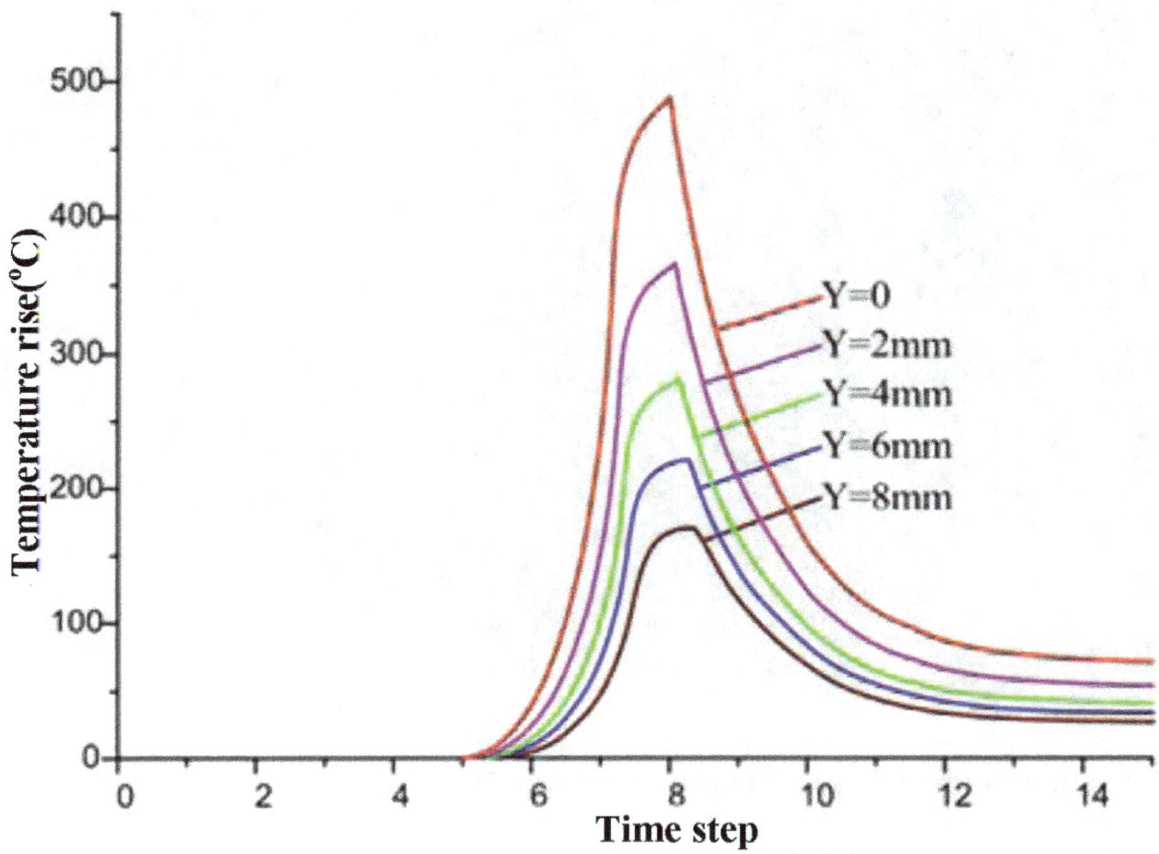

on workpiece surface, and internal temperature rise in the workpiece becomes smaller and smaller. Time-dependent temperature rising trend caused at the eighth analysis step along the direction of grinding depth (Y direction) under the influence of heat source is shown in Fig. 11.

According to Fig. 11, temperature trends of workpiece material under different depths and the influence of heat source are identical, firstly increasing and then decreasing. Maximum temperature rise occurs on workpiece surface, and it gradually reduces as the depth increases. Fig. 10 and Fig. 11 combined, temperature gradients of the eighth analysis step exist under different depths, where the temperature gradient is the maximum when maximum temperature rise is reached. The greater the depth from the workpiece surface, the smaller the generated gradient, mainly because of heat conductivity coefficient of the workpiece. As heat source is moving, the generated temperature field is also moving. Under the influence of heat conductivity of workpiece material, it takes a certain time for heat to be transferred from surface to different depths, so the time for temperature rise generation when workpiece is under a large depth is slightly delayed when compared with that when it is under a small depth. As there is no forced convective heat transfer condition inside the workpiece base, temperature trend becomes more and more gentle as the depth increases.

Figure 12. Temperature field cloud atlas of the eighth analysis step along direction of grinding width

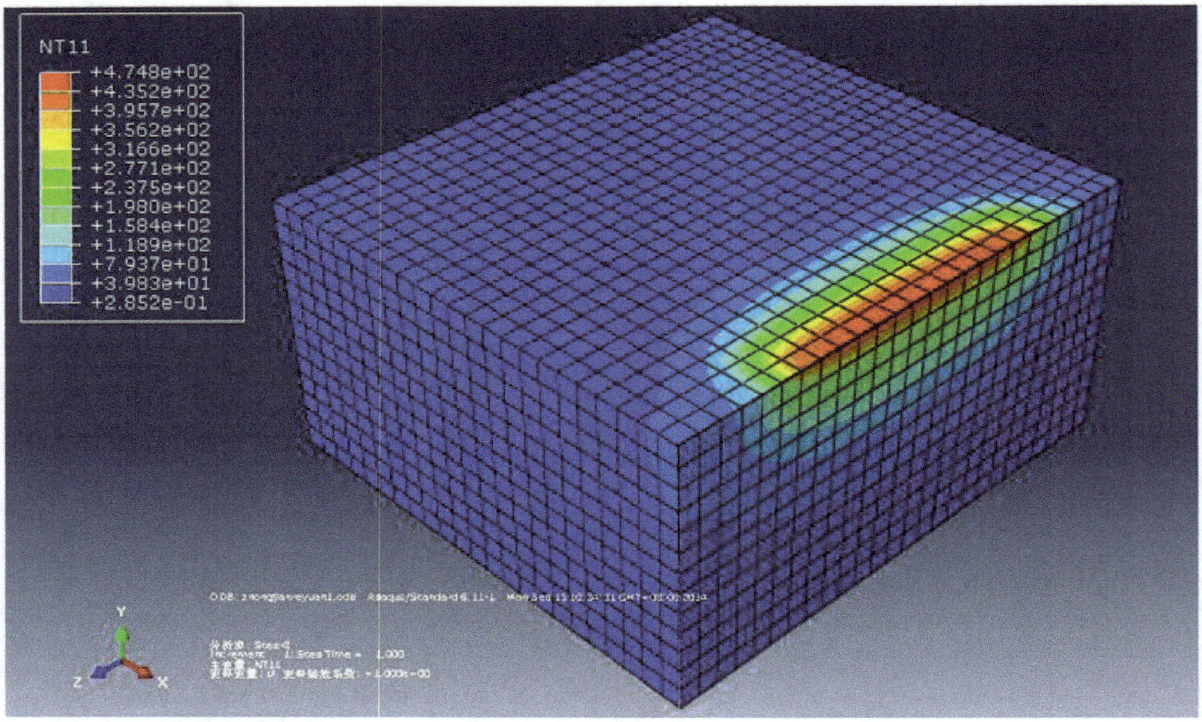

5.3.3 Temperature Analysis of Grinding Width Direction

When workpiece width is greater than width of grinding wheel, surface heat source formed at the grinding wheel/workpiece contact interface not only transfers heat to front and rear directions of heat source and direction of grinding depth but also to direction of grinding width. When heat source moves to the eighth analysis step, section view of workpiece model is got at the heat source along direction of grinding width, and cloud atlas of temperature field formed along direction of grinding width is shown in Fig. 12.

As shown in Fig. 12, the temperature formed at heat source along direction of grinding width is the highest and temperatures at two ends of heat source gradually decline. Temperature difference at the interface between grinding area and non-grinding area is large, and the farther it is from the grinding area, the smaller the temperature gradient. Temperature distribution trend along direction of grinding width when heat source is at different depths of the eighth analysis step is shown in Fig. 13.

According to Fig. 12 and Fig. 13, temperature rise generated in the workpiece/grinding wheel contact area is the highest along direction of grinding width, temperatures nearby two sides of the contact area rapidly decline, mainly because there is no heat input in this area, and temperature rise is only caused by heat conduction. Temperature difference between contact area and non-contact area is large, thus forming a large temperature gradient. Being farther away the contact area, temperature gradient gradually reduces while temperature rise gradually declines. At Y=2mm and Y=4mm depths from the surface, declining rate of the highest temperature gradually reduces. As there is temperature gradient in heat distribution along direction of grinding depth, temperature distribution trend formed along direction of grinding

Figure 13. Temperature distribution trend of the eighth analysis step along direction of grinding width

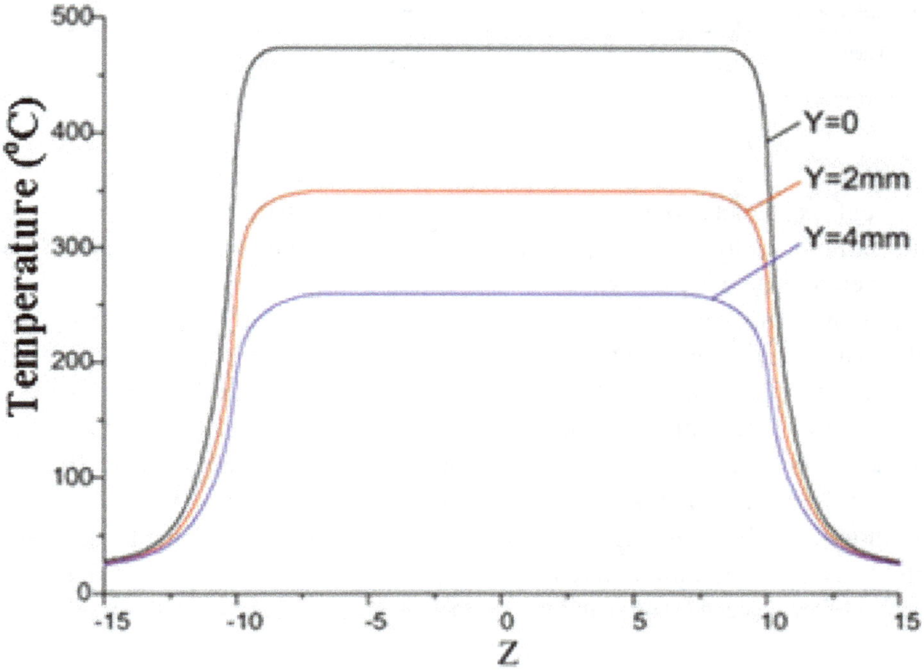

width is identical with that under Y=0 condition. Due to heat conduction and heat transfer effects, the highest temperature rise area is gradually reduced.

5.4 EFFECT OF CONVECTION HEAT TRANSFER ON GRINDING TEMPERATURE FIELD

Different cooling and lubrication conditions exert different influences on temperature field due to the difference of their heat transfer capacity, where a large quantity of heat is transferred to workpiece as there is no forced convective heat transfer of grinding fluid under dry grinding mode. However, under flood grinding, a large quantity of heat on workpiece surface is transferred to and carried away by grinding fluid in the form of forced convective heat transfer of grinding fluid. According to differences of grinding fluid type and cooling and lubrication form, convective heat transfer capacity of grinding fluid is changed so as to influence temperature field formed on workpiece surface.

5.4.1 Effect of Cooling and Lubrication Methods on Grinding Temperature Field

In order to profoundly study the influence of cooling and lubrication condition on grinding temperature field in the grinding process, simulated analysis of grinding machining of high-temperature nickel base alloy GH4169 material under different cooling and lubrication condition is carried out. In this simulation, four cooling and lubrication condition—dry grinding, flood grinding, MQL and NMQL—are successively

Table 5. Heat transfer coefficients of different cooling and lubrication condition

Cooling and lubrication mode	Dry grinding	Flood grinding	MQL	NMQL
Heat transfer coefficient h(W/ m^2K)	10	3×105	2.1×105	2.48×10^5

set. The same heat flux is loaded onto the surface of workpiece surface. According to simulation results of four cooling and lubrication condition, the influences of different cooling and lubrication condition on grinding temperature field are analyzed.

Different grinding fluid types or cooling condition result in different heat transfer coefficients of different cooling and lubrication condition, which are solved using experimental solving method, mathematical analysis method and numerical analysis method as shown in Table 5 (Eastman et al., 1996).

In the simulation of four cooling and lubrication condition, different heat transfer coefficients are set to simulate heat transfer capacity of four cooling and lubrication condition, and simulation program is operated. When the heat source is at the eighth analysis step (steady area) under four cooling and lubrication condition, grinding temperature field is obtained through simulation, the influences of different cooling and lubrication condition on temperature rise of workpiece are analyzed, and cloud atlas of temperature fields of workpiece formed under four cooling and lubrication condition is shown in Fig. 14.

(a) (b)

(c) (d)

As shown in Fig. 14, when heat source acts on the eighth analysis step, temperature fields formed under four cooling and lubrication condition are significantly different, indicating that different cooling and lubrication condition have different cooling effects on temperature field. Under dry grinding mode as shown in Fig. 14 (a), temperature field formed on workpiece model is large, high temperature formed when heat source at rear edge of grinding area on workpiece surface moves is not timely lowered, the whole rear edge is in the temperature field, and heat is transferred to workpiece base, which causes a large temperature rise of workpiece base; under flood grinding as shown in Fig. 14 (b), temperature field formed in workpiece model is small, temperature rise formed at rear edge of grinding area is timely lowered, heat transferred to workpiece base is reduced, and then temperature rise of workpiece base is small; under MQL mode as shown in Fig. 14 (c), temperature field formed in workpiece model is somehow reduced when compared with that under dry grinding, high-temperature area at rear edge of grinding area and the depth of workpiece base to which the heat is transferred are reduced, indicating that MQL mode has cooling effect on temperature field, but the effect is weaker than flood grinding, so MQL cooling effect is limited; under NMQL mode as shown in Fig. 14 (d), compared with MQL mode, temperature field is reduced, indicating that adding nanoparticles into the grinding fluid based on MQL can enhance cooling effect of grinding fluid, which is approximate to cooling effect of flood grinding, and moreover, it has advantages of energy conservation, high efficiency, greenness and environmental protection (Ivady et al., 2014; Wang et al., 2019; Yang et al., 2017) .

Temperature fields of workpiece surface generated under four cooling and lubrication condition are different in size (Dai et al., 2019; Tucek et al., 2011; Wang et al., 2016). Temperature rises caused at different analysis steps on workpiece surface along grinding direction (X direction) are different, and temperature rises on workpiece surface under four cooling and lubrication condition when heat source is at the eighth analysis step are shown in Fig. 15.

Figure 14. The temperature field cloud plot under four kinds of cooling and lubrication conditions: (a) dry, (b) flood, (c) MQL, (d) NMQL

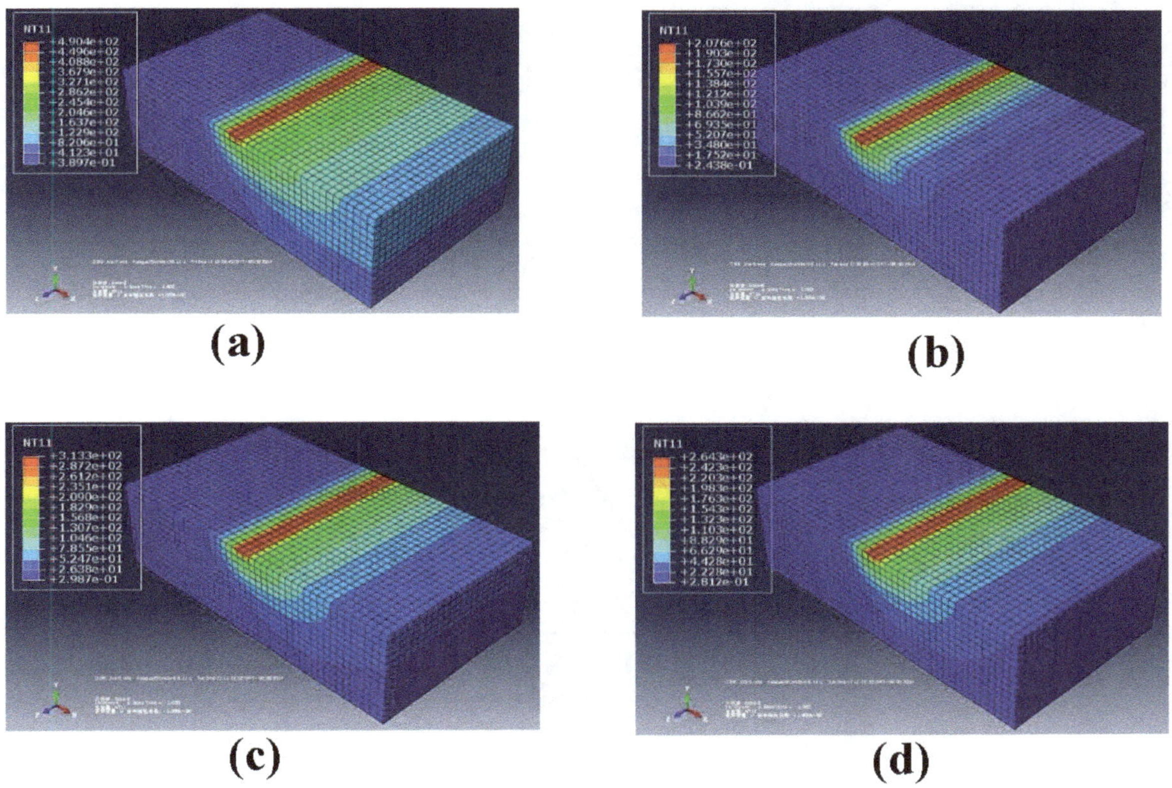

According to Fig. 15, under cooling effects of four cooling and lubrication condition in the grinding process, temperature rise difference on workpiece surface is quite obvious. Temperature rise on workpiece under dry grinding mode reaches 490°C which is the highest among four working conditions, heat transferred to workpiece can't be timely dissipated, and heat is aggregated on surface to form high temperature; temperature rise obtained under flood grinding is low, being 207°C, which reduces by at least 280°C when compared with that under dry grinding, and the heat transferred to workpiece is timely carried away by grinding fluid on workpiece surface in quantity, thus reaching the cooling effect; MQL grinding is between dry grinding and flood grinding, grinding fluid is atomized and sprayed to workpiece grinding area for cooling because of high-pressure gas, temperature rise on workpiece surface is 313°C which is between dry grinding and flood grinding and reduces by 177°C when compared with that under dry grinding, indicating that grinding fluid can exert cooling effect in the grinding process, but it still has a certain gap with flood grinding in the aspect of cooling effect; NMQL refers to adding nanoparticles in the grinding fluid based on MQL, temperature rise on workpiece surface obtained through grinding simulation is 264°C which reduces by about 50°C when compared with MQL, indicating that addition of nanoparticles enhances heat transfer capacity of grinding fluid and compensates for deficiencies of MQL cooling effect to a certain degree. The temperature rise is approximate to that obtained under flood grinding. When heat source is located at the eighth analysis step, temperature of workpiece surface in the

Figure 15. The temperature rise under four kinds of cooling and lubrication conditions

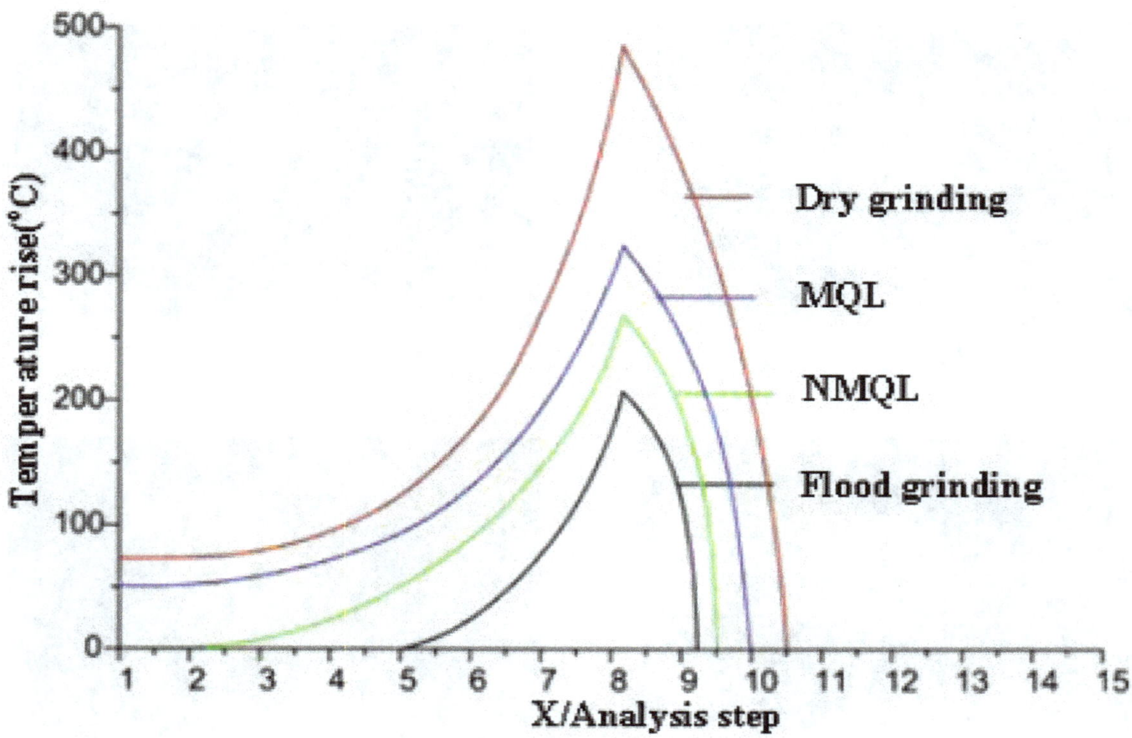

first to fifth analysis step areas at rear edge is already lowered to room temperature under flood grinding, temperature rise of workpiece in the first analysis step area is zero under NMQL condition, but a certain temperature rise exists on workpiece surface in the first analysis step area under MQL and dry grinding conditions, indicating that coverages of temperature fields are different and cooling effects of flood grinding and NMQL conditions are superior to those under MQL and dry grinding conditions.

Under cooling effects of four cooling and lubrication condition, when heat source moves to the eighth analysis step, temperature rise on workpiece surface along direction of width (Z direction) of grinding contact area is also different, and temperature rises on workpiece surface along direction of grinding width under four working conditions are shown in Fig. 16.

According to Fig. 16, the maximum temperature rises on workpiece surface are obviously different. MQL is between dry grinding and flood grinding and NMQL is between MQL and flood grinding, and this indicates that cooling effect of NMQL is approximate to that under flood grinding. Lengths of maximum temperature rise areas on workpiece surface are also different, it is the maximum under dry grinding and equal to length of heat source, that under flood grinding is the shortest, and lengths under MQL and NMQL are intermediate, indicating that their cooling effects at the boundary between grinding area and non-grinding area are different: under dry grinding mode, there is no cooling medium for heat dissipation, so heat transferred to workpiece is aggregated on surface to form high temperature; under flood grinding, grinding fluid enters the grinding interface to form liquid film, heat at the boundary

Figure 16. The temperature rise along the direction of the grinding width

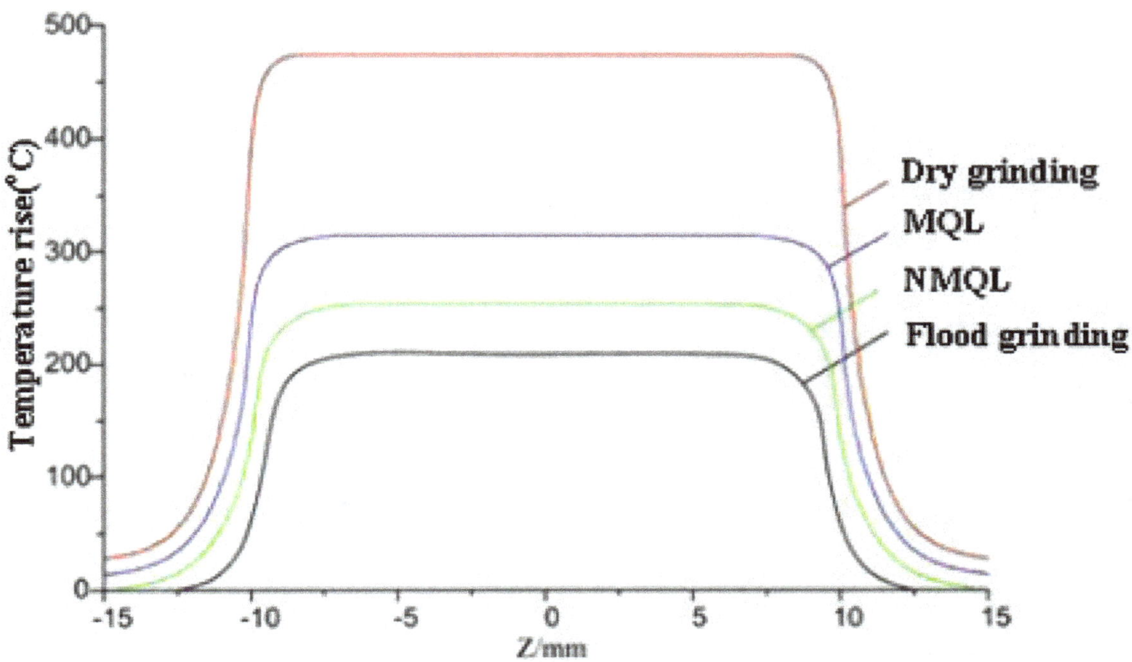

is carried away by grinding fluid due to its cooling effect, and on the other hand, heat is transferred to direction of grinding width, and temperature rise at the boundary is low with large temperature gradient. As the distance from grinding area increases, temperature rise on workpiece surface gradually declines under the convective heat transfer effect. Grinding fluid will experience side leakage phenomenon under flood grinding, which effectively reduces temperature rises in non-grinding areas at two sides. According to maximum temperature rise and cooling effect at boundary between grinding areas, cooling effects of four cooling and lubrication condition are compared along direction of grinding width, and it's obtained that cooling effect of NMQL is quite significant, which plays a significant role in lowering grinding temperature rise, and the cooling effect is approximate to that under flood grinding.

Under four cooling and lubrication condition, after grinding fluid carries away a part of heat, residual heat continues to be transferred into workpiece base. Temperature rise inside the workpiece base along direction of grinding depth (Y direction) in the grinding contact area is analyzed. Temperature rises along direction of grinding depth under four cooling and lubrication condition are shown in Fig. 17.

According to Fig. 17, different cooling and lubrication condition also have an influence on temperature rise inside the workpiece. Temperature rises at different depth positions of workpiece under four cooling and lubrication condition are different, indicating that heat quantities transferred to grinding fluid due to convective heat transfer effect of grinding fluid are different under four cooling and lubrication condition, and residual heat is transferred to workpiece base from workpiece surface, which causes temperature rise in workpiece base. Firstly, heat is accumulated on workpiece surface to form different temperature rises, and heat is dissipated in the form of heat conduction as the depth from workpiece surface increases.

Figure 17. The temperature rise along the direction of the grinding depth

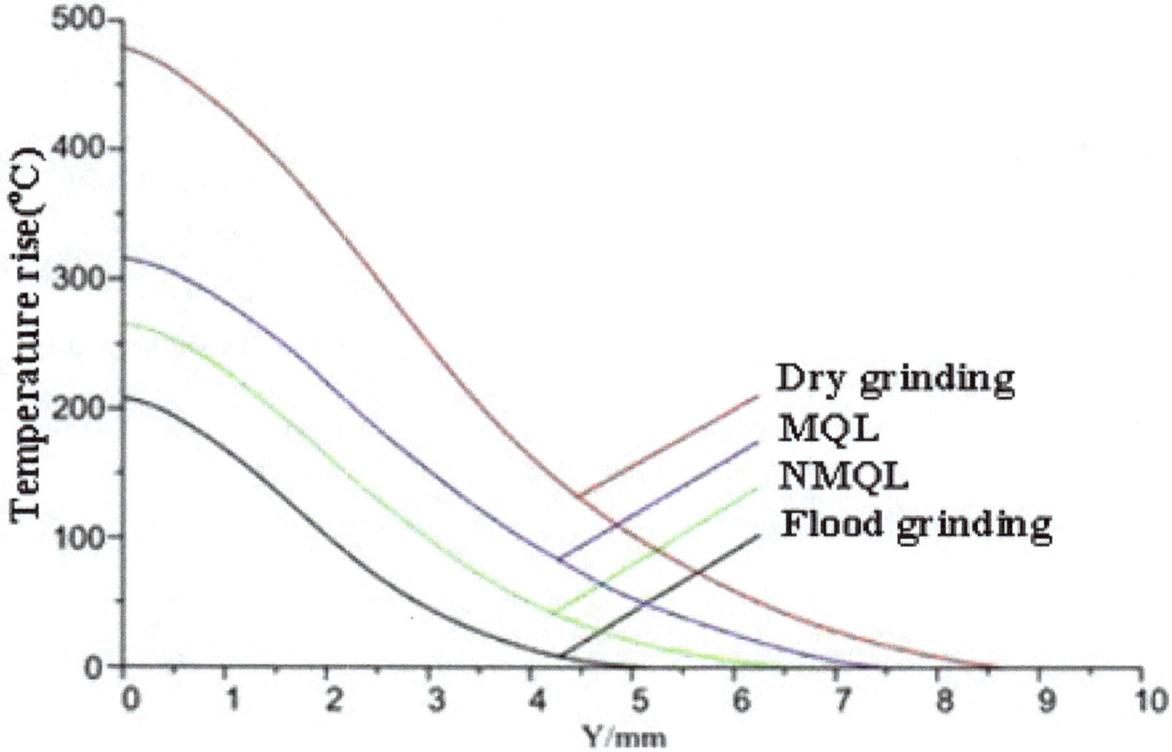

At the same depth from workpiece surface, temperature rises inside workpiece are different under four working conditions: the temperature rise under dry grinding mode is the highest, followed by MQL, NMQL and flood grinding, indirectly certifying that heat transfer capacity of four cooling and lubrication condition are different. Temperature rise formed inside workpiece gradually reduces along direction of grinding depth. When the depth from workpiece surface is small, heat conductivity of workpiece material nearby the surface is high due to the highest workpiece surface temperature. Transferred heat quantity is large at small depth. Temperature rise within a certain depth declines to a smaller degree compared with temperature rise on surface, temperature gradually declines, heat conductivity of workpiece material is gradually reduced, and temperature rise inside workpiece is gradually reduced at high speed. As the depth increases, transferred heat quantity is gradually reduced with small temperature gradient, and temperature rise slowly reduces. Under four cooling % lubrication condition, when heat source is at the eighth analysis step, the depth from maximum temperature rise to zero temperature rise along direction of grinding depth is 5mm under flood grinding, that under NMQL mode is 6.5mm, that under MQL mode is 7.5mm and that under dry grinding mode is the maximum, indicating that cooling and lubrication mode influences heat transferred into workpiece, this depth is not the maximum depth, but instead, as the heat source continues to move forward, it takes a certain time for heat to be transferred downward inside workpiece due to heat conductivity, maximum depth will increase. Influences of four cooling and lubrication condition on temperature rise along direction of grinding depth within the same time step can be compared, and comparison results indicate that flood grinding and NMQL mode can reduce the thermal influence of surface temperature rise on workpiece base in direction of grinding depth, reduce temperature rise of workpiece base and exert a favorable cooling effect.

As shown in Fig. 14, Fig. 15, Fig. 16 and Fig. 17, under four cooling and lubrication condition, heat transfer coefficients are different due to different heat transfer capacity of grinding fluid, much heat will be carried away when heat transfer capacity of grinding fluid is strong so that temperature rise on workpiece surface is reduced. Due to difference of heat transfer coefficients under four cooling and lubrication condition, maximum temperature rises on workpiece surface are obviously different. Temperature rises along directions of grinding width and grinding depth have obvious differences.

Among four cooling and lubrication condition, a large quantity of grinding fluid is casted between grinding wheel and workpiece for cooling under flood grinding with the optimal cooling effect, but as a large quantity of grinding fluid is used under flood grinding, the price and treatment expense of grinding fluid are high, and leakage and evaporation of grinding fluid will harm environment and workers; dry grinding is a relatively ideal green machining mode, but its cooling effect is not that ideal. When grinding temperature exceeds a limiting value, it can easily generate grinding burning, which causes thermal injury, residual stress and crack on workpiece surface, thus influencing workpiece material performance and service life; under MQL mode between dry grinding and flood grinding, grinding fluid is atomized and sprayed into grinding area for cooling with the help of high-pressure gas, its cooling effect is far poorer than flood grinding, but compared with dry grinding, cooling effect is obvious, and moreover, it reduces use level of grinding fluid by a large margin. Selecting degradable grinding fluid can reduce pollution hazards; according to heat transfer enhancement theory of solid, after nanoparticles are added in the grinding fluid based on MQL, cooling effect can be greatly improved, thus compensating for cooling deficiency of MQL, and its cooling performance is approximate to that under flood grinding. Therefore, NMQL mode can be the optimal cooling and lubrication mode among the four condition in grinding machining in aspects of energy conservation, environmental protection and high efficiency.

5.4.2 Influence of Nanofluids Minimum Quantity Lubrication on Grinding Temperature Field

As for the influence of cooling performance of NMQL on grinding temperature field, past studies mostly concentrate on jet angle, jet velocity and jet distance (Jia et al., 2014; Mao et al., 2012; Tsinober et al., 2006). However, improving heat transfer performance of nanofluid is also the main NMQL heat transfer enhancement technology. Nanofluid energy transfer performance is closely related to nanoparticles suspended in the grinding fluid (Cong et al., 2009).

Nanoparticles change structure of base fluid. Nanoparticles are uniformly suspended and distributed in the base fluid under ideal state. In fact, nanoparticles continuously do irregular Brownian movement due to collision of liquid molecules and VDW force. According to the theory of Brownian movement, motion velocity of nanoparticles influence frequency of energy exchange and transfer of nanoparticles with surrounding base fluid, and the smaller the nanoparticle size, the higher the motion velocity, and the higher the enhanced heat transfer performance of nanofluid (Wang et al., 2017; Zhang et al., 2018). According to heat transfer enhancement theory of solid, adding solid particles can improve heat conductivity coefficient and convective heat transfer performance of base fluid (Mao et al., 2013; Yi et al., 2019). Heat conductivity coefficient of solid is far greater than those of liquid and gas, so addition of solid nanoparticles can improve thermal diffusivity coefficient of nanofluid so as to enhance convective heat transfer capability. Therefore, heat transfer enhancement of nanofluid can be ascribed to random motion and heat diffusion of nanoparticles and elevation of heat conductivity coefficient of base fluid by nanoparticles (Krasyuk et al., 2012; Wen et al., 2009).

Selection of base fluid in nanofluid has a bearing on heat conductivity coefficient of nanofluid. It's theoretically and experimentally verified that increasing proportion of heat conductivity coefficient of grinding fluid after nanoparticles are added into oil-based fluid is larger than water-based fluid, and moreover, oil-based fluid contributes more to suspension stability (Alabi et al., 2017;Zhang et al., 2016). Differences of nanoparticle types added into the base grinding fluid and volume concentrations of nanoparticles forming the nanofluid will directly affect heat conductivity coefficient of nanofluid and convective heat transfer capacity of NMQL mode (Bense et al., 2016; Peres et al., 2008).

(1) The Influence of Nanoparticle Type on Cooling Performance of Nanofluid

In recent years, several international research groups have experimentally measured heat conductivity coefficients of nanofluids of different types. According to experimental data, adding nanoparticles in the base fluid significantly increases heat conductivity coefficient of liquid, but increasing amplitudes of heat conductivity coefficients of nanofluids added with different nanoparticles are different. Four nanoparticle types—molybdenum disulfide (MoS_2), zirconium oxide (ZrO_2), carbo nanotube (CNT) and diamond (ND) solid particles—are selected and added into the grinding fluid in the simulation to study convective heat transfer effect. MoS_2 is mainly used as lubricant. As its molecular structure is stratified structure, it is of easy shear, low friction coefficient, good lubricating effect but low heat conductivity coefficient; heat conductivity coefficient of carbon nanotube is high, reaching 3000-6000W/(m·K) with good heat-conducting property, and its stiffness is equivalent to diamond; ZrO_2 and diamond are hard materials, but diamond is selected as micron-order particles to be compared with other three nanoparticles.

After four types of nanoparticles of different properties are added into the base fluid, nanofluid with 1% volume concentration is formed. Convective heat transfer occurs on workpiece surface in the form of NMQL, and temperature rising trend of the eighth analysis step on workpiece surface within the whole time step is shown in Fig. 18.

As shown in Fig. 18, heat conductivity coefficient of nanofluid formed after nanoparticles are added is significantly elevated, but elevation effects of heat conductivity coefficients are different due to different heat conductivity coefficients of different nanoparticles. When these nanoparticles are applied to convective heat transfer on workpiece surface, grinding temperature rise of the eighth analysis step on workpiece surface within the whole time step is obtained. It's found that grinding temperature rise laws caused by four working conditions are identical and temperature differences are small. Compared with MQL-induced temperature rise, decreasing amplitudes of temperature rises are within 5%-18%, indicating that nanoparticles added into the grinding fluid can exert an important effect on lowering grinding temperature, where temperature rise obtained by adding carbon nanotube is the lowest, being about 245°C and embodying good heat transfer performance, followed by temperature rise after MoS_2 and ZrO_2 are added, but the temperature rise after addition of MoS_2 is partially low, possibly because good lubricating performance of MoS_2 reduces heat generated by friction. Temperature rise obtained by addition of diamond micro-powders is the highest, possibly because compared with nanoparticles, their particle size is large, thus reducing specific surface area and reducing heat transfer capacity. This indicates that heat transfer capacity of micron-order particles is different from that of nanoparticles. As for temperature rises obtained through simulation by adding four solid particles into the grinding fluid,

Figure 18. Influences of different nanoparticles on grinding temperature rise

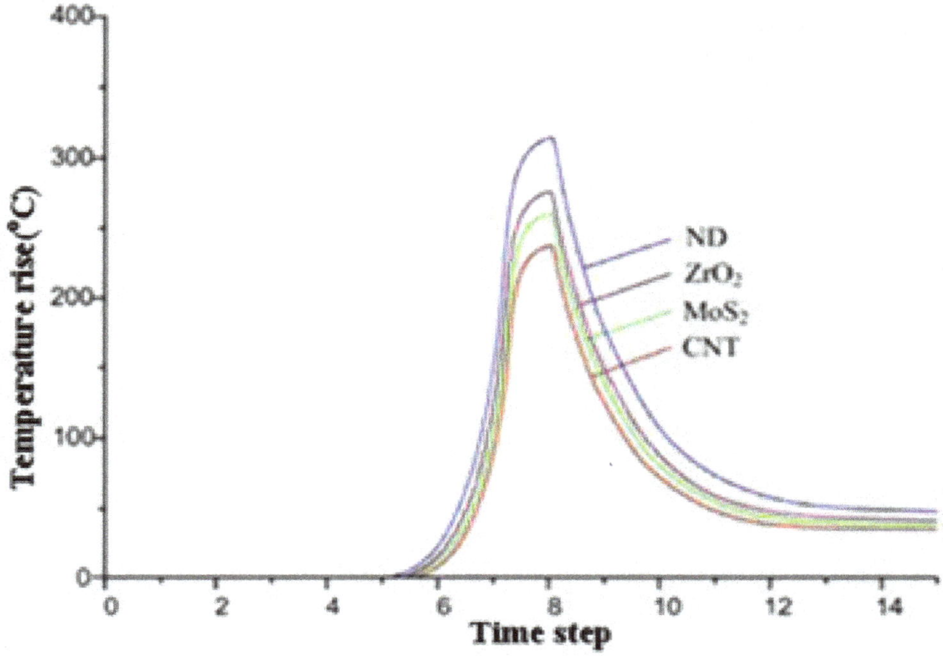

the grinding temperature rises obtained by adding MoS_2 and carbon nanotube particles are the lowest, so their cooling effects are satisfying.

(2) The Influence of Nanoparticle Volume Concentration on Cooling Performance of Nanofluid

Volume concentration of nanoparticles added into the base grinding fluid also has a certain influence on heat conductivity coefficient of the nanofluid. MoS_2 and carbon nanotube particles with good cooling effects are selected for an analysis. Ratios of heat conductivity coefficients of nanofluids formed by adding nanoparticles with different volume concentrations to heat conductivity coefficient of base grinding fluid are shown in Fig. 19.

It can be known from Fig. 19 that as volume concentration φ of nanoparticles added into the grinding fluid increases, heat conductivity coefficient of the nanofluid gradually increases, mainly because heat conductivity coefficient of nanoparticles is far greater than that of base fluid according to the heat transfer theory of solid. According to equation (1-1), increasing volume concentration of nanoparticles results in increase of heat conductivity coefficient of nanofluid. Theoretically calculated heat transfer coefficient increases with heat conductivity coefficient. When volume concentration of MoS_2 nanoparticles increases from 1% to 2%, heat transfer coefficient of nanofluid increases from 2.48×10^5W/(m^{2}K) to 2.85×10^5W/(m^{2}K). MoS_2 and carbon nanotube nanoparticles with volume concentrations being 1%, 2% and 3% are selected to simulate cooling performance of nanofluid. Influences of volume concentra-

Figure 19. Nanoparticle volume concentration effect on the heat conductivity

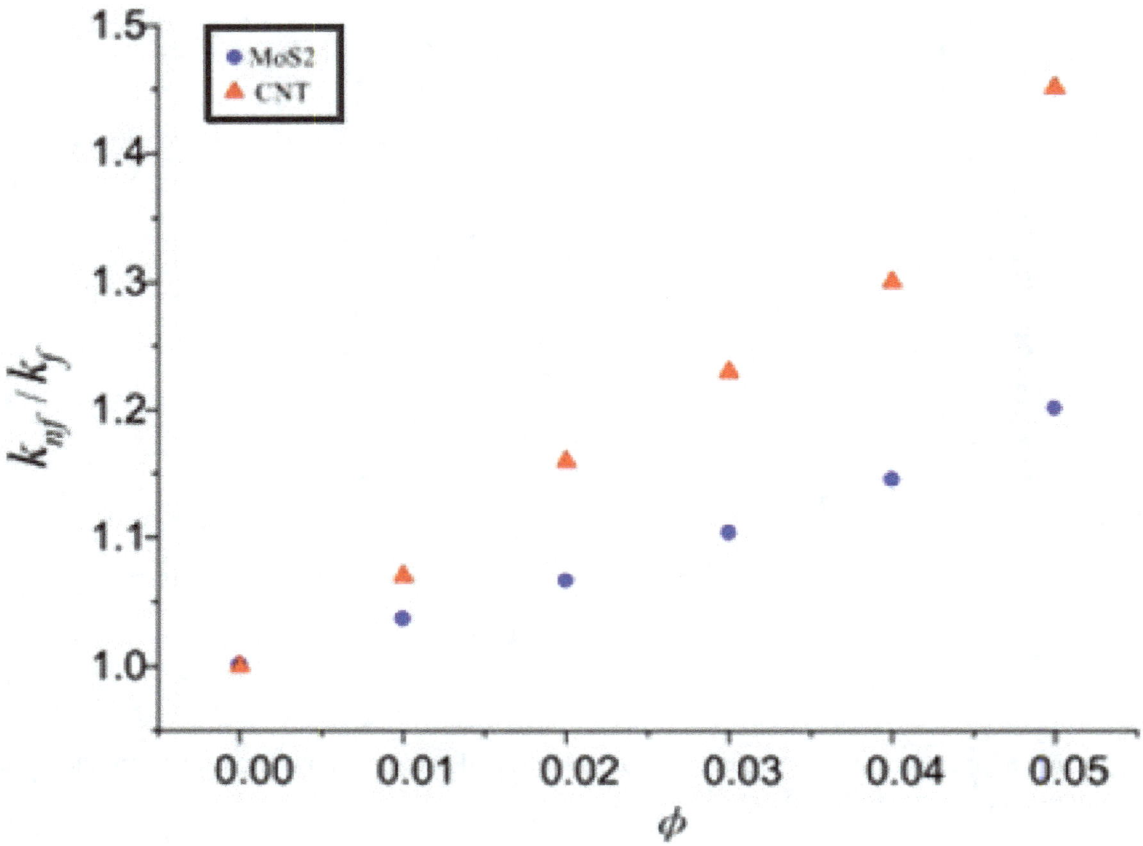

tions of two types of nanoparticles at the eighth analysis step within the whole time step on grinding temperature rise are shown in Fig. 20.

As shown in Fig. 20, grinding temperature rises caused by the two types of nanoparticles with the change of volume concentration present identical features. From grinding temperature rises obtained through the two nanoparticles, cooling performance of carbon nanotube is superior to that of MoS_2, but temperature difference is small and maximum temperature difference is about 50°C. Grinding temperature rises obtained by both nanoparticles are the maximum under 1% volume concentration and they are the maximum under 3% volume concentration. As volume concentration of added nanoparticles increases, convective heat transfer coefficient of grinding fluid increases, heat transfer capacity of workpiece surface is enhanced, and the obtained grinding temperature rise is lower, indicating that cooling and lubricating effect of nanofluid is better as the volume concentration increases.

Through a combination of influences of different types and volume concentrations of nanoparticles on cooling and lubricating effect of the nanofluid, heat conductivity coefficients of nanofluids formed by adding different types of nanoparticles are all elevated when compared with the base fluid. As heat conductivity coefficients of different types of nanoparticles can elevate heat conductivity coefficients of the nanofluid to different degrees, the obtained heat transfer coefficients are also different. Different

Figure 20. Nanoparticle volume concentration effect on the grinding temperature rise

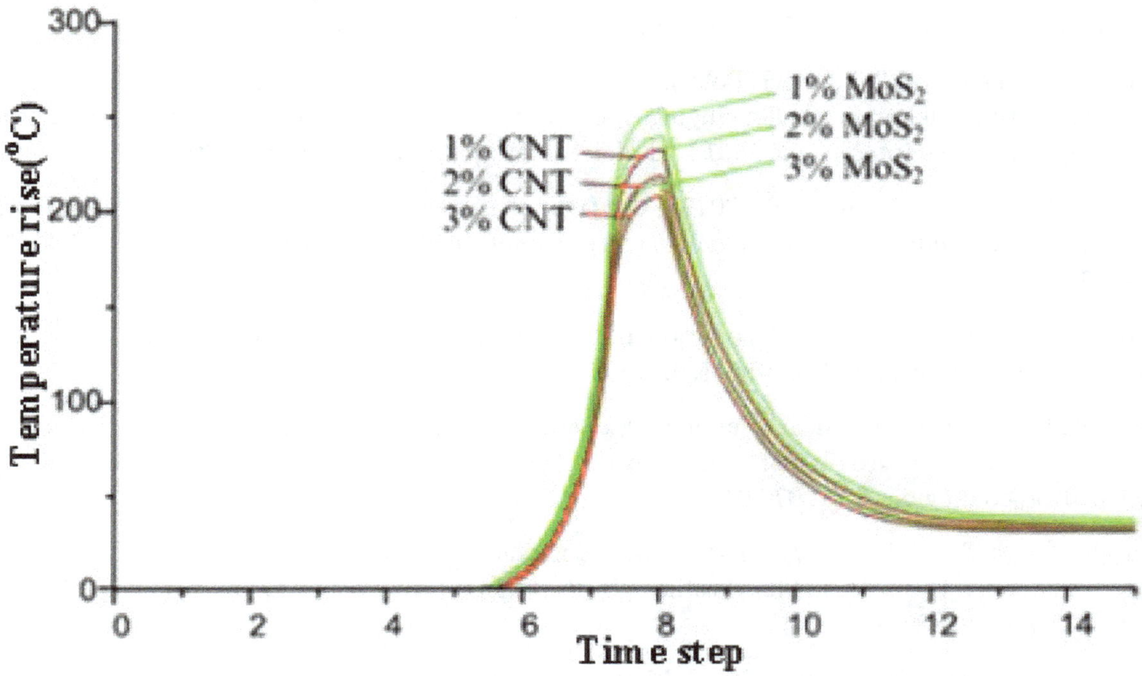

nanoparticles, when applied to NMQL, have different cooling effects on grinding temperature field. Through a comparison of four types of nanoparticles, cooling and lubricating effects of MoS_2 and carbon nanotube are good; as volume concentration of nanoparticles increases, heat conductivity coefficient of the nanofluid will increase according to the heat transfer enhancement theory of solid, so will heat transfer coefficient, and the obtained grinding temperature rise gradually reduces, indicating that enlarging volume concentration of nanoparticles can elevate heat transfer coefficient of the nanofluid, enhance convective heat transfer effect and gradually lower grinding temperature rise.

REFERENCES

Alabi, A. A., Akinyemi, O. D., Olowofela, J. A., Salako, F. K., Ajiboye, G. A., & Olurin, O. T. (2017). Temporal variation of ground temperature at depths 2cm to 200cm in an experimental field in Abeokuta, South-Western, Nigeria. *Arabian Journal of Geosciences*, *10*(19). doi:10.100712517-017-3201-z

Bense, V. F., Read, T., & Verhoef, A. (2016). Using distributed temperature sensing to monitor field scale dynamics of ground surface temperature and related substrate heat flux. *Agricultural and Forest Meteorology*, *220*, 207–215. doi:10.1016/j.agrformet.2016.01.138

Chen, Y. J., & Wang, X. (2018). Influences of the Temperature on the Ground State of the Coupling Polaron in an Asymmetric Quantum Pseudodot Under the External Magnetic Field. *International Journal of Theoretical Physics*, *57*(11), 3540–3549. doi:10.100710773-018-3868-2

Cong, T. N., Galanis, N., Polidori, G., Fohanno, S., Popa, C. V., & Bechec, A. L. (2009). An experimental study of a confined and submerged impinging jet heat transfer using AlO-water nanofluid. *International Journal of Thermal Sciences*, *48*(2), 401–411. doi:10.1016/j.ijthermalsci.2008.10.007

Dai, S. J., Li, X. Q., & Zhang, H. B. (2019). Research on temperature field of non-uniform heat source model in surface grinding by cup wheel. *Advances in Manufacturing*, *7*(3), 326–342. doi:10.100740436-019-00272-3

Eastman, J. A., Choi, U. S., Li, S., Thompson, L. J., & Lee, S. (1996). Enhanced Thermal Conductivity Through the Development of Nanofluids. *Mrs Proceedings, 457*.

Gavriliev, R. I. (2012). Analytical solution for the periodic temperature field in frozen ground with account for variation in mean daily air temperature. *Cold Regions Science and Technology*, *83-84*, 110–114. doi:10.1016/j.coldregions.2012.07.001

Gonzalez-Santander, J. L. (2017). Efficient Series Expansions of the Temperature Field in Dry Surface Grinding for Usual Heat Flux Profiles. *Mathematical Problems in Engineering*, *2017*, 1–13. doi:10.1155/2017/1856523

Gonzalez-Santander, J. L., Isidro, J. M., & Martin, G. (2015). An analysis of the transient regime temperature field in wet grinding. *Journal of Engineering Mathematics*, *90*(1), 141–171. doi:10.100710665-014-9713-6

Gonzalez-Santander, J. L., & Monreal, L. (2019). Efficient temperature field evaluation in wet surface grinding for arbitrary heat flux profile. *Journal of Engineering Mathematics*, *116*(1), 101–122. doi:10.100710665-019-10004-y

Gonzalez-Santander, J. L., Perez, J., de Cordoba, P. F., & Isidro, J. M. (2010). An analysis of the temperature field of the workpiece in dry continuous grinding. *Journal of Engineering Mathematics*, *57*(3), 165–174. doi:10.100710665-009-9335-6

Gonzalez-Santander, J. L., Placeres, J. M. V., & Isidro, J. M. (2011). Exact Solution for the Time-Dependent Temperature Field in Dry Grinding: Application to Segmental Wheels. *Mathematical Problems in Engineering*, *2011*, 1–28. doi:10.1155/2011/927876

Gu, R. J., Shillor, M., Barber, G. C., & Jen, T. (2004). Thermal analysis of the grinding process. *Mathematical and Computer Modelling, 39*(9–10), 991–1003. doi:10.1016/S0895-7177(04)90530-4

Hou, Y. L., Li, C. H., Han, Z. L., & Li, J. Y. (2011). Examination of the material removal mechanisms during the abrasive jet finishing of 45 steel. *Advanced Science Letters, 4*(4), 1478-1484. doi:10.1166/asl.2011.1304

Huang, L. (2011). Secondary development functions and applications of Abaqus/CAE. *Computer-Aided Engineering*.

Ivady, V., Simon, T., Maze, J. R., Abrikosov, I. A., & Gali, A. (2014). Pressure and temperature dependence of the zero-field splitting in the ground state of NV centers in diamond: A first-principles study. *Physical Review B, 90*(23). doi: B.90.235205 doi:10.1103/PhysRe

Jia, D. Z., Li, C. H., Zhang, D. K., Zhang, Y. B., & Zhang, X. W. (2014). Experimental verification of nanoparticle jet minimum quantity lubrication effectiveness in grinding. *Journal of Nanoparticle Research, 16*(12), 1–15. doi:10.100711051-014-2758-7

Jia, J., Leeb, W. L., & Cheng, Y. D. (2019). Field demonstration of a first constant-temperature thermal response test with both heat injection and extraction for ground source heat pump systems. *Applied Energy, 249*, 79–86. doi:10.1016/j.apenergy.2019.04.145

Jiang, J. L., Ge, P. Q., Sun, S. F., Wang, D. X., Wang, Y. L., & Yang, Y. (2016). From the microscopic interaction mechanism to the grinding temperature field: An integrated modelling on the grinding process. *International Journal of Machine Tools & Manufacture, 110*, 27–42. doi:10.1016/j.ijmachtools.2016.08.004

Jin, T., & Stephenson, D. J. (2004). Three dimensional finite element simulation of transient heat transfer in high efficiency deep grinding. *CIRP Annals - Manufacturing Technology, 53*(1), 259–262. doi:10.1016/S0007-8506(07)60693-3

Krasyuk, A. M., Lugin, I. V., & P'yankova, A. Y. (2012). Temperature fields in surrounding ground of shallow tube stations. *Journal of Mining Science, 48*(3), 465–473. doi:10.1134/S1062739148030094

Lan, S. L., & Jiao, F. (2019). Modeling of heat source in grinding zone and numerical simulation for grinding temperature field. *International Journal of Advanced Manufacturing Technology, 103*(5-8), 3077–3086. doi:10.100700170-019-03662-w

Li, C. F., Mao, J. F., Peng, X., Mao, W., Xing, Z. L., & Zhu, G. D. (2019). Analysis of intermittent operation on the ground temperature recovery of borehole heat exchanger fields. *International Journal of Green Energy*. doi:10.1080/15435075.2019.1671416

Li, C. H., Han, Z. L., & Li, J. Y. (2010). Investigation into fluid velocity field of wedge-shaped gap in grinding. *Applied Mechanics and Materials, 37-38*, 593-598.

Li, C. H., Li, J. Y., Wang, S., & Zhang, Q. (2013). Modeling and Numerical Simulation of the Grinding Temperature Field with Nanoparticle Jet of MQL. *Advances in Mechanical Engineering*. doi:10.1155/2013/986984

Li, C. H., Li, J. Y., & Zhao, H. Y. (2010). Modeling and simulation of hydrodynamic pressure in cutting zone. *Advanced Materials Research, 148-149*, 186-190.

Li, Q. J., Zhao, Z. Y., Zhou, H. D., Ke, W. P., Wang, X. M., Fan, C., . . . Sun, X. F. (2012). Paramagnetic ground state with field-induced partial order in Nd3Ga5SiO14 probed by low-temperature heat transport. *Physical Review B, 85*(17). doi: B.85.174438 doi:10.1103/PhysRev

Mao, C., Zhou, Z. X., Guo, K. X., & Li, X. Y. (2010). Numerical Simulation and Experimental Validation of the Temperature Field in Surface Grinding. *Machining Science and Technology, 14*(3), 344–364. doi:10.1080/10910344.2010.512198

Mao, C., Zhou, Z. X., Ren, Y. H., & Zhang, B. (2010). Analysis and FEM Simulation of Temperature Field in Wet Surface Grinding. *Materials and Manufacturing Processes*, *25*(6), 399–406. doi:10.1080/10426910903124811

Mao, C., Zou, H., Huang, Y., Li, Y., & Zhou, Z. (2013). Analysis of heat transfer coefficient on workpiece surface during;minimum quantity lubricant grinding. *International Journal of Advanced Manufacturing Technology*, *66*(1-4), 363–370. doi:10.100700170-012-4330-x

Mao, C., Zou, H. F., Huang, X. M., Zhang, J., & Zhou, Z. X. (2012). The influence of spraying parameters on grinding performance for;nanofluid minimum quantity lubrication. *International Journal of Advanced Manufacturing Technology, 64*(9-12), 1791-1799. doi:10.100700170-012-4143-y

Nguyen, T., & Zhang, L. C. (2010). Grinding-hardening using dry air and liquid nitrogen: Prediction and verification of temperature fields and hardened layer thickness. *International Journal of Machine Tools & Manufacture*, *50*(10), 901–910. doi:10.1016/j.ijmachtools.2010.06.002

Parente, M. P. L., Jorge, R. M. N., Vieira, A. A., & Baptista, A. M. (2012). Experimental and numerical study of the temperature field during creep feed grinding. *International Journal of Advanced Manufacturing Technology*, *61*(1-4), 127–134. doi:10.100700170-011-3711-x

Parente, M. P. L., Jorge, R. M. N., Vieira, A. A., Santos, S., & Baptista, A. M. (2013). Experimental and Numerical Study on the Temperature Field during Surface Grinding of a Ti-6Al-4V Titanium Alloy. *Mechanics of Advanced Materials and Structures*, *20*(5), 397–404. doi:10.1080/15376494.2011.627641

Peres, L. F., Sobrino, J. A., Libonati, R., Jimenez-Munoz, J. C., Dacamara, C. C., & Romaguera, M. (2008). Validation of a temperature emissivity separation hybrid method from airborne hyperspectral scanner data and ground measurements in the SEN2FLEX field campaign. *International Journal of Remote Sensing*, *29*(24), 7251–7268. doi:10.1080/01431160802036532

Sailer, E., Taborda, D. M. G., & Zdravkovic, L. (2018). A new approach to estimating temperature fields around a group of vertical ground heat exchangers in two-dimensional analyses. *Renewable Energy*, *118*, 579–590. doi:10.1016/j.renene.2017.11.035

Song, M. C., Pu, Y. S., Liu, Q., Wang, R., & Wang, B. Q. (2019). Numerical Simulation and Experimental Verification of the Quenching Temperature Field of Grinding Balls. *Materiali in Tehnologije*, *53*(4), 515–520. doi:10.17222/mit.2018.226

Su, J. X., Ke, Q. X., Deng, X. Z., & Ren, X. Z. (2018). Numerical simulation and experimental analysis of temperature field of gear form grinding. *International Journal of Advanced Manufacturing Technology*, *97*(5-8), 2351–2367. doi:10.100700170-018-2079-6

Tsinober, A., Kit, E., & Dracos, T. (2006). Experimental investigation of the field of velocity gradients in turbulent flows. *Journal of Fluid Mechanics*, *242*(242), 169–192.

Tucek, J., Ohkoshi, S., & Zboril, R. (2011). Room-temperature ground magnetic state of epsilon-Fe2O3: In-field Mossbauer spectroscopy evidence for collinear ferrimagnet. *Applied Physics Letters*, *99*(25), 253108. doi:10.1063/1.3671114

Vieira, A. A., Baptista, A. M., Parente, M. P. L., & Jorge, R. M. N. (2009). On the temperature field during superficial grinding: An experimental study. *International Journal of Advanced Manufacturing Technology*, *40*(11-12), 1084–1092. doi:10.100700170-008-1442-4

Wang, R. Q., Dai, S. J., Zhang, H. B., & Dong, Y. F. (2017). The temperature field study on the annular heat source model in large surface grinding by cup wheel. *International Journal of Advanced Manufacturing Technology*, *93*(9-12), 3261–3273. doi:10.100700170-017-0732-0

Wang, X. Q., Chen, Y. J., & Xiao, J. L. (2019). Effects of temperature and magnetic field on the ground state binding energy of the strong coupling magneto-polaron in an RbC1 asymmetrical semi-exponential quantum well. *International Journal of Modern Physics B*, *33*(21), 1950239. doi:10.1142/S0217979219502394

Wang, X. Z., Yu, T. B., Sun, X., Shi, Y., & Wang, W. S. (2016). Study of 3D grinding temperature field based on finite difference method: Considering machining parameters and energy partition. *International Journal of Advanced Manufacturing Technology*, *84*(5-8), 915–927. doi:10.100700170-015-7757-z

Wang, Z. X., Li, Y., Yu, T. B., Zhao, J., & Wen, P. H. (2019). Prediction of 3D grinding temperature field based on meshless method considering infinite element. *International Journal of Advanced Manufacturing Technology*, *100*(9-12), 3067–3084. doi:10.100700170-018-2801-4

Wang, Z. X., Yu, T. B., Wang, X. Z., Zhang, T. Q., Zhao, J., & Wen, P. H. (2019). Grinding temperature field prediction by meshless finite block method with double infinite element. *International Journal of Mechanical Sciences*, *153*, 131–142. doi:10.1016/j.ijmecsci.2019.01.037

Wen, D., Lin, G., Vafaei, S., & Kai, Z. (2009). Review of nanofluids for heat transfer applications. *Particuology*, *7*(2), 141–150. doi:10.1016/j.partic.2009.01.007

Yang, M., Li, C. H., Zhang, Y. B., Wang, Y. G., Li, B. K., Jia, D. Z., ... Li, R. Z. (2017). Research on microscale skull grinding temperature field under different cooling conditions. *Applied Thermal Engineering*, *126*, 525–537. doi:10.1016/j.applthermaleng.2017.07.183

Yi, J., Jin, T., & Deng, Z. H. (2019). The temperature field study on the three-dimensional surface moving heat source model in involute gear form grinding. *International Journal of Advanced Manufacturing Technology*, *103*(5-8), 3097–3108. doi:10.100700170-019-03752-9

Zhang, J. C., Li, C. H., Zhang, Y. B., Yang, M., Jia, D. Z., Hou, Y. L., & Li, R. Z. (2018). Temperature field model and experimental verification on cryogenic air nanofluid minimum quantity lubrication grinding. *International Journal of Advanced Manufacturing Technology*, *97*(1-4), 209–228. doi:10.100700170-018-1936-7

Zhang, Z. Y., Shang, W., Ding, H. H., Guo, J., Wang, H. Y., Liu, Q. Y., & Wang, W. J. (2016). Thermal model and temperature field in rail grinding process based on a moving heat source. *Applied Thermal Engineering*, *106*, 855–864. doi:10.1016/j.applthermaleng.2016.06.071

Zhou, L., Huang, S. T., & Zhang, C. Y. (2013). Numerical and experimental studies on the temperature field in precision grinding of SiCp/Al composites. *International Journal of Advanced Manufacturing Technology*, *67*(5-8), 1007–1014. doi:10.100700170-012-4543-z

Chapter 6
Experimental Research on Minimum Quantity Lubrication Surface Grinding With Different Cooling and Lubrication Conditions

ABSTRACT

Given the increasing attention to environmental and health problems caused by machining, the development of an environmentally friendly grinding fluid has become an urgent task. The cooling and lubricating properties of different cooling and lubricating conditions were analyzed. The influence mechanism of nanofluids minimum quantity lubrication (NMQL) on cooling and lubricating effect was revealed with different nanoparticles (MoS2, CNT, ZrO2) and different volume concentrations of MoS2 nanofluids (1%, 2%, 3%). The experimental results showed that the temperature rise (258 °C) and grinding force (Fn=70 N, Ft=27 N) obtained under NMQL grinding were the closest to the flood grinding. The specific grinding energy of MoS2 nanofluids was the lowest, which was 47 J/mm3. When the volume concentration was 2%, the best cooling and lubricating effect was obtained. The surface roughness of the workpiece was the lowest (Ra = 0.283 μm; Rz = 0.424 μm).

6.1 INTRODUCTION

Surface grinding, which is a very important machining force in the grinding technology, has high requirements for machining precision and surface integrity of workpiece surface, so it belongs to precision machining technology (Jawahir et al., 2011; Li & Zhao, 2003). There are two major parameters influencing workpiece surface quality in the grinding process namely grinding force and grinding temperature. Cooling and lubricating conditions at the grinding interface influences grinding force and grinding temperature. When lubrication is sufficient, friction coefficient will be reduced, good surface

DOI: 10.4018/978-1-7998-1546-4.ch006

quality and superior cooling effect will be finally formed and maximum temperature rise on workpiece surface can be lowered, thus avoiding grinding burning and guaranteeing workpiece surface integrity. Therefore, studying the influence of cooling and lubricating effect on workpiece surface quality in the grinding process will be of great significance (Ding et al., 2014; Guitouni, Chaieb, Ben Rhouma, & Ben Fredj, 2016; Krolczyk, Nieslony, Maruda, & Wojciechowski, 2017; Setti, Sinha, Ghosh, & Rao, 2015). This chapter is mainly about grinding experimental study for different cooling and lubricating conditions with the emphasis laid on studying NMQL conditions. Evaluation parameters like energy proportionality coefficient, specific grinding energy and workpiece surface quality are used to verify simulation results based on experimental results. An experimental program is formulated and the best cooling and lubricating conditions is selected according to experimental results.

6.2 EXPERIMENTAL EQUIPMENT AND MATERIALS

6.2.1 Experimental Equipment

Equipment used in the grinding experiment is shown in Fig. 1, where experimental platform is K-P36 NC surface grinder produced by German Schleifring. Built-in grinding program and grinding wheel dressing program are used in the grinding experiment. Main parameters include: maximum power output by spindle is 4.5 kW, range of spindle speed is 45~4800 r/min, transverse feed speed of workbench is 30~30000 mm/min, longitudinal feed speed of workbench is 4~4000 mm/min and workbench size is 1000×950 mm. Grinding wheel used in the experiment is 80# white corundum ceramic bond grinding wheel, grinding wheel model is WA80MV12P and its dimensions are 300×20×76.2 mm.

Liquid supply forms used in the grinding experiment include flood grinding and MQL as shown in Fig. 1 (a) and Fig. 1 (b). MQL liquid supply device in this experiment is Bluebe MQL liquid supply system produced by Fuji company. An impulse is used to convey grinding fluid, regulate air pressure, control flow quantities of conveyed oil fluid and compressed air and regulating ratio of gas to liquid sprayed from the nozzle for the sake cooling and lubricating. As a general rule, air pressure is set as 4 bar~6.5 bar, compressed air and MQL grinding fluid are sprayed out of nozzle and enter machining area through high-pressure air after being mixed and atomized, and quantity of grinding fluid consumed is 30-100 ml/h per unit width of grinding wheel(HEISEL, LUTZ, WASSMER, & WALTER, 1998).

6.2.2 Experimental Materials

Experimental materials used in the experiment mainly include: workpiece material, water-based grinding fluid, MQL oil and nanoparticles.

(1) Workpiece Material

Workpiece material selected in this experiment is high-temperature nickel base alloy GH4169 and close marks include Inconel718 (US) and NC19FeNb (France). GH4169 nickel base alloy has favorable comprehensive performance within the temperature range of -253~700 °C, it is mainly applied to aerospace, nuclear energy, petroleum industries and extrusion dies, and it belongs to difficult-to-machine material(Niu, Qu, & Li, 2018; Zeng, Qin, Chang, & Luo, 2018). Grinding performance of high-temperature nickel base

Figure 1. The experimental equipment in surface grinding process: 1-NC surface grinder, 2-grinding wheel, 3-operation panel, 4-MQL supply system, 5-workpiece, 6-three-way dynamometer, 7-nozzle, 8-thermal IR imager, 9-data acquisition unit, 10-data amplifier, 11-computer

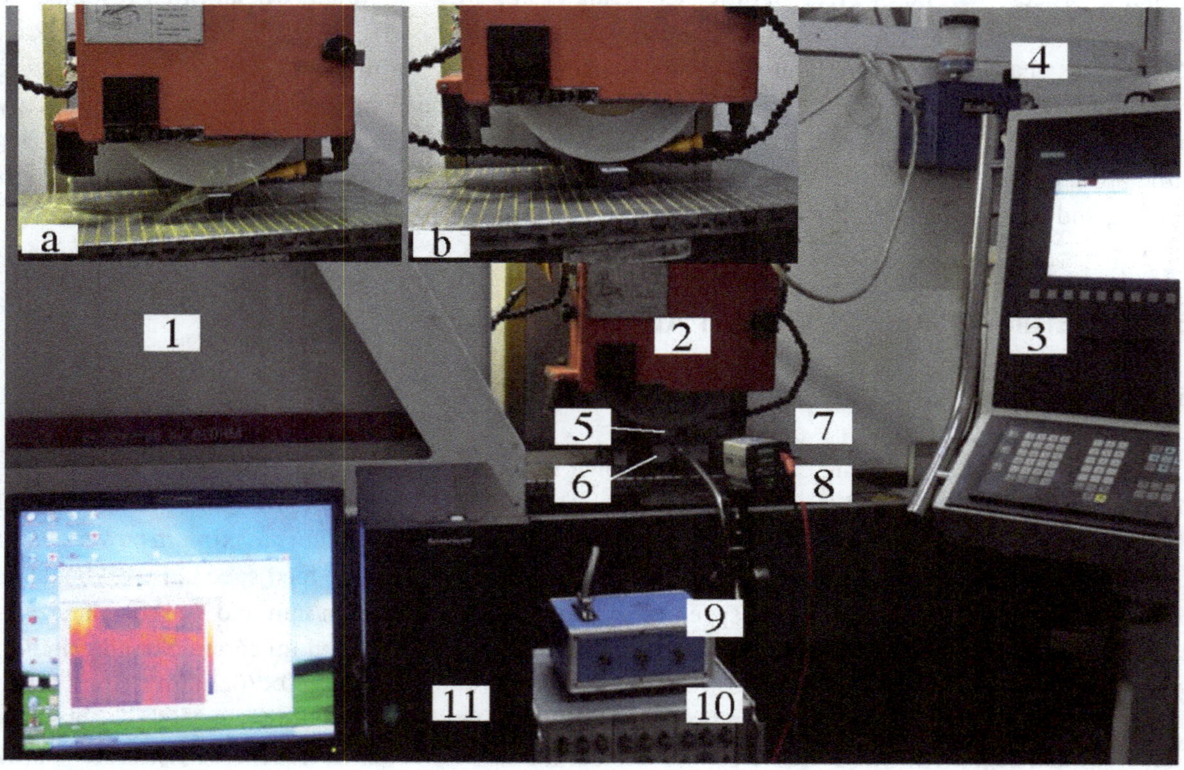

alloy GH4169 is studied through simulation and experiment(Zhang et al., 2016). According to the difference of contents of chemical elements, its components are divided into standard components, superior components and high-purity components. According to material anti-fatigue, anti-radiation, oxidation resistance and corrosion resistance, it has been applied to different fields. Chemical compositions of high-temperature nickel base alloy GH4169 are shown in Table 1 (GB/T 14992-2005).

Table 1. The chemical component of GH4169

Element	Ni	C	Cr	Nb	Mo	Ti	Co	Al	Mn	Cu	Fe
Component (%)	50-55	0.08	17-21	5.25	3.1	0.96	1.0	0.95	0.35	0.3	Margin

For the convenience of grinding experiment and result measurement of workpiece material, workpiece dimensions used in this experiment are selected as 40×30×30 mm and sample blocks are shown in Fig 2.

Figure 2. The sample piece of workpiece

(2) Water-Based Grinding Fluid

Water-based grinding fluid and oil-based grinding fluid are grinding fluids commonly used I grinding machining (Mechiri, Vasu, & Gopal, 2017; Ni, Yang, & Wu, 2019; Virdi, Chatha, & Singh, 2019). When water-based cutting fluid is used in the flood cooling process, convective heat transfer coefficient is about $3\times10^5 W/m^2$ K and it is about 2.3×10^4 W/m^2 K when oil-based cutting fluid is used, both reaching good cooling and lubricating effects (Incropera, 1990; Jin, Stephenson, & Rowe, 2003). Grinding fluid used in this experiment is CCF-04T water-soluble synthetic grinding fluid produced by Talent Chemicals Co., Ltd. Mass fraction of the grinding fluid is 4%, satisfying matching requirement for water-soluble grinding fluid (3~5%).

(3) MQL Oil

Grinding fluid used in the MQL experiment is Accu-lube/lb-2000 vegetable cutting oil(Simoes et al., 2014) as shown in Fig. 3. This oil has superior lubricating property and good cooling performance, lengths service life of cutting tool, keeps workpiece surface smooth and clean and eliminates stress generated by temperature change (Abdulquadir & Adeyemi, 2008; Kulkarni, Nadakatti, Patil, & Kulkarni, 2017; ManojKumar & Ghosh, 2016). As this cutting oil is vegetable oil with advantages of non-toxicity, no harm, non-volatility, easy decomposition and low cost, it will not result in environmental pollution or harm worker health, and moreover, it meets requirements for green and environmental protection, high efficiency and energy conservation (Awale, Srivastava, Vashista, & Yusufzai, 2019; Dambatta et al., 2019).

(4) Nanoparticles

NMQL grinding refers to adding nanoparticles with a certain volume concentration in the grinding fluid based on MQL condition. MQL lubricating oil is mixed with nanoparticles to form the nanofluid which is sprayed into grinding interface through high-pressure gas for cooling and lubricating(Huang, Liu, Tsai, & Chou, 2018; Rabiei, Rahimi, & Hadad, 2017; H. Singh, Sharma, Singh, & Dogra, 2019).

There are generally two nanofluid preparation methods: single-step method and two-step method(Bönnemann, Botha, Bladergroen, & Linkov, 2010), where the latter is applicable to preparation

Figure 3. MQL oil

of most nanofluids. Given this, two-step method is selected in this experiment for nanofluid preparation, base oil with a certain capacity is directly mixed with nanoparticles of a certain volume, and in order to prevent agglomeration of nanoparticles in the base oil, dispersing agent dimethyl sulfate ($C_2H_6O_4S$) is added and ultrasonic vibration is carried out to finally prepare the nanofluid.

Selected nanoparticle type and particle size are shown in Table 2. Samples are shown in Fig. 4, respectively being molybdenum disulfide, zirconium oxide, carbon nanotube and diamond micropowder.

6.3 EXPERIMENTAL SCHEME

The research purpose of this experiment is to discuss about cooling and lubricating conditions at the grinding wheel/workpiece interface under different cooling and lubricating conditions as well as their influences on workpiece surface quality, formulate a set of cooling and lubricating conditions for selection of experimental program and determine a cooling and lubricating way suitable to this working condition according to experimental results.

Univariate method is used in this experiment without consideration of influences of grinding parameters. Only influences of different cooling and lubricating conditions on grinding process are studied, so grinding parameters are unchanged in the whole experimental process as seen in Table 3.

Currently used cooling and lubricating conditions include: flood grinding, dry grinding, MQL and NMQL. As flooding has some defects, existing cooling and lubricating conditions are combined through the experiment, and then a cooling and lubricating condition, which can replace flood grinding, with good cooling and lubricating effect, green and environmental protection and high economical efficiency

Table 2. The species and size of nanoparticles

Nanoparticle	Molybdenum disulfide	Zirconium oxide	Carbon nanotube	(ND)Diamond
Particle size	50 nm	10 nm	10 nm	10 μm

Figure 4. The sample of nanoparticles

Table 3. The parameters of grinding process

Grinding environment	Parameter setting
Grinding ways	Surface grinding and up-grinding
Grinding equipment	K-P36
Grinding wheel	Al_2O_3 (WA80MV12P)
Grinding wheel linear velocity /v_s	35 m/s
Feed speed /v_w	0.05 m/s
Grinding depth /a_p	20 μm

is found out according to experimental results. An experimental program is formulated according to the structural order: from the overall cooling and lubricating conditions to single modes and then continuous detailing, and the experimental program is as follow:

Experiment I

According to existing cooling and lubricating conditions, cooling and lubricating effects in the grinding process under dry grinding, flood grinding, MQL and NMQL grinding are explored to verify cooling and lubricating effect of flood grinding. A cooling and lubricating conditions with good cooling and lubricating effect is obtained through the comparison of experimental results. In this group of four experiments, different cooling and lubricating conditions are used and their cooling and lubricating effects at the grinding interface are compared, and parameters of these cooling and lubricating conditions are shown in Table 4.

Four cooling and lubricating conditions are adopted in this experimental group, where grinding fluid is not used under dry grinding, CCF-04T water-based grinding fluid with 4% mass fraction is used under flood grinding, and cooling conditions are shown in Fig.1 (a). Accu-lube/lb-2000 vegetable oil is used as base oil under both MQL and NMQL conditions, but the former uses pure oil while the latter adds nanoparticles in the base fluid to prepare MoS_2 nanofluid with 1% volume concentration. MQL and NMQL

Table 4. The parameters of cooling and lubricating conditions

Cooling and lubricating conditions	Performance parameters
Dry grinding	No grinding fluid
Flood grinding	Water-based grinding fluid CCF-04T,Fluid supply 100L/min Water-based grinding fluid CCF-04T and liquid supply 100 L/min
MQL	Accu-lube/lb-2000 vegetable oil Liquid supply 30 ml/h and air pressure 0.5 MPa
NMQL	MoS_2 particles, particle size 50 nm Accu-lube/lb-2000 vegetable oil Volume concentration 1%, liquid supply 30 ml/h and air pressure 0.5 MPa

cooling conditions are shown in Fig. 1 (b). Nanofluid is sprayed out of nozzle through high-pressure gas (0.5MPa) to form nanoparticles, which are sprayed on workpiece surface for cooling and lubricating.

Experiment II

Cooling and lubricating effects on the grinding interface under four cooling and lubricating conditions are obtained through Experiment I. It's verified that flood grinding has good cooling and lubricating effect. A cooling and lubricating conditions with effect only second to flood grinding is selected according to experimental results. Based on theories of heat transfer and enhanced heat transfer of solid, it can be derived that NMQL has favorable cooling and lubricating effect. The emphasis is laid on studying cooling and lubricating effect under NMQL condition. Influences of added different kinds of nanoparticles on cooling and lubricating effect are further studied. MoS_2, CNT, ZrO_2 nanoparticles and ND micropowder are selected in this experimental group to mix with base fluid—Accu-lube/lb-2000 vegetable oil—to prepare grinding fluid with 1% volume concentration for the grinding experiment. Parameters for preparation of grinding fluid in this experimental group are shown in Fig. 5. According to influence of nanoparticle jet flow on workpiece quality and experimental results, a kind of nanoparticles with good cooling and lubricating effect when added to the grinding fluid is selected.

Experiment III

Based on results of Experiment II, the influence of volume concentration of nanoparticles, which are selected according to comparison results of cooling and lubricating effect, on cooling and lubricating effect is further studied. They are mixed with Accu-lube/lb-2000 vegetable oil to prepare grinding fluids

Table 5. The parameters of grinding fluid composition in nano particle jet MQL

Nanoparticle	Particle size/nm	Base oil of grinding fluid	Volume concentration
MoS_2	50nm	Accu-lube/lb-2000 vegetable oil	1%
CNT	10nm		
ZrO_2	10nm		
ND	10μm		

Figure 5. Grinding dynamometer and thermal imager used in experiment

with 1%, 2% and 3% volume concentrations for the grinding experiment. Cooling and lubricating effect at the grinding interface is investigated according to grinding temperature rise and tangential grinding force generated in the grinding process. Volume concentration of nanoparticles with good cooling and lubricating effect is obtained according to the comparison of workpiece surface quality after grinding.

6.4 EXPERIMENTAL MEASUREMENT

Physical quantities which are mainly measured in the grinding process include grinding force and grinding temperature. After grinding is completed, measured physical quantities of machined workpiece surface include surface roughness and surface morphology.

Grinding force in the grinding process is measured using YDM-III99 three-way piezoelectric grinding dynamometer developed by Dalian University of Technology. This dynamometer is made of stainless steel material and it is non-conducting magnetic material. As surface grinder workbench used in this experiment is a magnetic workbench, the dynamometer is directly placed on workbench but not fixed on it, and workpiece is placed but not fixed on the dynamometer. Therefore, a set of measuring technological equipment of grinding force is set, which on the one hand is convenient for stationary use of this dynamometer (indirectly fixed on magnetic workbench) on magnetic workbench, and on the other hand fixes workpiece on dynamometer for grinding and measuring grinding force. Six degrees of freedom of workpiece are restricted by this equipment and workpiece is clamped using this equipment.

Grinding temperature in the grinding process is measured using Thermovision A20 thermal IR imager produced by US FLIR. This compact-type thermal IR imager is mainly used to measure temperature in the field of industrial automation, which is non-contact type temperature measurement with temperature measuring range of -20 °C—900 °C and accuracy of ±2 °C. This thermal imager can be used to measure temperature at the end face of workpiece in the grinding experiment. According to the IR imaging principle, movement of heat generated by grinding wheel under the grinding effect on workpiece surface and the heat transfer process towards workpiece base can be clearly observed from end face of workpiece.

Grinding force dynamometer used in the experiment is shown in Fig. 5.

During the experimental process, grinding dynamometer and thermal IR imager can be used simultaneously to measure grinding force and grinding temperature and experimental measurement schematic is shown in Fig. 6.

Figure 6. The schematic diagram of experiment measurement and the measured figures: 1-grinding wheel, 2-workpiece, 3-three-way dynamometer, 4-workbench, 5-thermal IR imager, 6-date amplifier, 7-date acquisition unit I, 8- date acquisition II, 9-computer, 10-nozzle

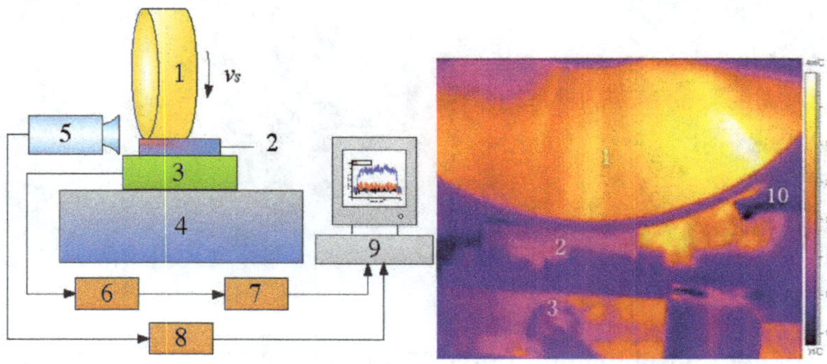

Grinding dynamometer and thermal IR imager are used in the experimental process to realize monitoring of grinding process as shown in Fig. 6. The workpiece is fixed on dynamometer through technological equipment, dynamometer is placed on magnetic workbench of grinder, and grinding wheel is close to end face of workpiece in tool setting. During the grinding process, grinding force measured through the dynamometer is amplified using the data amplifier and acquired by the data acquisition unit, and measured grinding force is displayed on dynamic measurement and display system of grinding force (DFMSG) of the computer. Grinding temperature rise of workpiece is monitored under sharp focusing state of thermal IR imager in the grinding process, temperature calibration is conducted for the imager before temperature measurement, 5 calibration points of different temperatures are selected for measurement and adjustment, detected IR images are acquired through the data acquisition unit in the measurement process, acquired IR data are processed and analyzed through post-processing analysis software Therma CAMResearcher on computer, one point on workpiece surface is measured for 5-10 times, average value of measured results is solved, and grinding temperature rise of workpiece is finally obtained.

After workpiece grinding is completed, the quality of already machined workpiece surface is measured and evaluated. TIME3220 contact pin-type surface roughness measuring instrument produced by Times Group Beijing TIME High Technology Co., Ltd is used in this experiment. Range of measured workpiece is set as 50 μm, Gauss function is used for linear smoothing and filtering of measuring result, workpiece surface is measured and measuring result is displayed, and profile curve of workpiece surface and corresponding surface roughness R_a value and R_z value are obtained; S-3500N SEM is used to scan workpiece surface, and workpiece surface morphology is obtained for observation and comparative analysis. TIME3220 roughness measuring instrument and S-3500N SEM used in the experimental measurement are shown in Fig. 7.

6.5 RESULT AND DISCUSSION

According to the above experimental program, surface grinding experiment is implemented, and main study objects are different cooling and lubricating conditions, including flood grinding, dry grinding, MQL and NMQL. The emphasis is laid on studying NMQL condition, including the influences of differ-

Figure 7. TIME3220 roughness measuring instrument and S-3500N scanning electron microscope

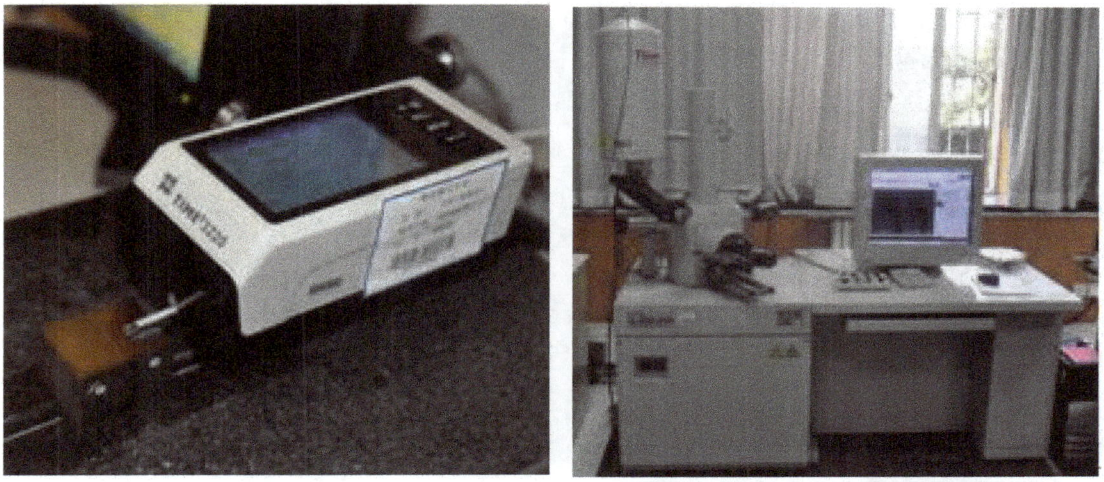

ent types of nanoparticles and their volume concentrations. Two important physical quantities—grinding force and grinding temperature—are measured in the grinding experimental process, energy proportionality coefficient and specific grinding energy are calculated(Li et al., 2016; Wang et al., 2016), surface morphology and surface roughness of machined surface are combined to comparatively analyze cooling and lubricating states under different cooling and lubricating conditions, and a cooling and lubricating conditions with good cooling and lubricating effect is selected.

6.5.1 Influences of Cooling-Lubricating Conditions on Cooling and Lubrication Property

For experimental program I, cooling and lubricating effects of four cooling and lubricating conditions—flood grinding, dry grinding, MQL and NMQL—are experimentally studied and grinding experiment is carried out under four cooling and lubricating conditions. By monitoring the experimental process, influences of different cooling and lubricating conditions on cooling and lubricating effect at the grinding interface in the grinding process are comparatively studied.

In the grinding process, grinding motion of grinding wheel converts consumed mechanical power into heat which is transferred to workpiece, and grinding temperature field in formed on workpiece. Thermal IR imager is used to monitor a grinding travel of grinding motion, and measured temperature rise and temperature field distribution on workpiece are shown in Fig. 6-8.

According to Fig. 8, a grinding travel under dry grinding environment is shown, including the whole grinding process from cut-in to cut-out of grinding wheel. Temperature field formed on workpiece changes with grinding motion of grinding wheel. In cut-in area (Fig. 8 (1)), grinding wheel starts contacting workpiece material, temperature field formed on workpiece is small, the caused temperature rise is small, and maximum temperature rise is 180 °C, mainly because volume of workpiece material participating in the grinding action is small and consumed grinding power is low. With transverse feed motion of workpiece, grinding wheel moves on workpiece surface, temperature field formed on workpiece also changes (Fig. 8 (2) and (3)), temperature rise caused in the grinding wheel/workpiece grind-

Figure 8. The grinding temperature measured figure: (1) cut-in area, (2) steady state, (3) steady state, (4) cut-out area

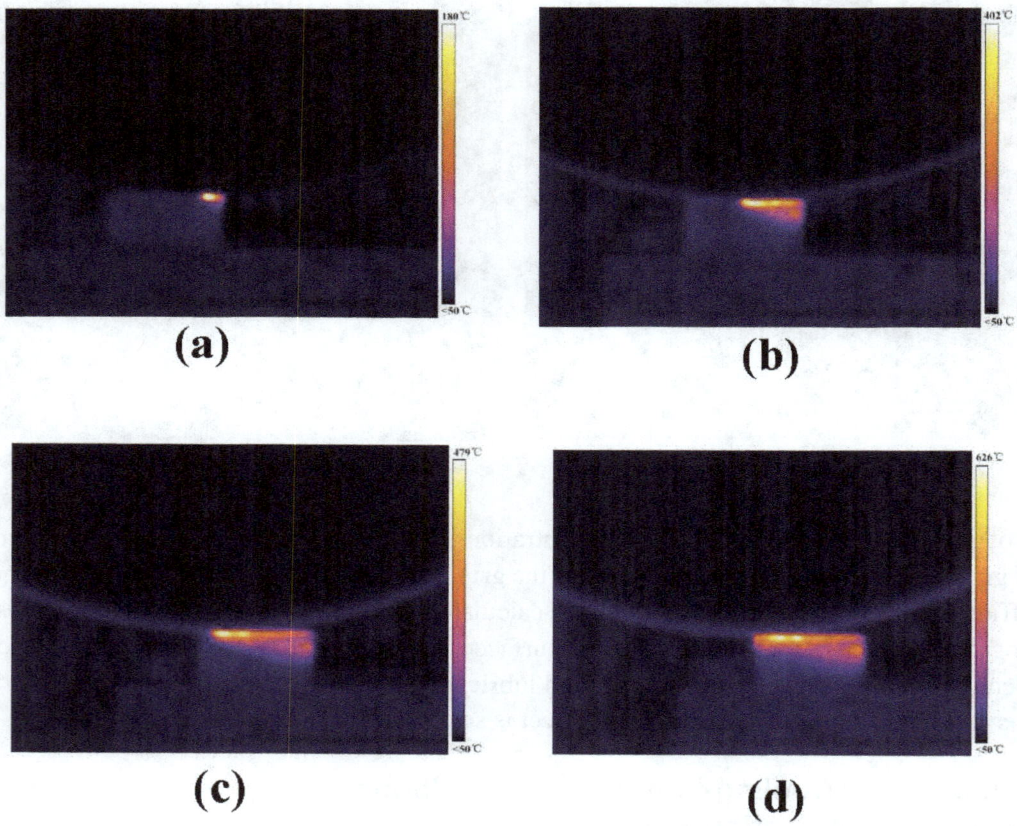

ing contact area increases, heat on workpiece surface at rear edge of grinding area is transferred into workpiece base, formed temperature field area is gradually enlarged from grinding area to rear edge of grinding area, maximum temperature rise is concentrated in grinding area, and maximum temperature rise reaches 402 °C. As the grinding motion proceeds, length of temperature field area is gradually enlarged, temperature rise at rear edge gradually reduces due to heat dissipation, but temperature rise in grinding area increases slightly and maximum temperature rise reaches 479 °C, and this is mainly caused by abrasive grain-induced blunting on grinding wheel or pore blocking between abrasive grains as well as increasing heat quantity generated by increasing power consumed by grinding. In cut-out area (Fig. 8 (4)), grinding temperature rise rapidly increases when grinding wheel is about to leave workpiece surface, and maximum temperature rise in grinding area reaches 626 °C, mainly because the quantity of workpiece materials is small in cut-out area and heat quantity transferred into unit volume of materials increases. Heat on workpiece surface is transferred to the base, temperature rise on workpiece surface fails to decline under dry grinding environment, and consequently, the whole superficial layer of workpiece is in the temperature field area.

Compared with grinding simulation results, temperature rising trend obtained through experiment accords with the simulation result, but the obtained maximum temperature rise is slightly different.

Figure 9. The grinding temperature rise under four kinds of cooling and lubrication conditions

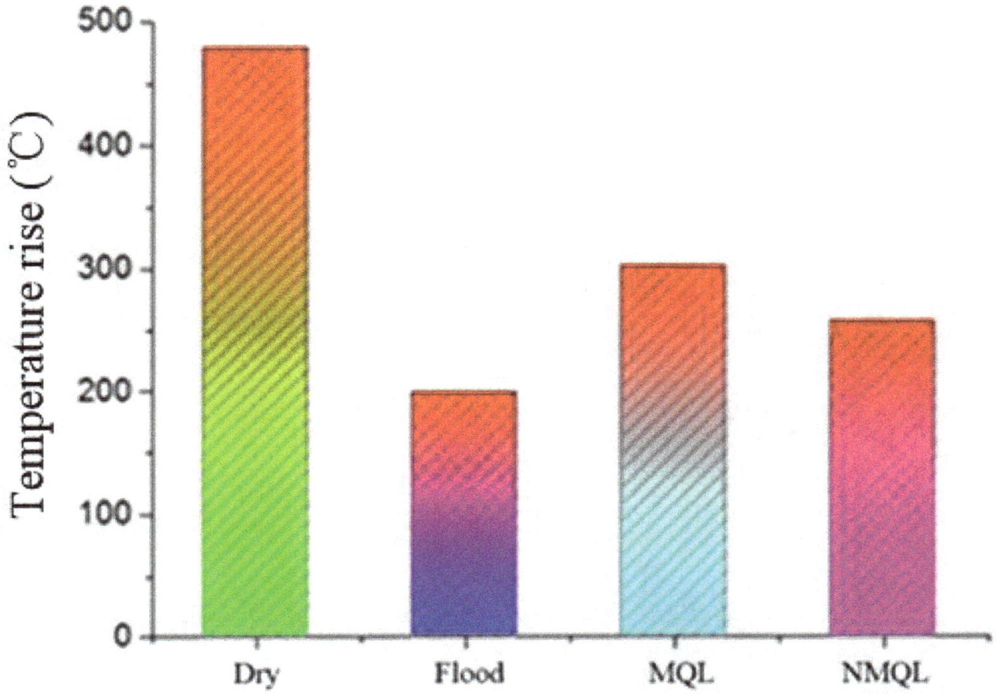

Maximum temperature in steady area tends to be stable in the simulation, but maximum temperature rise measured through experiment has increasing trend, which conforms to actual situation. Phenomena like grinding wheel wear and grinding wheel pore blocking by grinding debris are neglected in the simulation. Relative to simulation result, maximum temperature rise measured through experiment is slightly smaller, mainly because the part of heat which is transferred to and carried away by grinding debris is neglected in the simulation.

Different cooling and lubricating conditions have different influences on temperature field on the workpiece. Average temperature rises in steady area on workpiece surface obtained under four cooling and lubricating conditions—dry grinding, flood grinding, MQL and NMQL grinding—are shown in Fig. 9(Li et al., 2016).

It can be known from Fig. 9 that grinding temperature rise is the highest under dry grinding condition while those obtained under flood grinding, MQL and NMQL (semi-dry) conditions are obviously reduced, indicating that grinding fluid exerts a good cooling effect on the grinding interface and carries away a large quantity of heat so that temperature rise on workpiece surface is reduced. Temperature rise obtained under flood grinding is 198 °C which is the lowest among four cooling and lubricating conditions and reduces by 280 °C when compared with that under dry grinding, and this indicates that cooling effect is the optimal; temperature rise obtained under MQL grinding is 302 °C which declines by 180 °C relative to that under dry grinding, so a certain cooling effect is exerted on the grinding interface, but the cooling effect is far poorer than that under flood grinding; temperature rise obtained under NMQL grinding is 258 °C, cooling effect is improved after nanoparticles are added based on MQL, and

Figure 10. The grinding force measured under four kinds of cooling and lubrication conditions

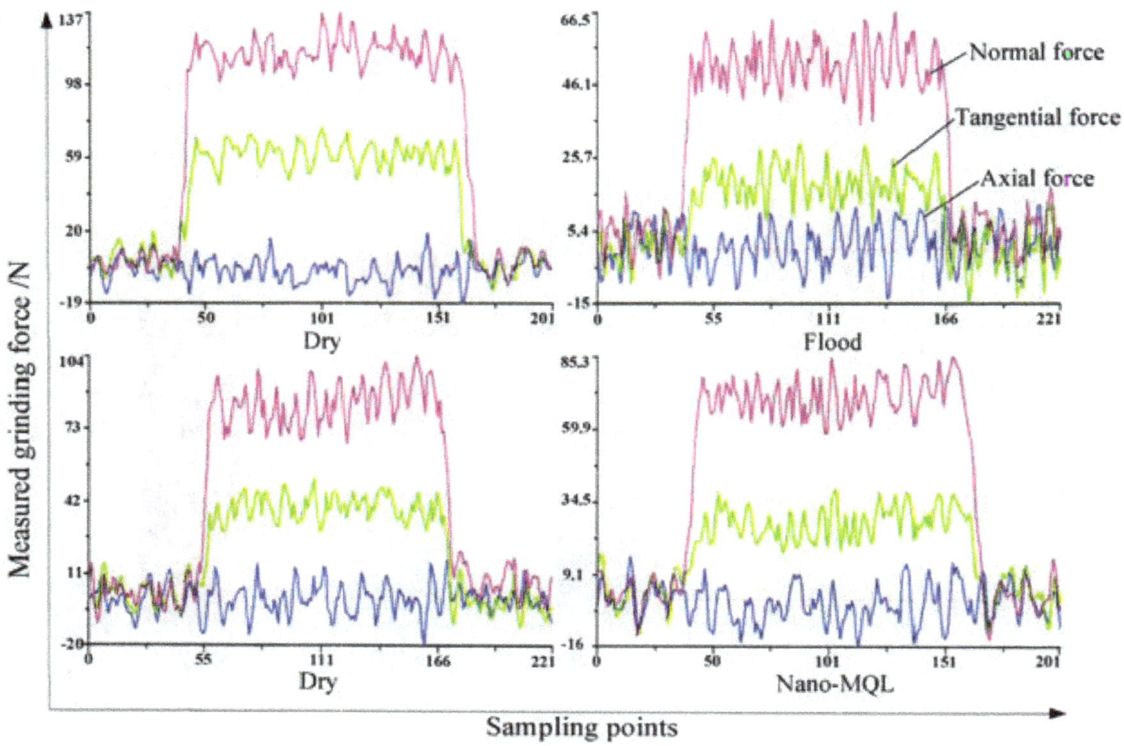

the temperature difference from flood condition is only 60 °C, indicating that NMQL cooling effect is approximate to that under flood grinding under four cooling and lubricating conditions.

Workpiece is placed on grinding dynamometer workbench using a technological equipment. Grinding dynamometer and thermal imager can simultaneously monitor grinding force and grinding heat in the grinding process (Fig. 1). Grinding forces obtained in the grinding process under four cooling and lubricating conditions are shown in Fig. 10.

According to Fig. 10, grinding forces obtained under different cooling and lubricating conditions are significantly different. Grinding force obtained under dry grinding is the highest among four cooling and lubricating conditions, normal force and average tangential force reach 115 N and 60 N respectively under grinding state, and the two reach 50 N and 18 N under flood grinding state. Based on a comparison between dry grinding and flood grinding, grinding fluid exerts a lubricating effect on the grinding interface so that grinding force is reduced. Normal force and tangential force obtained under MQL grinding are 82 N and 40 N respectively, and compared with flood, grinding force increases after use level of grinding fluid is lowered. The two figures are 70 N and 27 N respectively under NMQL grinding, compared with MQL grinding nanoparticles enhance the lubricating effect on the grinding interface, so grinding force is effectively reduced, and the difference between grinding force and flood grinding is the minimum, and this indicates that the lubricating effect is approximate to that under flood. According to equation (3-26), specific grinding energies under four cooling and lubricating conditions

Figure 11. Specific grinding energy and friction coefficient under four kinds of cooling and lubricating conditions

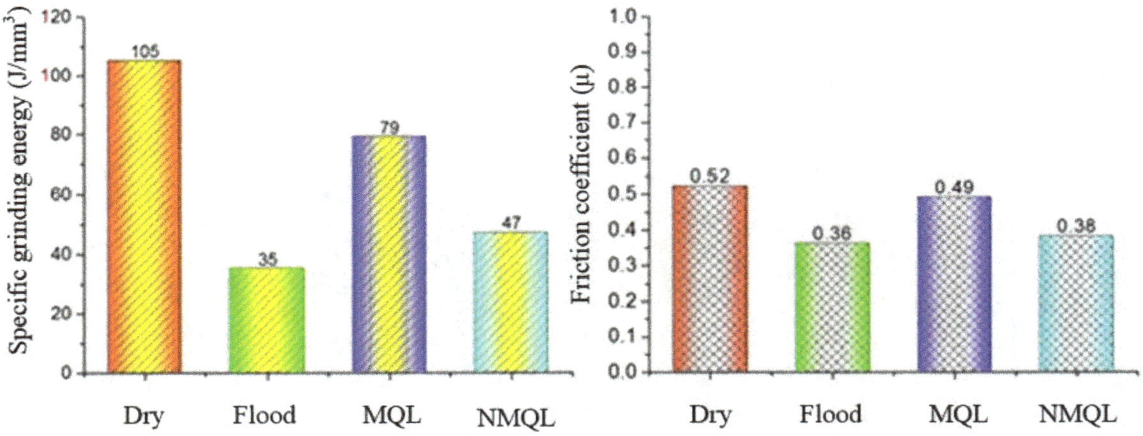

can be obtained through measured grinding forces. Lubricating effect on the grinding interface can be analyzed through specific grinding energy and friction coefficient(Wang et al., 2016). Specific grinding energies and friction coefficient under four cooling and lubricating conditions are shown in Fig. 11(Wang et al., 2016).

As shown in Fig. 11, specific grinding energy E_e and friction coefficient μ on the grinding interface obtained under four cooling and lubricating conditions are obviously different, indicating that four cooling and lubricating conditions exert different lubricating effects. Specific grinding energy obtained under dry grinding is the highest, indicating that the energy needed to consume unit volume of material is large. Specific grinding energy in the flood grinding process is significantly reduced under lubricating effect of grinding fluid. Compared with flood, under NMQL grinding, specific grinding energy E_e and friction coefficient μ are further reduced, indicating that lubricating effect is enhanced after nanoparticles are added.

Grinding forces and grinding temperature rises measured through the experiment are substituted into equations (3-27) and (4-51), and energy q_{wb} transferred to workpiece can be obtained with the help of theoretical formulas of Guo and Malkin. Proportionality coefficient of energy transferred to workpiece is obtained according to $q_{wb}/q_{total}=R$ (GUO & MALKIN, 1992). Energy proportionality coefficients under four cooling and lubricating conditions are shown in Fig. 12(Wang et al., 2016).

According to Fig. 12, proportionality coefficients of energies transferred to workpiece are different under different cooling and lubricating conditions, which causes different grinding temperature rises on workpiece, energy proportionality coefficient is 56% under dry grinding, and compared with simulation results, 4% of the heat is carried away by grinding debris, and energy proportionality coefficient under flood grinding is the lowest and the corresponding temperature rise is the lowest. Compared with MQL, energy proportionality coefficient is reduced somehow under NMQL, further certifying heat transfer enhancement effect of nanoparticles. Energy proportionality coefficient obtained under NMQL grinding is approximate to that under flood grinding, so their cooling effects are close.

Figure 12. Energy partition coefficient under four kinds of cooling and lubrication conditions

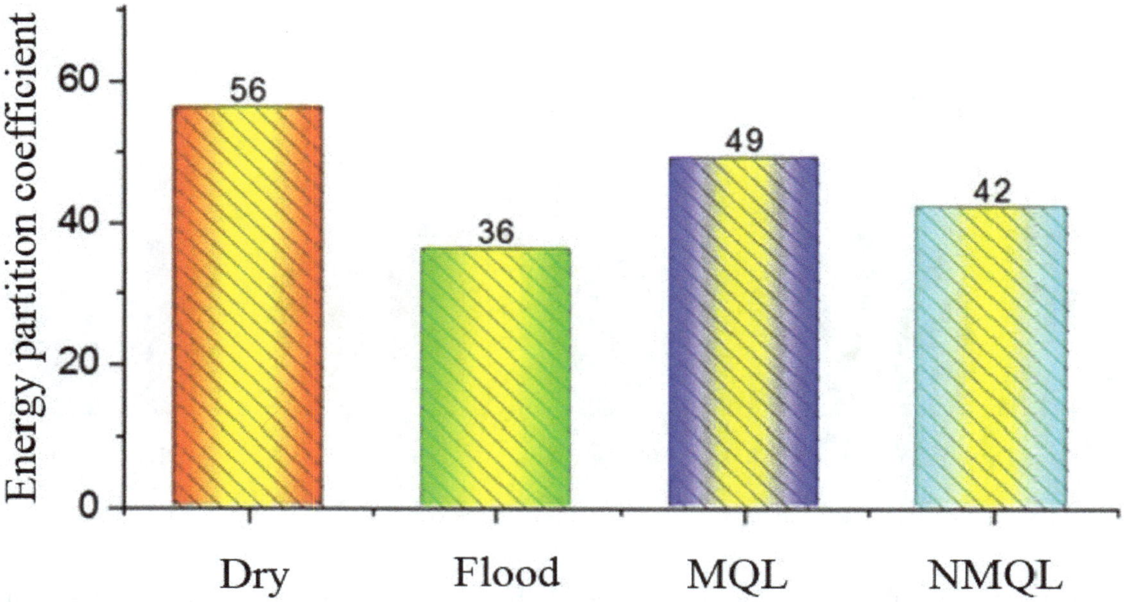

Cooling effects and lubricating effects under four cooling and lubricating conditions combined, workpiece surface quality is an assertive evidence used to authenticate cooling and lubricating effect. Workpiece surface morphologies of four workpieces under SEM after grinding is completed are shown in Fig. 13(Wang et al., 2016).

From Fig. 13, workpiece surface quality obtained under different cooling and lubricating effects due to four cooling and lubricating conditions is manifested by surface morphology. Workpiece surface after dry grinding is of irregular texture with material adhesion and stacking phenomena; workpiece surface texture after MQL grinding becomes regular, material adhesion and stacking phenomena are remitted, and the workpiece surface morphology is still different from that under flood grinding. However, under NMQL grinding, workpiece surface texture is clear with surface quality improved, and the workpiece surface quality is as satisfactory as that under flood grinding.

According to grinding temperature rise, grinding force and workpiece surface quality obtained through the grinding experiment under four cooling and lubricating conditions, cooling and lubricating effects under four cooling and lubricating conditions are analyzed. In full consideration of requirements for green and environmental protection, energy conservation and high efficiency, NMQL can replace flood cooling and lubricating conditions(Nam & Lee, 2018; Singh, Dureja, Singh, & Bhatti, 2019; Singh, Dureja, Dogra, & Bhatti, 2018).

6.5.2 Influences of Nanofluids Minimum Quantity Lubrication on Cooling and Lubrication Property

According to results of Experiment I and through comparison and screening from multiple aspects, NMQL is believed as a good cooling and lubricating conditions with satisfactory cooling and lubricat-

Figure 13. Surface morphology under four kinds of cooling and lubrication conditions

ing effect. Therefore, NMQL will be further investigated in this section. Experimental program II is designed to study influences of different types of added nanoparticles on cooling and lubricating effect.

In this experimental group, MoS_2, CNT, ZrO_2 nanoparticles and ND micropowder are selected. ND micropowder used in the experiment is micron-order solid particle while other three kinds are nanoscale solid particles. Four kinds of solid particle are added to the same base grinding fluid (Accu-lube/lb-2000 vegetable oil) to prepare grinding fluid with 1% volume concentration for the grinding experiment. Grinding temperature rises and grinding forces are measured through multiple experiments, and maximum temperature rise and tangential grinding force in the grinding area are shown in Table 6 (Li et al., 2016).

Experimental data of grinding temperature rise and tangential grinding force are substituted into equation (4-51) to obtain proportionality coefficients R of energy transferred to workpiece under MQL grinding state with four types of nanofluids as shown in Fig. 6-14.

Table 6. Grinding temperature rise and tangential grinding force in nano particle jet MQL

Nanoparticle type	MoS_2	CNT	ZrO_2	ND
Grinding temperature rise θ_{max} / °C	258	235	267	287
Tangential grinding force F_t /N	27	32	36	43

Figure 14. Energy partition coefficient in nano particle jet MQL

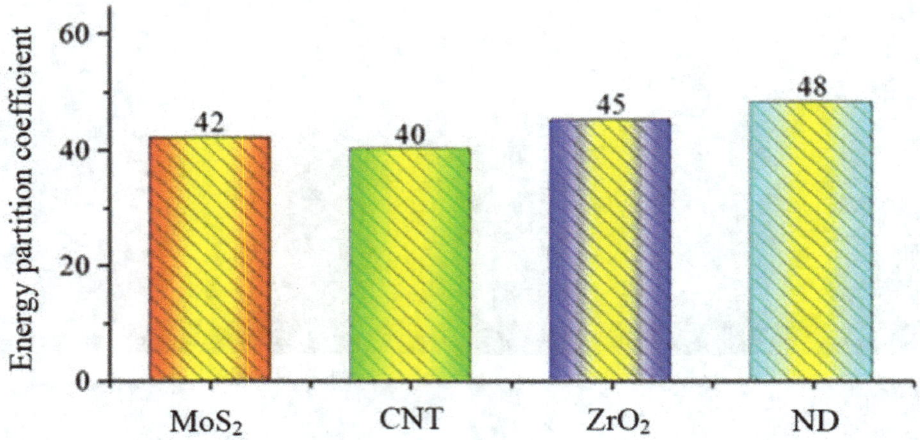

According to Fig. 14, among four kinds of solid particle, energy proportionality coefficient obtained through the grinding experiment added with carbon nanotube is the minimum, being 40%, indicating that the grinding fluid added with carbon nanotube has the best convective heat transfer effect mainly due to good heat conductivity of carbon nanotube(Choi, Jung, Choi, Lee, & Oh, 2011; Haque et al., 2018; Zhu et al., 2018), and then heat transfer performance of nanofluid is improved when it is added to the base grinding fluid; followed by MoS_2 and ZrO_2; energy proportionality coefficient obtained by the grinding fluid added with ND micropowder is the highest, indicating that heat transfer enhancement effect when micron-order solid particles are added into the grinding fluid is not as obvious as that when nanoparticles are added.

Nanoparticles not only exert cooling effect with base grinding fluid on the grinding interface, and according to lubrication mechanism of nanoparticles, they also exert lubricating effect(Naskar, Singh, Choudhary, & Paul, 2018; Sen, Mia, Krolczyk, Mandal, & Mondal, 2019; Uysal, 2017). Specific grinding energy in the grinding process reflects lubrication state at the grinding interface. Specific grinding energy E_e of MQL grinding with four different nanofluids are calculated as shown in Fig. 15 according to tangential grinding force and grinding parameters in the grinding force.

From Fig. 15, specific grinding energy in the MQL grinding process with MoS_2 nanofluid is the lowest, being 47 J/mm^3. Energy needed to consume unit volume of material under the effect of MoS_2 nanoparticles at the grinding interface is the lowest, indicating that the lubricating effect is the best, mainly because MoS_2 nanoparticles have low friction coefficient as a solid lubricating material and they can improve lubrication and anti-friction characteristics(Koshy, Rajendrakumar, & Thottackkad, 2015; SHEN, MALSHE, KALITA, & SHIH, 2008; Verma, Jiang, Abu Safe, Brown, & Malshe, 2008); specific grinding energy of carbon nanotube and ZrO_2 nanoparticles namely their lubricating effects are second and third to MoS_2 nanoparticles; specific grinding energy obtained through ND micropowder is the highest, indicating that particle size of added solid particles can exert an influence on lubricating effect at the grinding interface. Grinding force is large when micron-order solid particles are used due to their large particle size, so its lubricating effect on the grinding interface is not as obvious as nanoparticles(Choi et al., 2011; Hwang, Park, Lee, & Jung, 2006).

Figure 15. Specific grinding energy in nano particle jet MQL

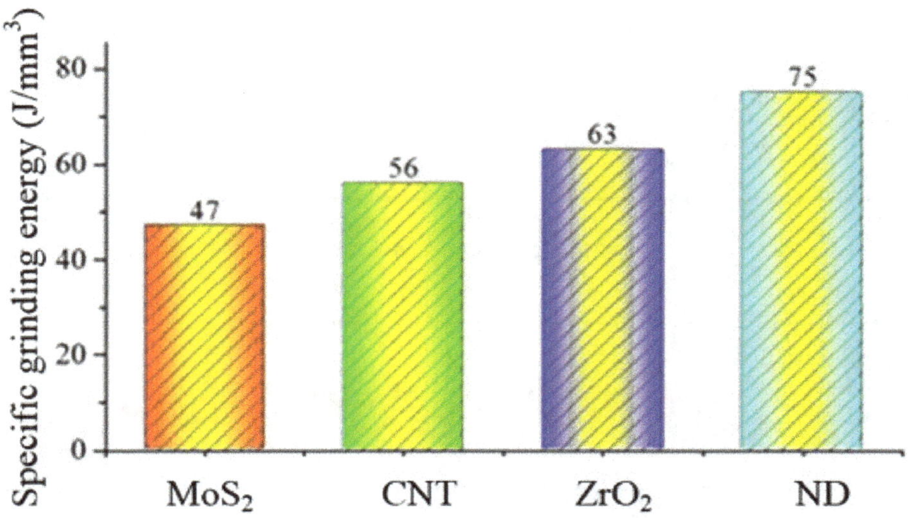

After grinding machining, workpiece surface quality is analyzed. Surface morphologies measured through macroscopic observation are not much different, so profile curves of workpiece surface after grinding are observed from a microscopic scale using a roughness measuring instrument and then compared. Surface profile curves of four workpieces under MQL grinding conditions with different nanofluids are shown in Fig. 16.

Grinding textures can be clearly observed from surface profile curves measured along the direction perpendicular to grinding direction. From an analysis of texture fluctuation, crest and valley distribution on surface profile curve (Fig. 16 (a)) obtained using MoS_2 nanoparticles is quite uniform, microscopic unevenness of each profile unit fluctuates little and surface profile is quite smooth; crest and valley on surface profile curves (Fig. 16 (b) and (c)) obtained through MQL grinding with carbon nanotube and ZrO_2 nanoparticles become gradually distinct, indicating that lubricating effects of carbon nanotube and ZrO_2 nanoparticles on the grinding interface ae not as good as that of MoS_2 nanoparticles; surface profile curve (Fig. 16 (d)) obtained using ND micropowder experiences large fluctuation, surface quality seriously degrades when compared with other three kinds of nanofluid, and this means that lubricating effect of ND micropowder on the grinding interface is not as good as nanoparticles. Based on obtained surface profile curves, surface roughness values of four workpieces can be calculated as shown in Fig. 17.

Workpiece surface roughness R_a value and R_z value can directly reflect workpiece surface quality(Abdulquadir & Adeyemi, 2008; Adibi, Esmaeili, & Rezaei, 2018; Alves, Barros, Trajano, Ribeiro, & Moura, 2013). According to Fig. 17, surface roughness R_a value and R_z value obtained using MoS_2 nanoparticles are both minimum values among four solid particles, respectively being 0.3μm and 2.45 μm; surface roughness obtained using carbon nanotube is larger than that using MoS_2, indicating that the workpiece surface is not as smooth as that using MoS_2; workpiece surface roughness obtained using ZrO_2 is larger than those using the abovementioned two solid particles, so different types of nanoparticles have different influences on workpiece surface quality; surface roughness obtained using

Figure 16. Surface profile curve in nano particle jet MQL

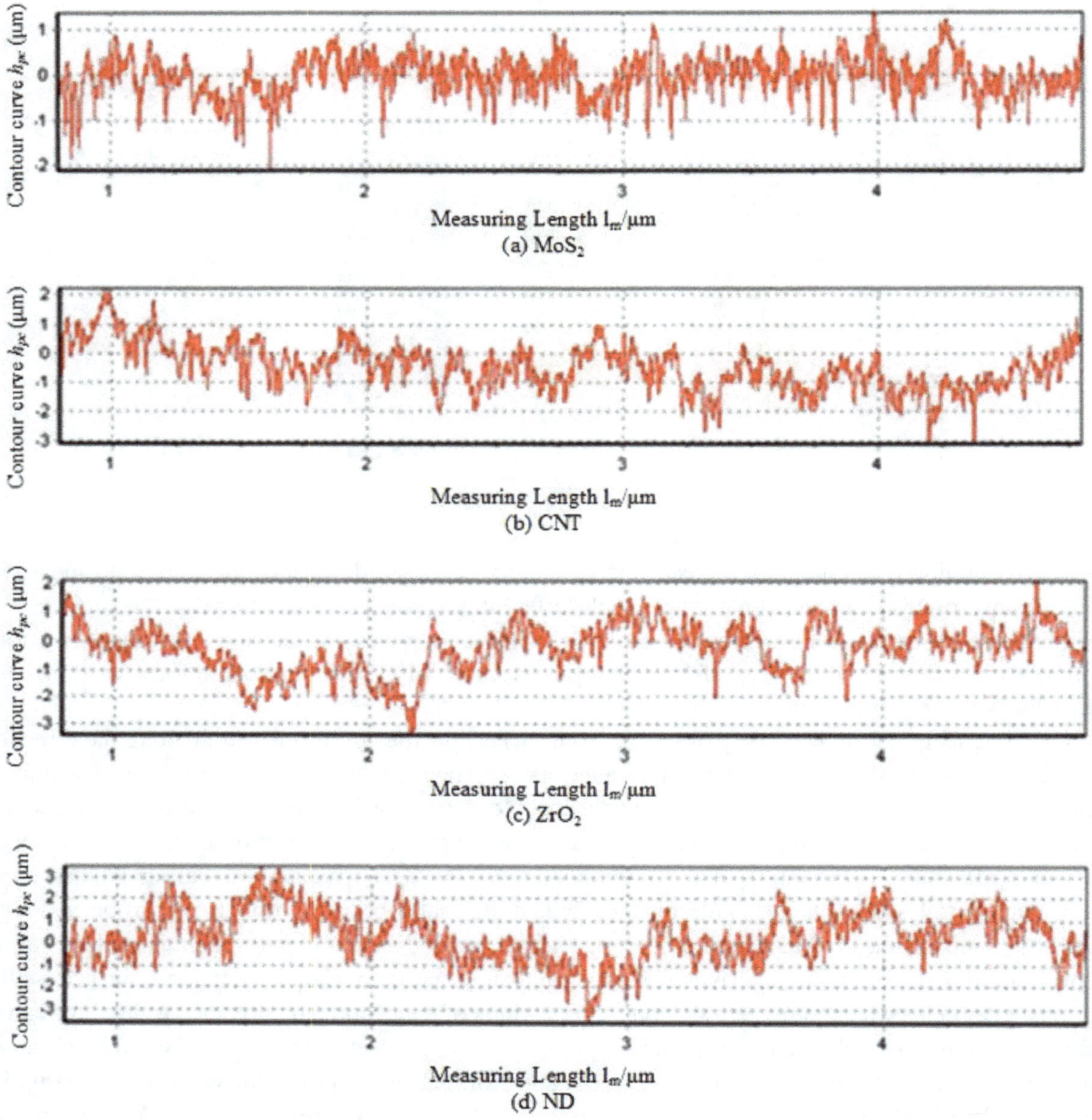

ND micropowder is the maximum, re-verifying that lubricating effect of micron-order solid particles is not as good as nanoparticles.

According to grinding experimental results when four kinds of solid particle are added into the grinding fluid, ND micropowder taken as the control in this experimental group, cooling and lubricating effects of micron-order and nanoscale particles are experimentally verified. In the aspect of cooling effect, heat transfer enhancement performances of three kinds of nanoparticle are ranked as carbon nanotube,

Figure 17. Surface roughness in nano particle jet MQL

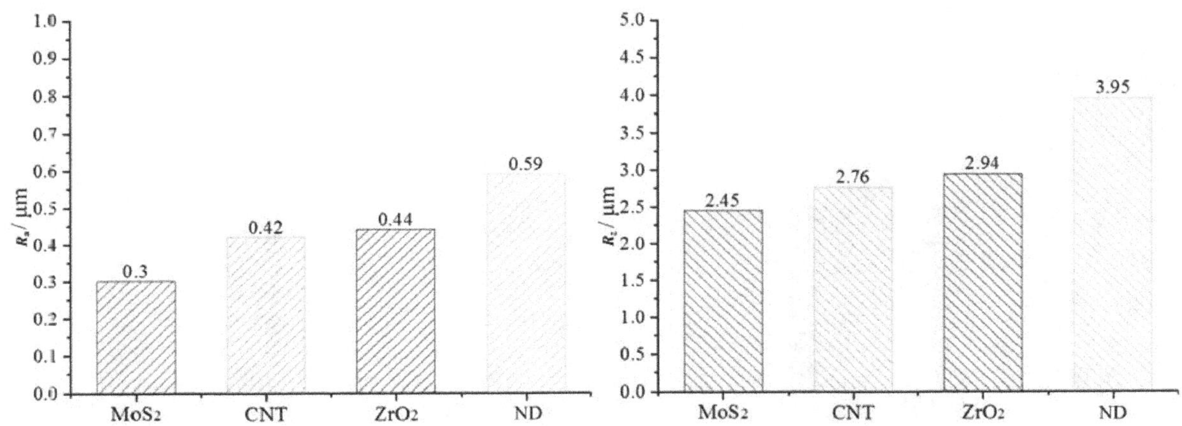

MoS_2 and ZrO_2 in a descending order; in the aspect of lubricating effect, lubricating properties of three kinds of nanoparticle are ranked as MoS_2, carbon nanotube and ZrO_2 in a descending order. Cooling and lubricating properties of MoS_2 and carbon nanotube are comprehensively evaluated by combining workpiece surface quality. It's believed that MoS_2 can obtain higher surface quality while enhancing heat transfer, so cooling and lubricating effect of MoS_2 nanoparticles is the optimal in this experimental group.

6.5.3 Influences of Nanoparticles Volume Concentration on Cooling and Lubrication Property

According to conclusions of Experiment II, further experimental study and analysis of influence of MoS_2 nanoparticles on cooling and lubricating effect in the NMQL grinding are conducted based on design of experimental program III. According to schematic design of Experiment III, MoS_2 nanoparticles are divided into three portions and added into the same grinding fluid (Accu-lube/lb-2000 vegetable oil) to prepare nanofluids with 1%, 2% and 3% volume concentrations for the grinding experiment. Same as experimental methods of Experiment I and Experiment II, maximum temperature rise and tangential grinding force are measured when a steady state is reached in the grinding process. Average values measured for multiple times are taken, and measured results are shown in Table 7.

According to maximum temperature rise θ_{max} and tangential grinding force F_t measured through the experiment, volume concentration of MoS_2 nanoparticles increases from 1% to 3%, temperature rise at the grinding interface doesn't decline as nanoparticle volume concentration increases, but instead, it firstly declines and then increases, indicating that nanofluids of different volume concentrations have

Table 7. Grinding temperature rise and tangential grinding force in MoS_2 nano particle jet MQL

MoS_2 Volume concentration	**1%**	**2%**	**3%**
Grinding temperature rise θ_{max} / °C	258	239	268
Tangential grinding force F_t /N	27	25	30

Figure 18. Energy proportionality coefficient under three nanoparticle volume concentrations

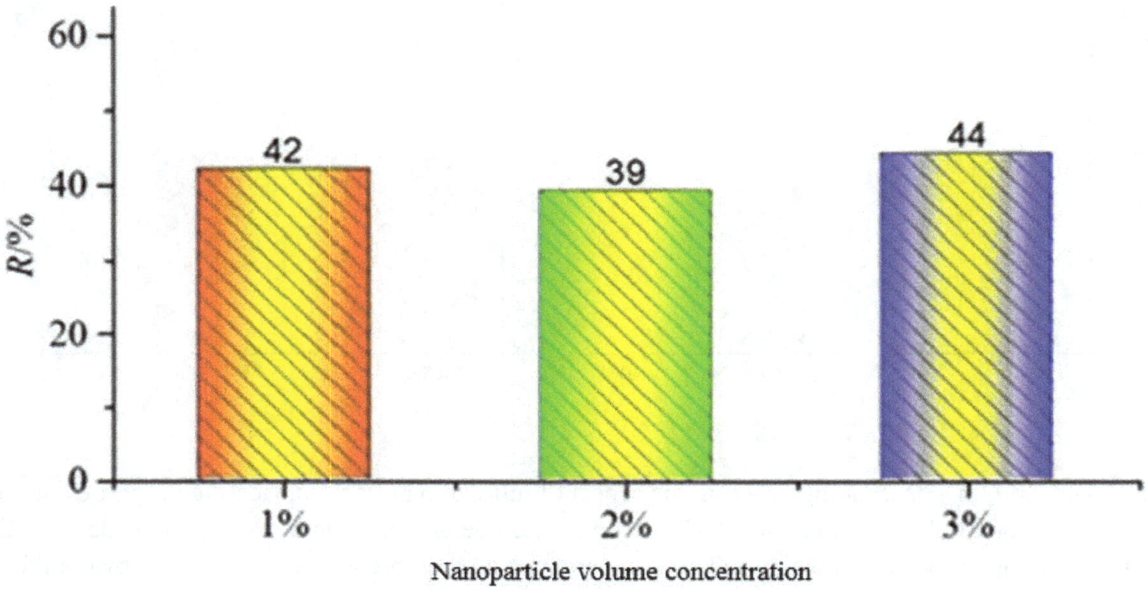

different heat transfer enhancement performances and different influences on cooling and heat transfer on the grinding interface, so temperature rises on workpiece surface are also different. Maximum temperature rise θ_{max} and tangential grinding force F_t are substituted into equation (4-51) to obtain energy proportionality coefficients R of MoS_2 nanoparticles with three different volume concentrations are shown in Fig. 18.

As shown in Fig. 18, energy proportionality coefficients are different under the influence of nanoparticle volume concentration. As volume concentration increases from 1% to 3%, energy proportionality coefficient firstly declines and then increases but the fluctuation is minor. Energy proportionality coefficient obtained under 2% volume concentration is the minimum, indicating that heat quantity under convective heat transfer effect of nanofluid is the largest with the best heat transfer enhancement effect. When volume concentration increases from 2% to 3%, energy proportionality coefficient increases, indicating that heat transfer coefficient of nanofluid declines as the concentration increases, mainly because nanofluid viscosity presents nonlinear growth as the volume concentration increases. Too high viscosity weakens nanofluid mobility on the grinding interface, reduces heat quantity transferred to nanofluid, lowers the heat transfer coefficient and inhibits heat transfer enhancement performance of nanofluid.

MoS_2 nanofluids of three volume concentrations are used for grinding, and power consumption and lubrication conditions on the grinding interface are measured. Specific grinding energy E_e and friction coefficient μ are calculated and compared as shown in Fig. 18 according to measured grinding force.

According to Fig. 19, as nanoparticle volume concentration increases from 1% to 3%, both specific grinding energy E_e and friction coefficient μ present firstly declining and then rising trend. When volume concentration is 2%, specific grinding energy E_e is the lowest, indicating that when nanoparticle volume concentration is 2%, energy consumed to remove unit volume of material at the grinding interface is the

Figure 19. Specific grinding energy and friction coefficient under three volume concentrations

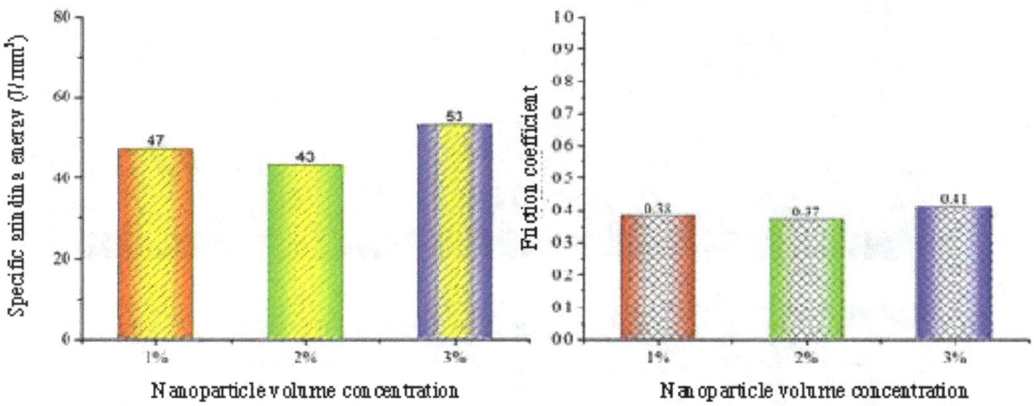

least. The workpiece is easy to grind and friction coefficient μ is the lowest, so lubricating state at the grinding interface is good.

The influence of nanoparticle volume concentration on cooling and lubricating effect is directly reflected by workpiece surface quality after grinding. Workpiece surface morphologies after grinding when volume concentrations of nanoparticles in the grinding fluid are 1%, 2% and 3%, measured through SEM, are shown in Fig. 20.

Influence results of different nanoparticle volume concentrations on workpiece surface morphology are observed from Fig. 20 and all surface textures are clear and orderly. Based on the observation of surface morphologies of three workpieces, texture is clear with smooth surface in Fig. 20 (b). Textures in Fig. 20 (a) and (c) are rough with slightly large texture unevenness, indicating that surface quality is influenced by nanoparticle volume concentration, and the surface quality is the highest under 2% volume concentration. Surface profile curves obtained through grinding under 1%, 2% and 3% nanoparticle volume concentrations, measured using roughness measuring instrument, are shown in Fig. 21.

It can be observed from Fig. 21 that fluctuation range of workpiece surface profile curves after grinding under three nanoparticle volume concentrations is small, where surface profile curve obtained under 2% nanoparticle volume concentration has higher evenness than other two planes. There are many crests and valleys on profile curves of the other two planes with slightly declining surface quality, indicating that surface quality under 2% nanoparticle volume concentration is relatively high, and obtained workpiece surface roughness values under three nanoparticle volume concentrations are shown in Fig. 22.

According to the analysis of surface roughness values, measured R_a value and R_z value are small and difference roughness values among three workpiece surfaces are small, indicating that surface quality obtained through NMQL grinding is high. Different nanoparticle volume concentrations have a minor influence on workpiece surface quality. Surface roughness R_a value obtained under 2% nanoparticle volume concentration is 0.283 μm which reduces by 5.3% and 14.5% when compared with those under 1% and 3% nanoparticle volume concentrations, and surface roughness R_z value is 2.424 μm, which reduces by 0.9% and 16.7% respectively. Therefore, the workpiece surface quality obtained through NMQL grinding under 2% nanoparticle volume concentration is the highest.

Figure 20. Surface morphologies of workpieces under three nanoparticle volume concentrations: (a) 1%; (b) 2%; (c) 3%

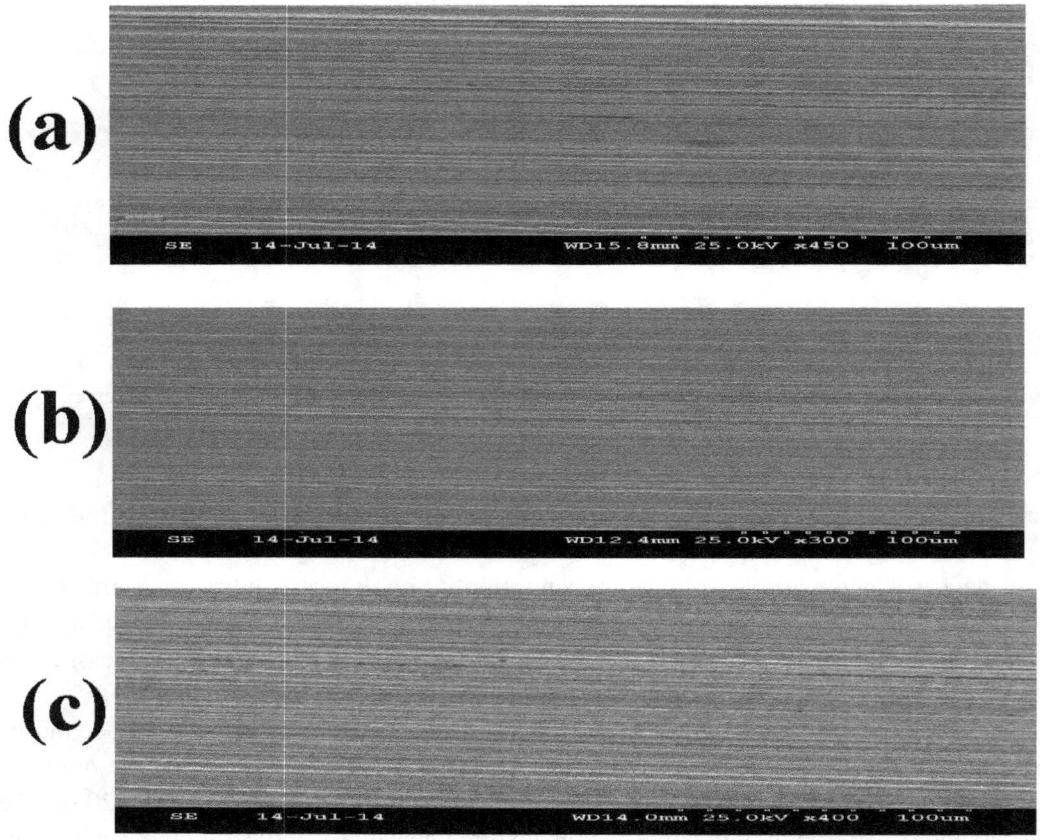

NMQL grinding experiment is carried out under 1%, 2% and 3% volume concentrations of MoS_2 nanoparticles. Three working conditions are experimented and analyzed from three aspects: cooling effect, lubricating effect and workpiece surface quality, and the following conclusions are drawn: heat transfer enhancement and lubricating performances of nanofluid are not strengthened as nanoparticle volume concentration increases with a certain deviation from simulation result, mainly because the influence of nanofluid viscosity on heat transfer effect is neglected. Analysis results are verified according to measured workpiece surface qualities. It's believed that nanoparticle volume concentration has a bearing on cooling and lubricating effect, and the effect is satisfactory under 2% MoS_2 volume concentration and the acquired workpiece surface quality is the highest. Volume concentration with the optimal cooling and lubricating effect can be further refined and investigated.

Figure 21. Surface profile curves under three nanoparticle volume concentrations

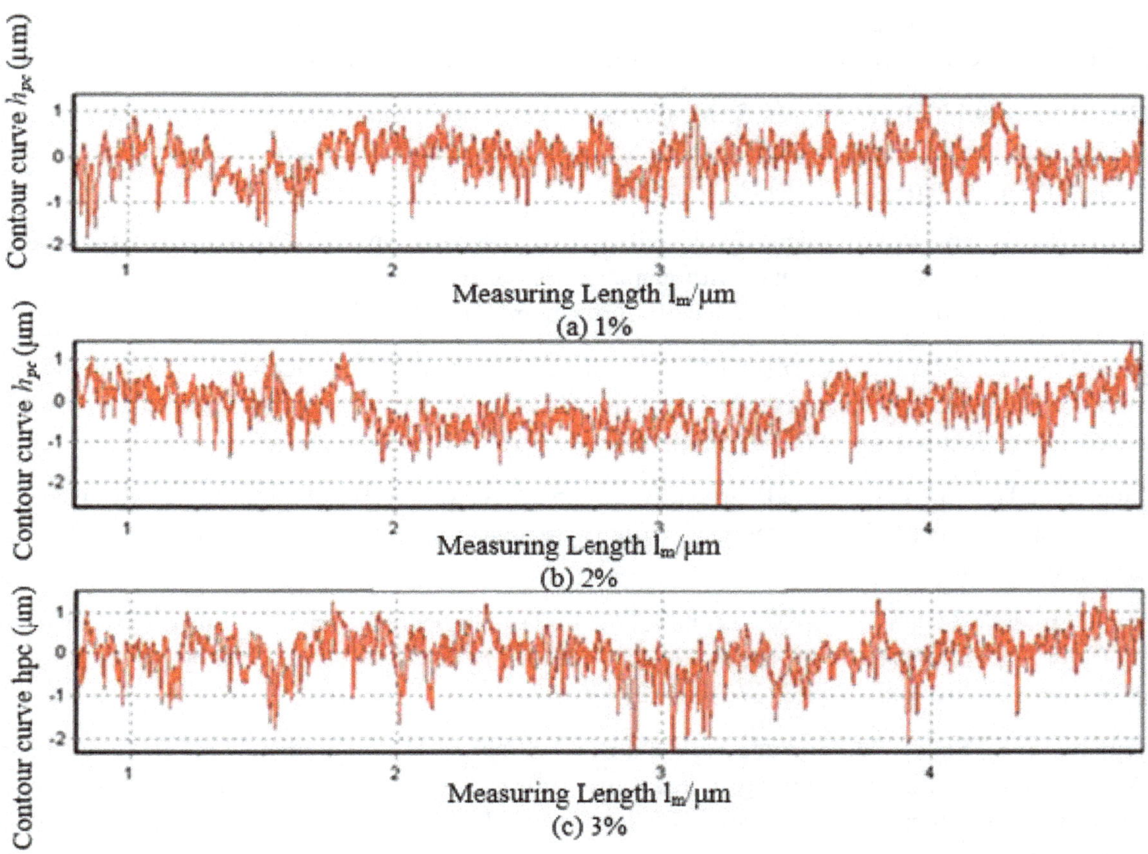

Figure 22. The surface roughness values under three nanoparticle volume concentrations

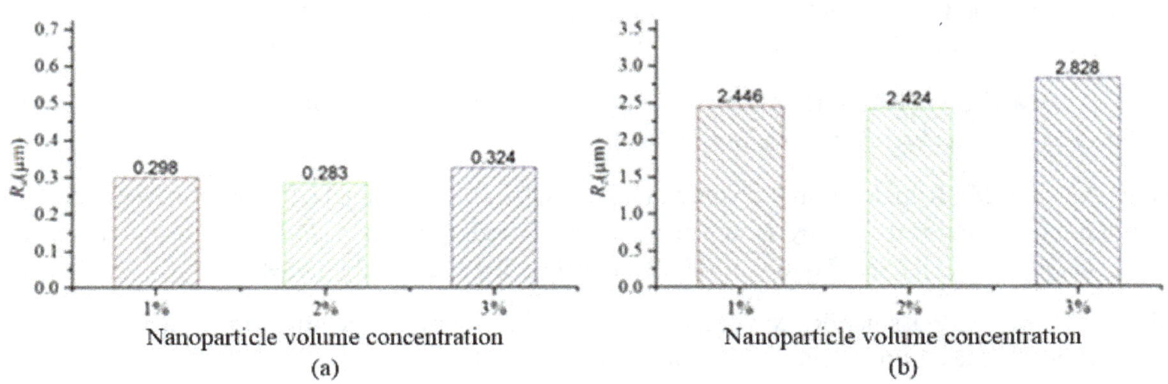

REFERENCES

Abdulquadir, B. L., & Adeyemi, M. B. (2008). Evaluations of vegetable oil-based as lubricants for metal-forming processes. *Industrial Lubrication and Tribology*, *60*(5), 242–248. doi:10.1108/00368790810895178

Adibi, H., Esmaeili, H., & Rezaei, S. M. (2018). Study on minimum quantity lubrication (MQL) in grinding of carbon fiber-reinforced SiC matrix composites (CMCs). *International Journal of Advanced Manufacturing Technology*, *95*(9-12), 3753–3767. doi:10.100700170-017-1464-x

Alves, S. M., Barros, B. S., Trajano, M. F., Ribeiro, K. S. B., & Moura, E. (2013). Tribological behavior of vegetable oil-based lubricants with nanoparticles of oxides in boundary lubrication conditions. *Tribology International*, *65*, 28–36. doi:10.1016/j.triboint.2013.03.027

Awale, A. S., Srivastava, A., Vashista, M., & Yusufzai, M. Z. K. (2019). Influence of minimum quantity lubrication on surface integrity of ground hardened H13 hot die steel. *International Journal of Advanced Manufacturing Technology*, *100*(1-4), 983–997. doi:10.100700170-018-2777-0

Bönnemann, H., Botha, S. S., Bladergroen, B., & Linkov, V. M. (2010). Monodisperse copper- and silver-nanocolloids suitable for heat-conductive fluids. *Applied Organometallic Chemistry*, *19*(6), 768–773. doi:10.1002/aoc.889

Choi, C., Jung, M., Choi, Y., Lee, J., & Oh, J. (2011). Tribological Properties of Lubricating Oil-Based Nanofluids with Metal/Carbon Nanoparticles. *Journal of Nanoscience and Nanotechnology*, *11*(1), 368–371. doi:10.1166/jnn.2011.3227 PMID:21446457

Dambatta, Y. S., Sayuti, M., Sarhan, A. A. D., Bin Ab Shukor, H., Derahman, N. A. B., & Manladan, S. M. (2019). Prediction of specific grinding forces and surface roughness in machining of AL6061-T6 alloy using ANFIS technique. *Industrial Lubrication and Tribology*, *71*(2), 309–317. doi:10.1108/ILT-03-2018-0098

Ding, W. F., Zhao, B. A., Xu, J. H., Yang, C. Y., Fu, Y. C., & Su, H. H. (2014). Grinding behavior and surface appearance of (TiCp+TiBw)/Ti-6Al-4V titanium matrix composites. *Chinese Journal of Aeronautics*, *27*(5), 1334–1342. doi:10.1016/j.cja.2014.08.006

Guitouni, A., Chaieb, I., Ben Rhouma, A., & Ben Fredj, N. (2016). Effects of Jet Pressure on the Ground Surface Quality and CBN Wheel Wear in Grinding AISI 690 Nickel-Based Superalloy. *Journal of Materials Engineering and Performance*, *25*(11), 5055–5064. doi:10.100711665-016-2330-y

Guo, C., & Malkin, S. (1992). Heat transfer in grinding. In *Transport phenomena in food processing, first international conference proceedings*. CRC Press.

Haque, A. K. M. M., Kim, S., Kim, J., Noh, J., Huh, S., Choi, B., ... Jeong, H. (2018). Surface Modification of Graphene Nanoparticles by Acid Treatment and Grinding Process. *Journal of Nanoscience and Nanotechnology*, *18*(1), 645–650. doi:10.1166/jnn.2018.13928 PMID:29768890

Heisel, U., Lutz, D., Wassmer, R., & Walter, U. (1998). The minimum quantity lubricant technique and its application in the cutting process. *Machines and Metals Magazine*, *386*, 22–38.

Huang, W. T., Liu, W. S., Tsai, J. T., & Chou, J. H. (2018). Multiple Quality Characteristics of Nanofluid/ Ultrasonic Atomization Minimum Quality Lubrication for Grinding Hardened Mold Steel. *IEEE Transactions on Automation Science and Engineering*, *15*(3), 1065–1077. doi:10.1109/TASE.2017.2726000

Hwang, Y., Park, H. S., Lee, J. K., & Jung, W. H. J. C. A. P. (2006). *Thermal conductivity and lubrication characteristics of nanofluids*. Academic Press.

Incropera, F. P., Lavine, A. S., Bergman, T. L., & DeWitt, D. P. (2007). *Fundamentals of heat and mass transfer*. Wiley.

Jawahir, I. S., Brinksmeier, E., M'Saoubi, R., Aspinwall, D. K., Outeiro, J. C., Meyer, D., ... Jayal, A. D. (2011). Surface integrity in material removal processes: Recent advances. *Cirp Annals-Manufacturing Technology*, *60*(2), 603–626. doi:10.1016/j.cirp.2011.05.002

Jin, T., Stephenson, D. J., & Rowe, W. B. (2003). Estimation of the convection heat transfer coefficient of coolant within the grinding area. *Proceedings of the Institution of Mechanical Engineers. Part B, Journal of Engineering Manufacture*, *217*(3), 397–407. doi:10.1243/095440503321590550

Koshy, C. P., Rajendrakumar, P. K., & Thottackkad, M. V. (2015). Evaluation of the tribological and thermo-physical properties of coconut oil added with MoS_2 nanoparticles at elevated temperatures. *Wear*, *330*, 288–308. doi:10.1016/j.wear.2014.12.044

Krolczyk, G. M., Nieslony, P., Maruda, R. W., & Wojciechowski, S. (2017). Dry cutting effect in turning of a duplex stainless steel as a key factor in clean production. *Journal of Cleaner Production*, *142*, 3343–3354. doi:10.1016/j.jclepro.2016.10.136

Kulkarni, H. B., Nadakatti, M. M., Patil, M. S., & Kulkarni, R. M. (2017). A Review on Nanofluids for Machining. *Current Nanoscience*, *13*(6), 634–653. doi:10.2174/1573413713666170623094121

Li, B. K., Li, C. H., Zhang, Y. B., Wang, Y. G., Jia, D. Z., & Min, Y. (2016). Grinding temperature and energy ratio coefficient in MQL grinding of high-temperature nickel-base alloy by using different vegetable oils as base oil. *Chinese Journal of Aeronautics*, *29*(4), 1084–1095. doi:10.1016/j.cja.2015.10.012

Li, B. M., & Zhao, B. (2003). *Modern grinding technology*. Beijing: China Machine Press.

ManojKumar, K., & Ghosh, A. (2016). Assessment of cooling-lubrication and wettability characteristics of nano-engineered sunflower oil as cutting fluid and its impact on SQCL grinding performance. *Journal of Materials Processing Technology*, *237*, 55–64. doi:10.1016/j.jmatprotec.2016.05.030

Mechiri, S. K., Vasu, V., & Gopal, A. V. (2017). Investigation of thermal conductivity and rheological properties of vegetable oil based hybrid nanofluids containing Cu-Zn hybrid nanoparticles. *Experimental Heat Transfer*, *30*(3), 205–217. doi:10.1080/08916152.2016.1233147

Nam, J., & Lee, S. W. (2018). Machinability of Titanium Alloy (Ti-6Al-4V) in Environmentally-Friendly Micro-Drilling Process with Nanofluid Minimum Quantity Lubrication Using Nanodiamond Particles. *International Journal of Precision Engineering and Manufacturing-Green Technology*, *5*(1), 29–35. doi:10.100740684-018-0003-z

Naskar, A., Singh, B. B., Choudhary, A., & Paul, S. (2018). Effect of different grinding fluids applied in minimum quantity cooling-lubrication mode on surface integrity in cBN grinding of Inconel 718. *Journal of Manufacturing Processes*, *36*, 44–50. doi:10.1016/j.jmapro.2018.09.023

Ni, J., Yang, Y. F., & Wu, C. (2019). Assessment of water-based fluids with additives in grinding disc cutting process. *Journal of Cleaner Production*, *212*, 593–601. doi:10.1016/j.jclepro.2018.12.066

Niu, S., Qu, N. S., & Li, H. S. (2018). Investigation of electrochemical mill-grinding using abrasive tools with bottom insulation. *International Journal of Advanced Manufacturing Technology*, *97*(1-4), 1371–1382. doi:10.100700170-018-2043-5

Rabiei, F., Rahimi, A. R., & Hadad, M. J. (2017). Performance improvement of eco-friendly MQL technique by using hybrid nanofluid and ultrasonic-assisted grinding. *International Journal of Advanced Manufacturing Technology*, *93*(1-4), 1001–1015. doi:10.100700170-017-0521-9

Sen, B., Mia, M., Krolczyk, G. M., Mandal, U. K., & Mondal, S. P. (2019). Eco-Friendly Cutting Fluids in Minimum Quantity Lubrication Assisted Machining: A Review on the Perception of Sustainable Manufacturing. *International Journal of Precision Engineering and Manufacturing-Green Technology*. doi:10.100740684-019-00158-6

Setti, D., Sinha, M. K., Ghosh, S., & Rao, P. V. (2015). Performance evaluation of Ti-6Al-4V grinding using chip formation and coefficient of friction under the influence of nanofluids. *International Journal of Machine Tools & Manufacture*, *88*, 237–248. doi:10.1016/j.ijmachtools.2014.10.005

Shen, B., Malshe, A. P., Kalita, P., & Shih, A. J. (2008). *Performance of novel MoS_2 nanoparticles based grinding fluids in minimum quantity lubrication grinding*. Paper presented at the SME Technical Conference.

Simoes, R. P., Bianchi, E. C., Oikawa, M. H., de Aguiar, P. R., Droppa, R., & Canarim, R. C. (2014). Evaluation of Different Methods of Cooling-lubrication in Cylindrical Grinding of Advanced Ceramic Dip. *Materials Research-Ibero-American Journal of Materials*, *17*(5), 1201–1212. doi:10.1590/1516-1439.250713

Singh, H., Sharma, V. S., Singh, S., & Dogra, M. (2019). Nanofluids assisted environmental friendly lubricating strategies for the surface grinding of titanium alloy: Ti6Al4V-ELI. *Journal of Manufacturing Processes*, *39*, 241–249. doi:10.1016/j.jmapro.2019.02.004

Singh, P., Dureja, J. S., Singh, H., & Bhatti, M. S. (2019). Performance evaluation of coated carbide tool during face milling of AISI 304 under different cutting environments. *Materials Research Express*, *6*(5), 056546. doi:10.1088/2053-1591/ab0696

Singh, T., Dureja, J. S., Dogra, M., & Bhatti, M. S. (2018). Environment Friendly Machining of Inconel 625 under Nano-Fluid Minimum Quantity Lubrication (NMQL). *International Journal of Precision Engineering and Manufacturing*, *19*(11), 1689–1697. doi:10.100712541-018-0196-7

Uysal, A. (2017). Surface Roughness in Nanofluid Minimum Quantity Lubrication Milling of AISI 430 Ferritic Stainless Steel. *Journal of Testing and Evaluation*, *45*(3), 933–939. doi:10.1520/JTE20150417

Verma, A., Jiang, W. P., Abu Safe, H. H., Brown, W. D., & Malshe, A. P. (2008). Tribological Behavior of Deagglomerated Active Inorganic Nanoparticles for Advanced Lubrication. *Tribology Transactions*, *51*(5), 673–678. doi:10.1080/10402000801947691

Virdi, R. L., Chatha, S. S., & Singh, H. (2019). Experiment evaluation of grinding properties under Al_2O_3 nanofluids in minimum quantity lubrication. *Materials Research Express*, *6*(9), 096574. doi:10.1088/2053-1591/ab301f

Wang, Y., Li, C., Zhang, Y., Li, B., Yang, M., Zhang, X., ... Liu, G. (2016). Experimental evaluation of the lubrication properties of the wheel/workpiece interface in MQL grinding with different nanofluids. *Tribology International*, *99*, 198–210. doi:10.1016/j.triboint.2016.03.023

Wang, Y. G., Li, C. H., Zhang, Y. B., Yang, M., Li, B. K., Jia, D. Z., ... Mao, C. (2016). Experimental evaluation of the lubrication properties of the wheel/workpiece interface in minimum quantity lubrication (MQL) grinding using different types of vegetable oils. *Journal of Cleaner Production*, *127*, 487–499. doi:10.1016/j.jclepro.2016.03.121

Zeng, Q. R., Qin, Y., Chang, W. L., & Luo, X. C. (2018). Correlating and evaluating the functionality-related properties with surface texture parameters and specific characteristics of machined components. *International Journal of Mechanical Sciences*, *149*, 62–72. doi:10.1016/j.ijmecsci.2018.09.044

Zhang, X. P., Li, C. H., Zhang, Y. B., Jia, D. Z., Li, B. K., Wang, Y. G., ... Zhang, X. W. (2016). Performances of Al_2O_3/SiC hybrid nanofluids in minimum-quantity lubrication grinding. *International Journal of Advanced Manufacturing Technology*, *86*(9-12), 3427–3441. doi:10.100700170-016-8453-3

Zhu, D. H., Qi, Y., Yu, W., Chen, L. F., Wang, M. Z., & Xie, H. Q. (2018). Enhanced Thermal Conductivity for Graphene Nanoplatelets/Epoxy Resin Composites. *Journal of Thermal Science and Engineering Applications*, *10*(1), 011011. doi:10.1115/1.4036796

Chapter 7
Experimental Research on Grinding Temperature and Energy Ratio Coefficient in MQL Grinding Using Different Types of Vegetables Oils

ABSTRACT

Vegetable oil can be used as a base oil in minimal quantity of lubrication (MQL). This study compared the performances of MQL grinding by using castor oil, soybean oil, rapeseed oil, corn oil, sunflower oil, peanut oil, and palm oil as base oils. Results revealed that 1) castor oil-based MQL grinding yields the lowest grinding force but exhibits the highest grinding temperature and energy ratio coefficient; 2) palm oil-based MQL grinding generates the second lowest grinding force but shows the lowest grinding temperature and energy ratio coefficient; 3) viscosity significantly influences grinding force and grinding temperature to a greater extent than fatty acid varieties and contents in vegetable oils. Palm oil is the optimum base oil of MQL grinding, and this base oil yields 26.98 N tangential grinding force, 87.10 N normal grinding force, 119.6 °C grinding temperature, and 42.7% energy ratio coefficient.

7.1 INTRODUCTION

Grinding is one of the most basic and important techniques in the machining industry. The final precision and surface quality of most machine parts are determined by grinding technique. Grinding is currently the only machining technique applicable to most difficult-to-process materials(Jia, Li, Zhang, Zhang, & Zhang, 2014; Li & Zhao, 2003). During grinding, high energy consumption is necessary to eliminate unit material volume; thus, a high amount of heat is generated in the grinding zone. Grinding depth is also small; as a result, a larger specific grinding energy is produced by grinding than by cutting and milling(Li, Chen, & Chen, 2011). Heat disperses to cuttings, tools, and workpieces. However, only a small

DOI: 10.4018/978-1-7998-1546-4.ch007

proportion of grinding heat is removed by abrasive dust; heat is mostly transferred to grinding wheels and workpieces(Khan & Wuyi; Lefebvre, Lipinski, Vieville, & Lescalier, 2008; Vieira, Baptista, Parente, & Jorge, 2009). Grinding heat likely affects the surface quality and usability of workpieces. In particular, an excessively high energy density of a workpiece surface burns the workpiece and deteriorates surface integrity(Belentani et al., 2014; Chen, Fu, He, Zhu, & Zhang, 2017; Ding, Xu, Chen, Su, & Fu, 2010; Malkin & Guo, 2007). During grinding, the grinding zone is often cooled and lubricated by traditional cooling via flooding and lubrication. A significant amount of grinding fluid is poured into the grinding zone(Jin & Stephenson, 2008) at usually 60 L/min per unit width of the grinding wheel.

With advances in grinding technology, cooling lubrication approaches have also been developed, improved, and optimized in terms of various aspects, such as energy conservation, emission reduction, eco-friendliness, and high efficiency(Zhang, Li, Zhang, Jia, & Zhang, 2015). However, flood grinding cannot comply with green manufacturing and sustainable development in terms of environmental protection, low carbon emission, processability, and economical efficiency. Therefore, environmentally friendly, highly efficient, and low-energy-consuming grinding fluid or new cooling lubrication techniques and equipment should be developed to achieve various technical effects of cooling lubrication, high workpiece processing quality and precision, prolonged service life of grinding wheel, and low environmental pollution caused by grinding fluid(Jia, Li, Wang, Zhang, & Hou, 2014; Zhang, Li, Jia, Zhang, & Zhang, 2015). Researchers proposed a technique. In dry grinding, grinding fluid is used, but high efficiency, high processing quality, prolonged grinding wheel service life, and a reliable grinding process can be maintained. Considering grinding wheel performance, machining tools, grinding dosage, and processing mode, we should integrate manufacturing techniques and materials science, as well as other subjects, such as information technology, electronics, and management. However, dry grinding is characterized by several disadvantages, such as high grinding force and grinding temperature; as a consequence, low processing precision, short grinding wheel service life, geometric errors in workpieces, and poor processing quality are obtained(Azarhoushang, Daneshi, & Lee, 2017; Huang, Liu, & Wu, 2016; Simoes et al., 2014).

In addition to dry grinding, MQL grinding is a green machining technique(Adibi, Esmaeili, & Rezaei, 2018). Hadad et al. (2012) investigated 100Cr6 grinding by using MQL technique with an overhead thermocouple to measure grinding temperature. Hadad et al. (2012) found that MQL consumes 7%–10% less energy than dry grinding. Dhar et al. (2007) reported that MQL generates 5%–15% smaller grinding force than dry grinding; thus, the service life of cutters is prolonged. Gaitondea et al.(2008) found that MQL achieves a higher workpiece surface processing quality than dry grinding. Mao et al. (2013) studied heat transfer on the workpiece surface during MQL grinding; in this method, the grinding zone is divided into four regions according to different microdroplet heat transfer mechanisms of workpiece surface temperature: non-boiling heat transfer region, nucleate-boiling heat transfer region, transition-boiling heat transfer region, and stable film-boiling heat transfer region. The heat transfer mechanism on the workpiece surface during MQL grinding is more credible than that of flood grinding. Tawakoli et al. (2009) discussed the effect of grinding parameters on workpiece surface quality. With optimum grinding fluid dosage and feed liquid parameters, MQL grinding achieves a greater workpiece surface quality but a lower tangential grinding force and specific grinding energy than flood grinding. Barczak et al. (2010) compared the grinding power, grinding force, grinding temperature, and surface roughness of MQL grinding with those of flood grinding and dry grinding. Silva et al. (2007) compared workpiece surface integrity, specific grinding energy, and grinding wheel wear via dry grinding, flood grinding, and MQL grinding.

Although MQL grinding exhibits some advantages over flood grinding, the former produces abundant oil mists in air; as a result, human health is threatened. In a previous study(Silva et al., 2007), the leakage and volatilization of grinding fluid cause pollution and inflict hazards to the environment and the human body. Therefore, vegetable oils have been used as base oils in MQL. Vegetable oils are characterized by a higher boiling point and molecular weight than mineral oils; thus, the wastes of the former are significantly reduced through atomization and gasification. Cetin et al. (2011) analyzed the grinding surface roughness of AISI 304L steel by using sunflower oil and rapeseed oil as base oils. Elmunafi et al. (2015) used castor oil as the base oil of MQL grinding to grind hardened stainless steel and subsequently explored grinding performances, such as surface roughness and grinding force. Le et al. (2013) used peanut oil as the base oil of MQL grinding to cut the difficult-to-process 9CrSi steel; Le et al. (2013) found that the processed workpiece is characterized by a relatively integrated surface roughness. Jain et al. (2010) aimed to replace the cutting fluid with non-edible vegetable mineral oils, such as rapeseed oil and Karanja tree oil, in metal cutting. Rahim et al. (2011a, 2011b, 2011c) investigated metal cutting by using palm oil and synthetic ester as MQL base oils and compared their corresponding grinding performances.

Although studies(Parash Kalita, Malshe, Kumar, Yoganath, & Gurumurthy, 2012; Khan, Mithu, & Dhar, 2009) have discussed and confirmed that vegetable oils can be used as base oils of MQL grinding and can increase production efficiency, service life of cutters, and workpiece surface processing quality(Guo et al., 2017; Kalita, Malshe, & Rajurkar, 2012; Lawal, Choudhury, & Nukman, 2013; Li et al., 2016; Mao et al., 2014; Rahim & Sasahara, 2011a, 2011b; Wang et al., 2016), only few researches have involved categories of vegetable oils, not to mention comparative and analytical researches on multiple kinds of vegetable oil under the same grinding conditions. This chapter mainly carries out an experimental study of multiple kinds of common vegetable oil. Multiple kinds of vegetable oil are used as base oils for MQL grinding of high-temperature nickel base alloy. Theoretical analysis and experimental study of grinding temperature and energy proportionality coefficient will be implemented, and then vegetable oils which are suitable to be MQL base oils will be selected out.

7.2 EXPERIMENTAL DESIGN

7.2.1 Experimental Equipment

Grinder used in the experiment is precision surface NC grinder KP-36, and its main technical parameters include spindle power 4.5 kW; maximum revolving speed 2000 r/min; workbench drive motor power 5 kW; grinding range 600×300 mm; corundum grinding wheel is used with dimensions of 300×20×76.2 mm and granularity of 80#, and the highest circular velocity is 50 m/s; Bluebe minimal quantity of oil supply system is used to atomize the nanofluid and input it into the grinding area; the experiment is carried out using certain grinding parameters as shown in Table 1.

Grinding force in the experimental process is measured using YDM-III99 three-way dynamometer and sampling frequency is 1 kHz. YDM-III99 dynamometer is placed at the proper position of magnetic workbench. The workpiece is clamped through a clamp installed on the dynamometer, grinding force generated by the workpiece grinded by grinding wheel is acquired through the dynamometer and loaded to computer for processing after passing through charge amplifier and signal acquisition unit, and 100 effective data points are randomly selected for calculation of average grinding force. Grinding tempera-

Table 1. Grinding process parameters

Grinding technological parameter	Value
Grinding way	**Surface grinding**
Grinding wheel velocity v_s (m/s)	30
Feed velocity v_w (mm/s)	50
Cutting depth a_p (μm)	10
MQL flow rate (mL/h)	50
MQL nozzle distance (mm)	12
MQL nozzle angle (°)	15
MQL air pressure (Mpa)	0.6

ture is measured using the clamp-type thermocouple method and thermocouple wire is clamped in the rectangular groove of two workpieces. In order to prevent the workpiece from affecting heat-conducting property of thermocouple wire, thermocouple wire is separated from the workpiece using a sheet mica in the rectangular groove, and the other end of the thermocouple wire is connected to signal acquisition unit which is connected to the computer. Temperature is measured in the whole grinding process and grinding temperature curve is drawn, and 100 data points are selected and their average grinding temperature is calculated. Schematic diagram of the whole measuring instrument is shown in Fig. 1. One-way analysis of variance is adopted for statistical data analysis, and confidence interval (CI) of cause and effect detection is 95%(Sullivan et al., 1998). Errors of grinding force and grinding temperature are 4 N and 5 °C respectively.

In each grinding experiment, in order to ensure measurement accuracy and controllability and ensure that each experimental condition is identical, grinding wheel will be dressed once data are measured for 60 times. Grinding wheel dressing parameters are shown in Table 2.

7.2.2 Experimental Material

Workpiece material used in the experiment is high-temperature nickel base alloy GH4169 as shown in Fig. 2, which has been mainly applied to fields like ocean, energy, aviation and mechanical and electrical products by virtue of good comprehensive performance. As high-temperature nickel base alloy GH4169 has low heat conductivity coefficient, grinding heat generated in the grinding process is large, which seriously impacts quality of the machined surface, so it's very important to carry out a grinding experimental study on high-temperature nickel base alloy GH4169. Researchers have carried out a large quantity of researches on nickel base alloy. Ding et al.(2013) studied residual stresses of difficult-to-machine materials including structural metallic structures like nickel base alloy in the grinding process; Dai et al.(2015) studied influences of high grinding speed of nickel base alloy GH4169 on chip formation and undeformed chip thickness; Chen et al.(2015) conducted an experimental study of high-efficiency deep grinding of nickel base alloy. MQL base oils are divided into castor oil, soybean oil, rapeseed oil, corn oil, peanut oil, palm oil and sunflower oil. Table 3 lists chemical compositions of high-temperature nickel base alloy GH4169 and Table 4 lists its performance parameters, where viscosities of seven vegetable oils are measured using brookfield DV2T viscosimeter under 25 °C, their contact angles are measured using a contact angle measuring instrument, and the two measuring instruments are shown in Fig. 3.

Figure 1. The experimental equipment in surface grinding process

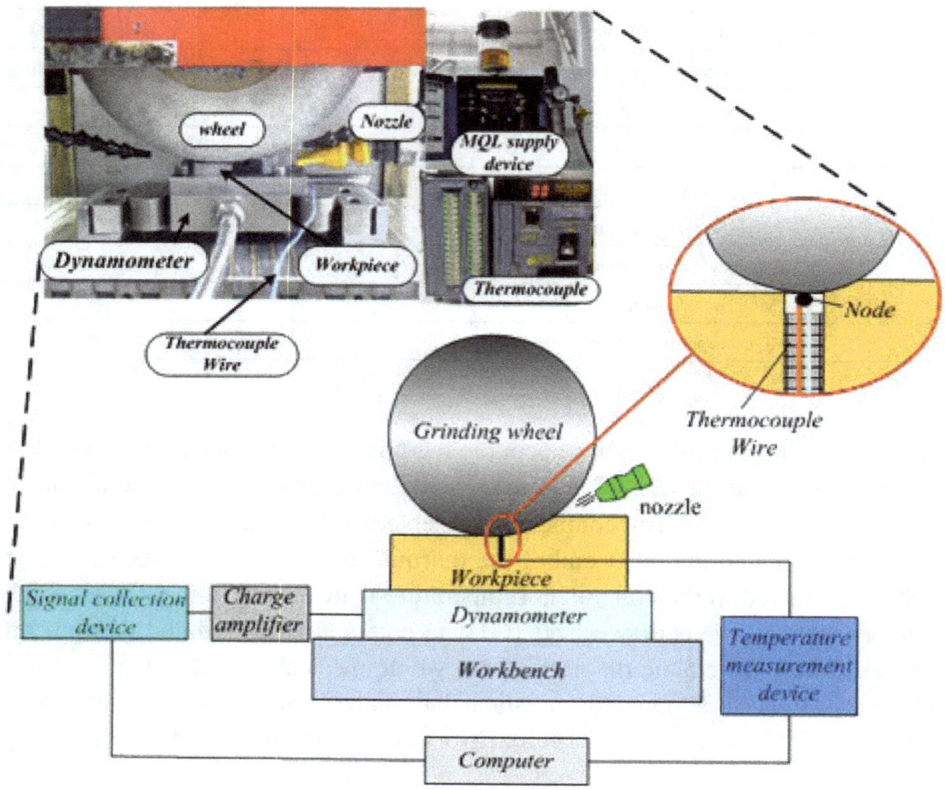

Table 5 lists basic properties of seven kinds of MQL base oil. Chip shapes and workpiece surface morphologies generated during the grinding process are shot using a scanning electron microscope (SEM) which is shown in Fig. 4.

Table 5 displays basic properties of seven kinds of vegetable oil used in MQL grinding. Viscosity of castor oil is 535.39 cp which is nearly 10 times of other vegetable oils, and convent of ricinoleic acid reaches as high as 90.85%, so castor oil separately classified into the first group; in addition, it can be found that soybean oil, corn oil, sunflower oil and peanut oil contain oleic acid, linoleic acid, palmitic acid and stearic acid, their total contents are basically approximate, no fatty acid has too high content, and their viscosity values are not that difference, so the four kinds of vegetable oil are classified into the second group; erucic acid content is high in rapeseed oil while contents of oleic acid, linoleic acid and stearic acid are relatively low, so rapeseed oil is classified into the third group; palmitic acid content in

Table 2. Dressing parameters of grinding wheel

Dressing item	Fixed PCD dresser of K-P36 grinder
Single-point dressing quantity (mm)	0.01
Transverse feed rate (mm/rev)	0.5
Number of stroke	25

Figure 2. Workpiece material

Table 3. Chemical compositions of GH4169

Element	Ni	C	Cr	Nb	Mo	Ti	Co	Al	Mn	Cu	Fe
Component (%)	50-55	0.08	17-21	5.25	3.1	0.96	1.0	0.95	0.35	0.3	Margin

Table 4. Performance parameters of GH4169 material

Temperature (°C)	25	100	200	300	400	500	600	700
Heat conductivity λ (W/m·K)	13.4	14.7	15.7	17.8	18.3	19.6	21.2	22.8
Specific heat c (J/kg·K)	435	458	467	481	494	515	539	573
Thermal expansion coefficient α (10^{-5}/K)	—	1.18	1.30	1.35	1.41	1.44	1.48	1.54
Density ρ(kg/m^3)	8240							
Elasticity modulus (GPa)	200							
Poisson's ratio	0.3							
Yield strength (MPa)	550							

palm oil is higher than those in the abovementioned vegetable oils, so is its viscosity value, so palmitic acid is separately classified into the fourth group.

7.2.3 Experimental Scheme

Seven vegetable oils—soybean oil, corn oil, peanut oil, sunflower oil, rapeseed oil, palm oil and castor oil—are used as base oils in MQL grinding experiment. Based a comparison of grinding force, grinding

Figure 3. Measuring instruments of viscosity and contact angle

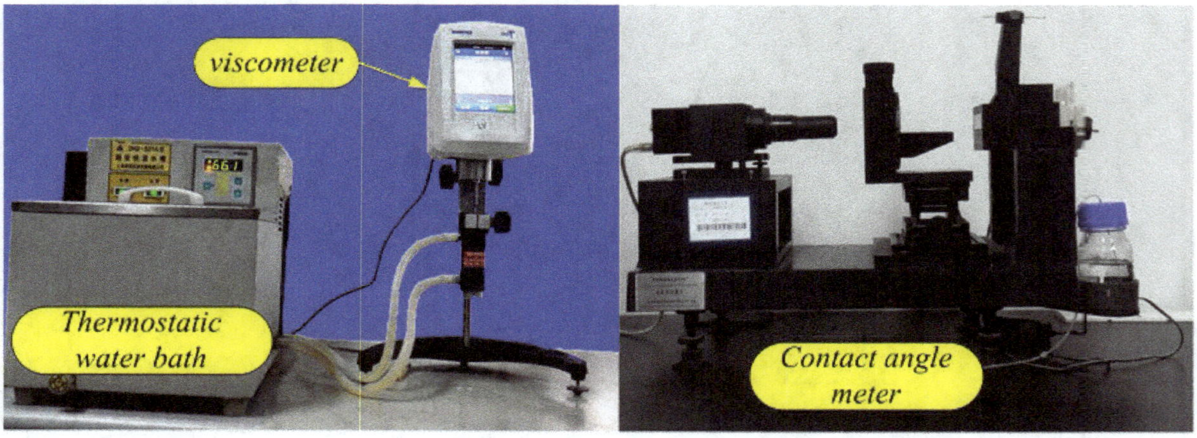

Table 5. The basic nature of the seven-trace lubricating base oils

Parameter	Castor oil	Soybean oil	Rapeseed oil	Corn oil	Peanut oil	Sunflower oil	Palm oil
Oleic acid (%)	0.82	23	5.92	32.55	40.84	28.76	36.8
Linoleic acid (%)	3.74	52.4	23.63	51.4	34.54	50.48	10.2
Palmitic acid (%)	0.72	8.9	2.63	1.94	11.92	11.24	45.1
Stearic acid (%)	0.64	3.8	11.72	13.27	4.3	6.48	4.8
Eicosenoic acid (%)	—	—	2.92	—	1.38	—	—
Erucic acid (%)	—	—	43.63	—	—	—	—
Vaccenic acid (%)	—	—	4.99	—	—	—	—
Linolenic acid (%)	—	10.6	3.34	—	—	—	—
Ricinoleic acid (%)	90.85	—	—	—	—	—	—
Saponification value (%)	181	192	193	190	192	189	193
Saturated fatty acid (%)	52	15	6	14	21	11	35
Monounsaturated fatty acid (%)	37	24	38	29	49	19	15
Polyunsaturated fatty acid (%)	11	61	36	54	30	70	50
Viacosity (cp)	540	420	51	47	53	51	62

temperature and energy proportionality coefficient, MQL grinding performances of the seven vegetable oils are investigated. The experimental program is shown in Table 6.

Figure 4. Chip and workpiece surface morphology shooting device

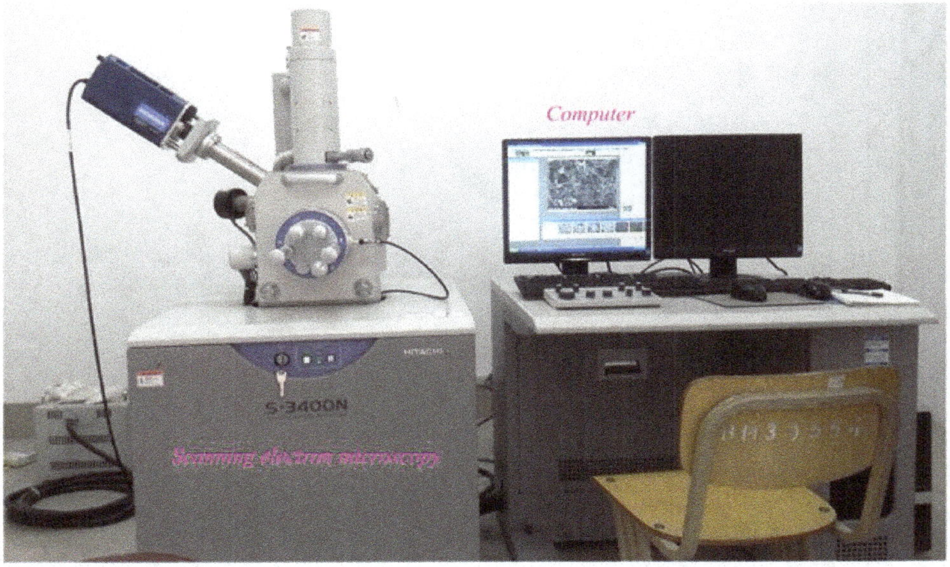

7.3 EXPERIMENTAL RESULTS

7.3.1 Grinding Force

Fig. 5 (Li et al., 2016) lists filtering images of grinding forces under MQL grinding with four kinds of vegetable oil, respectively being castor oil, peanut oil, rapeseed oil and palm oil.

As a very important technical parameter in the grinding process, grinding force can not only reflect grinding state in the machining process but also reflect lubricating state. Under certain grinding, normal cutting force, tangential cutting force and normal sliding frictional force on grinding surface are all constant values, so sliding frictional force will decline and tangential grinding force will present proportional decline when lubricating state changes.

Table 6. Grinding experimental program with seven kinds of vegetable oil

Experiment No.	MQL lubricating oil	Workpiece material	Cooling and lubricating conditions
1	Castor oil	Nickel base alloy	MQL
2	Soybean oil	Nickel base alloy	MQL
3	Rapeseed oil	Nickel base alloy	MQL
4	Peanut oil	Nickel base alloy	MQL
5	Corn oil	Nickel base alloy	MQL
6	Sunflower oil	Nickel base alloy	MQL
7	Palm oil	Nickel base alloy	MQL

Figure 5. Four kinds of vegetable oil MQL grinding force measurement signal image

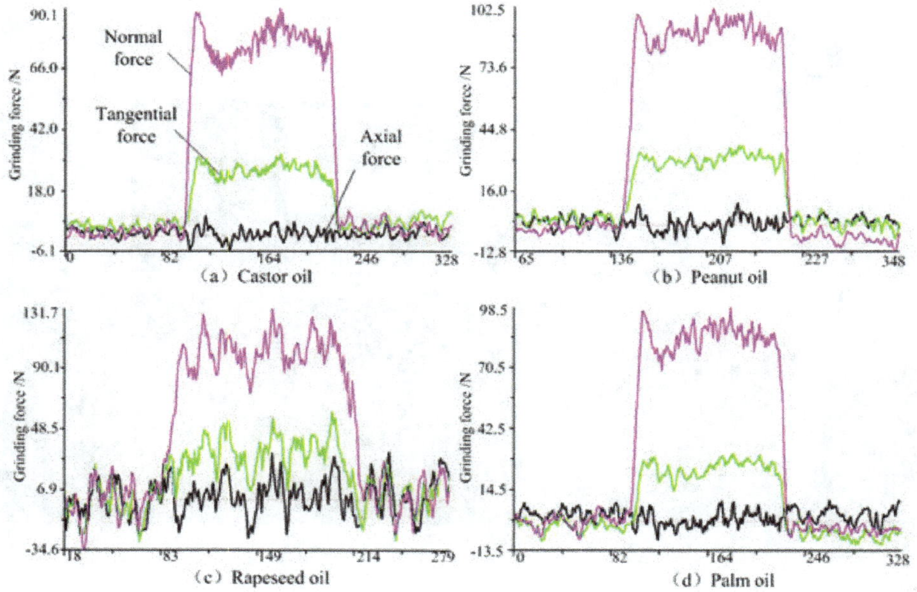

Fig. 6 (Li et al., 2016) shows experimental results of grinding forces under MQL grinding with seven kinds of vegetable oil. It can be seen that both tangential grinding force and normal grinding force of castor oil are the minimum, being 24.33 N and 76.83 N respectively; followed by palm oil, respectively being 26.98 N and 87.10 N; those of rapeseed oil are 31.88 N and 102.53 N; tangential grinding forces of soybean oil, sunflower oil, corn oil and peanut oil are 31.54 N, 31.34 N, 33.91 N and 31.10 N respectively and their normal grinding forces are 104.24 N, 101.88 N, 99.85 N and 91.37 N respectively. Compared with castor oil, tangential grinding forces and normal grinding forces of palm oil and rapeseed oil increase 11.0% and 13.4% and 33.5% and 131.0% respectively; normal grinding forces of soybean oil, sunflower oil, corn oil and peanut oil increase by 35.7%, 32.6%, 30.0% and 18.9% respectively, and their tangential grinding forces increase by 29.6%, 28.8%, 39.4% and 27.8% respectively; relative to palm oil, tangential grinding forces of rapeseed oil, soybean oil, sunflower oil, corn oil and peanut oil are elevated by 18.2%, 16.9%, 16.2%, 25.7% and 15.3% respectively, and their normal grinding forces are elevated by 17.7%, 19.7%, 17.0%, 14.6% and 4.9% respectively.

7.3.2 Grinding Temperature

Grinding temperature generated in the grinding process has an important influence on the machined workpiece, so it should be effectively controlled. Among factors generating grinding temperature, grinding parameters can exert significant influences on grinding temperature, but effective cooling and heat transfer effect of grinding fluid is also an essential factor.

Fig. 7 (Li et al., 2016) shows change curves of grinding temperature with time in the MQL grinding process with seven kinds of vegetable oil. It can be seen that the highest grinding temperature is

Figure 6. MQL grinding force for seven kinds of vegetable oil

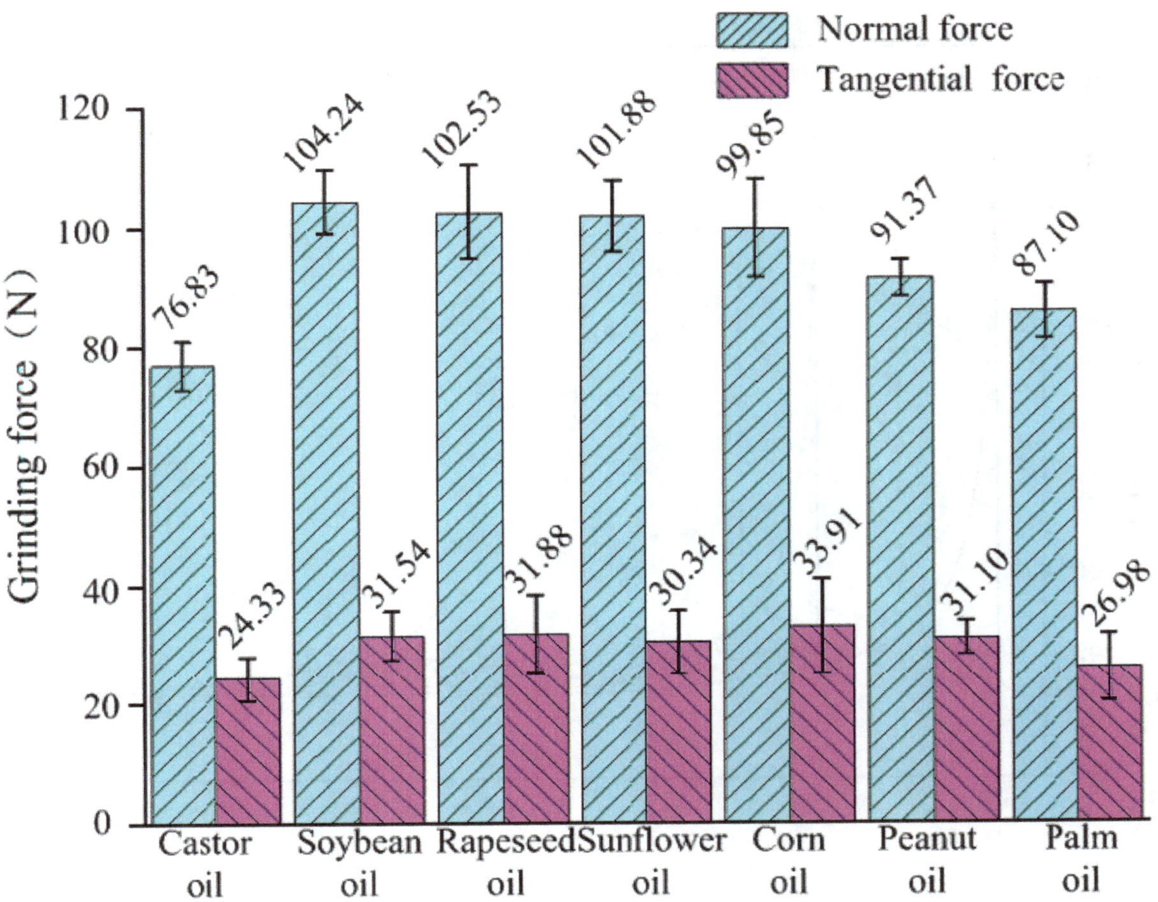

obtained when castor oil is used for MQL grinding while the lowest grinding temperature is obtained when palm oil is used.

Fig. 8 (Li et al., 2016) shows the highest temperature temperatures under MQL grinding with seven kinds of vegetable oil. Obviously, the grinding temperature under MQL grinding with castor oil is the highest, being 176 °C; the lowest grinding temperature under MQL grinding with palm oil is 119.6 °C; grinding temperature of rapeseed oil is 143.4 °C; grinding temperatures of soybean oil, corn oil, sunflower oil and peanut oil are 143.5 °C, 139.6 °C, 139.3 °C and 138.4 °C respectively.

7.3.3 Thermal Distribution Ratio Transferred into Workpieces

Energy distribution in workpiece is a critical factor calculating grinding temperature and controlling thermal injury. Studying energy proportionality coefficient of workpiece is also an important basis for determining transfer of grinding temperature and cooling effect of grinding fluid.

Proportionality coefficients of energy transferred to workpiece under vegetable oil MQL conditions are shown in Fig. 9 (Li et al., 2016). Castor oil obtains the highest energy proportionality coefficient,

Figure 7. Seven kinds of vegetable oil grinding temperature graph

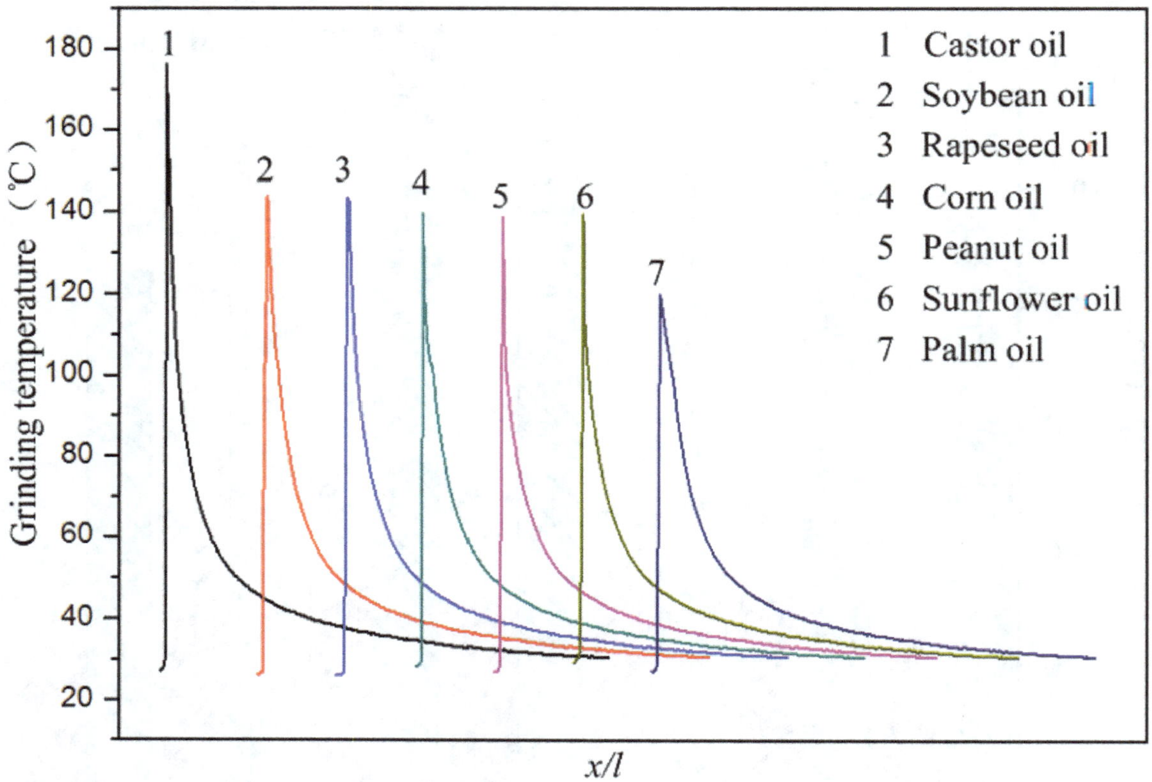

being 69.3%; followed by soybean oil, being 49.3% which decreases by 20% relative to castor oil; that of palm oil is the lowest, only being 42.7% which declines by 26.6% relative to castor oil and by 6.6% relative to soybean oil; those of rapeseed oil, sunflower oil, peanut oil and corn oil are 49.1%, 48.8%, 48.1% and 47.8% respectively, which decrease by 20.2%, 20.5%, 21.2% and 21.5% respectively relative to castor oil, by 0.2%, 0.5%, 1.2% and 1.5% respectively relative to soybean oil, and by 6.4%, 6.1%, 5.4% and 5.1% respectively relative to palm oil.

7.4 RESULTS AND DISCUSSION

7.4.1 Grinding Force

According to data of tangential and normal grinding forces, lubricating effect is satisfactory when castor oil is used for MQL grinding, because lubricating oil form formed by castor oil in the grinding area has good anti-friction & anti-wear characteristics and bearing capacity as shown in Fig. 10. This phenomenon can be explained from three aspects:

Figure 8. Seven kinds of vegetable oil MQL grinding maximum grinding temperature

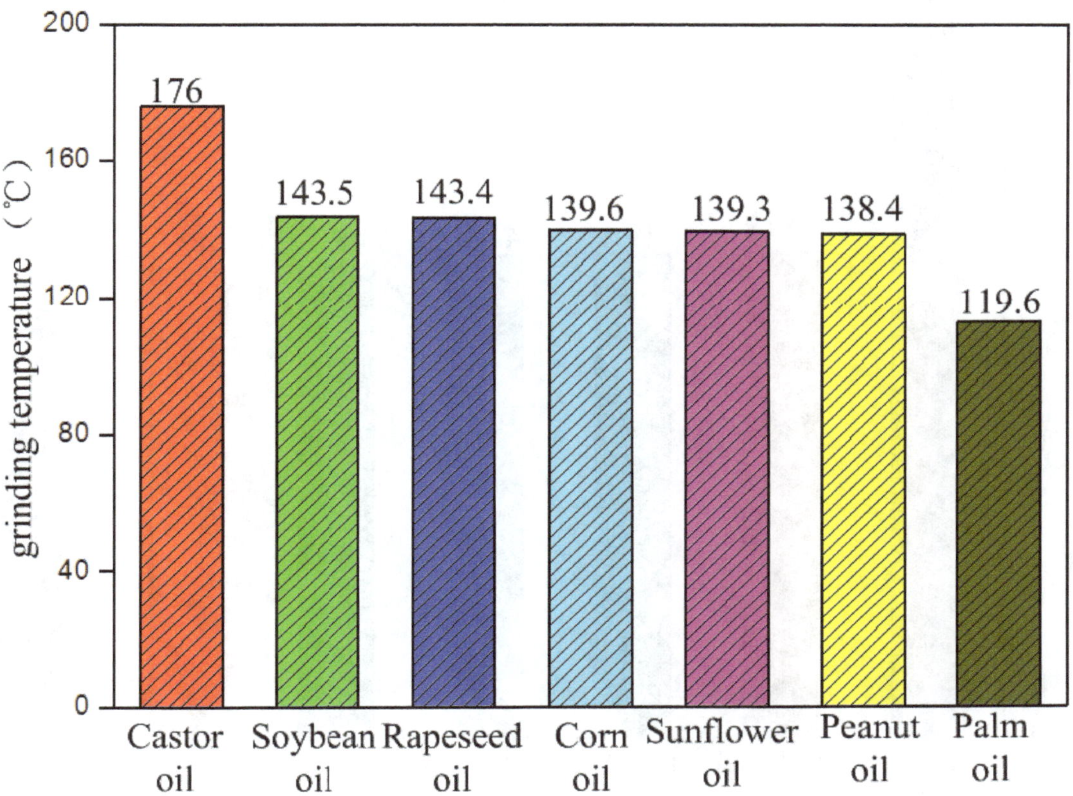

(1) Influence of vegetable oil viscosity. Viscosity of castor oil is 535.39 cp under 25 °C, indicating that castor oil is very viscous. In the MQL grinding process, mobility between high-viscosity castor oil molecules is poor, a compact protective film is formed after castor oil is sprayed into the grinding area, and this protective film has high anti-friction & anti-wear and bearing capacity, reduces friction between grinding wheel and workpiece and exerts good lubricating effect(Bhaumik, Datta, & Pathak, 2017; Boran & Tamas, 2014; Wang, Wan, Lu, Zhang, & Tang, 2018), so grinding force is greatly reduced.

(2) The influence of physical adsorption film formed by fatty acid. According to formation principle of lubricating oil film and separation degree of surfaces of friction pairs, lubricating state can be divided into boundary lubrication, mixture lubrication and hydrodynamic lubrication. Grinding fluid is kept under boundary lubricating state due to the effect of impulsive load in the grinding process. Formation of physical adsorption film is related to vegetable oil property and carbon chain length in fatty acid. On the one hand, some polar atoms like S, O, N and P or polar groups like OH, -COOH, -COOR, -COR, -CN, -CHO, -NCS, $-NH_2$, $-NHCH_3$, $-NH_3$ and -NROH are contained in vegetable oil, which have strong affiliative activity with workpiece surface and go through physical adsorption with workpiece surface molecules through VDW force to form physical adsorption film, and grinding force is reduced as physical adsorption film has anti-friction and anti-wear effects; on the other hand, vegetable oil contains various fatty acids, and carbon chain length of fatty acid molecules can exert an important influence on adsorption characteristics of physical adsorption film: the longer the carbon chain length, the better the

Figure 9. Seven kinds of vegetable oil MQL grinding energy ratio coefficient

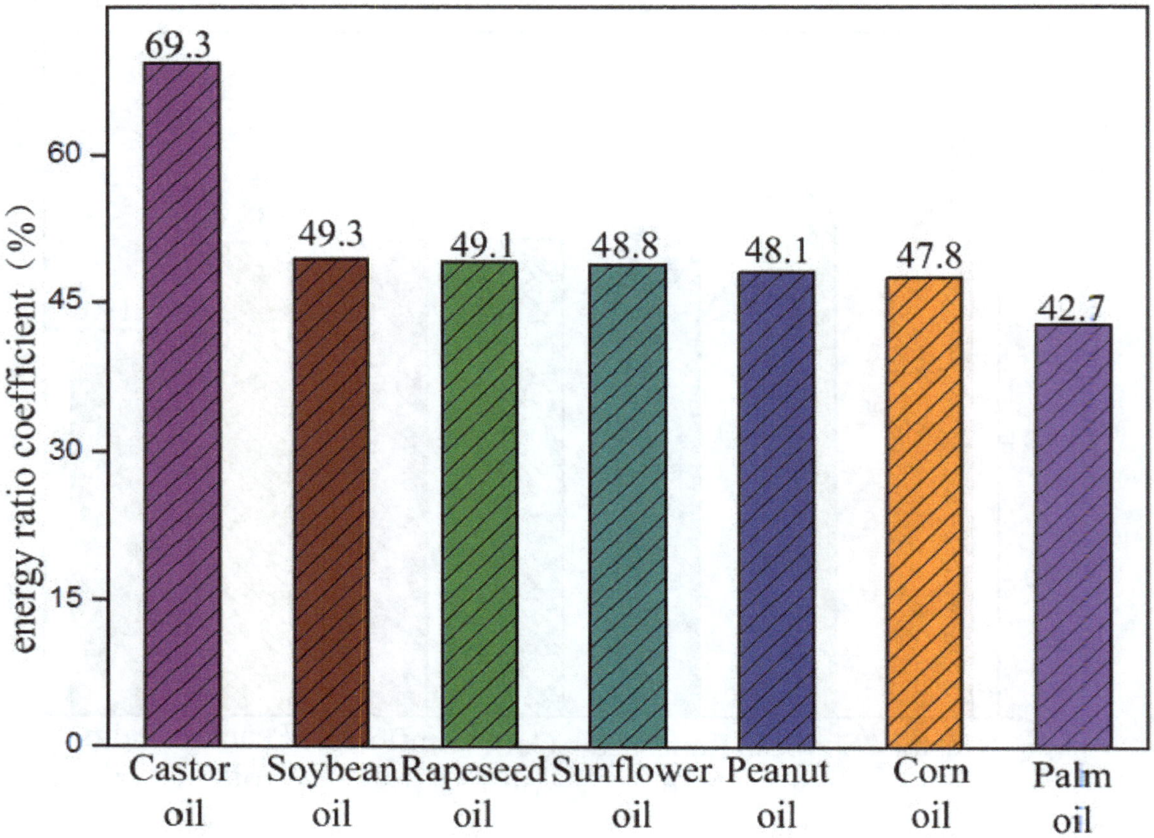

Figure 10. Lubricant film antiwear and load carrying capacity schematic: 1-workpiece 2-grinding wheel 3-oil drop 4-nozzle 5-abrasive grain 6-lubricant film

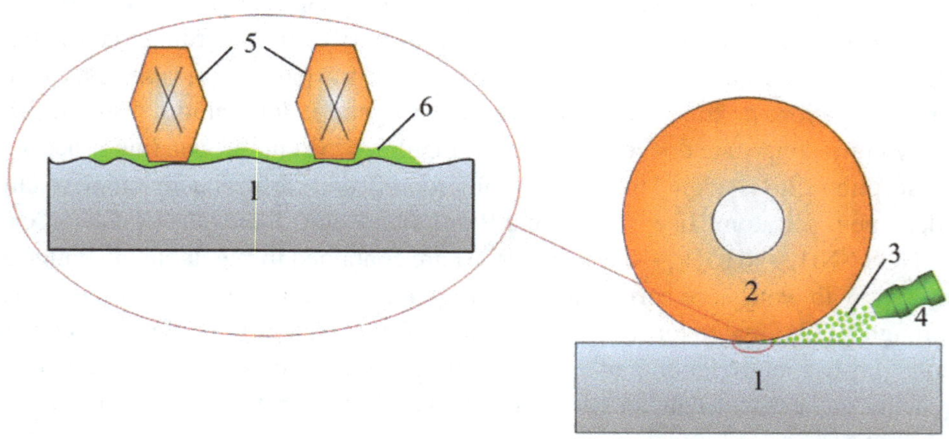

lubricating property of adsorption film(Zhang, Li, Jia, Zhang, & Zhang, 2015). Ricinoleic acid content reaches as high as 90.85% in castor oil, and it contains two polar groups (-OH and -COOH)(Choudhary, Naskar, & Paul, 2018; Hayat, Wen, Li, & Cao, 2015), so strength and durability of physical adsorption film formed in the grinding area are good, and then the minimum grinding force is obtained when castor oil is used for MQL grinding. In addition, Zhang et al. (2015) found that carbon chain length is not the main factor influencing lubricating property of physical adsorption film. This can also be verified according to abovementioned experimental results of grinding forces under MQL grinding with seven kinds of vegetable oil. For instance, erucic acid content in rapeseed oil is 43.63% while palmitic acid content in palm oil is 45.1%, so contents of the two fatty acids are basically identical, and moreover, other fatty acids in the two kinds of vegetable oil are also similar with largely identical contents. It can be found from Fig. 11 that carbon chain length of erucic acid is longer than that of palmitic acid. Based on a theoretical analysis, grinding force obtained when rapeseed oil is used for MQL grinding is smaller than that when palm oil is used for MQL grinding. However, experimental result is not so: grinding force of palm oil is smaller than that of rapeseed oil, proving that carbon chain length is not the main factor influencing lubricating property of physical adsorption film to a certain degree.

(3) The influence of metallic saponification. Saturated fatty acids in vegetable oil will go through chemical reaction with metal on workpiece surface so as to generate fatty acid saponification-induced adsorption film with monomolecular layer or polymolecular layers. This layer of adsorption film will be adsorbed to workpiece surface to form a compact lubricating oil film so that lubricating property between abrasive grain of grinding wheel and workpiece surface is improved and friction and wear of workpiece surface are reduced. Generally speaking, adsorption capacity and strength of adsorption film will be strengthened as carbon number of saturated fatty acids increases. When adsorption film reaches the maximum compactness and the strongest adsorption capacity, carbon number of saturated fatty acids will also reach a critical value. However, carbon number has no bearing on physical adsorption firm formed by saturated fatty acids. Nevertheless, carbon number must be greater than 16. For instance: C_{18} octadecanoic acid and C_{22} behenic acid have identical anti-friction and anti-wear effects. However, fatty acids of vegetable oil contain polar unsaturated bonds. Under the effect of ethylenic bond, compactness of physical and chemical adsorption films is reduced, so its lubricating property becomes poor. At the time, transverse cohesion (mutual attraction) between fatty acid molecules plays a dominant role. It's already known that transverse cohesion is in direct proportion of carbon number of fatty acids, so lubricating property of cetoleic acid is better than that of octadecenoic acid. From Table 5, the maximum content of saturated fatty acids in castor oil is 52%, so compactness of saponification-induced adsorption film formed on workpiece surface when castor oil is used for MQL grinding is better than those when other vegetable oils are used, and carbon chain length of ricinoleic acid is long so that the strength of the formed fatty acid saponification-induced adsorption film is strong; in the meantime, under the same carbon chain length, lubricating property of unsaturated fatty acids is not as good as that of saturated fatty acids, so the grinding force of castor oil is the lowest. It's noteworthy that content of saturated fatty acids in palm oil is 35%, which is higher than saturated fatty acid contents in other five kinds of vegetable oil, and this may be the reason why grinding force of palm oil is lower than those of other five kinds of vegetable oil.

Figure 11. The molecular formula of several kinds of fatty acids: a. structure of castor acid molecular structure; b. structure of palmitic acid molecular; c. structure of stearic acid molecular
d. structure of CIS oleic acid molecular; e. structure of trans oleic acid molecular; f. structure of linoleic acid molecular; g. structure of molecular structure of erucic acid
a. molecular structural formula of ricinoleic acid; b. molecular structural formula of palmitic acid; c. molecular structural formula of stearic acid; d. molecular structural formula of cis-form oleic acid; e. molecular structural formula of anti-form oleic acid; f. molecular structural formula of linoleic acid; g. molecular structural formula of erucic acid

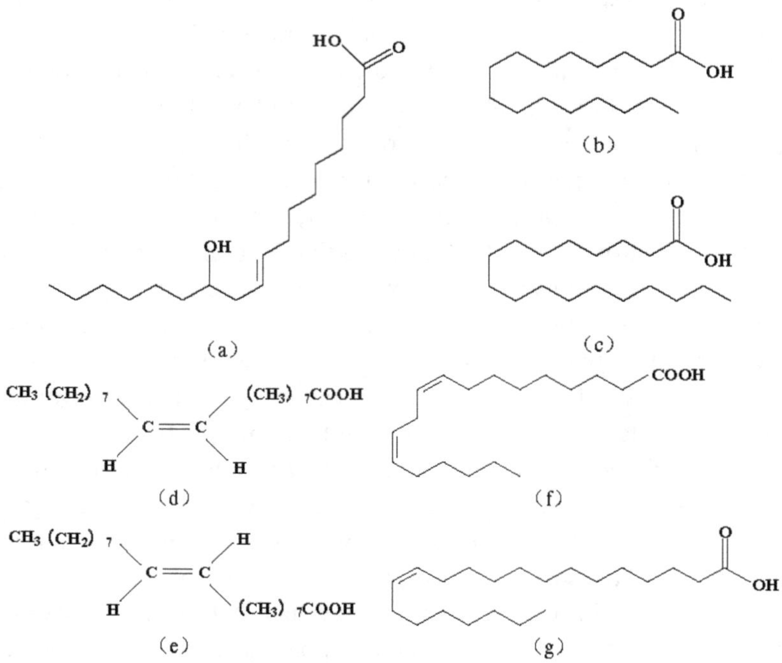

7.4.2 Grinding Temperature

As shown in Fig. 8, compared with castor oil, grinding temperatures is lowered by 32.0% when palm oil is used for MQL grinding; grinding temperatures of both soybean oil and rapeseed oil reduce by 18.5%; grinding temperatures of corn oil and sunflower oil reduce by 20.7% and 20.9% respectively; grinding temperature of peanut oil reduces by 21.4%; when compared with palm oil, grinding temperatures of both soybean oil and rapeseed oil increase by 20% and those of corn oil, sunflower oil and peanut oil increase by 16.7%, 16.5% and 15.7% respectively. Meanwhile, it can be noted that grinding temperatures when soybean oil, corn oil, sunflower oil and peanut oil are used for MQL grinding are basically identical. According to fatty acid contents and viscosity values of four kinds of vegetable oil in Table. 5, the following conclusion can be obtained: fatty acid contents in vegetable oils are not much different and their viscosity values are basically approximate, and the highest grinding temperatures when they are used for MQL grinding are also basically similar, which means that their cooling effects are basically the same. Furthermore, soybean oil has the minimum viscosity value, being 41.8 cp and its difference value from other vegetable oils is nearly 10 cp, but grinding temperature is approximate to those obtained using other three kinds of vegetable oil, and this may be related to mobility between vegetable oil molecules

Figure 12. Different viscosities vegetable oil heat transfer capacity schematic: 1-nozzle, 2-grinding wheel, 3- Heat transfer fluid before grinding,4- Heat transfer fluid after grinding, 5-workpiece

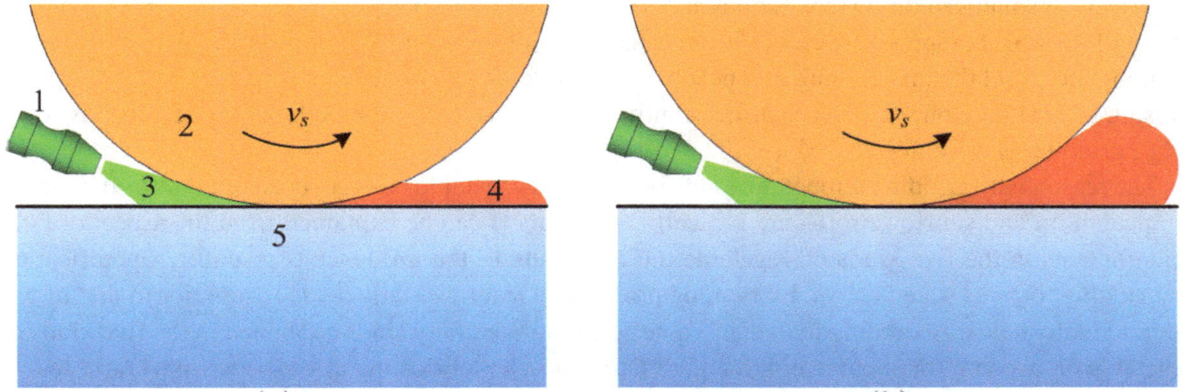

(a) Low viscosity vegetable oil heat capacity schematic

(b) High viscosity vegetable oil heat capacity schematic

under high temperature; when vegetable oil is sprayed into the grinding area together with compressed air, grinding process is proceeding and high grinding temperature is generated, viscosity of vegetable oil declines somehow and its mobility is enhanced, and meanwhile, compressed air can accelerate its flow, no compact and effective protective film is formed on the grinding surface between grinding wheel and workpiece to protect its operation, and thus relatively high grinding temperature is generated; it's also found that mobilities of soybean oil, corn oil, sunflower oil and peanut oil with basically identical viscosity values are basically identical under MQL grinding state. Literature(Zhang et al., 2015) points out that carbon chain length and polar groups in vegetable oil influence viscosity of vegetable oil: the longer the carbon chain length, the greater the viscosity of vegetable oil, and viscosity of vegetable oil can exert an important influence on grinding temperature: the greater the viscosity of vegetable oil, the higher the grinding temperature(Vishwanathan, Singh, & Subramanian, 2011). Erucic acid content in rapeseed oil is 43.63% and its carbon chain length is great, but viscosity of rapeseed oil is not that high, possibly because rapeseed oil also contains other kinds of fatty acids, total content of which is higher than erucic acid content. Therefore, erucic acid in rapeseed oil has a great effect on its viscosity, and moreover, its grinding temperature is basically identical with those obtained when the above four kinds of vegetable oil are used.

Through a comparison of tangential grinding force and normal grinding force, it can be found that tangential grinding force and normal grinding force of castor oil are the minimum but it obtains the highest grinding temperature. Meanwhile, grinding forces of soybean oil, rapeseed oil, corn oil, peanut oil and sunflower oil are high, and their grinding temperatures are low as expected, which may be closely related to property of vegetable oil: ricinoleic acid content contained in castor oil is 90.85%, ricinoleic acid contains two polar groups with a large carbon number, so ricinoleic acid exerts an extreme effect on viscosity of castor oil. Viscosity of castor oil is the maximum among 7 kinds of vegetable oil. The greater the viscosity, the poorer the mobility between molecules. Viscosity values of other vegetable oils are low, and oil fluid leaves grinding area with flow of compressed air after absorbing grinding heat in the grinding area as shown in Fig. 12 (a), its heat transfer capacity is better than that of castor oil, so the obtained grinding temperature is lower than castor oil. Under MQL grinding state, minimal quantity of castor oil sprayed in the grinding area forms a lubricating oil film on workpiece with good lubricating effect, so small grinding force is obtained. However, as it has poor mobility, it can't leave grinding area with flow of compressed air after heat exchange, high-temperature oil fluid is accumulated in the grinding area, and then its cooling and heat exchange effects become poor as shown in Fig. 12 (b). As a result, temperature on workpiece surface continuously rises, so grinding temperature under castor oil MQL grinding is the highest.

The reason why grinding temperatures of five kinds of vegetable oil—soybean oil, rapeseed oil, corn oil, peanut oil and sunflower oil—are basically approximate can be explained from the aspect of fatty acid contents in the five kinds of vegetable oil: according to the influence of metallic saponification on grinding force as stated above, lubricating property of saturated fatty acids is superior to that of unsaturated fatty acids. According to data in Table 5, unsaturated fatty acids contained in the five kinds of vegetable oil account for about over 80% of all fatty acids, so lubricating effect between workpiece and grinding wheel surface is poor in the grinding process and grinding force is large. Interaction between grinding wheel and workpiece is fierce for castor oil and palm oil, so the generated grinding temperature is relatively high. From the comparison of castor oil with five kinds of vegetable oil, the influence of viscosity on grinding force and grinding temperature is higher than the influence of fatty acid type and content in vegetable oils on grinding force and grinding temperature.

From a comparison between Fig. 6 and Fig. 8, it can be found that both grinding force and grinding temperature under palm oil MQL grinding are relatively low. On the whole, viscosity of palm oil is 61.62 cp which is higher than viscosity values of soybean oil, rapeseed oil, corn oil, sunflower oil and peanut oil. In addition, saturated fatty acid content in palm oil is higher than those in other five kinds of vegetable oil, and its lubricating effect is higher than other vegetable oil except for castor oil, so its grinding force is between castor oil and other five kinds of vegetable oil; meanwhile, as viscosity of palm oil is not that different from those of soybean oil, rapeseed oil, corn oil, sunflower oil and peanut oil, their mobilities will not be that different, either. Partial grinding heat generated in the grinding process is carried away by grinding fluid. Through a combination of several aspects of effects, grinding temperature obtained under palm oil MQL grinding is relatively low.

7.4.3 Energy Partition Ratio of Workpieces

Experimental results of energy proportionality coefficients are basically identical with expected results. Energy proportionality coefficient of castor oil is the highest while that of palm oil is the lowest, and from the aspects of viscosity values of vegetable oils and fatty acid type and content in vegetable oils: as castor oil is of extremely high viscosity, ricinoleic acid of high content and long carbon chain contributes to formation of an effective lubricating oil film between grinding wheel and workpiece surface in the grinding process, which reduces grinding force, but grinding temperature soars up and a large quantity of heat is generated, so the reason why its energy proportionality coefficient is the highest is that high grinding temperature results in falling of physical adsorption film and metallic saponification film, the formed adsorption film can't completely and effectively protect workpiece surface in the grinding area, and meanwhile, due to extremely high viscosity of castor oil, its mobility is poor and the generated grinding heat can't be timely dissipated. Consequently, a large quantity of grinding heat is transferred to workpiece, so energy proportionality coefficient is higher than other vegetable oils. The possible reasons for the lowest energy proportionality coefficient of palm oil are: on the one hand, grinding heat generated in the grinding process is low due to good cooling & lubricating effect of palm oil, so heat quantity transferred to workpiece is small, and energy proportionality coefficient is small; on the other hand, the difference values of palm oil from soybean oil, rapeseed oil, corn oil, sunflower oil and peanut oil are small in the aspect of viscosity, their mobilities are basically identical under high temperature in the grinding area, but its energy proportionality coefficient is lower than those of the five kinds of vegetable oil, possibly because palmitic acid has the shortest carbon chain, but short carbon chain can enhance heat transfer capacity in the grinding area and contribute to heat dissipation by palm oil grinding fluid, and as a result, grinding heat transferred to workpiece is reduced and the obtained energy proportionality coefficient is low.

REFERENCES

Adibi, H., Esmaeili, H., & Rezaei, S. M. (2018). Study on minimum quantity lubrication (MQL) in grinding of carbon fiber-reinforced SiC matrix composites (CMCs). *International Journal of Advanced Manufacturing Technology*, *95*(9-12), 3753–3767. doi:10.100700170-017-1464-x

Azarhoushang, B., Daneshi, A., & Lee, D. H. (2017). Evaluation of thermal damages and residual stresses in dry grinding by structured wheels. *Journal of Cleaner Production*, *142*, 1922–1930. doi:10.1016/j.jclepro.2016.11.091

Barczak, L. M., Batako, A. D. L., & Morgan, M. N. (2010). A study of plane surface grinding under minimum quantity lubrication (MQL) conditions. *International Journal of Machine Tools & Manufacture*, *50*(11), 977–985. doi:10.1016/j.ijmachtools.2010.07.005

Belentani, R. D., Funes, H., Canarim, R. C., Diniz, A. E., Hassui, A., Aguiar, P. R., & Bianchi, E. C. (2014). Utilization of Minimum Quantity Lubrication (MQL) with Water in CBN Grinding of Steel. *Materials Research-Ibero-American Journal of Materials*, *17*(1), 88–96. doi:10.1590/S1516-14392013005000165

Bhaumik, S., Datta, S., & Pathak, S. D. (2017). Analyses of Tribological Properties of Castor Oil With Various Carbonaceous Micro- and Nano-Friction Modifiers. *Journal of Tribology-Transactions of the Asme*, *139*(6), 061802. doi:10.1115/1.4036379

Boran, S., & Tamas, A. (2014). Rheological behaviour of castor oil mixed with different pyromellitic esters. *Journal of the Serbian Chemical Society*, *79*(2), 241–251. doi:10.2298/JSC130129062B

Cetin, M. H., Ozcelik, B., Kuram, E., & Demirbas, E. (2011). Evaluation of vegetable based cutting fluids with extreme pressure and cutting parameters in turning of AISI 304L by Taguchi method. *Journal of Cleaner Production*, *19*(17), 2049–2056. doi:10.1016/j.jclepro.2011.07.013

Chen, J. J., Fu, Y. C., He, Q. S., Zhu, Y. B., & Zhang, W. (2017). Experimental investigation on high-efficiency grinding of Inconel 718 with heat pipe grinding wheel. *Machining Science and Technology*, *21*(1), 86–102. doi:10.1080/10910344.2016.1260431

Chen, Z. Z., Xu, J. H., Ding, W. F., Ma, C. Y., & Fu, Y. C. (2015). Grinding temperature during high-efficiency grinding Inconel 718 using porous CBN wheel with multilayer defined grain distribution. *International Journal of Advanced Manufacturing Technology*, *77*(1-4), 165–172. doi:10.100700170-014-6403-5

Choudhary, A., Naskar, A., & Paul, S. (2018). An investigation on application of nano-fluids in high speed grinding of sintered alumina. *Journal of Manufacturing Processes*, *35*, 624–633. doi:10.1016/j.jmapro.2018.09.013

Dai, J. B., Ding, W. F., Zhang, L. C., Xu, J. H., & Su, H. H. (2015). Understanding the effects of grinding speed and undeformed chip thickness on the chip formation in high-speed grinding. *International Journal of Advanced Manufacturing Technology*, *81*(5-8), 995–1005. doi:10.100700170-015-7265-1

Dhar, N. R., Ahmed, M. T., & Islam, S. (2007). An experimental investigation on effect of minimum quantity lubrication in machining AISI 1040 steel. *International Journal of Machine Tools & Manufacture*, *47*(5), 748–753. doi:10.1016/j.ijmachtools.2006.09.017

Ding, W. F., Xu, J. H., Chen, Z. Z., Su, H. H., & Fu, Y. C. (2010). Grindability and Surface Integrity of Cast Nickel-based Superalloy in Creep Feed Grinding with Brazed CBN Abrasive Wheels. *Chinese Journal of Aeronautics*, *23*(4), 501–510. doi:10.1016/S1000-9361(09)60247-8

Ding, W. F., Xu, J. H., Chen, Z. Z., Yang, C. Y., Song, C. J., & Fu, Y. C. (2013). Fabrication and performance of porous metal-bonded CBN grinding wheels using alumina bubble particles as pore-forming agents. *International Journal of Advanced Manufacturing Technology*, *67*(5-8), 1309–1315. doi:10.100700170-012-4567-4

Elmunafi, M. H. S., Kurniawan, D., & Noordin, M. Y. (2015).. . *Use of Castor Oil as Cutting Fluid in Machining of Hardened Stainless Steel with Minimum Quantity of Lubricant Procedia CIRP*, *26*, 408–411.

Gaitonde, V., Karnik, S. R., & Davim, J. P. (2008). Selection of optimal MQL and cutting conditions for enhancing machinability in turning of brass. *Journal of Materials Processing Technology*, *204*(1), 459–464. doi:10.1016/j.jmatprotec.2007.11.193

Guo, S. M., Li, C. H., Zhang, Y. B., Wang, Y. G., Li, B. K., Yang, M., ... Liu, G. T. (2017). Experimental evaluation of the lubrication performance of mixtures of castor oil with other vegetable oils in MQL grinding of nickel-based alloy. *Journal of Cleaner Production*, *140*, 1060–1076. doi:10.1016/j.jclepro.2016.10.073

Hadad, M. J., Tawakoli, T., Sadeghi, M. H., & Sadeghi, B. (2012). Temperature and energy partition in minimum quantity lubrication-MQL grinding process. *International Journal of Machine Tools & Manufacture*, *54-55*, 10–17. doi:10.1016/j.ijmachtools.2011.11.010

Hayat, M. D., Wen, G., Li, T., & Cao, P. (2015). Compatibility improvement of Ti-MIM feedstock using liquid surfactant. *Journal of Materials Processing Technology*, *224*, 33–39. doi:10.1016/j.jmatprotec.2015.04.027

Huang, W. T., Liu, W. S., & Wu, D. H. (2016). Investigations into lubrication in grinding processes using MWCNTs nanofluids with ultrasonic-assisted dispersion. *Journal of Cleaner Production*, *137*, 1553–1559. doi:10.1016/j.jclepro.2016.06.038

Jain, A., & Bisht, R. P. S. (2010). Metalworking emulsions from industrial vegetable oils. *Lubrication Science*, *25*(3), 87–94. doi:10.1002/jsl.51

Jia, D., Li, C., Wang, S., Zhang, Q., & Hou, Y. (2014). Advances and Patents about Grinding Equipments with Nano-Particle Jet Minimum Quantity Lubrication. *Recent Patents on Nanotechnology*, *8*(3), 215–229. doi:10.2174/1872210508666141022113257 PMID:25336175

Jia, D., Li, C., Zhang, D., Zhang, Y., & Zhang, X. (2014). Experimental verification of nanoparticle jet minimum quantity lubrication effectiveness in grinding. *Journal of Nanoparticle Research*, *16*(12), 2758. doi:10.100711051-014-2758-7

Jin, T., & Stephenson, D. J. (2008). A study of the convection heat transfer coefficients of grinding fluids. *CIRP Annals - Manufacturing Technology, 57*(1), 367-370. doi:10.1016/j.cirp.2008.03.074

Kalita, P., Malshe, A. P., Kumar, S. A., Yoganath, V. G., & Gurumurthy, T. (2012). Study of specific energy and friction coefficient in minimum quantity lubrication grinding using oil-based nanolubricants. *Journal of Manufacturing Processes*, *14*(2), 160–166. doi:10.1016/j.jmapro.2012.01.001

Kalita, P., Malshe, A. P., & Rajurkar, K. P. (2012). Study of tribo-chemical lubricant film formation during application of nanolubricants in minimum quantity lubrication (MQL) grinding. *Cirp Annals-Manufacturing Technology*, *61*(1), 327–330. doi:10.1016/j.cirp.2012.03.031

Khan, A. W., & Wuyi, C. (n.d.). Systematic Geometric Error Modeling for Workspace Volumetric Calibration of a 5-axis Turbine Blade Grinding Machine. *Chinese Journal of Aeronautics*.

Khan, M. M. A., Mithu, M. A. H., & Dhar, N. R. (2009). Effects of minimum quantity lubrication on turning AISI 9310 alloy steel using vegetable oil-based cutting fluid. *Journal of Materials Processing Technology*, *209*(15-16), 5573–5583. doi:10.1016/j.jmatprotec.2009.05.014

Lawal, S. A., Choudhury, I. A., & Nukman, Y. (2013). A critical assessment of lubrication techniques in machining processes: A case for minimum quantity lubrication using vegetable oil-based lubricant. *Journal of Cleaner Production*, *41*, 210–221. doi:10.1016/j.jclepro.2012.10.016

Lefebvre, A., Lipinski, P., Vieville, P., & Lescalier, C. (2008). Experimental analysis of temperature in grinding at the global and local scales. *Machining Science and Technology*, *12*(1), 1–14. doi:10.1080/10910340701873489

Li, B. K., Li, C. H., Zhang, Y. B., Wang, Y. G., Jia, D. Z., & Min, Y. (2016). Grinding temperature and energy ratio coefficient in MQL grinding of high-temperature nickel-base alloy by using different vegetable oils as base oil. *Chinese Journal of Aeronautics*, *29*(4), 1084–1095. doi:10.1016/j.cja.2015.10.012

Li, B. M., & Zhao, B. (2003). *Modern grinding technology*. Beijing: China Machine Press.

Li, X., Chen, Z., & Chen, W. (2011). Suppression of Surface Burn in Grinding of Titanium Alloy TC4 Using a Self-inhaling Internal Cooling Wheel. *Chinese Journal of Aeronautics*, *24*(1), 96–101. doi:10.1016/S1000-9361(11)60012-5

Malkin, S., & Guo, C. (2007). Thermal Analysis of Grinding. *CIRP Annals - Manufacturing Technology, 56*(2), 760-782.

Mao, C., Zou, H. F., Huang, Y., Li, Y. F., & Zhou, Z. X. (2013). Analysis of heat transfer coefficient on workpiece surface during minimum quantity lubricant grinding. *International Journal of Advanced Manufacturing Technology*, *66*(1-4), 363–370. doi:10.100700170-012-4330-x

Mao, C., Zou, H. F., Zhou, X., Huang, Y., Gan, H. Y., & Zhou, Z. X. (2014). Analysis of suspension stability for nanofluid applied in minimum quantity lubricant grinding. *International Journal of Advanced Manufacturing Technology*, *71*(9-12), 2073–2081. doi:10.100700170-014-5642-9

Nguyen, V. C., Le, T. S., Tran, M. D., & Nguyen, D. B. (2013). An Investigation on Effect of Characteristics of the Peanut oil MQL on Tool life in Hard turning 9CrSi steel. *International Journal of Machining & Machinability of Materials*, *13*(4), 428–438. doi:10.1504/IJMMM.2013.054275

Rahim, E. A., & Sasahara, H. (2011a). An Analysis of Surface Integrity When Drilling Inconel 718 Using Palm Oil and Synthetic Ester under Mql Condition. *Machining Science and Technology*, *15*(1), 76–90. doi:10.1080/10910344.2011.557967

Rahim, E. A., & Sasahara, H. (2011b). Investigation of tool wear and surface integrity on MQL machining of Ti-6AL-4V using biodegradable oil. *Proceedings of the Institution of Mechanical Engineers. Part B, Journal of Engineering Manufacture*, *225*(9B9), 1505–1511. doi:10.1177/0954405411402554

Rahim, E. A., & Sasahara, H. (2011c). A study of the effect of palm oil as MQL lubricant on high speed drilling of titanium alloys. *Tribology International*, *44*(3), 309–317. doi:10.1016/j.triboint.2010.10.032

Silva, L. R. D., Bianchi, E. C., Fusse, R. Y., Catai, R. E., França, T. V., & Aguiar, P. R. (2007). Analysis of surface integrity for minimum quantity lubricant—MQL in grinding. *International Journal of Machine Tools & Manufacture*, *47*(2), 412–418. doi:10.1016/j.ijmachtools.2006.03.015

Simoes, R. P., Bianchi, E. C., Oikawa, M. H., de Aguiar, P. R., Droppa, R., & Canarim, R. C. (2014). Evaluation of Different Methods of Cooling-lubrication in Cylindrical Grinding of Advanced Ceramic Dip. *Materials Research-Ibero-American Journal of Materials*, *17*(5), 1201–1212. doi:10.1590/1516-1439.250713

Sullivan, P. A., Eisen, E. A., Woskie, S. R., Kriebel, D., Wegman, D. H., Hallock, M. F., ... Monson, R. R. (1998). Mortality studies of metalworking fluid exposure in the automobile industry: VI. A case-control study of esophageal cancer. *American Journal of Industrial Medicine*, *34*(1), 36–48. doi:10.1002/(SICI)1097-0274(199807)34:1<36::AID-AJIM6>3.0.CO;2-O PMID:9617386

Tawakoli, T., Hadad, M. J., Sadeghi, M. H., Daneshi, A., Stockert, S., & Rasifard, A. (2009). An experimental investigation of the effects of workpiece and grinding parameters on minimum quantity lubrication-MQL grinding. *International Journal of Machine Tools & Manufacture*, *49*(12-13), 924–932. doi:10.1016/j.ijmachtools.2009.06.015

Vieira, A. A., Baptista, A. M., Parente, M. P. L., & Jorge, R. M. N. (2009). On the temperature field during superficial grinding: An experimental study. *International Journal of Advanced Manufacturing Technology*, *40*(11-12), 1084–1092. doi:10.100700170-008-1442-4

Vishwanathan, K. H., Singh, V., & Subramanian, R. (2011). Wet grinding characteristics of soybean for soymilk extraction. *Journal of Food Engineering*, *106*(1), 28–34. doi:10.1016/j.jfoodeng.2011.04.002

Wang, Y. G., Li, C. H., Zhang, Y. B., Yang, M., Li, B. K., Jia, D. Z., ... Mao, C. (2016). Experimental evaluation of the lubrication properties of the wheel/workpiece interface in minimum quantity lubrication (MQL) grinding using different types of vegetable oils. *Journal of Cleaner Production*, *127*, 487–499. doi:10.1016/j.jclepro.2016.03.121

Wang, Y. N., Wan, Z. P., Lu, L. S., Zhang, Z. H., & Tang, Y. (2018). Friction and wear mechanisms of castor oil with addition of hexagonal boron nitride nanoparticles. *Tribology International*, *124*, 10–22. doi:10.1016/j.triboint.2018.03.035

Zhang, D., Li, C., Jia, D., Zhang, Y., & Zhang, X. (2015). Specific grinding energy and surface roughness of nanoparticle jet minimum quantity lubrication in grinding. *Chinese Journal of Aeronautics*, *28*(2), 570–581. doi:10.1016/j.cja.2014.12.035

Zhang, D., Li, C., Zhang, Y., Jia, D., & Zhang, X. (2015). Experimental research on the energy ratio coefficient and specific grinding energy in nanoparticle jet MQL grinding. *International Journal of Advanced Manufacturing Technology*, *78*(5-8), 1275–1288. doi:10.100700170-014-6722-6

Zhang, Y., Li, C., Jia, D., Zhang, D., & Zhang, X. (2015). Experimental evaluation of MoS_2 nanoparticles in jet MQL grinding with different types of vegetable oil as base oil. *Journal of Cleaner Production*, *87*, 930–940. doi:10.1016/j.jclepro.2014.10.027

Zhang, Y. B., Li, C. H., Jia, D. Z., Zhang, D. K., & Zhang, X. W. (2015). Experimental evaluation of MoS_2 nanoparticles in jet MQL grinding with different types of vegetable oil as base oil. *Journal of Cleaner Production*, *87*, 930–940. doi:10.1016/j.jclepro.2014.10.027

Chapter 8
Experimental Research on Heat Transfer Performance in MQL Grinding With Different Nanofluids

ABSTRACT

An investigation into the effect of nanofluid minimum quantity lubrication (MQL) on the temperatures in surface grinding is presented and discussed. Six types of nanoparticles, namely molybdenum disulfide (MoS2), zirconium dioxide (ZrO2), carbon nanotube (CNT), polycrystalline diamond, aluminum oxide (Al2O3), and silica dioxide (SiO2), are considered to mix individually with a pollution-free palm oil in preparing the nanofluids. A commonly used Ni-based alloy was chosen as the workpiece material. It is shown that CNT nanofluid results in the lowest grinding temperature of 110.7°C and the associated energy proportionality coefficient of 40.1%. The relevant physical properties of the nanofluids such as the coefficient of thermal conductivity, viscosity, surface tension, and the contact state between the droplets and workpiece surface (contact angle) were discussed to shine a light on their effect on the cooling performance. A mathematical model for convective heat transfer coefficient was then developed based on the boundary layer theories.

8.1 INTRODUCTION

The grinding process to remove a unit volume of metal materials has a large energy input. Almost all energy transformations occur in the grinding zone. Most of the energies are converted into heat, so the temperature of the grinding zone increases. Grinding temperature influences the surface quality of the processed workpiece and the cutting performance of abrasive particles on the grinding wheel. A high grinding temperature will change the metallographic structure of the surface and cause grinding burn (Hadad, Tawakoli, Sadeghi, & Sadeghi, 2012). A high transient temperature will result in thermal stress during grinding and cooling; this thermal stress will in turn cause considerable residual stress and even

DOI: 10.4018/978-1-7998-1546-4.ch008

cracks on the processed surface. This problem is aggravated in metallic materials with low thermal conductivity (e.g., titanium alloy). A high grinding temperature will also affect the dimensional and shape accuracies of parts. Additionally, the hardness of abrasive particles declines significantly at a high temperature. The bond wear and diffusion wear between abrasive particles and grinding materials passivate abrasive particles so fast that they lose their cutting capability (Chen, Rowe, & Mccormack, 2000). Therefore, studying grinding temperature is important.

According to the requirements of sustainable development and environmental protection, traditional pouring grinding is unfit for the development in this new century. Existing green processing technologies, such as minimum quantity lubrication (MQL) and nanofluid MQL, can replace traditional pouring cooling lubrication and have shown significant advantages and development prospects (Malkin & Guo, 2007). Hadad et al. (2012) investigated 100Cr6 grinding through MQL and used an overhead thermocouple to measure the grinding temperature. They found that MQL consumes 7%–10% less energy than dry grinding. Barczak et al. (2010) compared the grinding power, grinding force, grinding temperature, and surface roughness of MQL grinding with those of pouring grinding and dry grinding. Their results revealed that MQL grinding is superior to pouring grinding in terms of grinding force and grinding power but inferior in terms of workpiece surface roughness and residual stress at an appropriate material removal rate. Although MQL grinding has more technological advantages than dry grinding and pouring grinding, the research indicated that it has insufficient heat transfer capability.

Researchers have explored nanofluid MQL machines according to the heat transfer enhancement of nanofluids. Shen et al. (2011) prepared nanofluids by adding Al_2O_3 and diamond nanoparticles to water to grind cast irons. They found that nanofluid MQL significantly reduces grinding force and surface roughness and effectively eliminates workpiece burning. A high nanoparticle concentration leads to a high G ratio. Lee et al. (2011) prepared nanofluids by adding diamond nanoparticles to liquid paraffin and studied the grinding performances of dry grinding, liquid paraffin MQL, and nanofluid MQL. The results showed that nanofluid MQL has a much lower grinding force and surface roughness than the others. Moreover, a small nanoparticle size leads to good surface quality. Kalita et al. (2012) conducted an experimental study on nanofluid grinding of cast iron and EN24 alloy steel by adding MoS_2 nanoparticles to paraffin and soybean oils. They found that MoS_2 nanofluids can reduce grinding force, grinding force ratio, and specific grinding energy and increase the G ratio of MQL grinding significantly. Kalita et al. (2012) also conducted an experimental study on spheroidal graphite cast iron 100-70-30 grinding by using different concentrations of MoS_2 nanofluids and found that 8% MoS_2 nanofluids has the lowest grinding temperature. Furthermore, they reported that nanofluid MQL grinding has good wear resistance. Mao et al. (2012)tested dry grinding, pouring grinding, MQL grinding, and Al_2O_3 nanofluid MQL grinding and discovered that nanofluid MQL has a lower grinding force and grinding temperature than the others and excellent workpiece surface quality. The researchers also discussed the effect of nanofluid MQL parameters and concluded that the lowest grinding temperature is achieved at 15° nozzle angle, 20 mm distance between the nozzle and the grinding zone, and 0.6 MPa air pressure (Mao et al., 2013). Zhang et al. (2015) ground quenching 45 steel with different nanofluids, including soybean oil, ZrO_2, and CNT. The volume fraction of all nanofluids was 1%. They analyzed the energy proportionality coefficient and specific grinding energy during the grinding process and reported that CNT nanofluids contribute lower grinding temperature and energy proportionality coefficient than the other two nanofluids. According to the thermal conductivity of CNT nanoparticles, CNT nanofluids possess excellent heat conduction and reduce the temperature of the grinding zone dramatically.

In MQL grinding, minimal quantity of lubricating oil is sprayed into the grinding area to improve lubricating and heat transfer properties at the workpiece/grinding wheel interface (Shen et al., 2011). However, ideal heat transfer property can't be reached yet (Zhang, Li, Zhang, Jia, & Zhang, 2015). In consideration of insufficient cooling effect of MQL technology and advent of nanotechnology, NMQL grinding has become a feasible method to solve this difficulty(Singh, Dureja, Dogra, & Bhatti, 2018; Zhang et al., 2018). Based on heat transfer enhancement theory of solid, adding nanoparticles into the base fluid can significantly elevate heat conductivity coefficient of the base fluid and strengthen heat transfer capacity in the grinding area (Rabiei, Rahimi, & Hadad, 2017). Scholars from various countries have carried out a large quantity of researches on NMQL grinding (Li, et al., 2017; Shabgard, Seyedzavvar, & Mohammadpourfard, 2017), and results indicate that performance of NMQL grinding is superior to that of MQL grinding (Lee, Kim, & Lee, 2018; Paul & Ghosh, 2019). However, these researches only direct at very few kinds of nanofluids with a lack of systematic studies on MQL grinding heat transfer with different kinds of vegetable oil-based nanofluids (Shabgard et al., 2017). Therefore, an experimental study of palm oil-based NMQL grinding with addition of aluminum oxide, diamond, molybdenum disulfide, carbon nanotube, zirconium oxide and silicon dioxide nanoparticles, mainly aiming at exploring into the heat transfer mechanism of the nanofluids in the grinding process.

8.2 EXPERIMENTAL MATERIALS

8.2.1 Workpiece Material

Workpiece material used in the experiment is high-temperature nickel base alloy GH4169, which has been mainly applied to fields like ocean, energy, aviation and mechanical and electrical products by virtue of good comprehensive performance (Du, Lu, Deng, & Bi, 2015). However, its heat conductivity coefficient is low (Akhtar, Sun, & Chen, 2016; Xi et al., 2017), so grinding heat generated in the grinding process is large, which seriously impacts quality of the machined surface (X. Li, Ma, & Meng, 2015). Therefore, it will be important to conduct a grinding experimental study of high-temperature nickel base alloy GH4169 (Wang et al., 2016). Chemical compositions and characteristic parameters of high-temperature nickel base alloy GH4169 are shown in Table. 3 and Table. 4 (Wu, Zhang, & Yao, 2018; Ye, Gong, Zhang, Tu, & Zhang, 2017). For the convenience of grinding experiment and data measurement, dimensions of used sample pieces of the workpiece are 40×30×30 mm.

8.2.2 Nanofluids

NMQL grinding is adopted in the experiment and base fluid is palm oil (Hussein, Lingenthiran, Kadirgamma, Noor, & Aik, 2018). The reason why palm oil is selected as base oil is: it's found in the study in Chapter 3 that palm oil has relatively high viscosity and triglyceride it contains has good polarity, so it can not only realize good lubricating property in the grinding area but also has superior heat transfer property (Li, Li, Zhang, Wang, Yang, et al., 2017; Zhang et al., 2019); furthermore, Wang et al. (2016) found in their study that although castor oil had better lubricating property than palm oil, its viscosity was much greater than that of palm oil. Mobility of castor oil was poor in the grinding process, so grinding heat could not be timely transferred out of the grinding area but accumulated in the grinding area (Seyedzavvar, Shabgard, & Mohammadpourfard, 2019; Wang et al., 2018), and then grinding temperature rose,

Figure 1. Six kinds of nanofluids and palm oil-based fluid

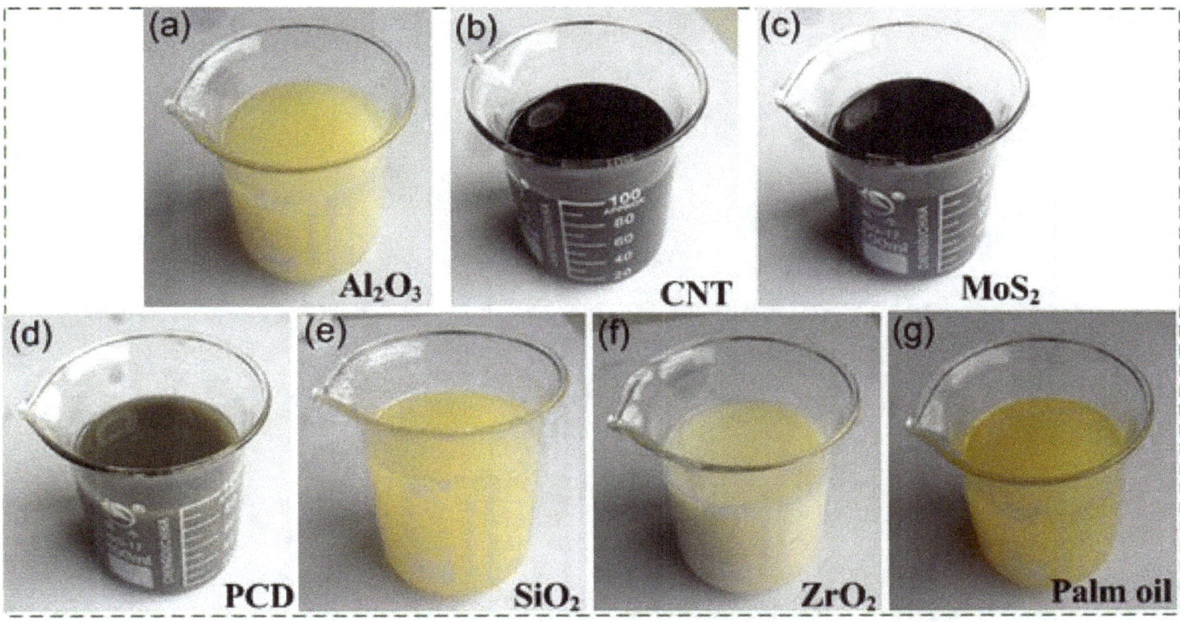

thus affecting machining quality and precision of the experimental workpiece. Meanwhile, according to the experiment specified in Chapter 7, castor oil has poor atomization effect, which seriously influences its lubricating and cooling performances, so in full consideration of grinding performances of castor oil and palm oil, palm oil is more suitable to be base oil of nanofluid (Javed et al., 2018). Used nanoparticles are MoS_2, ZrO_2, CNT, ND, Al_2O_3 and SiO_2 respectively, their particle sizes are all 50nm, and average length of carbon nanotube is 10-30μm; nanofluid is prepared using two-step method. Base oil of a certain mass is directly mixed with nanoparticles of certain mass, and mass fraction of nanofluid is 6%. In order to prevent agglomeration phenomenon of nanoparticles in the base oil, 0.3% Polysorbate 80 is added as dispersing agent (Ilyas, Pendyala, & Narahari, 2017) and the mixture is oscillated in the ultrasonic oscillator (model: KQ3200DB) for 1h (Sharma, Tiwari, & Dixit, 2016). The prepared six nanofluids are shown in Fig. 8-1. Viscosity and contact angle of nanofluids are measured using Brookfield DV2T viscosimeter and contact angle measuring instrument which are shown in Fig. 7-2.

8.2.3 Experimental Scheme

Grinder and measuring instruments used in the experiment are shown in Fig. 7-1. Experimental parameters and measuring methods are identical with those in Chapter 3. The abovementioned 6 nanofluids are used in this experiment for NMQL grinding of nickel base alloy GH4169. Grinding force and grinding temperature are measured and heat transfer conditions in MQL grinding using different nanofluids are investigated. Experimental program is shown in Table. 1.

Figure 2. Force signal diagram of MQL grinding with six nanofluids

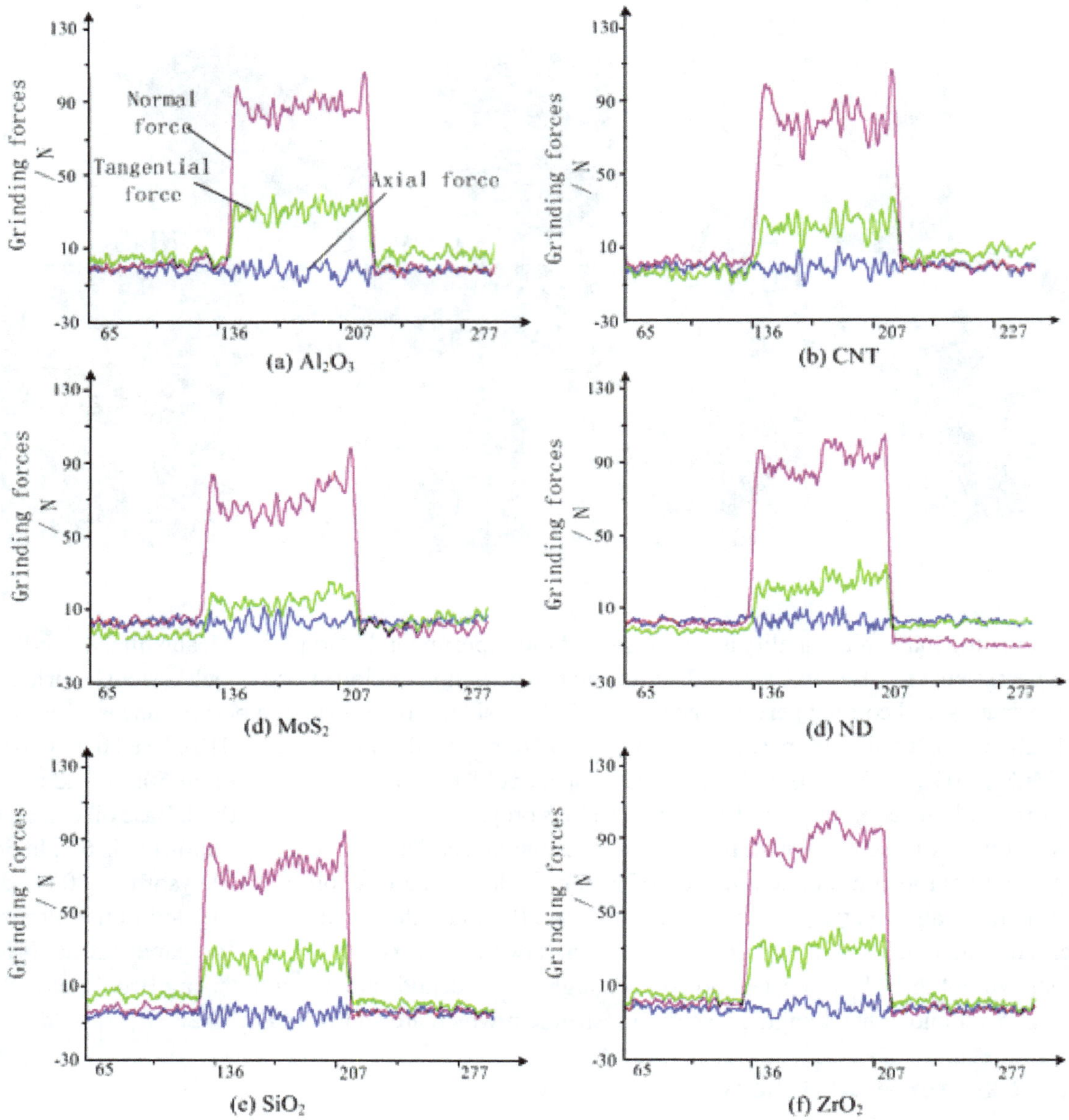

8.3 EXPERIMENTAL RESULTS

8.3.1 Grinding Force Ratio

During the grinding process, grinding force is an important parameter which reflects grinding state and lubricating state in the grinding process. Fig. 2 shows filtering signal images of grinding forces cf 6

Table 1. Different nanofluids minimal quantities of lubricant grinding experiment table

No.	Nanoparticles	Base oil	Nanofluid mass fraction	Cooling & lubrication condition
1	**MoS_2**	Palm oil	6%	MQL
2	ZrO_2			
3	CNT			
4	ND			
5	Al_2O_3			
6	SiO_2			

kinds of nanofluids.

Surface quality of machined workpiece is influenced by lubricating property in the grinding area. Parameters used to evaluate lubricating property in the grinding area can be expressed by grinding force ratio: the lower the grinding force ratio, the better the lubricating property. Fig. 3 shows grinding force ratios under MQL grinding using 6 kinds of nanofluids.

Figure 3. Friction coefficients of MQL grinding with six nanofluids (95% CI)

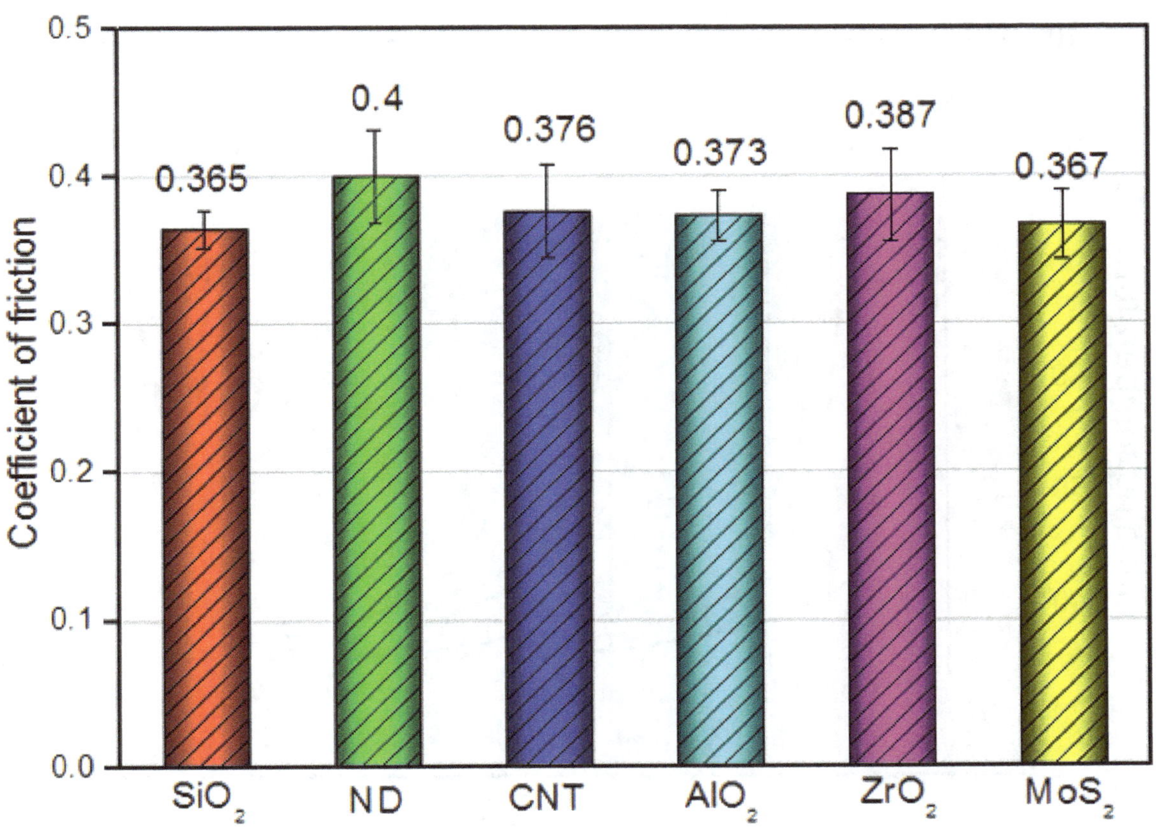

According to grinding force ratios under MQL grinding using 6 kinds of nanofluids, maximum grinding force ratio under ND NMQL grinding is 0.40, followed by ZrO_2 nanoparticles, being 0.387 which reduces by 3.25% relative to ND nanoparticles; grinding force ratio is the lowest under SiO_2 NMQL grinding, being 0.365 and grinding force ratio of MoS_2 nanoparticles is 0.367, which decrease by 8.75% and 8.25% respectively relative to ND nanoparticles. This indicates that lubricating effects under ND and ZrO_2 NMQL grinding states are not as good as those under SiO_2 and MoS_2 NMQL grinding. It can be seen from the figure that grinding forces under CNT and Al_2O_3 NMQL grinding are 0.376 and 0.373 respectively, which increase by 2.92% and 2.14% respectively relative to SiO_2 nanoparticles. The figure shows that grinding force ratio of SiO_2 nanoparticles is the lowest, indicating that it has the best lubricating property.

8.3.2 Grinding Temperature

Grinding temperature can exert an important influence on workpiece machining quality (Setti, Sinha, Ghosh, & Rao, 2015). Too high grinding temperature will give rise to burning of workpiece surface, aggravation of workpiece surface roughness and lowering of machining precision. Fig. 4 shows grinding temperature curve chart under MQL grinding using 6 kinds of nanofluids, where *x/l* is a dimensionless

Figure 4. Graph of grinding temperature for the six nanofluids (95% CI)

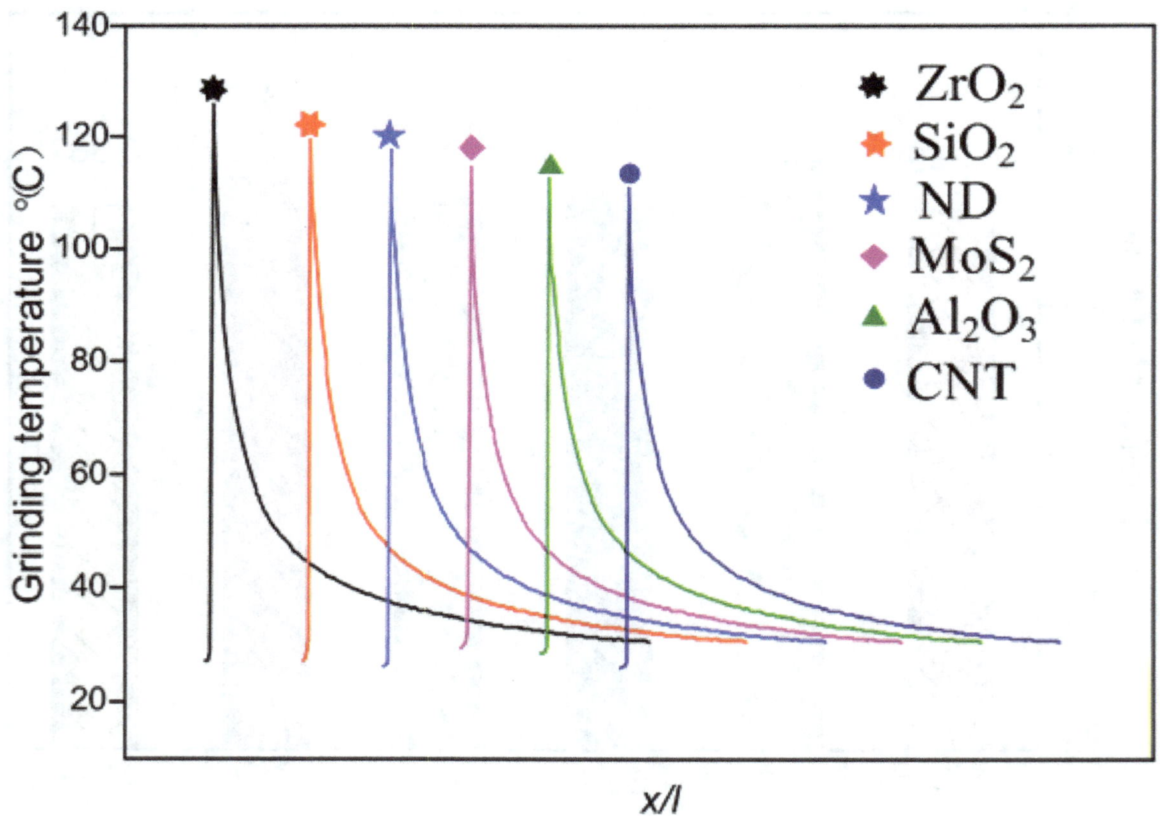

quantity. It can be seen from the figure that when grinding wheel is not grinding the workpiece, workpiece is under normal temperature; after it grinds the workpiece, working temperature rapidly rises and slowly declines to room temperature under the cooling effect of nanofluid.

Fig. 5 shows grinding temperatures under MQL grinding using 6 kinds of nanofluids. ZrO_2 nanofluid obtains the highest grinding temperature, being 125.7 °C. CNT obtains the lowest grinding temperature, being 110.7 °C which reduces by 11.9% relative to ZrO_2 nanoparticles; grinding temperatures obtained using ND and SiO_2 nanoparticles are slightly lower than that obtained using ZrO_2 nanofluid, respectively being 117.7 °C and 119.3 °C which decrease by 6.4% and 5.1% respectively; Al_2O_3 and MoS_2 nanofluids obtain relatively low grinding temperatures, respectively being 112.4 °C and 114.7 °C which increase by 1.5% and 3.6% respectively relative to CNT NMQL grinding. According to the analysis of experimental results, CNT nanoparticles have not ideal lubricating effect but obtain the lowest grinding temperature, indicating that CNT nanoparticles have good heat transfer performance.

8.3.3 Energy Partition of Workpieces

During the grinding process, the more the energy transferred to workpiece, the higher the workpiece temperature. When workpiece temperature reaches one critical value, it will cause serious damage to the workpiece, including surface burning, too large residual stress, etc. Therefore, using the grinding fluid

Figure 5. Grinding temperatures of the six nanofluids (95% CI)

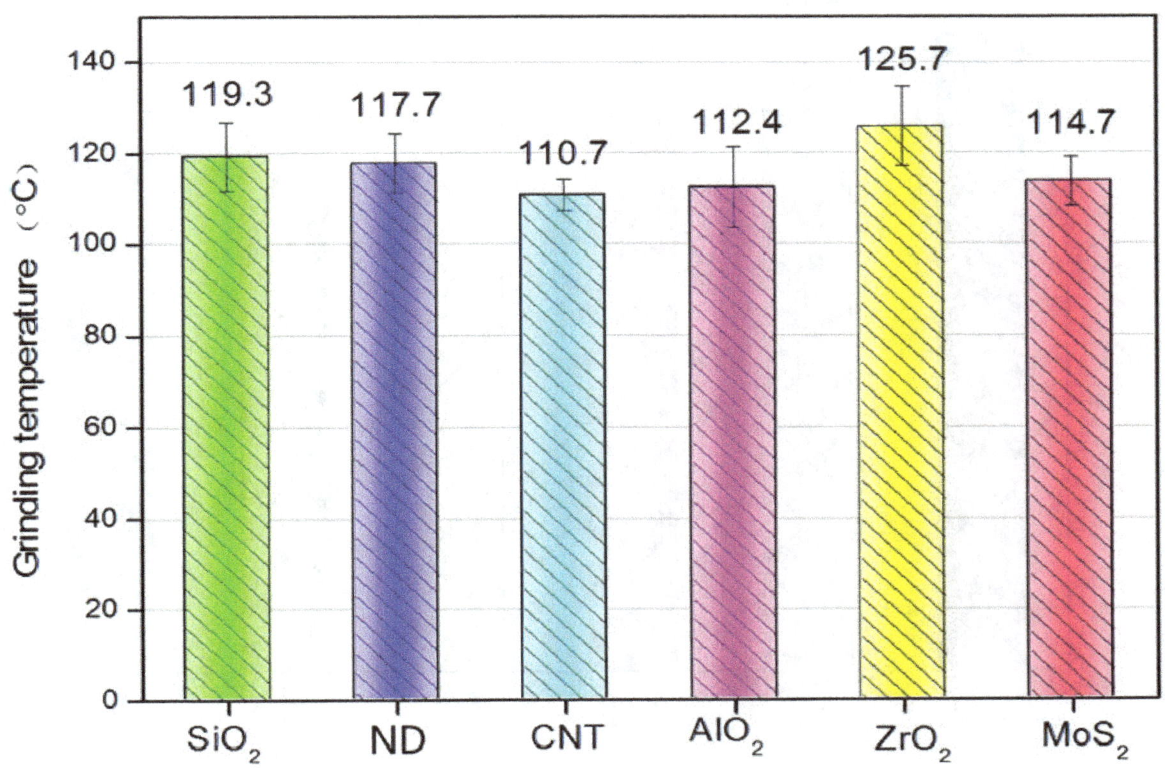

with high heat transfer efficiency is to reduce energy transferred to workpiece so as to reduce workpiece temperature and obtain ideal machining quality and precision.

Fig. 6 shows energy proportionality coefficients under MQL grinding using 6 kinds of nanofluids. The lowest proportionality coefficient of energy transferred to workpiece under CNT NMQL grinding is 40.1%, followed by Al_2O_3 NMQL grinding, being 41.0%; energy proportionality coefficient of MoS_2 nanoparticles is slightly higher than those of CNT and Al_2O_3 nanoparticles (by 11% and 8.5% respectively), being 44.5%; the difference between SiO_2 and ZrO_2 nanoparticles is not large in energy proportionality coefficient, respectively being 46.4% and 46.5% respectively, which increase by 15.7% and 16% respectively relative to CNT nanoparticles. Energy proportionality coefficient of ND nanoparticles is 43.9% which increases by 9.5% relative to CNT nanoparticles. According to proportionality coefficients of energy transferred to workpiece, it can be further proved that CNT nanoparticles have good heat transfer performance. Based on the analysis of grinding temperatures and proportionality coefficients of energy transferred to workpiece under MQL grinding using 6 kinds of nanofluids, it can be known that they have very different heat transfer performances, so influences of physical characteristics of nanofluids on their heat transfer performances will be analyzed and discussed.

Figure 6. Ratio coefficients of energy of the six nanofluids (95% CI)

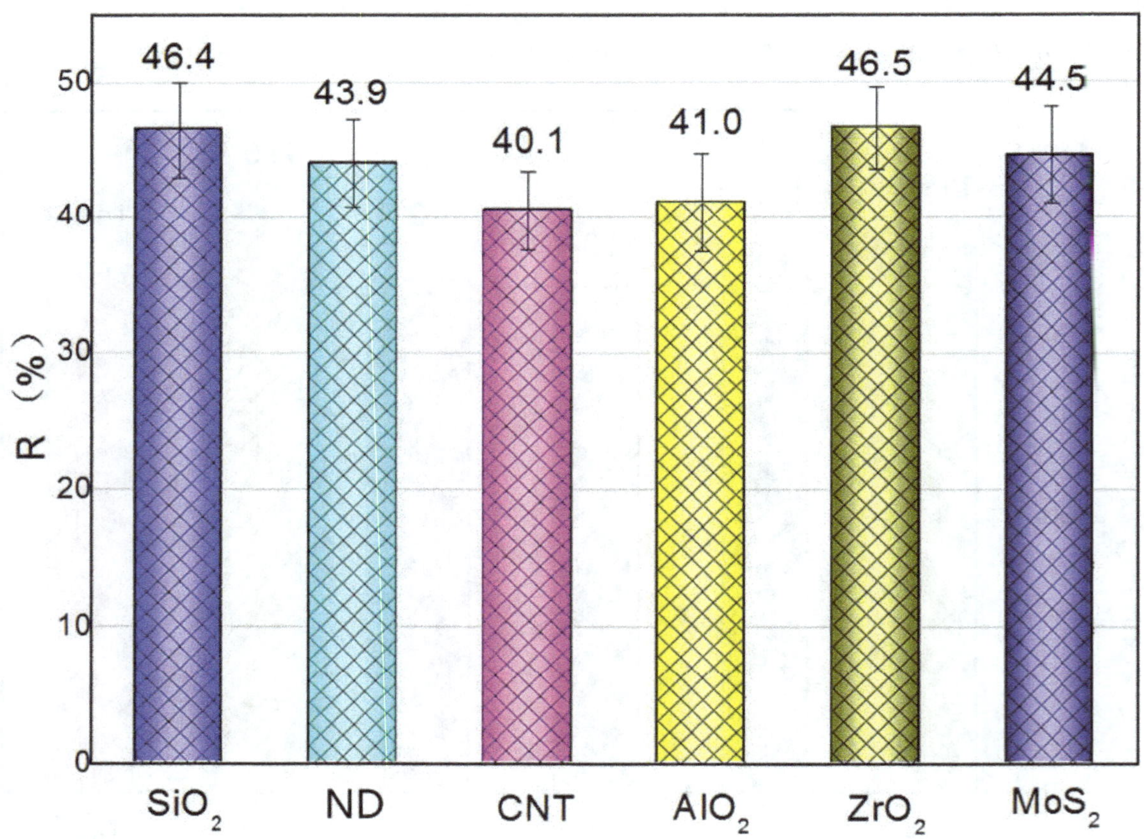

8.4 RESULTS AND DISCUSSION

8.4.1 Effects of Thermal Conductivity of Nanoparticles on Heat Transfer Performance

It's well known that adding nanoparticles into the base fluid can significantly elevate its heat conductivity coefficient and enhance its convective heat transfer enhancement performance (Zhang, C. H. Li, D. Z. Jia, et al., 2015), because nanofluid is a solid-liquid two-phase fluid in essence. Nanoparticles conduct intense Brownian movement in the base fluid (Michaelides, 2015), which is not only conducive to dispersing performance of nanoparticles in the base fluid (Mao et al., 2014) but also to interaction between nanoparticles and that between nanoparticles and base fluid. Therefore, nanofluid mobility is enhanced, so is its heat transfer rate; nanoparticles continuously do irregular micromovement in the base fluid and this micromovement results in weak convection phenomenon between nanoparticles and base fluid, which can accelerate energy transfer of the nanofluid (Dambatta, Sarhan, Sayuti, & Bin Abd Shukor, 2019; Yang et al., 2019), so heat conductivity of the nanofluid increases. Moreover, irregular movement also causes transfer of energy that nanoparticles have, and this transferred heat significantly enhances energy transfer inside the nanofluid. Heat transfer coefficient of nanoparticles can exert an important effect on heat transfer performance of the nanofluid in the grinding process. It's generally believed that nanoparticles with high conductivity coefficient have good heat transfer performance (Pryazhnikov, Minakov, Rudyak, & Guzei, 2017). Table. 2 shows heat conductivity coefficients of 6 kinds of nanofluids. Heat conductivity coefficient of CNT nanoparticles is the highest, being 3000 $W(m{\cdot}K)^{-1}$, followed by that (2300 $W(m{\cdot}K)^{-1}$) of ND nanoparticles, both of which are far higher than those of other 4 kinds of nanoparticles. It can be inferred that heat transfer performance of CNT is better than those of other 5 kinds of nanoparticles. This inference can be verified by Fig. 6: proportionality coefficient of energy transferred to workpiece under CNT NMQL grinding is 40.1%, indicating that heat generated in the grinding area is effectively dissipated and transferred under the effect of CNT nanoparticles; in addition, it can be found that heat conductivity coefficient of ND nanoparticles is dozens of times of that of Al_2O_3 nanoparticles, possibly because Al_2O_3 nanofluid has favorable lubricating property in the grinding process and reduces energy input in the grinding process. The generated grinding heat is fewer than that under ND nanofluid grinding, which can be verified by Fig. 5. Obviously, grinding temperature of ND nanofluid is higher than that of Al_2O_3 nanofluid, so proportionality coefficient of energy transferred to workpiece is small.

The possible reason why lubricating property of Al_2O_3 nanofluid is better than that of ND nanofluid is that hardness of Al_2O_3 nanoparticles is far lower than that of ND nanoparticles, and Al_2O_3 nanoparticles are deformed due to extrusion by grinding wheel in the grinding process. On the other hand, soft Al_2O_3 nanoparticles are melted on workpiece surface under the local high-temperature effect in the grinding area, and a layer of nanoscale protective film is formed, which reduces friction coefficient and grinding force and exerts good lubricating effect by virtue of its superior anti-friction and anti-wear characteristics (P. H. Lee, Nam, Li, & Lee, 2012; Yin et al., 2018). In addition, as shown in Fig. 3, that friction coefficient of ND nanofluid is higher than those of other nanofluids is related to its high hardness, and ND nanoparticles have relatively stable structure, so they will not be easily deformed or experience chemical reaction in the grinding process. Therefore, some ND nanoparticles are very likely to be filled into pores of grinding wheel and abrasive grain gaps, these nanoparticles act as abrasive grains of grinding

Table 2. Heat conductivity coefficients of the six nanoparticles

Nanoparticles	ZrO_2	SiO_2	Al_2O_3	MoS_2	ND	CNT
Heat conductivity coefficient/ $W(m{\cdot}K)^{-1}$	< 2	7.6	40	138	2300	3000

wheel and grind the workpiece with high-speed revolution of grinding wheel, "two-body machining" phenomenon occurs, so ND nanofluid has high friction coefficient.

8.4.2 Effects of Viscosity of Nanofluids on Heat Transfer Performance

Adding nanoparticles in the base fluid can not only increase heat conductivity coefficient of the base fluid and improve its enhanced heat transfer performance but also elevate viscosity of the base fluid. Nanofluid viscosity has a great bearing on heat transfer in the grinding process. Study by Wang et al. (2016) indicated that the higher the viscosity, the weaker the mobility of nanofluid at the grinding interface, the lower the heat quantity carried away by the nanofluid and the lower the heat transfer coefficient, and this would inhibit enhanced heat transfer performance of the nanofluid.

This viewpoint can be explained from the aspect of boundary-layer theory (Castilho, Fuller, Sane, & Liu, 2019): according the hydromechanical principle, when the fluid with a certain viscosity flows through the solid wall surface, fluid velocity close to the wall surface becomes zero due to existence of wall surface and under the influence of fluid viscosity, thus satisfying the boundary condition for non-slipping on wall surface. From zero velocity on the wall surface, frictional resistance hindering fluid flow will be generated between wall surface and fluid and transferred into the fluid due to fluid viscosity along the normal direction, but as the distance from the wall surface increases, the effect of wall surface is gradually weakened, fluid velocity is gradually recovered, a thin fluid layer with obvious velocity change is formed nearby the wall surface, and this thin layer is called velocity boundary layer; in the meantime, when the fluid passes through the wall surface with different temperature, this is similar to non-slipping boundary condition on the wall surface, fluid temperature close to the wall surface is surely equal to temperature on the wall surface, and one part of the heat transferred from the wall surface to fluid is carried downstream by the fluid in way of flowing, so the heat transferred along the normal direction of the wall surface becomes less and less, fluid temperature change becomes smaller and smaller, an area with abrupt temperature change is formed nearby the wall surface due to internal heat transfer in the fluid, and this area is called temperature boundary layer (Sheikholeslami, Jafaryar, et al., 2018; Sheikholeslami, Shehzad, & Li, 2018). From physical scale, velocity boundary layer reflects fluid momentum diffusivity, the thicker the velocity boundary namely the farther the fluid is from the influence area of surface frictional effect on fluid velocity, the greater the moment diffusivity coefficient of the fluid, and the stronger the momentum diffusion capacity of the fluid; temperature boundary layer reflects heat diffusivity of the fluid: the thicker the temperature boundary layer namely the farther the fluid is from the influence area of surface heating or cooling on fluid temperature, the stronger the heat diffusivity. Thermal diffusivity is a mark of heat diffusivity of the fluid, and the fluid with higher thermal diffusivity has thicker temperature boundary layer (Sheikholeslami, Shehzad, Abbasi, & Li, 2018; Subramani, Nagarajan, Mahian, & Sathyamurthy, 2018).

Figure 7. Schematic of the nanofluid boundary in the grinding area

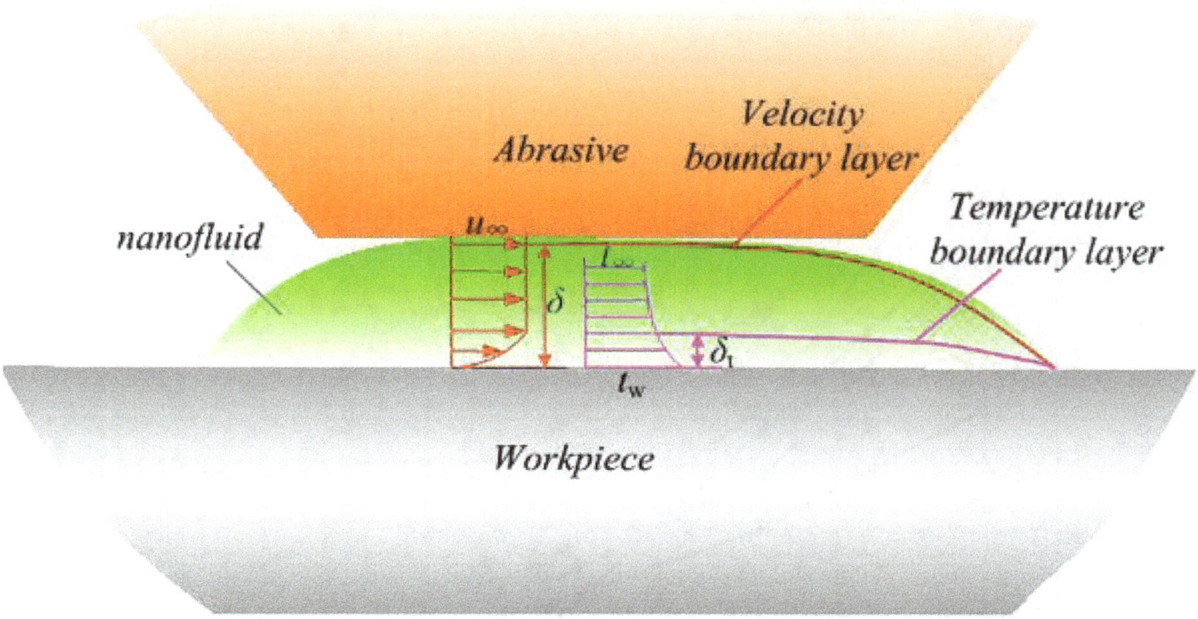

According to boundary-layer theory, as the nanofluid is viscous fluid, velocity boundary layer and temperature boundary layer will appear on workpiece surface after the nanofluid is sprayed onto workpiece surface through the MQL nozzle as shown in Fig. 8-7. Different nanofluids have different viscosity values. The nanofluid with large viscosity has thin nanofluid velocity boundary layer, so its momentum diffusivity is weak and mobility becomes poor; thickness of the temperature boundary layer is large, thermal diffusivity is strong, so the heat accumulated in the grinding area is higher than nanofluids with small viscosity values, and then its heat transfer coefficient is small, which weakens its enhanced heat transfer performance; thickness of velocity boundary layer of the nanofluid with low viscosity is the same as or slightly lower than that of the temperature boundary layer, so its heat transfer ability is favorable.

Fig. 8 shows curve chart of nanofluid viscosity change with temperature. Viscosity values of 6 nanofluids are greatly different under 40°C, but they are basically identical under100°C; meanwhile, it can be found that viscosity of CNT nanofluid is relatively large while that of MoS_2 nanofluid is relatively small. According to the above analysis, velocity boundary layer formed by CNT nanofluid on workpiece surface in the grinding process is smaller than that formed by MoS_2 nanofluid and temperature boundary layer formed by CNT nanofluid is thicker than that formed by MoS_2 nanofluid, so according to the analysis of thickness values of boundary layers, heat transfer performance of CNT nanofluid is poorer than that of MoS_2 nanofluid. However, this analysis doesn't accord with the obtained experimental results, which may be related to heat transfer coefficients of boundary layers formed by the 6 nanofluids.

According to the boundary layer theory, thickness values of velocity boundary layer and temperature boundary layer formed when viscous fluid flows through the wall surface can be expressed as follows (Amanbaev, 2018; Huang., 2014):

Figure 8. Relation curves of the temperature and viscosity of the six nanofluids

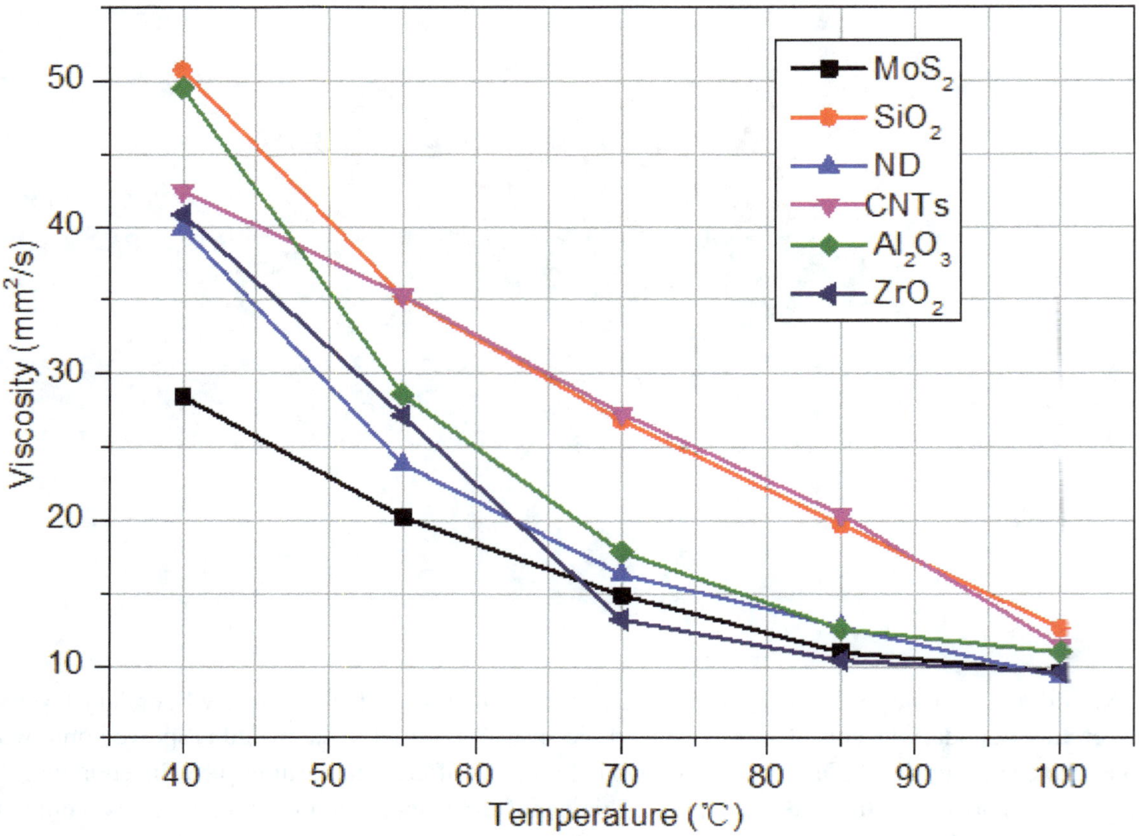

$$\delta = 4.64x\,\mathrm{Re}_x^{-1/2} \tag{1}$$

$$\xi = \frac{\delta_t}{\delta} = \frac{1}{1.026}\mathrm{Pr}^{-1/3} \tag{2}$$

where δ is velocity boundary layer; δ_t is temperature boundary layer; Re_x is Reynolds number of x point; Pr is Prandtl number.

$$\mathrm{Re}_x = \frac{\mu_\infty x}{v} = \frac{\mu_\infty x \rho}{\eta} \tag{3}$$

where μ_∞ is outlet flow velocity of nanofluid; ρ is nanofluid density; ν is dynamic viscosity of nanofluid; η is dynamic viscosity of nanofluid.

$$\Pr = \frac{v}{\alpha} = \frac{C_p \eta}{k} \tag{4}$$

where α is thermal diffusivity of nanofluid; C_p is isobaric specific heat capacity; k is heat conductivity.

Base oil of nanofluid is palm oil and order of magnitudes of Pr is 10^3. According to formula 98-2), thickness of the temperature boundary layer formed when nanofluid is sprayed into the grinding area is smaller than that of velocity boundary layer. If the velocity boundary layer is thick, its momentum diffusivity is high, so its heat transfer capability in the grinding area is better, and its heat transfer coefficient can be calculated through the following formula:

Heat transfer coefficient of local surface of boundary layer is:

$$h_x = 0.332 \frac{\lambda}{L} \mathrm{Re}_x^{1/2} \Pr^{1/3} \tag{5}$$

where λ is nanofluid heat conductivity coefficient, and its range is $Re \leq 5\times10^5$in the formula.

Workpiece length is set as L, then average heat transfer coefficient on the whole grinded workpiece surface is $h = \frac{1}{L} \int_0^L h_x dx$, and average heat transfer coefficient is obtained through integration:

$$h = 0.664 \frac{\lambda}{L} \mathrm{Re}_L^{1/2} \Pr^{1/3} \tag{6}$$

The curve of average heat transfer coefficients of boundary layers formed by 6 nanofluids is shown in Fig. 9:

According to the figure, average heat transfer coefficient of boundary layer formed by CNT nanofluid is the highest, being 1.3×10^4 W/(m·K), and that formed by ZrO_2 nanofluid is the lowest, being 0.08×10^4 W/(m·K). Through a comparison, the boundary layer formed by CNT nanofluid has the best heat transfer capability.

8.4.3 Effects of Nanofluid Contact Angle

During the MQL grinding process, nanofluid is sprayed from the nozzle into the grinding area between workpiece and grinding wheel in the form of droplets to exert cooling and lubricating effect. Therefore, the state of droplets falling on workpiece decides cooling and lubricating effect, which refers to the contact angle between droplets and workpiece. Zhang et al. (2015) investigated the influence of contact angle on lubricating effect of MQL grinding and found that contact angle had a significant influence on workpiece surface quality: the smaller the contact angle, the smaller the roughness value on workpiece surface. The smaller the contact angle of droplets in the grinding process, the better the obtained cooling and lubricating effect, because small contact angle of droplets represented large infiltration area of droplets (effective lubricating area of MQL grinding fluid), and the larger the infiltration area, the better the wettability, and the more sufficient the cooling and lubricating effect of MQL grinding fluid. Contact angle of nanofluid droplets sprayed onto workpiece not only had an important influence on lubricating effect but also on cooling effect, which was mainly manifested by wettability of droplets.

Figure 9. Mean heat conductivity curves of the boundary layers of the six nanofluids

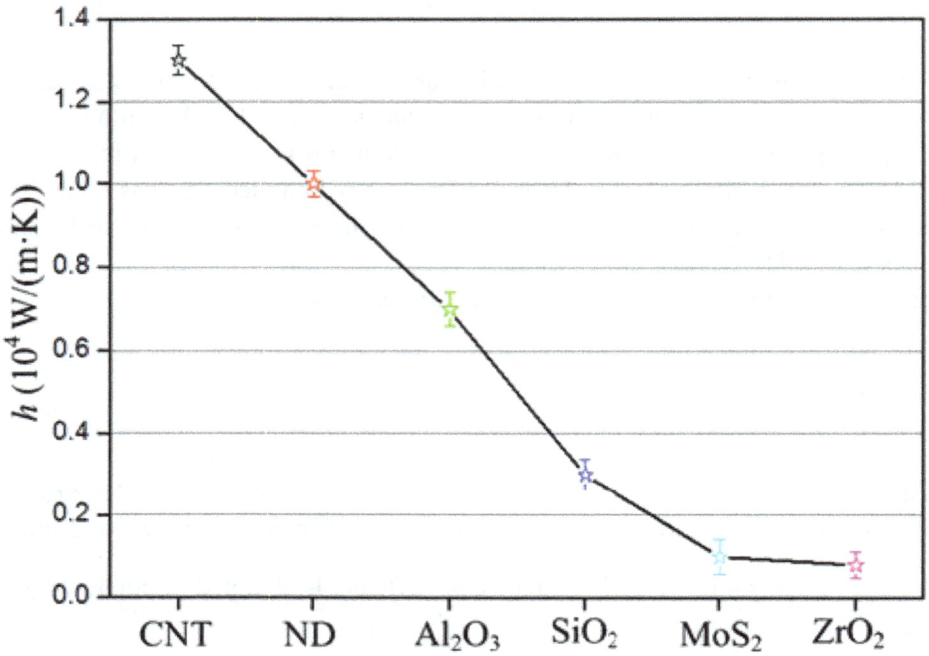

Fig. 10 shows contact angles of 6 nanofluids and the surface of nickel base alloy workpiece, $\theta(SiO_2)$=49.2°, θ(CNT)=47.5°, $\theta(MoS_2)$=46°, $\theta(Al_2O_3)$=45.5°, θ(ND)=43.5° and $\theta(ZrO_2)$=41.5°, and it can be seen that 6 nanofluids are all hydrophilic, so when sprayed into the grinding area, they can exert favorable lubricating effect and have good heat transfer performance. According to the above theory, ZrO_2 nanofluid has the best wettability on workpiece surface, so its heat transfer performance should be the best, but the proportionality coefficient of energy transferred to workpiece indicates that heat transfer performance of CNT nanofluid is better than that of ZrO_2 nanofluid, meaning that superior heat transfer performance is attributed to multiple factors. Contact angle, though having a great bearing on heat transfer performance, is not a decisive factor.

In addition, the relationship between convective heat transfer coefficient and contact angle can be expressed by $h \propto (1-\cos\theta)$, so heat transfer coefficients of another 5 nanofluids can be sorted as CNT>MoS_2>Al_2O_3>ND>ZrO_2 according to their contact angles. Therefore, in the heat transfer process, heat transfer coefficient of CNT nanofluid is higher than those of the 4 nanofluids, so the more the heat transferred out, the fewer the heat transferred into workpiece, which accords with experimental results of proportionality coefficient of energy transferred to workpiece.

8.4.4 Effects of Nanofluid Surface Tension

Das et al. (2003) measured nanofluid surface tension, and experimental results showed that adding nanoparticles in the base fluid to prepare the nanofluid could lower the surface tension of the base fluid, which then could enhance its heat transfer performance (2016), lower the temperature in the grinding

Figure 10. Contact angles between the MQL droplets of the six nanofluids and the workpiece

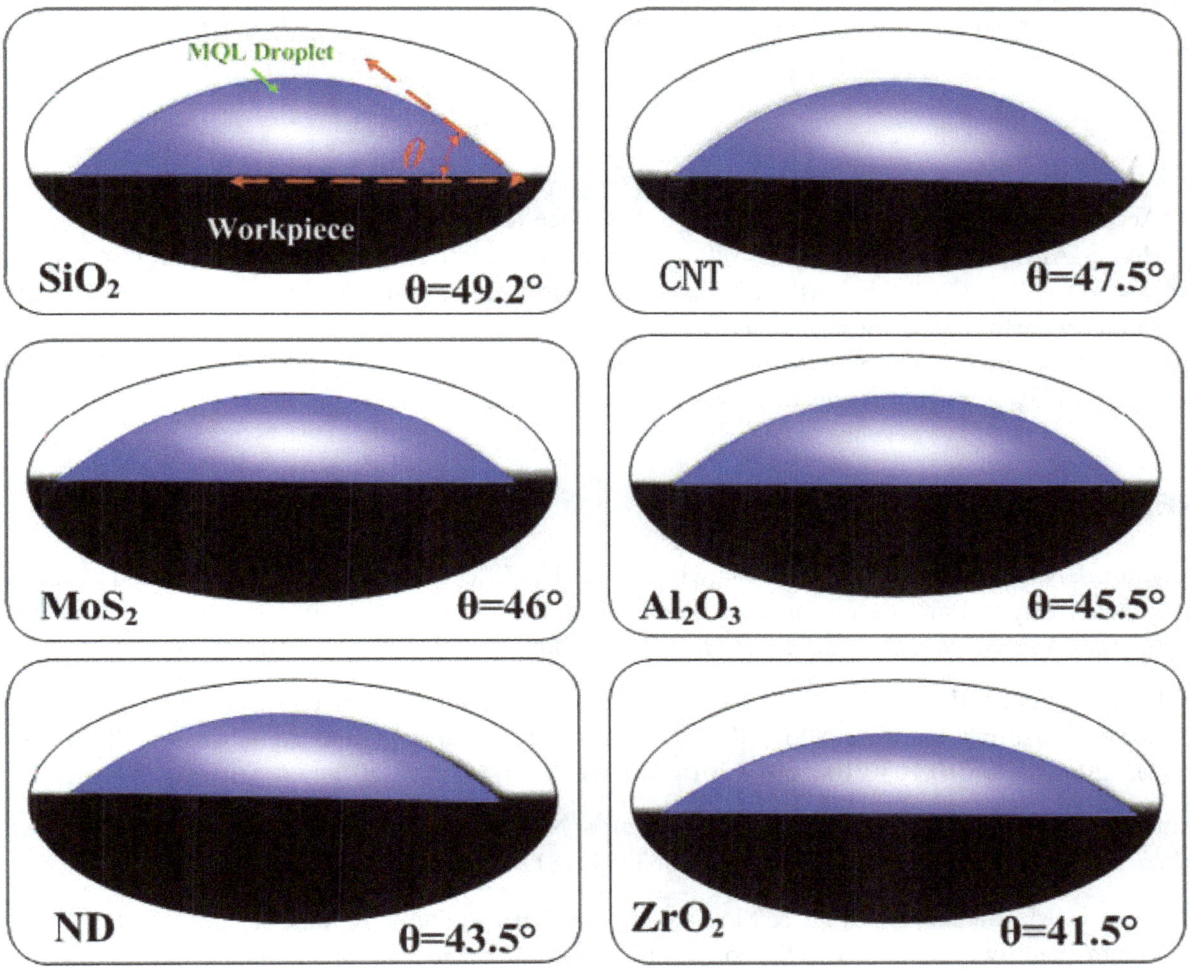

area, avoid workpiece burning and improve machining quality of workpiece. Meanwhile, the relationship between surface tension and heat transfer coefficient is:

$$h \propto 1/\sigma^3 \tag{7}$$

where σ is nanofluid surface tension.

Nanofluid surface tension can be calculated using Szyszkowski equation (Xu & Jia, 2000):

$$\frac{\sigma_0 - \sigma}{\sigma_0} = b\ln(\frac{\varnothing}{a} + 1) \tag{8}$$

where σ_0 is surface tension of base fluid; ϕ is volume fraction of nanoparticles; a and b are constants.

Table 3. Density of the six nanoparticles

Nanoparticles	ZrO_2	MoS_2	Al_2O_3	ND	SiO_2	CNT
Density g/cm³	5.8	4.8	3.7	3.45	2.2	2.1

According to formulas (8-5) and Table. 8-3, surface tensions of 6 nanofluids are sorted as $ZrO_2 > MoS_2 > Al_2O_3$>ND> SiO_2>CNT, so their heat transfer coefficients are sorted as $ZrO_2 < MoS_2 < Al_2O_3$<ND< SiO_2<CNT. Obviously, CNT nanofluid has the highest heat transfer coefficient with good heat transfer performance, the heat quantity transferred out in the grinding process is large while the heat quantity transferred into workpiece is small, so the energy proportionality coefficient is small, which accords with experimental results of energy proportionality coefficients very well.

REFERENCES

Ahammed, N., Asirvatham, L. G., & Wongwises, S. (2016). Effect of volume concentration and temperature on viscosity and surface tension of graphene-water nanofluid for heat transfer applications. *Journal of Thermal Analysis and Calorimetry*, *123*(2), 1399–1409. doi:10.100710973-015-5034-x

Akhtar, W., Sun, J. F., & Chen, W. Y. (2016). Effect of Machining Parameters on Surface Integrity in High Speed Milling of Super Alloy GH4169/Inconel 718. *Materials and Manufacturing Processes*, *31*(5), 620–627. doi:10.1080/10426914.2014.994769

Amanbaev, T. R. (2018). Modeling and Calculation of Nanofluid Flows in a Boundary Layer. *Technical Physics*, *63*(11), 1582–1589. doi:10.1134/S1063784218110026

Barczak, L. M., Batako, A. D. L., & Morgan, M. N. (2010). A study of plane surface grinding under minimum quantity lubrication (MQL) conditions. *International Journal of Machine Tools & Manufacture*, *50*(11), 977–985. doi:10.1016/j.ijmachtools.2010.07.005

Chen, X., Rowe, W. B., & Mccormack, D. F. (2000). Analysis of the transitional temperature for tensile residual stress in grinding. *Journal of Materials Processing Technology*, *107*(1-3), 216–221. doi:10.1016/S0924-0136(00)00692-0

Dambatta, Y. S., Sarhan, A. A. D., Sayuti, M. B., & Bin Abd Shukor, M. H. (2019). Fuzzy logic method to investigate grinding of alumina ceramic using minimum quantity lubrication. *International Journal of Applied Ceramic Technology*, *16*(4), 1668–1683. doi:10.1111/ijac.13219

Das, S. K., Putra, N., & Roetzel, W. (2003). Pool boiling characteristics of nano-fluids. *International Journal of Heat and Mass Transfer*, *46*(5), 851–862. doi:10.1016/S0017-9310(02)00348-4

de Castilho, C. J. B., Fuller, M. E., Sane, A., & Liu, J. T. C. (2019). Nanofluid Flow and Heat Transfer in Boundary Layers: The Influence of Concentration Diffusion on Heat Transfer Enhancement. *Heat Transfer Engineering*, *40*(9-10), 725–737. doi:10.1080/01457632.2018.1442298

Du, J. H., Lu, X. D., Deng, Q., & Bi, Z. B. (2015). Progress in the Research and Manufacture of GH4169 Alloy. *Journal of Iron and Steel Research International*, *22*(8), 657–663. doi:10.1016/S1006-706X(15)30054-6

Hadad, M. J., Tawakoli, T., Sadeghi, M. H., & Sadeghi, B. (2012). Temperature and energy partition in minimum quantity lubrication-MQL grinding process. *International Journal of Machine Tools & Manufacture*, *54-55*, 10–17. doi:10.1016/j.ijmachtools.2011.11.010

Hadada, M. J., Sadeghi, M. H., & Sadeghi, B. (2012). Temperature and energy partition in minimum quantity lubrication-MQL grinding process. *International Journal of Machine Tools & Manufacture*, *55*(3), 10–17. doi:10.1016/j.ijmachtools.2011.11.010

Huang, S. B. (2014). *Heat Transfer Theory*. Petroleum University Press.

Hussein, A. M., Lingenthiran, Kadirgamma, K., Noor, M. M., & Aik, L. K. (2018). Palm oil based nanofluids for enhancing heat transfer and rheological properties. *Heat and Mass Transfer*, *54*(10), 3163–3169. doi:10.100700231-018-2364-9

Ilyas, S. U., Pendyala, R., & Narahari, M. (2017). Rheological behavior of mechanically stabilized and surfactant-free MWCNT-thermal oil-based nanofluids. *International Communications in Heat and Mass Transfer*, *87*, 250–255. doi:10.1016/j.icheatmasstransfer.2017.07.015

Javed, M., Shaik, A. H., Khan, T. A., Imran, M., Aziz, A., Ansari, A. R., & Chandan, M. R. (2018). Synthesis of stable waste palm oil based CuO nanofluid for heat transfer applications. *Heat and Mass Transfer*, *54*(12), 3739–3745. doi:10.100700231-018-2399-y

Kalita, P., Malshe, A. P., Kumar, S. A., Yoganath, V. G., & Gurumurthy, T. (2012). Study of specific energy and friction coefficient in minimum quantity lubrication grinding using oil-based nanolubricants. *Journal of Manufacturing Processes*, *14*(2), 160–166. doi:10.1016/j.jmapro.2012.01.001

Kalita, P., Malshe, A. P., & Rajurkar, K. P. (2012). Study of tribo-chemical lubricant film formation during application of nanolubricants in minimum quantity lubrication (mql) grinding. *CIRP Annals - Manufacturing Technology*, *61*(1), 327-330. doi:10.1016/j.cirp.2012.03.031

Lee, P. H., Kim, J. W., & Lee, S. W. (2018). Experimental characterization on eco-friendly micro-grinding process of titanium alloy using air flow assisted electrospray lubrication with nanofluid. *Journal of Cleaner Production*, *201*, 452–462. doi:10.1016/j.jclepro.2018.07.307

Lee, P. H., Nam, J. S., Li, C., & Lee, S. W. (2012). An Experimental Study on Micro-Grinding Process with Nanofluid Minimum Quantity Lubrication (MQL). *International Journal of Precision Engineering and Manufacturing*, *13*(3), 331–338. doi:10.100712541-012-0042-2

Lee, P. H., Nam, T. S., Li, C., & Sang, W. L. (2011). *Environmentally-Friendly Nano-fluid Minimum Quantity Lubrication (MQL) Meso-scale Grinding Process Using Nano-diamond Particles*. Paper presented at the International Conference on Manufacturing Automation.

Li, B. K., Li, C. H., Zhang, Y. B., Wang, Y. G., Jia, D. Z., Yang, M., ... Sun, K. (2017). Heat transfer performance of MQL grinding with different nanofluids for Ni-based alloys using vegetable oil. *Journal of Cleaner Production*, *154*, 1–11. doi:10.1016/j.jclepro.2017.03.213

Li, B. K., Li, C. H., Zhang, Y. B., Wang, Y. G., Yang, M., Jia, D. Z., ... Ding, W. F. (2017). Numerical and experimental research on the grinding temperature of minimum quantity lubrication cooling of different workpiece materials using vegetable oil-based nanofluids. *International Journal of Advanced Manufacturing Technology*, *93*(5-8), 1971–1988. doi:10.100700170-017-0643-0

Li, X., Ma, S., & Meng, F. J. (2015). Surface integrity of GH4169 affected by cantilever finish grinding and the application in aero-engine blades. *Chinese Journal of Aeronautics*, *28*(5), 1539–1545. doi:10.1016/j.cja.2015.06.023

Mao, C., Tang, X., Zou, H., Huang, X., & Zhou, Z. (2012). Investigation of grinding characteristic using nanofluid minimum quantity lubrication. *International Journal of Precision Engineering and Manufacturing*, *13*(10), 1745–1752. doi:10.100712541-012-0229-6

Mao, C., Zhang, J., Huang, Y., Zou, H., Huang, X., & Zhou, Z. (2013). Investigation on the effect of nanofluid parameters on MQL grinding. *Advanced Manufacturing Processes*, *28*(4), 436–442. doi:10.1080/10426914.2013.763970

Mao, C., Zou, H. F., Zhou, X., Huang, Y., Gan, H. Y., & Zhou, Z. X. (2014). Analysis of suspension stability for nanofluid applied in minimum quantity lubricant grinding. *International Journal of Advanced Manufacturing Technology*, *71*(9-12), 2073–2081. doi:10.100700170-014-5642-9

Michaelides, E. E. (2015). Brownian movement and thermophoresis of nanoparticles in liquids. *International Journal of Heat and Mass Transfer*, *81*, 179–187. doi:10.1016/j.ijheatmasstransfer.2014.10.019

Paul, S., & Ghosh, A. (2019). Suitability of aqueous MoS_2 nanofluid for small quantity cooling lubrication-assisted diamond grinding of WC-Co cermets. *Proceedings of the Institution of Mechanical Engineers. Part B, Journal of Engineering Manufacture*, *233*(2), 426–442. doi:10.1177/0954405417739037

Pryazhnikov, M. I., Minakov, A. V., Rudyak, V. Y., & Guzei, D. V. (2017). Thermal conductivity measurements of nanofluids. *International Journal of Heat and Mass Transfer*, *104*, 1275–1282. doi:10.1016/j.ijheatmasstransfer.2016.09.080

Rabiei, F., Rahimi, A. R., & Hadad, M. J. (2017). Performance improvement of eco-friendly MQL technique by using hybrid nanofluid and ultrasonic-assisted grinding. *International Journal of Advanced Manufacturing Technology*, *93*(1-4), 1001–1015. doi:10.100700170-017-0521-9

Setti, D., Sinha, M. K., Ghosh, S., & Rao, P. V. (2015). Performance evaluation of Ti-6Al-4V grinding using chip formation and coefficient of friction under the influence of nanofluids. *International Journal of Machine Tools & Manufacture*, *88*, 237–248. doi:10.1016/j.ijmachtools.2014.10.005

Seyedzavvar, M., Shabgard, M., & Mohammadpourfard, M. (2019). Investigation into the performance of eco-friendly graphite nanofluid as lubricant in MQL grinding. *Machining Science and Technology*, *23*(4), 569–594. doi:10.1080/10910344.2019.1575403

Shabgard, M., Seyedzavvar, M., & Mohammadpourfard, M. (2017). Experimental investigation into lubrication properties and mechanism of vegetable-based CuO nanofluid in MQL grinding. *International Journal of Advanced Manufacturing Technology*, *92*(9-12), 3807–3823. doi:10.100700170-017-0319-9

Sharma, A. K., Tiwari, A. K., & Dixit, A. R. (2016). Effects of Minimum Quantity Lubrication (MQL) in machining processes using conventional and nanofluid based cutting fluids: A comprehensive review. *Journal of Cleaner Production*, *127*, 1–18. doi:10.1016/j.jclepro.2016.03.146

Sheikholeslami, M., Jafaryar, M., Saleem, S., Li, Z. X., Shafee, A., & Jiang, Y. (2018). Nanofluid heat transfer augmentation and exergy loss inside a pipe equipped with innovative turbulators. *International Journal of Heat and Mass Transfer*, *126*, 156–163. doi:10.1016/j.ijheatmasstransfer.2018.05.128

Sheikholeslami, M., Shehzad, S. A., Abbasi, F. M., & Li, Z. X. (2018). Nanofluid flow and forced convection heat transfer due to Lorentz forces in a porous lid driven cubic enclosure with hot obstacle. *Computer Methods in Applied Mechanics and Engineering*, *338*, 491–505. doi:10.1016/j.cma.2018.04.020

Sheikholeslami, M., Shehzad, S. A., & Li, Z. X. (2018). Water based nanofluid free convection heat transfer in a three dimensional porous cavity with hot sphere obstacle in existence of Lorenz forces. *International Journal of Heat and Mass Transfer*, *125*, 375–386. doi:10.1016/j.ijheatmasstransfer.2018.04.076

Shen, B., Shih, A. J., & Xiao, G. X. (2011). A Heat Transfer Model Based on Finite Difference Method for Grinding. *Journal of Manufacturing Science and Engineering-Transactions of the Asme*, *133*(3), 031001. doi:10.1115/1.4003947

Singh, T., Dureja, J. S., Dogra, M., & Bhatti, M. S. (2018). Environment Friendly Machining of Inconel 625 under Nano-Fluid Minimum Quantity Lubrication (NMQL). *International Journal of Precision Engineering and Manufacturing*, *19*(11), 1689–1697. doi:10.100712541-018-0196-7

Subramani, J., Nagarajan, P. K., Mahian, O., & Sathyamurthy, R. (2018). Efficiency and heat transfer improvements in a parabolic trough solar collector using TiO_2 nanofluids under turbulent flow regime. *Renewable Energy*, *119*, 19–31. doi:10.1016/j.renene.2017.11.079

Wang, Y. G., Li, C. H., Zhang, Y. B., Yang, M., Li, B. K., Dong, L., & Wang, J. (2018). Processing Characteristics of Vegetable Oil-based Nanofluid MQL for Grinding Different Workpiece Materials. *International Journal of Precision Engineering and Manufacturing-Green Technology*, *5*(2), 327–339. doi:10.100740684-018-0035-4

Wang, Y. G., Li, C. H., Zhang, Y. B., Yang, M., Li, B. K., Jia, D. Z., ... Mao, C. (2016). Experimental evaluation of the lubrication properties of the wheel/workpiece interface in minimum quantity lubrication (MQL) grinding using different types of vegetable oils. *Journal of Cleaner Production*, *127*, 487–499. doi:10.1016/j.jclepro.2016.03.121

Wu, D. X., Zhang, D. H., & Yao, C. F. (2018). Effect of Turning and Surface Polishing Treatments on Surface Integrity and Fatigue Performance of Nickel-Based Alloy GH4169. *Metals*, *8*(7), 549. doi:10.3390/met8070549

Xi, M. Z., Zhou, W., Shang, J. Y., Lv, C., Wu, Z. H., & Gao, S. Y. (2017). Effect of Heat Treatment on Microstructure and Mechanical Properties of Consecutive Point-Mode Forging and Laser Rapid Forming GH4169 Alloy. *Chin Shu Hsueh Pao*, *53*(2), 239–247. doi:10.11900/0412.1961.2016.00356

Xu, J. Y., & Jia, D. N. (2000). *Boiling Heat Transfer and Gas-liquid Flow*. Beijing: Atomic Energy Press.

Yang, M., Li, C. H., Zhang, Y. B., Jia, D. Z., Li, R. Z., Hou, Y. L., & Cao, H. J. (2019). Effect of friction coefficient on chip thickness models in ductile-regime grinding of zirconia ceramics. *International Journal of Advanced Manufacturing Technology*, *102*(5-8), 2617–2632. doi:10.100700170-019-03367-0

Ye, S., Gong, J. G., Zhang, X. C., Tu, S. T., & Zhang, C. C. (2017). Effect of Stress Ratio on the Fatigue Crack Propagation Behavior of the Nickel-based GH4169 Alloy. *Acta Metallurgica Sinica. English Letters*, *30*(9), 809–821. doi:10.100740195-017-0567-6

Yin, Q. A., Li, C. H., Dong, L., Bai, X. F., Zhang, Y. B., Yang, M., ... Li, R. Z. (2018). Effects of the physicochemical properties of different nanoparticles on lubrication performance and experimental evaluation in the NMQL milling of Ti-6Al-4V. *International Journal of Advanced Manufacturing Technology*, *99*(9-12), 3091–3109. doi:10.100700170-018-2611-8

Zhang, D. K., Li, C. H., Jia, D. Z., Zhang, Y. B., & Zhang, X. W. (2015). Specific grinding energy and surface roughness of nanoparticle jet minimum quantity lubrication in grinding. *Chinese Journal of Aeronautics*, *28*(2), 570–581. doi:10.1016/j.cja.2014.12.035

Zhang, D. K., Li, C. H., Zhang, Y. B., Jia, D. Z., & Zhang, X. W. (2015). Experimental research on the energy ratio coefficient and specific grinding energy in nanoparticle jet MQL grinding. *International Journal of Advanced Manufacturing Technology*, *78*(5-8), 1275–1288. doi:10.100700170-014-6722-6

Zhang, J. C., Li, C. H., Zhang, Y. B., Yang, M., Jia, D. Z., Liu, G. T., ... Cao, H. J. (2018). Experimental assessment of an environmentally friendly grinding process using nanofluid minimum quantity lubrication with cryogenic air. *Journal of Cleaner Production*, *193*, 236–248. doi:10.1016/j.jclepro.2018.05.009

Zhang, X. P., Li, C. H., Jia, D. Z., Gao, T., Zhang, Y. B., Yang, M., ... Ji, H. J. (2019). Spraying parameter optimization and microtopography evaluation in nanofluid minimum quantity lubrication grinding. *International Journal of Advanced Manufacturing Technology*, *103*(5-8), 2523–2539. doi:10.100700170-019-03642-0

Zhang, Y. B., Li, C. H., Jia, D. Z., Zhang, D. K., & Zhang, X. W. (2015). Experimental evaluation of the lubrication performance of MoS_2/CNT nanofluid for minimal quantity lubrication in Ni-based alloy grinding. *International Journal of Machine Tools & Manufacture*, *99*, 19–33. doi:10.1016/j.ijmachtools.2015.09.003

Chapter 9 Experimental Evaluation on the Effect of Nanofluids Physical Properties With Different Concentrations on Grinding Temperature

ABSTRACT

This chapter is proposed to solve the insufficient MQL cooling and heat transfer capability based on the heat transfer enhancement theory of solid. Adding nanoparticles into the base fluid can significantly elevate heat conductivity coefficient of the base fluid and enhance convective heat transfer capability of the grinding area. Researchers have carried out numerous experimental studies on nanofluids with different concentrations. However, the scientific nature of MQL cooling has not been explained. Degradable, nontoxic, low-carbon, and environmentally friendly green grinding fluid, palm oil taken as the base fluid, grinding force, grinding temperature and proportionality coefficient of energy transferred to workpiece of nanofluids with different volume fractions, are investigated in this chapter. Based on the analysis of the influence of physical characteristics of nanofluids on experimental results, cooling and heat transfer mechanism of NMQL grinding is studied. The experimental study can provide a certain technical guidance for industrial machining.

9.1 INTRODUCTION

Grinding is the final procedure in machining to achieve high machining accuracy (Li, Wang, & Wu, 2008). Given that grinding requires an extremely high energy input, a large portion of energy will be converted in the grinding zone. Most is converted to heat, thereby increasing the temperature in the grinding zone (Barczak, Batako, & Morgan, 2010). Grinding temperature can significantly affect the surface quality of the workpiece and the cutting performance of abrasive particles of the grinding wheel (Shen, Shih, &

DOI: 10.4018/978-1-7998-1546-4.ch009

Tung, 2009). Higher grinding temperature will change the superficial metallographic structure of parts, causing grinding burns (Rowe, 2014). Moreover, high temperature will decrease the hardness of abrasive particles considerably and cause bonding wear and diffusive wear between abrasive particles and the workpiece, inducing rapid passivation of abrasive particles (Miyashita, 1990). Therefore, grinding temperature is important toin machining accuracy and quality. Grinding temperature should be studied.

With the technological progress and proposed further higher requirements on processing technology, MQL grinding is widely accepted by researchers. Nanofluid MQL grinding was proposed based on the theory of solid heat transfer enhancement to overcome the poor heat transfer of MQL cooling. The addition of nanoparticles into the base fluid can significantly increase the thermal conductivity of the base fluid and enhance the heat transfer capacity of the grinding zone (Ozerinc, Kakac, & Yazicioglu, 2010). Vajjha and Das added Al_2O_3 nanoparticles into the base fluid and prepared a 6 vol.% nanofluid. The tandem tested the thermal conductivity of the nanofluid through experiments and found that the prepared 6 vol.% nanofluid achieved 22.4% higher thermal conductivity compared with the base fluid (Vajjha & Das, 2012). Choi et al. carried out an experimental study on the effective thermal conductivity of multi-walled carbon nanotube (MWCNT) nanofluid, and found that the thermal conductivity of MWCNT was markedly higher than the theoretical value. The measured thermal conductivity was 150% higher than that of the base fluid (Li, 2011) Hong et al. (2005) synthesized the Fe-–glycol nanofluid by chemical vapor condensation method and increased its dispersity by using ultrasonic and high-energy pulses. According to experimental results, thermal conductivity presents a linear positive correlation with the volume fraction of nanoparticles. The thermal conductivity of the nanofluid with 0.55 vol.% nanoparticles was 18% higher than that of the base fluid. Murshed et al. (2005) measured the thermal conductivity of rod-like and spherical TiO_2 water-based nanofluid by using a transient barretter mount and a new integral relational model. The experimental results reflected that thermal conductivity increases with the increase in volume fraction of nanoparticles. In addition, the shape and size of nanoparticles could affect thermal conductivity (Mintsa, Roy, Nguyen, & Doucet, 2009). When the volume concentration of nanofluids was fixed at 5%, the nanofluids with rod-like and spherical nanoparticles achieved 33% and 30% higher thermal conductivity than the base fluid.

On the basis of the fact that nanoparticles can significantly increase the heat transfer capacity of the base fluid significantly (Sadeghi, Haddad, Tawakoli, & Emami, 2009), researchers have shifted their attention to grinding and conducted numerous experimental studies. Lee et al. (2011) prepared nanofluids by adding diamond nanoparticles into liquid paraffin and conducted three grinding experiments by using dry grinding, liquid paraffin MQL grinding, and nanofluid MQL grinding. Results showed that nanofluid MQL considerably decreased grinding force and surface roughnes, and workpiece surface quality was improved with the reduction of nanoparticle size. Mao et al. (2012) carried out different grinding experiments involving dry grinding, pouring grinding, MQL grinding, and nanofluid MQL grinding, and discovered that nanofluid MQL grinding dattained lower grinding force and grinding temperature and more ideal workpiece surface quality compared with other grinding processes. Mao et al. (2012) further discussed the effect of nanofluid MQL parameters through experiments. The group concluded that the lowest grinding temperature is achieved at a nozzle angle wof 15°, distance between nozzle and the grinding zone of 20 mm, and air pressure of 0.6 MPa. ; moreover, nanofluid MQL technology resolves heat transfer in the grinding area while enhancing the lubricating property in the area (Mao et al., 2014). Sayuti et al. (2013) conducted an experimental study on the face milling of AL 6061-T6 alloy with a SiO_2 nanofluid, and observed that the SiO_2 nanofluid can contribute low cutting force, cutting temperature, and surface roughness. These studies indicate that grinding is improved by MQL technology. More importantly, the dosage of the grinding fluid significantly decreases, thereby reducing hazards to workers and the environment.

According to the good machining performance of nanofluid MQL cooling, researchers have studied nanofluids with different concentrations. Shen et al. (2008) prepared the volume fractions of 1%, 2.5%, and 4% nanofluids by using Al_2O_3 nanoparticles for cast-iron grinding. The team discovered in experiments that compared with dry grinding, pouring grinding, and MQL grinding, nanofluid MQL grinding gained significantly lower grinding force and surface roughness, eliminated workpiece burns effectively, and increased G ratio. Moreover, 1% nanofluid MQL grinding achieved the lowest grinding force and G ratio, indicating the best surface quality. Kalita et al. (2012) added MoS_2 nanoparticles into paraffin oil and soybean oil for nanofluid grinding of cast-iron and EN24 alloy steel. According to experimental results, MoS_2 nanofluid MQL grinding reduces grinding force, friction coefficient, and specific grinding energy, but increased G ratio significantly. Kalita et al. (2012) prepared nanofluids at 2% and 8% mass fractions by adding MoS_2 nanoparticles into paraffin oil and soybean oil, which were used in the nanofluid MQL grinding of cast-iron and EN24 alloy steel. The group found that 8% MoS_2 nanofluid MQL grinding achieved lower grinding force, friction coefficient, and specific grinding energy, but higher G ratio compared with 2% MoS_2 nanofluid MQL grinding. Furthermore, the 8% MoS_2 nanofluid MQL grinding showed lower grinding temperature and better wear resistance. Setti et al. (Setti, Sinha, Ghosh, & Rao, 2015) tested performances of MQL grinding and MQL grinding with different concentrations of water-based Al_2O_3 nanofluids on Ti-–6Al-–4V alloy. According to test results, water-based Al_2O_3 nanofluid MQL grinding got smaller surface roughness than MQL grinding. Besides, the 4 vol.% water-based Al_2O_3 nanofluid MQL grinding yielded better workpiece quality 1 vol.% water-based Al_2O_3 nanofluid MQL grinding. Zhang et al. (2015) added MoS_2 nanoparticles to soybean oil and prepared nanofluids with different mass fractions (2%, 4%, 6%, and 8%) for MQL grinding of 45 steel. The experimental results demonstrated that the mass fraction of 6% nanofluid achieved the lowest grinding force and surface roughness, reflecting the best surface quality. Zhang et al. (2015) used different concentrations of nanofluids for the nanofluid MQL grinding of quenched 45 steel and analyzed energy proportionality coefficient and specific grinding energy during grinding. In this experiment, they also used soybean oil-based CNT nanofluids with different volume fractions (1%, 2%, and 3%) and found that the volume fraction of 2% nanofluid MQL grinding yielded the lowest grinding temperature and energy proportionality coefficient. Viewed in terms of the thermal conductivity of CNT nanoparticles, the CNT nanofluid presents good thermal conductivity and could significantly decrease the temperature of the grinding zon. In addition, vegetable oil is available for use as MQL base oil and is advantageous because of its environment friendliness, non-hazardous effect ton operators' health, andreduction in the cost of grinding fluid compared with flood grinding (Mechiri, Vasu, & Gopal, 2017).

This literature review reveals that most of the existing literature focused on the influences of nanofluid parameters (e.g.such as nanoparticle type, nanofluid concentration, and base fluid type), jet parameters (such as nozzle distance, nozzle angle, and air pressure) and grinding parameters (such as feed speed, grinding depth, circular velocity of the abrasive wheel, and workpiece type) on grinding force, grinding temperature and workpiece quality. They did not provide a relatively reasonable explanation for the experimental phenomena and disclose the scientific essence of MQL from the physical properties (such viscosity and surface tension) of nanofluids. In this paper, palm oil, a degradable, non-toxic, low-carbon, and environmental-friendly grinding fluid, was used as base oil. The grinding force, grinding temperature, and proportionality coefficient of MQL grinding based on nanofluids with different volume fractions were studied. The heat transfer mechanism of nanofluid MQL grinding was discussed by analyzing the influences of the physical properties of nanofluid on experimental results. Results can provide certain technical references for industrial machining and present important reference significance.

9.2 EXPERIMENTAL DESIGN

9.2.1 Experimental Materials

Workpiece material used in the experiment is high-temperature nickel base alloy GH4169, and its chemical compositions and characteristic parameters are shown in Table. 7-3 and Table. 7-4. High-temperature alloy has been widely applied to many fields because of its good comprehensive performance, but its heat conductivity coefficient is low. Therefore, a large quantity of heat is generated in the machining process, thus affecting its machining performance. It's necessary to conduct a more extensive and profound study on this material. For the convenience of grinding experiment and data measurement, dimensions of the workpiece sample are 40×30×30 mm.

NMQL grinding is used in the experiment. Experimental study in Chapter 7 indicates that palm oil has good lubricating and heat transfer characteristics, so it is the most suitable to be MQL base oil among seven kinds of vegetable oils. Zhang et al. (2015) experimented on MQL grinding of quenched 45 steel with nanofluids of different volume fractions and analyzed energy proportionality coefficient and specific grinding energy in the grinding process. Soybean oil-based carbon nanotube (CNT) nanofluids with volume fractions 1%, 2% and 3% are used in the experiment, and it's found that CNT NMQL grinding with 2% nanofluid volume fraction obtains the lowest grinding temperature and energy proportionality coefficient. According to heat conductivity coefficient of CNT nanoparticles, CNT nanofluid has good heat-conducting property and significantly lowers the temperature in the grinding area. Therefore, nanoparticles used in this experiment are carbon nanotubes (CNT), and particle size and average length of nanoparticles are 50 nm and 10-30 μm respectively; base oil and nanoparticles used in the experiment are shown in Fig. 1. The nanofluid is prepared using two-step method (Shen & Gradworks, 2008). Base oil is mixed with nanoparticles of certain masses, and volume fractions of prepared nanofluids are 0.5%, 1%, 1.5%, 2%, 2.5%, 3%, 3.5% and 4% respectively. Dispersing agent should be added into the base oil to prevent agglomeration phenomenon of nanoparticles. When the content of added dispersing agent is too small, nanoparticles may experience agglomeration. When the additive amount is too large, it will affect physiochemical characteristics of the nanofluid, so 0.3% Polysorbate80 is added as dispersing agent (Jia, Li, Zhang, Zhang, & Zhang, 2014), and oscillated in an ultrasonic oscillator (model: KQ3200DB) for 1h. Temperature-dependent variable value of the nanofluid is measured using Brookfield DV2T viscosimeter and DKB-501A super thermostatic water bath; nanofluid contact angle is measured using contact angle measuring instrument as shown in Fig. 7-2.

9.2.2 Experimental Scheme

Grinding equipment used in the experiment and grinding parameters are identical with those in Chapter 3 and Chapter 4. The same method is still used in the experiment to measure grinding force and grinding temperature. Eight kinds of palm oil-based CNT nanofluids with different concentrations are prepared for nanofluid plane grinding of high-temperature nickel base alloy. According to experimental results

Figure 1. Nanoparticles and base oil

and from physical characteristics of nanofluids like viscosity, surface tension and contact angle, heat transfer mechanism of eight nanofluids with different concentrations are studied. The experimental program is shown in Table. 1.

9.3 EXPERIMENTAL RESULTS

9.3.1 Grinding Force

Grinding force can be used to diagnose grinding conditions, including lubricating state, grinding wheel wear, etc., and it has an important influence on workpiece surface quality, specific grinding energy and grinding wheel quality, so it's an extremely important parameter in the grinding process. Fig. 2 shows grinding force signal images of eight different concentrations of nanofluids in the MQL grinding process.

Fig. 3 shows grinding forces of eight nanofluids with different volume fractions in the MQL grinding process. Under volume fractions of 0.5%, 1%, 1.5%, 2%, 2.5%, 3%, 3.5% and 4%, tangential grinding forces are 22.25 N, 23.12 N, 23.26 N, 21.93 N, 21.30 N, 22.60 N, 22.61 and 23.62 N respectively and normal grinding forces are 2.47 N, 71.65 N, 74.94 N, 49.50 N, 59.20 N, 71.22 N, 67.44 N and 68.99 N respectively. It can be seen that tangential grinding force and normal grinding force obtained under 2% and 2.5% nanofluid volume fractions are low, indicating that nanofluids with such volume fractions have satisfactory lubricating properties.

9.3.2 Grinding Temperature

Fig. 4 shows grinding temperature curves obtained through MQL grinding with 8 nanofluids; when grinding wheel grinds the workpiece, workpiece surface temperature will rapidly rise and reach the maximum value, and it will gradually decline under the effect of the nanofluid.

Table. 1 Different concentration nanofluids minimum quantity lubricant grinding experimental program

Experiment No.	Nanoparticles	Base oil	Volume concentration	Cooling & lubricating condition
1	CNT	Palm oil	0.5%	MQL
2	CNT	Palm oil	1%	MQL
3	CNT	Palm oil	1.5%	MQL
4	CNT	Palm oil	2%	MQL
5	CNT	Palm oil	2.5%	MQL
6	CNT	Palm oil	3%	MQL
7	CNT	Palm oil	3.5%	MQL
8	CNT	Palm oil	4%	MQL

Figure 2. Grinding force signal image of eight nanofluid

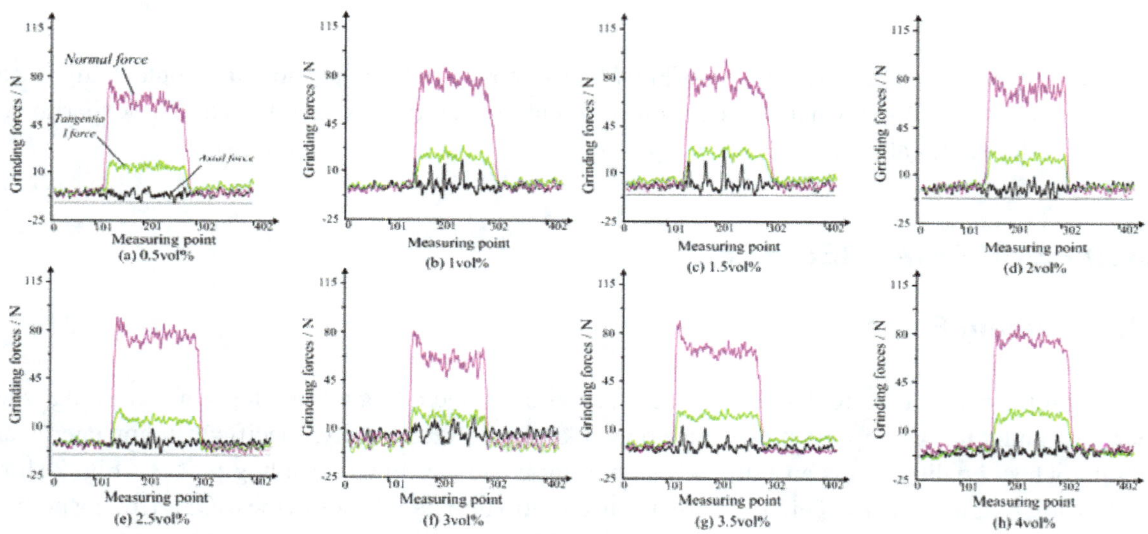

Fig. 5 shows grinding temperatures in the MQL grinding experiment with 8 nanofluids. Grinding temperatures in the MQL grinding with nanofluids of 0.5%, 1%, 1.5%, 2%, 2.5%, 3%, 3.5% and 4% volume concentrations are 109.4 °C, 112.4 °C, 117.8 °C, 108.9 °C, 110.3 °C, 114.2°C, 113.8 °C and 116.5 °C respectively. It can be seen that as the nanofluid volume fraction increases, grinding temperature presents firstly rising and then declining and rising again trendy, where nanofluid with 2% volume fraction has the lowest grinding temperature. The change trend of grinding temperature is related to physical properties of the nanofluid.

Figure 3. The results of grinding force (95% CI)

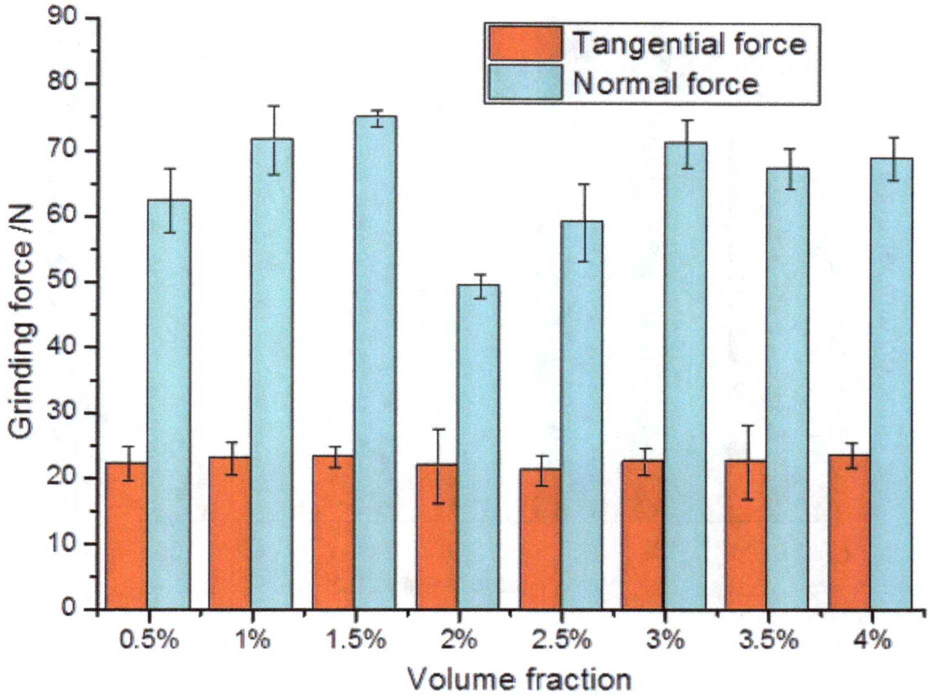

Figure 4. Grinding temperature curves (95% CI)

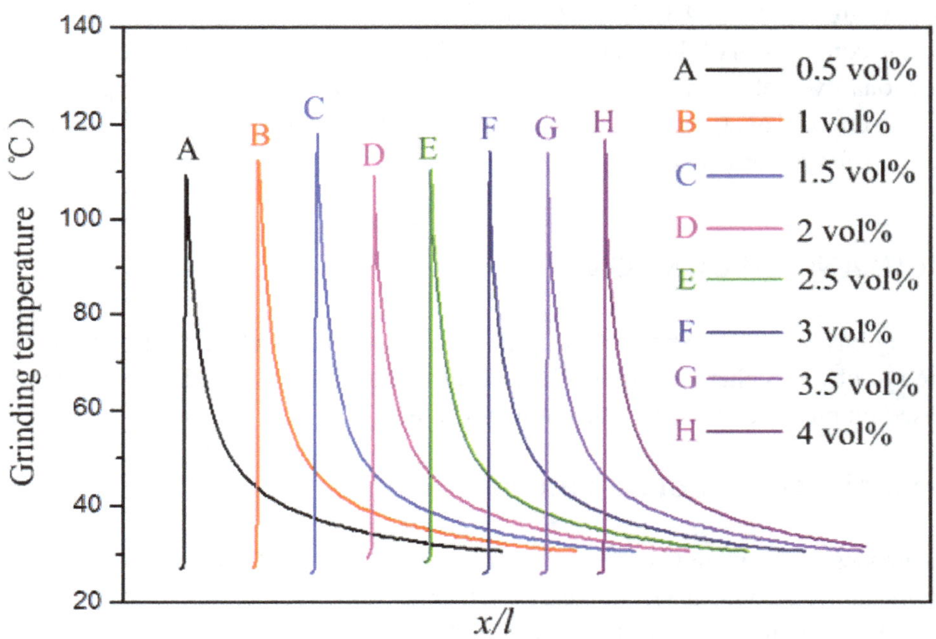

Figure 5. Grinding temperatures of eight kinds of nanofluids (95% CI)

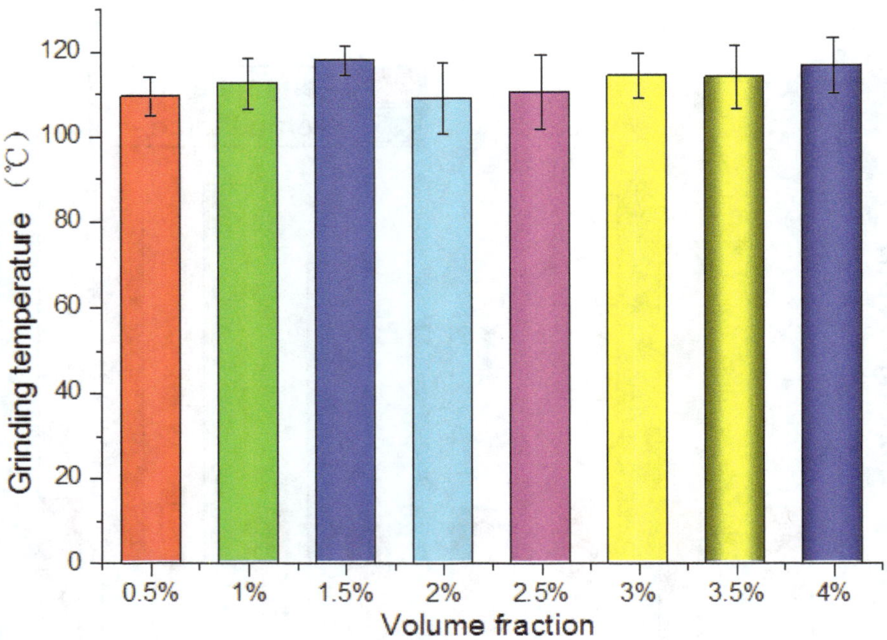

9.3.3 Energy Partition of Workpieces

Fig. 6 shows proportionality coefficients of energy transferred to workpiece obtained through the experiment. Energy proportionality coefficients obtained through MQL grinding with nanofluids of 0.5%-4% volume fractions are 43.76%, 44.15%, 43.7%, 42.7%, 46.6%, 45.4%, 46.5% and 44.38% respectively. According to the comparison of the obtained results, the nanofluid with 2% volume fraction obtains the lowest proportionality coefficient of energy transferred to workpiece. Through experimental results of grinding temperature and energy proportionality coefficient, the nanofluid with 2% volume fraction has superior heat transfer performance.

9.4 RESULTS AND DISCUSSION

NMQL grinding has advantages not owned under other cooling & lubricating conditions, and this is closely related to physical characteristics of nanofluids. Experimental results are hereby analyzed and discussed based on physical characteristics of nanofluids.

9.4.1 Effects of Viscosity of Nanofluids

Viscosity influences mobility and heat transfer characteristic of nanofluid as a very important physical property parameter. Heat conductivity of base fluid increases with addition of nanoparticles, so viscosity of base fluid will be enhanced. According to the study carried out in Chapter 7, castor oil MQL grind-

Figure 6. Energy ratio coefficients of heat energy transferred into workpieces of eight nanofluids (95% CI)

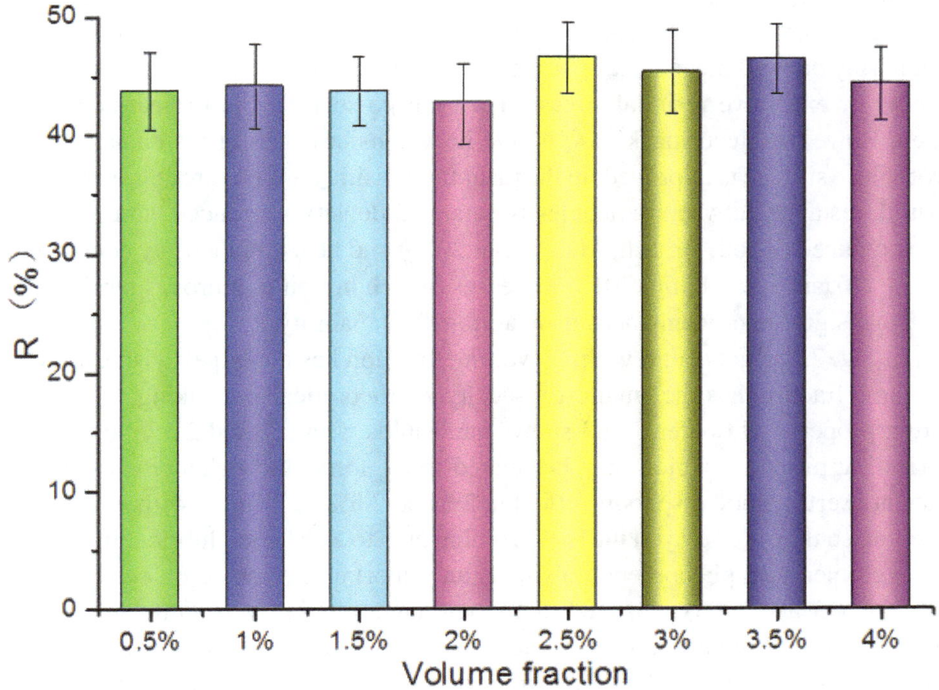

ing obtains the highest grinding temperature mainly because viscosity of castor oil is too high. Due to its poor mobility in the grinding area, convective heat transfer efficiency is reduced, and grinding heat can't be effectively dissipated; it's also found that other vegetable oils with relatively low viscosity like palm oil and soybean oil have satisfactory heat transfer performance, because vegetable oils with low viscosity have good mobility and can rapidly carry heat out of the grinding area under the scouring action of high-pressure air, thus avoiding heat accumulation in the grinding area and improving heat transfer performance. Therefore, nanofluid viscosity is a nonnegligible influence factor of convective heat transfer performance.

Fig. 7 shows temperature-dependent curves of eight nanofluids with different volume fractions. Nanofluid viscosity presents nonlinear growth trend as volume fraction of CNT nanoparticles increases due to enhanced interactions of nanoparticles in the palm oil, including interaction and agglomeration between particles (Chandrasekar, Suresh, & Senthilkumar, 2012). Wang et al. (2016) studied lubricating properties of different kinds of nanofluids for nickel base alloy grinding and found that nanofluid viscosity could exert an important influence on lubricating property in the grinding process, and high-viscosity nanofluid had better lubricating property, because high viscosity could prevent flowing trend of nanofluid, and then grinding fluid could stay in the grinding area for a longer time so as to improve lubricating property at grinding wheel/workpiece contact interface, reduce friction and inhibit rapid wear of abrasive grains; on the other hand, as viscosity increased, colloidal force, Brownian movement and viscous force between grinding fluid molecules correspondingly increased, which increased thickness and strength of adsorption film so that their lubricating properties were somehow improved. According to

Fig. 7, nanofluid viscosity gradually declines with temperature, possible because as the temperature rises, nanofluid volume is expanded, interaction between nanoparticles and palm oil molecules is weakened and then nanofluid viscosity declines. After nanofluid viscosity is reduced, its mobility is enhanced and the time for it to stay on workpiece surface is shortened, heat can be rapidly transferred out of workpiece surface, burning or excessive residual stress on workpiece surface due to heat accumulation is thus avoided (Chen, Rowe, & Mccormack, 2000), and heat transfer capability is enhanced and machining quality of workpiece surface is improved; in the meantime, rising temperature aggravates thermal motion in the nanofluid, distance between molecules is enlarged, density is reduced, and then surface tension of nanofluid is reduced (Zhou, Di, Liu, Ma, & Cai, 2016) and heat conductivity coefficient (Wusiman, Jeong, Tulugan, Afrianto, & Chung, 2013) increases, which not only improves lubricating property of nanofluid but also significantly enhances its heat transfer capability.

As shown in Fig. 7, the nanofluid with 4% volume fraction has the largest viscosity while nanofluid with 0.5% volume fraction has the smallest viscosity, so theoretically speaking, 4% nanofluid has the best lubricating property. However, Fig.3 shows nanofluids with 2% and 2.5% volume fractions have the best lubricating property, which may be related to agglomeration phenomenon of nanoparticles. Dispersing agent exerts a good dispersing effect in 2% and 2.5% nanofluids without serious agglomeration phenomenon, so they are sprayed into the grinding area to exert good lubricating effect. In addition, nanoparticle agglomeration phenomenon also has an important influence on convective heat transfer performance. Xuan et al. (2003) proposed that when nanoparticles were uniformly distributed in the nanofluid namely no agglomeration phenomenon, they would keep high degree of thermal motion state and temperature gradient inside the fluid became small so as to enhance heat transfer and then overall

Figure 7. The viscosity-temperature curve of eight kinds of nanofluids

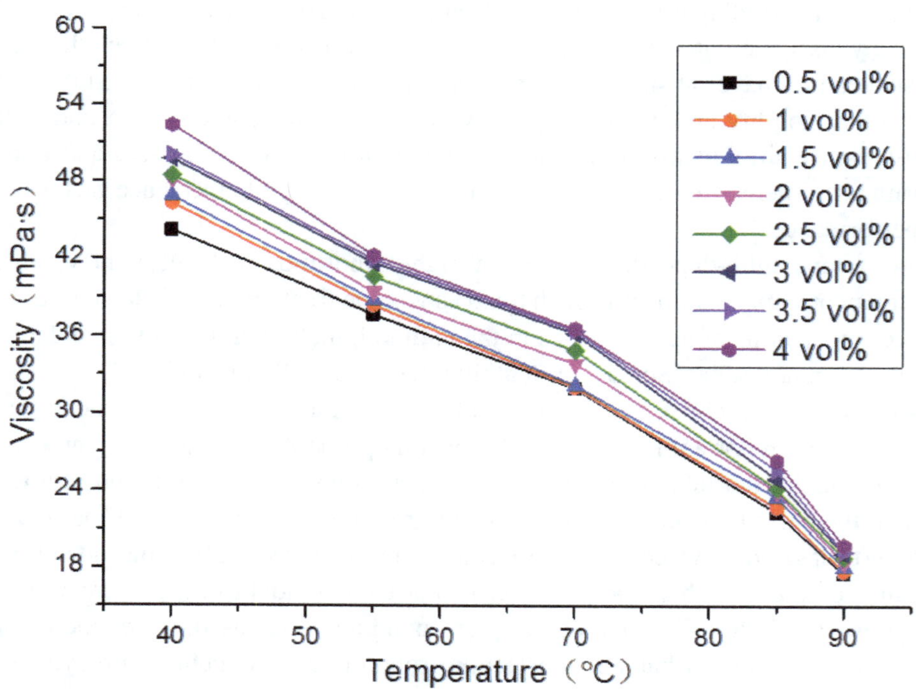

convective heat transfer performance was the best; Keblinsk et al. (2002) put forward that loose aggregation structure of nanoparticles was conducive to heat transfer, the area with large heat conductivity coefficient in the nanofluid was larger than the actual solid-phase area, and the looser the aggregation structure, the larger the effective heat transfer volume, and the strong the heat transfer capability; Xie et al. (2003) also proposed similar viewpoint

In order to analyze the influence of nanofluid viscosity in the grinding area on grinding temperature, nanofluid viscosity should be calculated. According to experimental data and existing models, Namburu et al. (2008) proposed the following formula:

$$\log(\mu_{ff}) = Ae^{-BT} \tag{1}$$

where A and B are constants. Kole and Dey (2010) used experimental results to verify the formula.

As temperature in the grinding area is higher than water bath heating temperature, formula (1) is fitted according to viscosity data of eight nanofluids with different volume fractions, and obtained values of constants A and B are shown in Table. 2.

Viscosity data of eight nanofluids under temperature in the grinding area are calculated as shown in Fig. 9.

Figure 8. Schematic of nanoparticles dispersed state

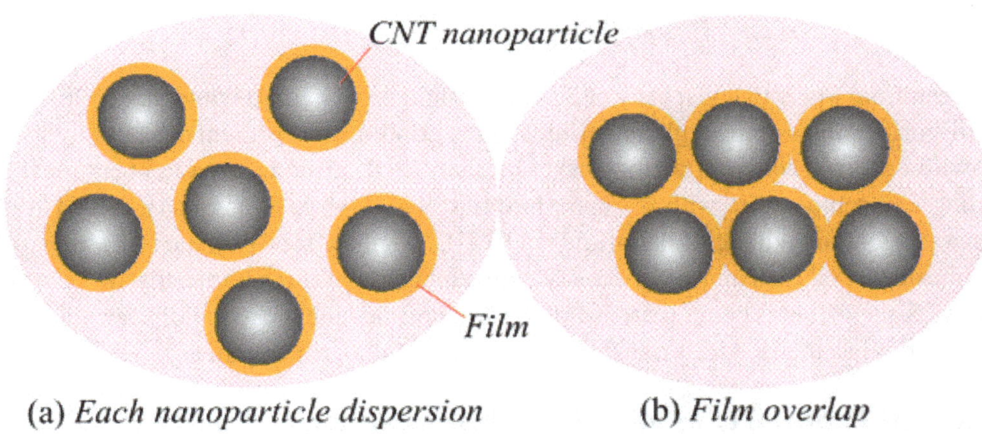

Table 2. Fitting numeric of constants A and B

	0.5%	1%	1.5%	2%	2.5%	3%	3.5%	4%
A($\times10^{-3}$)	7.1	6.4	6.95	11	6.8	7.1	5.67	6.13
B	2.27	2.2	2.36	3.4	2.23	2.57	2.34	2.4

Figure 9. Nanofluids viscosity values at the grinding area temperature

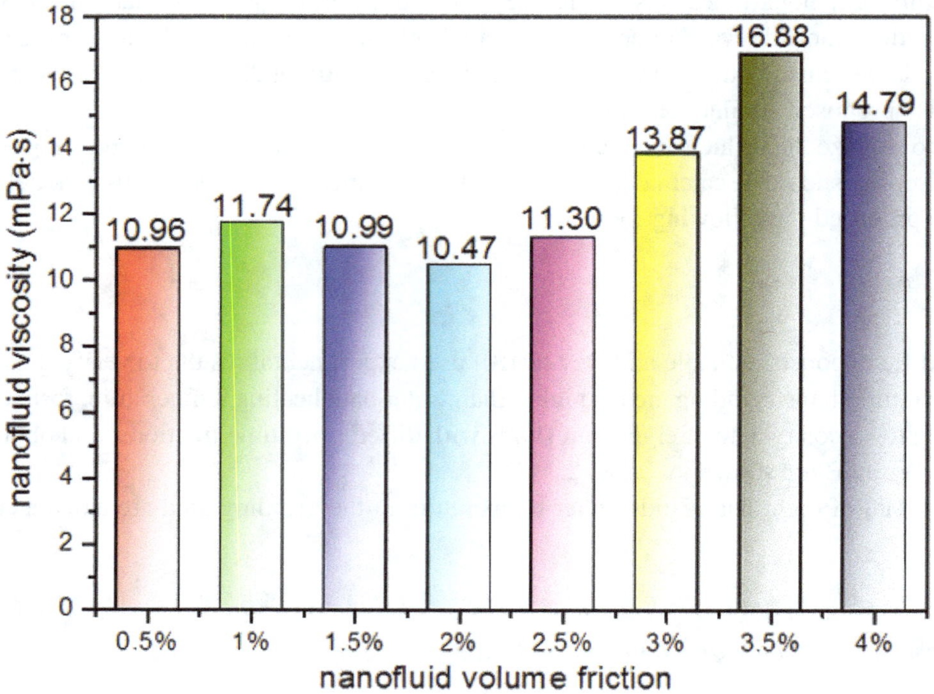

It can be seen from Fig. 9 that viscosity of 3.5% nanofluid is the highest while that of 2% nanofluid is the lowest; viscosity values of the eight nanofluids in the grinding area present different trends, because nanofluids with different volume fractions have different MQL grinding area temperatures; however, MQL grinding area temperature with 2% nanofluid is 109.4°C while its viscosity value is the lowest. Grinding area temperature with 3.5% nanofluid is 113.8°C, but the viscosity value is the highest, indicating that decline degree of 2% nanofluid viscosity with temperature is higher than that of 2.5% nanofluid, so mobility of 2% nanofluid is better than that of 2.5% nanofluid, and it can rapidly carry heat out of the grinding area and it has the best heat transfer performance in the grinding area.

9.4.2 Effects of Thermal Conductivity of Nanoparticles

It's well known that adding nanoparticles in the base fluid can significantly enhance heat conductivity performance of the base fluid, because small size effect and irregular motion of nanoparticles enlarge heat conductivity coefficient of the nanofluid; in addition, VDW force existing between nanoparticles contribute to their mutual attraction (Ruoff, Tersoff, Lorents, Subramoney, & Chan, 1993). If nanoparticles are not agglomerated, surface heat resistance will not be reduced just due to direct contact between nanoparticles, so its heat conductivity is greatly enlarged (Sundar, Sousa, & Singh, 2015). Heat conductivity coefficient of nanofluid will exert an important influence on energy transfer in the grinding process, and its heat conductivity can be calculated through a theoretical formula besides using the experimental measurement method, where the commonly used at Maxwell Model (Sundar et al., 2015) and H-C model (Maïga et al., 2005).

M-G model:

$$\frac{k_{eff}}{k_f} = \frac{k_p + 2k_f - 2f(k_f - k_p)}{k_p + 2k_f + f(k_f - k_p)} \tag{2}$$

H-C model:

$$\frac{k_{eff}}{k_f} = \frac{k_p + (n-1)k_f - (n-1)f(k_f - k_p)}{k_p + (n-1)k_f + f(k_f - k_p)} \tag{3}$$

Xue (2005) obtained heat conductivity of carbon nanotube particles in the base oil, and its model formula is expressed as:

$$\frac{k_{eff}}{k_f} = \frac{1 - f + 2f\dfrac{k_p}{k_p - k_f}\ln\dfrac{k_c + k_f}{2k_f}}{1 - f + 2f\dfrac{k_f}{k_p - k_f}\ln\dfrac{k_p + k_f}{2k_f}} \tag{4}$$

where k_{eff} is heat conductivity of nanofluid; k_f is heat conductivity of base fluid; k_p is heat conductivity of nanoparticles; f is volume fraction of nanofluid; n is shape factor, $n=3/\varphi$ according to empirical value obtained through the experiment, and φ is degree of sphericity of nanoparticles.

Luo et al. (2015) experimentally verified equation (4) and found that elevated heat conductivity of carbon nanotube-conduction oil nanofluid was highly correlated with shape factor. When $n=7$, H-C model obtains basically identical result with experimental data, indicating that nanotube particles experience agglomeration in the base oil, so shape of carbon nanotube particles is changed, which then gives rise to change of heat-conducting property of nanofluid; however. Heat conductivity model of Xue has a large deviation from the experimental result. Therefore, degree of sphericity of carbon nanotube nanofluid in the H-C model is selected as 7, and then H-C model can be simplified as:

$$\frac{k_{eff}}{k_f} = \frac{(1+6f)k_p + 6(1-f)k_f}{(1-f)k_p + (6+f)k_f} \tag{5}$$

Temperature can also greatly influence heat conductivity coefficient of nanofluid. After nanofluid is sprayed into the grinding area, its own temperature rises and enhanced irregular motion of nanoparticles gives rise to change of heat conductivity coefficient. Therefore, in consideration of temperature influence, formula (4) is optimized. Xuan et al. (Xuan, Qiang, & Hu, 2003) put forward nanoparticle motion-enhanced heat conductivity coefficient as:

$$k_{effp} = \frac{1}{2}\rho\varnothing c_p\sqrt{\frac{k_B T}{3\pi r\mu_{ff}}} \tag{6}$$

where k_B is Boltzmann constant, k_B=1.381×10^{-13} J/K; r is average nanoparticle size. It's noteworthy that the influences of nanoparticle agglomeration and heat resistance at nanoparticle/base fluid interface on heat conductivity coefficient are not considered in formula (6).

According to formulas (5) and (6), prediction formula of nanofluid heat conductivity coefficient is obtained as:

$$\frac{k_{eff}}{k_f} = \frac{(1+6f)k_p + 6(1-f)k_f}{(1-f)k_p + (6+f)k_f} + \frac{\rho\varnothing c_p}{2k_f}\sqrt{\frac{k_B T}{3\pi r\mu_{ff}}} \tag{7}$$

According to the experimental conditions, heat conductivity coefficients of 8 nanofluids under normal temperature are calculated through formula (5) as shown in Fig. 10. It can be seen that heat conductivity of carbon nanotube nanofluid increases with volume concentration and presents linear growth trend, indicating that carbon nanotube added into the base fluid—palm oil—significantly enhances heat conductivity of palm oil, and on the other hand, draw ratio of carbon nanotube particles is high, so their specific surface area is large, which enhances heat transfer performance and heat capacity of nanofluid in the grinding area.

Figure 10. The heat conductivity of nanofluids curve

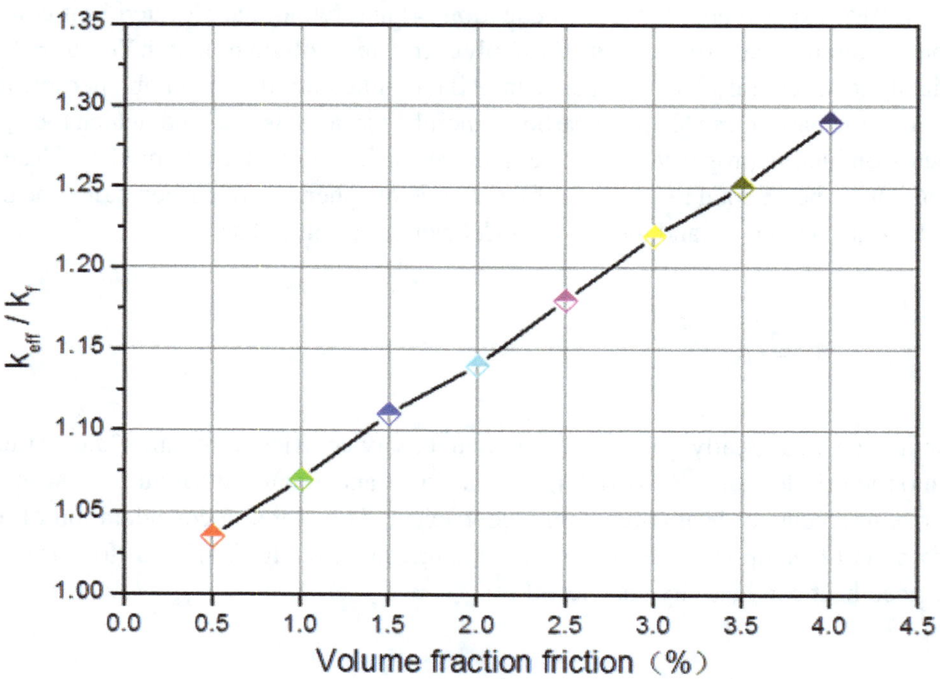

Fig. 11 shows growth rate curve of heat conductivity coefficients of nanofluids sprayed into the grinding area, and it can be seen that as volume fraction of nanofluid increases, heat conductivity coefficient tends to change gently after rapidly increasing. This phenomenon may be attributed to changes of physical properties of nanofluid with temperature rise like viscosity, specific heat and density, and meanwhile, enlargement of volume fraction results in more and more serious nanoparticle agglomeration phenomenon in the base fluid, average radius of particle clusters formed through agglomeration gradually increases, and for particle clusters with large radius, irregular motion intensity is reduced, so is transferred energy, and therefore, increasing amplitude of heat conductivity coefficient is reduced.

Heat transfer performance of nanofluid is enhanced as its heat conductivity increases. From theoretical analysis, 4% nanofluid has the best heat transfer performance, so its grinding temperature should be the lowest; however, according to experimental results of grinding temperature, 2% nanofluid obtains the lowest grinding temperature in MQL grinding, which doesn't accord with the theoretical analysis. Possible reason is that as the volume fraction of nanofluid increases, the quantity of carbon nanotube particles becomes larger and larger. As carbon nanotube has very large draw ratio (Yamamoto, Noda, & Kato, 2011), being about 2×10^5-6×10^5, carbon nanotubes will go through agglomeration easily when added into palm oil; when content of carbon nanotubes in the nanofluid is small, bending and twisting will not easily occur under the effect of dispersing agent, not to mention agglomeration phenomenon, but instead, they exist in the form of "independent microtube" in the suspension (Zhang, Li, Jia, Zhang, & Zhang, 2015b). With large draw ratio, carbon nanotube has high heat exchange performance in its axial direction, which is lower than that in its radial direction (Liu et al., 2013), so heat can be transferred

Figure 11. The heat conductivity of nanofluids growth curve

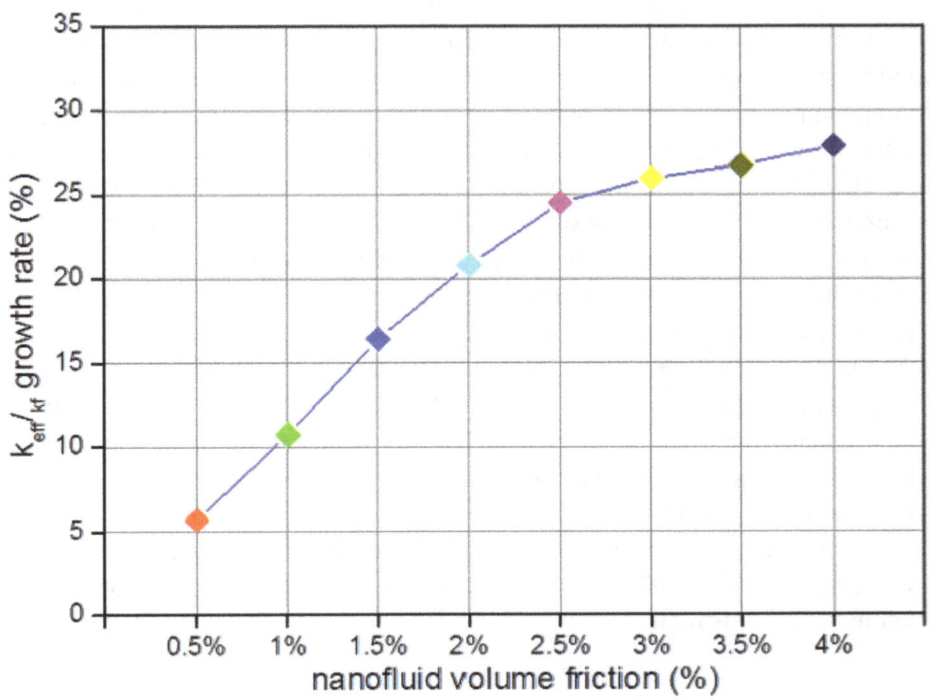

along axial direction of nanoparticles, which will improve heat conductivity performance of nanofluid; nevertheless, when content of carbon nanotubes in nanofluid is large, dispersing agent can't wrap carbon nanotubes so that strong VDW force exists between carbon nanotubes, which gives rise to agglomeration as shown in Fig. 12, where nanoparticles are intertwined but not existing in the form of "independent microtube", an irregular "tube cluster" structure is formed, they become particles with large diameters, and at the time, thermal contact resistance (Lianxiang, Chang, Qiu, Wang, & Amp, 2015) between tubes is adverse to heat transfer, thus degrading heat transfer performance. During the grinding process, if nanoparticles sprayed into the grinding area exist in the form of "independent microtube", heat can be transferred in axial direction of nanoparticles while not restricted by thermal contact resistance, so high convective heat transfer performance can be achieved; if nanoparticles exist in the form of "tube cluster", heat transfer is constrained by thermal contact resistance, which reduces convective heat transfer performance and elevates temperature in the grinding area. According to experimental results, when volume fraction of nanofluid is smaller than 1.5%, grinding temperature gradually rises as the volume fraction increases; Grinding temperature obtained through MQL grinding is low under 2% and 2.5% volume fractions of nanofluid; when volume fraction is higher than 3%, grinding temperature presents rising trend, possibly because under 0.5% volume fraction, heat conductivity coefficient of nanofluid is the lowest with the poorest convective heat transfer performance; when volume fraction increases to 1.5%, the obtained grinding force is large according to Fig. 3, so lubricating performance is poor with high grinding temperature, possibly because carbon nanotubes exist in the form of "independent microtube", existence of carbon-carbon bonds results in relatively low slippage performance without "bearing-like" effect in the grinding area, and high modulus and strength of carbon nanotubes make high friction coefficient in the grinding area, and lubricating property becomes poor; although 1.5% nanofluid doesn't go through serious agglomeration phenomenon and its heat conductivity coefficient is higher than that of 0.5% nanofluid, quantity of carbon nanotube particles is smaller than those of 2% and 2.5% nanofluids and the quantity of nanoparticles sprayed into the grinding area and then existing in the form of "independent microtube" is small while the quantities of carbon nanotube particles existing in the form of "independent microtube" in 2% and 2.5% nanofluids are large with good lubricating property and low grinding temperature; as agglomeration phenomenon is relatively serious in the nanofluid with above 3% volume fraction, numerous nanoparticles are intertwined to form "tube cluster" particles with large diameters. On the one hand, "tube cluster" particles are adhered around abrasive grains, thus reducing grinding efficiency, and on the other hand, "tube cluster" particles destruct integrity of lubricating oil film. This may be related to lubricating property in the grinding area. When volume fraction of nanofluid is higher than 3%, nanoparticles are agglomerated so that lubricating state in the grinding area is changed, lubricating property is reduced, grinding force gradually increases, and thermal contact resistance between nanoparticles increases, and as a result, heat transfer efficiency is reduced and temperature in the grinding area increases; in addition, it's pointed out in theory that draw ratio of carbon nanotube particles is an important factor influencing heat conductivity of carbon nanotube nanofluid: heat conductivity of carbon nanotube nanofluid changes with draw ratio of carbon nanotube particles. Generally speaking, this is manifested by reduced draw ratio and heat conductivity[26]. Agglomeration phenomenon of carbon nanotube particles in palm oil will result in relatively reduced draw ratio and poor heat-conducting property, so grinding temperature obtained through carbon nanotube NMQL grinding under high volume fraction is high.

Figure 12. Nanoparticle agglomeration

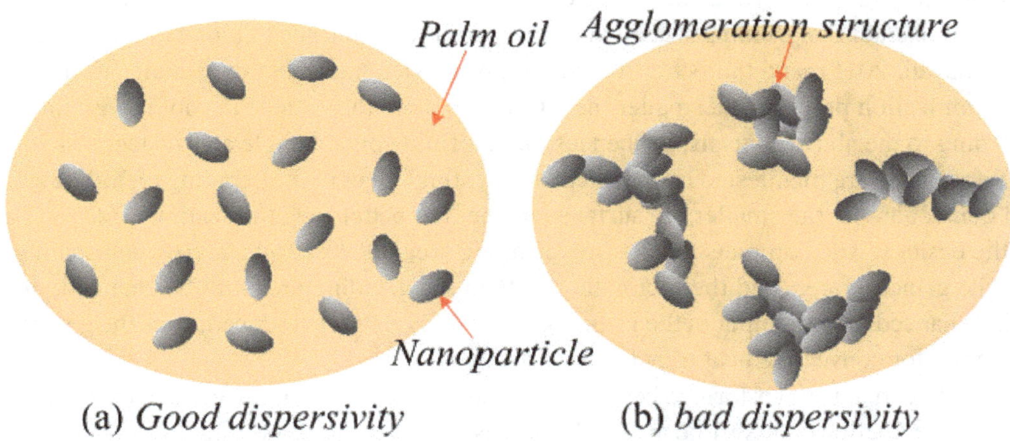

9.4.3 Effects of Nanofluid Contact Angle

Between workpiece and grinding wheel in the form of droplets during the MQL grinding process to exert the cooling & lubricating effect. Therefore, contact angle between droplets and workpiece decides its cooling & lubricating effect. Zhang et al. (2015a) studied the influence of contact angle on lubricating effect of MQL grinding and found that the influence was significant: the smaller the contact angle, the smaller the roughness value on workpiece surface, and the better the lubricating effect, because small contact angle of droplets represents a large infiltration area, and the larger the infiltration area, the better the wettability, and the more sufficient the cooling & lubricating effect of MQL grinding. This theory can be explained from the aspect of surface tension of nanofluid. Fig. 13 shows schematic diagram of contact angle formed by droplets sprayed onto workpiece surface. Workpiece, droplets and air are marked as medium 1, 2 and 3 respectively, surface tensions on contact surfaces of the three media are recorded as σ_{21}, σ_{23} and σ_{31} respectively. When droplets fall on workpiece surface, boundary surfaces of the three media may intersect at point p, contact line is influenced by tensions of the three surfaces, and in order that component force of resultant force of three surface tensions in these directions is zero, three boundary surfaces need to intersect at a certain angle, and this angle is called contact angle (Xia, 2015). Therefore, mutual relation of the three surface tensions is expressed as

$$\cos\theta = \frac{\sigma_{31} - \sigma_{21}}{\sigma_{23}},$$

from which it can be seen that contact angle is directly related to σ_{23}: contact angle is reduced when σ_{23} decreases, infiltration area is enlarged, and the wettability in the grinding area is better. Furthermore, surface tension σ_{23} will exert an influence on crushing and atomization effect of droplets: the smaller the surface tension, the easier the droplet crushing (Shavit & Chigier, 1995), and the better the atomization effect. In terms of NMQL grinding, when nanofluid is sprayed in the grinding area through the nozzle, the smaller the surface tension of nanofluid, the better the droplet crushing and atomization effect, and the larger the quantity of small droplets obtained through atomization. As the wedge-shaped space formed when grinding wheel grinds the workpiece is very narrow and small, the smaller the droplets, the easier for them to enter the grinding wheel/workpiece grinding interface (Chidambaram, Arunachalam, & Vijayaraghavan, 2015), and the better the cooling & lubricating effect they exert. Therefore, in the nanofluid atomization process, the smaller the surface tension, the better the cooling & lubricating effect. According to the above analysis, influences of nanofluid contact angle and surface tension on heat transfer performance are manifested by enhanced lubricating effect in the grinding area. The smaller the nanofluid contact angle, the smaller the surface tension, the better the atomization effect on nanofluid droplets, the easier it is for droplets to enter the grinding area, the larger the area of lubricating oil film formed in the grinding area, and the lubricating state in the grinding area will be changed, lubricating property is enhanced, energy input in the grinding process is relatively reduced, so is the generated heat, and this is good for convective heat transfer performance.

Fig. 14 shows contact angles formed between eight nanofluids and workpiece surface. It can be known that as the volume fraction of nanofluid gradually increases, contact angle presents gradual reducing trend, and contact angles of eight nanofluids are sorted as $\theta_{2\%}<\theta_{2.5\%}<\theta_{4\%}<\theta_{3.5\%}<\theta_{3\%}<\theta_{1.5\%}<\theta_{1\%}<\theta_{0.5\%}$. According to the above formula, surface tension keeps identical with contact angle, where 2% nanofluid has the smallest contact angle, followed by 2.5% nanofluid. Through the above analysis, the smaller the contact angle, the larger the infiltration area on workpiece surface, the smaller the surface tension, the better the atomization effect on nanofluid, the smaller the droplets sprayed into the grinding area, the easier it is for smaller droplets to enter the grinding area, so nanofluid can exert good lubricating effect in the grinding area. Therefore, 2% nanofluid has the best lubricating effect, followed by 2.5% nanofluid; according to Fig. 3, 2% and 2.5% nanofluids obtain the lowest tangential grinding force, certifying that their lubricating effect is ideal, and this accords with analysis result of contact angle. During the grinding process, the better the lubricating effect of nanofluid, the lower the grinding force, the lower the energy input, and generated grinding heat quantity is small, and this is good for enhancing heat transfer performance.

Figure 13. The schematic of the contact angle

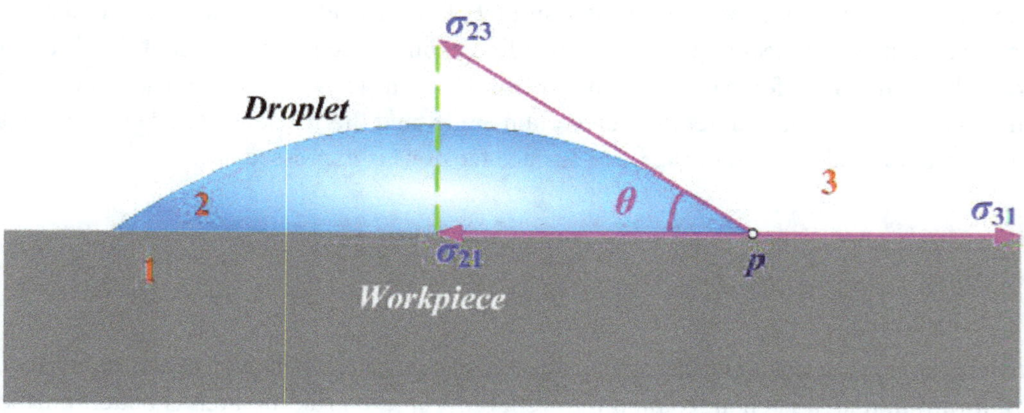

Figure 14. The contact angle of eight kinds of nanofluids

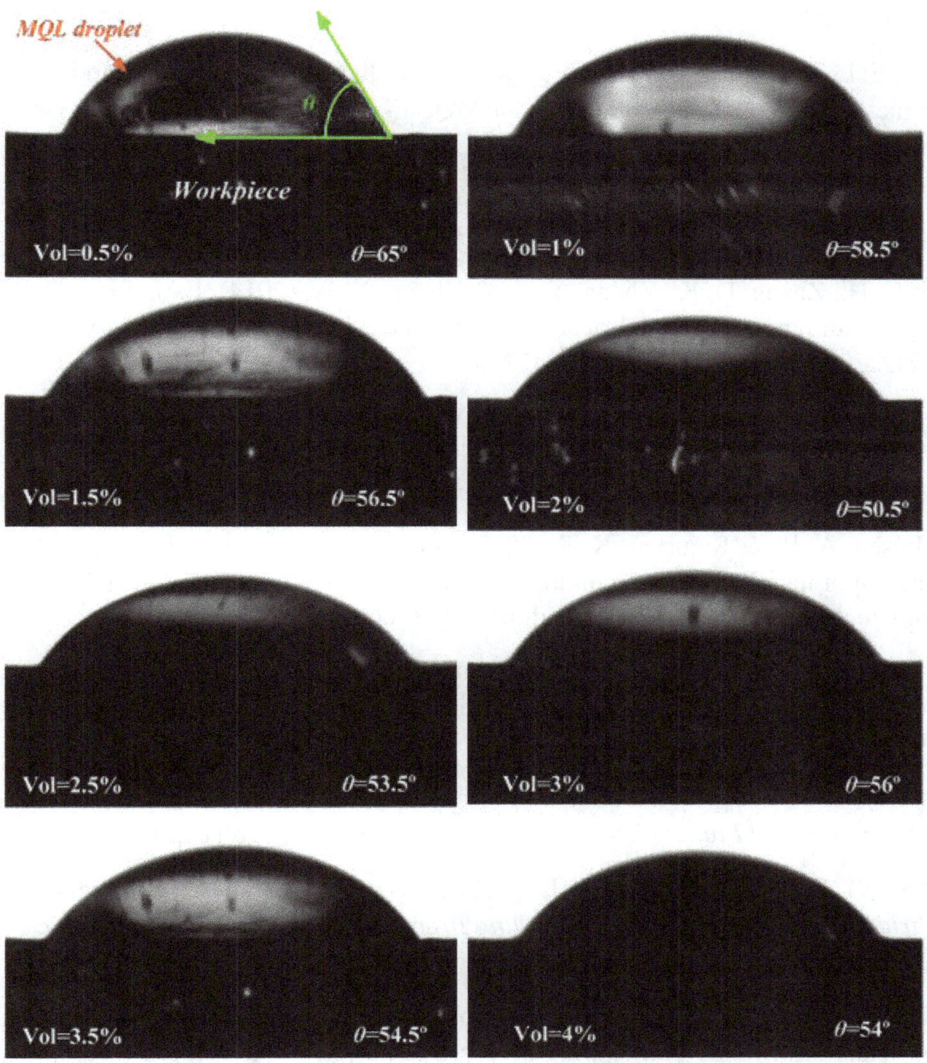

REFERENCES

Barczak, L. M., Batako, A. D. L., & Morgan, M. N. (2010). A study of plane surface grinding under minimum quantity lubrication (MQL) conditions. *International Journal of Machine Tools & Manufacture*, *50*(11), 977–985. doi:10.1016/j.ijmachtools.2010.07.005

Chandrasekar, M., Suresh, S., & Senthilkumar, T. (2012). Mechanisms proposed through experimental investigations on thermophysical properties and forced convective heat transfer characteristics of various nanofluids - A review. *Renewable & Sustainable Energy Reviews*, *16*(6), 3917–3938. doi:10.1016/j.rser.2012.03.013

Chen, X., Rowe, W. B., & Mccormack, D. F. (2000). Analysis of the transitional temperature for tensile residual stress in grinding. *Journal of Materials Processing Technology*, *107*(1), 216–221. doi:10.1016/S0924-0136(00)00692-0

Chidambaram, E. L., Arunachalam, N., & Vijayaraghavan, L. (2015). Analytical Model to Predict Sauter Mean Diameter in Air Assisted Atomizers for MQL in Machining Application. *Procedia Cirp*, *37*, 117–121. doi:10.1016/j.procir.2015.09.007

Hong, T. K., Yang, H. S., & Choi, C. J. (2005). Study of the Enhanced Thermal Conductivity of Fe Nanofluids. *Journal of Applied Physics*, *97*(6), 064311. doi:10.1063/1.1861145

Jia, D. Z., Li, C. H., Zhang, D. K., Zhang, Y. B., & Zhang, X. W. (2014). Experimental verification of nanoparticle jet minimum quantity lubrication effectiveness in grinding. *Journal of Nanoparticle Research*, *16*(12), 2758. doi:10.100711051-014-2758-7

Kalita, P., Malshe, A. P., Kumar, S. A., Yoganath, V. G., & Gurumurthy, T. (2012). Study of specific energy and friction coefficient in minimum quantity lubrication grinding using oil-based nanolubricants. *Journal of Manufacturing Processes*, *14*(2), 160–166. doi:10.1016/j.jmapro.2012.01.001

Kalita, P., Malshe, A. P., & Rajurkar, K. P. (2012). Study of tribo-chemical lubricant film formation during application of nanolubricants in minimum quantity lubrication (MQL) grinding. *CIRP Annals - Manufacturing Technology*, *61*(1), 327-330. doi:10.1016/j.cirp.2012.03.031

Keblinski, P., Phillpot, S. R., Choi, S. U. S., Eastman, J. A., & Transfer, M. (2002). Mechanisms of heat flow in suspensions of nano-sized particles (nanofluids). *International Journal of Heat*, *45*(4), 855–863. doi:10.10160017-9310(01)00175-2

Kole, M., & Dey, T. K. (2010). Viscosity of alumina nanoparticles dispersed in car engine coolant. *Experimental Thermal and Fluid Science*, *34*(6), 677–683. doi:10.1016/j.expthermflusci.2009.12.009

Lee, P. H., Nam, T. S., Li, C., & Sang, W. L. (2011). *Environmentally-Friendly Nano-fluid Minimum Quantity Lubrication (MQL) Meso-scale Grinding Process Using Nano-diamond Particles.* Paper presented at the International Conference on Manufacturing Automation.

Li, S. Y., Wang, Z., & Wu, Y. L. (2008). Relationship between subsurface damage and surface roughness of optical materials in grinding and lapping processes. *Journal of Materials Processing Technology*, *205*(1-3), 34–41. doi:10.1016/j.jmatprotec.2007.11.118

Li, Y., Zhou, J., Luo, Z., Tung, S., Schneider, E., Wu, J., & Li, X. (2011). Investigation on two abnormal phenomena about thermal conductivity enhancement of BN/EG nanofluids. *Nanoscale Research Letters*, *6*(1), 443–443. doi:10.1186/1556-276X-6-443 PMID:21740586

Lianxiang, M. A., Chang, Q., Qiu, J., Wang, Y., & Amp, M. (2015). *Experimental Study on Preparation and Thermal Properties of Water-based Carbon Nanotubes Nanofluids*. Materials Review.

Liu, J., Wu, S., He, C., Zhuo, H., Zhu, C., Li, C., & Zhang, Q. (2013). Structure, property and application of carbon nanotubes and carbon microtubes. *Journal of Shenzhen University Science Engineering*, *30*(1), 1–11. doi:10.3724/SP.J.1249.2013.01001

Liu, Z. Q., Guo, G. G., Zheng, X. H., An, Q. L., & Chen, M. (2011). Investigation of Grinding Ti-6Al-4V under Minimum Quantity Lubrication (MQL) Condition. *Key Engineering Materials, 487*, 84-88.

Luo, Z. Y., Yue-Qiong, W. U., Qian, H. U., Wang, T., & Ming-Jiang, N. I. (2015). Thermal and Rheological Properties of Carbon Nanotubes-Oil Nanofluid. *Journal of Chemical Engineering of Chinese Universities*, *29*(1), 35–42. doi:10.3969/j.issn.1003-9015.2015.01.005

Maïga, S. E. B., Palm, S. J., Cong, T. N., Roy, G., & Galanis, N. (2005). Heat transfer enhancement by using nanofluids in forced convection flows. *International Journal of Heat and Fluid Flow*, *26*(4), 530–546. doi:10.1016/j.ijheatfluidflow.2005.02.004

Mao, C., Tang, X., Zou, H., Huang, X., & Zhou, Z. (2012). Investigation of grinding characteristic using nanofluid minimum quantity lubrication. *International Journal of Precision Engineering and Manufacturing*, *13*(10), 1745–1752. doi:10.100712541-012-0229-6

Mao, C., Zou, H., Zhang, J., Huang, X., & Zhou, Z. (2012). The influence of spraying parameters on grinding performance for;nanofluid minimum quantity lubrication. *International Journal of Advanced Manufacturing Technology*, *64*(9-12), 1791–1799. doi:10.100700170-012-4143-y

Mao, C., Zou, H. F., Zhou, X., Huang, Y., Gan, H., & Zhou, Z. (2014). Analysis of suspension stability for nanofluid applied in minimum quantity lubricant grinding. *International Journal of Advanced Manufacturing Technology*, *71*(9-12), 2073–2081. doi:10.100700170-014-5642-9

Mechiri, S. K., Vasu, V., & Gopal, A. V. (2017). Investigation of thermal conductivity and rheological properties of vegetable oil based hybrid nanofluids containing Cu-Zn hybrid nanoparticles. *Experimental Heat Transfer*, *30*(3), 205–217. doi:10.1080/08916152.2016.1233147

Mintsa, H. A., Roy, G., Nguyen, C. T., & Doucet, D. (2009). New temperature dependent thermal conductivity data for water-based nanofluids. *International Journal of Thermal Sciences*, *48*(2), 363–371. doi:10.1016/j.ijthermalsci.2008.03.009

Miyashita, M. (1990). *Way to nanogrinding technology*. Academic Press.

Murshed, S. M. S., Leong, K. C., & Yang, C. (2005). Enhanced thermal conductivity of TiO_2—Water based nanofluids. *International Journal of Thermal Sciences*, *44*(4), 367–373. doi:10.1016/j.ijthermalsci.2004.12.005

Namburu, P. K., Kulkarni, D. P., Misra, D., & Das, D. K. (2008). Viscosity of copper oxide nanoparticles dispersed in ethylene glycol and water mixture. *Experimental Thermal and Fluid Science*, *32*(2), 397–402. doi:10.1016/j.expthermflusci.2007.05.001

Ozerinc, S., Kakac, S., & Yazicioglu, A. G. (2010). Enhanced thermal conductivity of nanofluids: A state-of-the-art review. *Microfluidics and Nanofluidics*, *8*(2), 145–170. doi:10.100710404-009-0524-4

Rowe, W. (2014). *Principles of Modern Grinding Technology*. Academic Press.

Ruoff, R. S., Tersoff, J., Lorents, D. C., Subramoney, S., & Chan, B. (1993). Radial deformation of carbon nanotubes by van der Waals forces. *Nature*, *364*(6437), 514–516. doi:10.1038/364514a0

Sadeghi, M. H., Haddad, M. J., Tawakoli, T., & Emami, M. (2009). Minimal quantity lubrication-MQL in grinding of Ti–6Al–4V titanium alloy. *International Journal of Advanced Manufacturing Technology*, *44*(5-6), 487–500. doi:10.100700170-008-1857-y

Sayuti, H., Sarhan, A. A. D., & Hamdi, M. (2013). An investigation of optimum SiO_2 nanolubrication parameters in end;milling of aerospace Al6061-T6 alloy. *International Journal of Advanced Manufacturing Technology*, *67*(1-4), 833–849. doi:10.100700170-012-4527-z

Setti, D., Sinha, M. K., Ghosh, S., & Rao, P. V. (2015). Performance evaluation of Ti-6Al-4V grinding using chip formation and coefficient of friction under the influence of nanofluids. *International Journal of Machine Tools & Manufacture*, *88*, 237–248. doi:10.1016/j.ijmachtools.2014.10.005

Shavit, U., & Chigier, N. (1995). The role of dynamic surface tension in air assist atomization. *Physics of Fluids*, *7*(1), 24–33. doi:10.1063/1.868725

Shen, B. (2008). *Minimum quantity lubrication grinding using nanofluids.* Academic Press.

Shen, B., Shih, A. J., & Tung, S. C. (2008). Application of Nanofluids in Minimum Quantity Lubrication Grinding. *Tribology Transactions*, *51*(6), 730–737. doi:10.1080/10402000802071277

Shen, B., Shih, A. J., & Tung, S. C. (2009). Application of Nanofluids in Minimum Quantity Lubrication Grinding. *Tribology & Lubrication Technology*, *65*(3), 73–80. doi:10.1080/10402000802071277

Sundar, L. S., Sousa, A. C. M., & Singh, M. K. (2015). Heat Transfer Enhancement of Low Volume Concentration of Carbon Nanotube-Fe3O4/Water Hybrid Nanofluids in a Tube With Twisted Tape Inserts Under Turbulent Flow. *Journal of Thermal Science and Engineering Applications*, *7*(2). doi:10.1115/1.4029622

Vajjha, R. S., & Das, D. K. (2012). A review and analysis on influence of temperature and concentration of nanofluids on thermophysical properties, heat transfer and pumping power. *International Journal of Heat and Mass Transfer*, *55*(15-16), 4063–4078. doi:10.1016/j.ijheatmasstransfer.2012.03.048

Wang, Y. G., Li, C. H., Zhang, Y. B., Li, B. K., Yang, M., Zhang, X. P., & Liu, G. T. (2016). Experimental evaluation of the lubrication properties of the wheel/workpiece interface in MQL grinding with different nanofluids. *Tribology International*, *99*, 198–210. doi:10.1016/j.triboint.2016.03.023

Wusiman, K., Jeong, H., Tulugan, K., Afrianto, H., & Chung, H. (2013). Thermal performance of multi-walled carbon nanotubes (MWCNTs) in aqueous suspensions with surfactants SDBS and SDS. *International Communications in Heat and Mass Transfer*, *41*, 28–33. doi:10.1016/j.icheatmasstransfer.2012.12.002

Xia, H. T. (2015). *Physical Chemistry*. Beijing: Chemical Industry Press.

Xie, H. Q. (2003). Study on the mechanism of heat conduction in nanofluid medium. *Wuli Xuebao*, *52*(6), 1444–1449. doi:0.1016/S0040-6090(03)00396-1

Xuan, Y., Qiang, L., & Hu, W. (2003). Aggregation structure and thermal conducting of nanofluids. *AIChE Journal. American Institute of Chemical Engineers*, *49*(4), 1038–1043. doi:10.1002/aic.690490420

Xue, Q. Z. (2005). Model for thermal conductivity of carbon nanotube-based composites. *Physica B, Condensed Matter*, *368*(1-4), 302–307. doi:10.1016/j.physb.2005.07.024

Yamamoto, T., Noda, S., & Kato, M. (2011). A simple and fast method to disperse long single-walled carbon nanotubes introducing few defects. *Carbon, 49*(10), 3179–3183. doi:10.1016/j.carbon.2011.03.040

Zhang, D., Li, C., Zhang, Y., Jia, D., & Zhang, X. (2015). Experimental research on the energy ratio coefficient and specific grinding energy in nanoparticle jet MQL grinding. *International Journal of Advanced Manufacturing Technology, 78*(5-8), 1275–1288. doi:10.100700170-014-6722-6

Zhang, D. K., Li, C. H., Zhang, Y. B., Jia, D. Z., & Zhang, X. W. (2015). Experimental research on the energy ratio coefficient and specific grinding energy in nanoparticle jet MQL grinding. *International Journal of Advanced Manufacturing Technology, 78*(5-8), 1275–1288. doi:10.100700170-014-6722-6

Zhang, Y., Li, C., Jia, D., Zhang, D., & Zhang, X. (2015). Experimental evaluation of MoS_2 nanoparticles in jet MQL grinding with different types of vegetable oil as base oil. *Journal of Cleaner Production, 87*(1), 930–940. doi:10.1016/j.jclepro.2014.10.027

Zhang, Y. B., Li, C. H., Jia, D. Z., Zhang, D. K., & Zhang, X. W. (2015a). Experimental evaluation of MoS_2 nanoparticles in jet MQL grinding with different types of vegetable oil as base oil. *Journal of Cleaner Production, 87*, 930–940. doi:10.1016/j.jclepro.2014.10.027

Zhang, Y. B., Li, C. H., Jia, D. Z., Zhang, D. K., & Zhang, X. W. (2015b). Experimental evaluation of the lubrication performance of MoS_2/CNT nanofluid for minimal quantity lubrication in Ni-based alloy grinding. *International Journal of Machine Tools & Manufacture, 99*, 19–33. doi:10.1016/j.ijmachtools.2015.09.003

Zhou, Z. Y., Di, Q. Q., Liu, B., Ma, X. Y., & Cai, B. H. (2016). *Experimental Study on the Surface Tension of Al_2O_3-H_2O Nanofluid.* Paper presented at the Materials Science Forum.

Chapter 10
Experimental Research on Grinding Temperature With Different Workpiece Materials

ABSTRACT

The results show that the nanofluids with a volume fraction of 2% have good comprehensive properties for lubrication. However, this study only concentrates on workpiece material—nickel base alloy. Based on literature analysis, researchers have carried out a large quantity of researches on NMQL grinding of different workpiece materials including hard steel HSS, quenched steel 100Cr6, nickel base alloy, Ti-6Al-4V alloy, soft steel CK45, and cast iron. They have mainly focused on grinding force, surface roughness, etc. The heat transfer state of nanofluids is neglected, and the research on the heat transfer mechanism of grinding temperature is lacking. Therefore, grinding temperatures obtained through MQL grinding with different concentrations of vegetable oils-based nanofluids as well as their heat transfer mechanism are investigated through simulation and experiment in this chapter.

10.1 INTRODUCTION

Removing the unit volume of metal through grinding requires high-energy input, especially for materials that are difficult to machine. However, almost all energy conversions are concentrated in the grinding zone;much energy is converted into heat and accumulates in the grinding zone (González-Santander, Fernández, Martín, & Arrazola, 2016). If the generated heat cannot be timely transmitted to the grinding zone, then the temperature in the grinding zone will sharply increase and cause various forms of thermal damage, such as burns, metallurgical changes, cracks, and residual stress, to the workpiece (Malkin & Guo, 1991). Hence, the vital link is an available cooling lubrication that can reduce the temperature and effectively improve the quality of the grinding process. With environmental issues becoming increasingly prominent, the conventional cooling lubrication called flood cooling no longer meets the needs of modern development. New green processing methods mainly include dry grinding, low-temperature cooling grinding, minimum quantity lubrication (MQL), and nanofluidMQL grinding. The lubrication

DOI: 10.4018/978-1-7998-1546-4.ch010

of dry grinding and low-temperature cooling grinding should be enhanced, but the limitations of these two methods can be improved by MQL. This technology provides effective lubrication by atomizing and injecting the lubricant liquid into the grinding zone while maintaining high-speed air flow throughout the grinding zone. Convective heat transfer occurs between the grinding wheel and the workpiece, thereby reducing the temperature of the workpiece. Hafenbraedl and Malkin (Hafenbraedl & Malkin, 2001) found that MQL technology can improve the performance of lubrication and reduce the specific grinding energy as compared with conventional processing methods. However, the limited cooling performance of compressed air affects the application of MQL technology in mechanical machining. With the development of nanotechnology, the in-depth study of nanofluids, and the consideration of the green environmental protection and energy saving requirements, nanofluid MQL has been regarded as an effective way for cooling and lubrication (Tasdelen, Thordenberg, & Olofsson, 2008).

MQL technology features excellent processability. Many studies reported about the MQL grinding of different workpiece materials. Tawakoli et al. (Tawakoli et al., 2009; Tawakoli, Westkämper, & Azarhoushang, 2009) studied the effect of workpiece materials on the performance of MQL grinding. The grinding forces, friction coefficient, and surface roughness of brittle hard 100Cr6 steel and 42CrMo4 soft steel were studied in their experiment. They discovered that MQL technology can significantly reduce the grinding force and friction coefficient; however, the brittle hard 100Cr6 steel has lower surface roughness than the 42CrMo4 soft steel. Tawakoli et al. (Tawakoli, Hadad, & Sadeghi, 2010) also explored the effect of jet parameters on the MQL grinding of brittle hard 100Cr6 steel and found that the grinding performance is affected by the jet angle, air pressure, and jet flux.Silva et al. (Silva, Catai, Fusse, França, & Aguiar, 2005) tested the effect of MQL grinding parameters on ABNT 4340 steel. They found that the grinding force and surface roughness are significantly declined, the grinding wheel wear is improved, the residual stress of workpieces is reduced, and the grinding performance is increased. Sadeghi et al. (Sadeghi, Hadad, Tawakoli, Vesali, & Emami, 2010) conducted an experimental study on the MQL grinding of AISI 4140 steel and found that MQL grinding achieves far lower grinding force and friction coefficient and significantly better workpiece surface quality than other cooling lubrication approaches. Barczak et al. (Barczak, Batako, & Morgan, 2010) conducted an experiment on the MQL grinding of EN8,M2, and EN31 steel types. Their results revealed that MQL technology can achieve better processability than flood grinding under some specific experimental conditions; however, MQL is inapplicable to the processing of very hard materials. Applications of MQL technology are limited to soft materials. However, Rabiei et al. (Rabiei, Rahimi, Hadad, & Ashrafijou, 2015) reported opposite conclusions. They conducted an experimental study on the dry grinding, flood grinding, and MQL grinding of hard steels HSS and 100Cr6 and soft steels CK45 and S305. Experimental results revealed that MQL grinding can reduce the grinding force and friction coefficient for the hard and soft steels. Furthermore,MQL grinding can improve the surface smoothness and surface quality of the hard steels but can worsen the surface smoothness of the soft steels compared with other grinding methods. Therefore,MQL technology is inapplicable to some soft steels. Hadad et al. (Hadada, Sadeghi, & Sadeghi, 2012) conducted an experimental study on the grinding temperature and energy distribution of MQL grinding of brittle hard 100Cr6 steel by using an aluminum oxide grinding wheel and a cubic boron nitride (CBN) grinding wheel; they found that the CBN grinding wheel can improve the shortcomings of MQL cooling. Moreover, special additives could improve the lubricating properties of the lubricant and reduce the grinding force and heat production in the grinding zone.

Nanotechnology has been used in many areas (Chen et al., 2016; Yang, Chen, Song, Zhang, & Chen, 2017; Yang et al., 2016),where nanofluid was further applied in grinding machining.Nanofluid is a suspension where the nanoparticles in MQL lubricating oil are evenly added and mixed. Adding few nanoparticles (1–5 vol%) can increase the thermal conductivity of the suspension to 20–30% (Fan & Wang, 2011). Choi et al. (Choi, Zhang, Yu, Lockwood, & Grulke, 2001; Das & Choi, 2009) studied the effective thermal conductivity of multi-walled carbon nanotube nanofluids and found that the actual measured thermal conductivity is 150% higher than that of the base solution. Wu et al. (Wu, Wu, & Cheng, 2009; Xin, Hui, Jian, & Zheng, 2008) discussed the thermal conductivities of water-based Al_2O_3 nanofluids with volume fractions of 0.15 and 0.26%; they concluded that the thermal conductivity increases with increasing volume fraction. Ogut et al. (Öğüt, 2009) analyzed the convective heat transfer coefficient of water-based Ag, Cu, TiO_2,andAl_2O_3 nanofluids, and they found that the convective heat transfer coefficient increases gradually with increasing volume fraction. Eastman et al. (Eastman, Choi, Li, Yu, & Thompson, 2001) prepared a nanofluid by adding the volume fraction of 0.3% Cu nanoparticles into ethylene glycol and studied its thermal conductivity. The thermal conductivity of this nanofluid is 40% higher than that of the base liquid. Nanofluid MQL grinding has been extensively studied due to the heat exchange performance of nanofluids. Shen et al. (Shen, Shih, & Tung, 2008) prepared nanofluids with 1, 2.5, and 4% volume fractions and used them in a MQL grinding experiment of cast iron. Results revealed that nanofluid MQL grinding achieves lower grinding force and surface roughness, fewer workpiece burns, and higher g-ratio than dry grinding, pouring grinding, and MQL cooling grinding. Among the nanofluids, that with 1% volume fraction exhibits the best surface quality. Kalitaet al. (Kalita, Malshe, & Rajurkar, 2012) added MoS_2 nanoparticles into paraffin oil and soybean oil and prepared nanofluids with mass fractions of 2 and 8%. These two nanofluids were subsequently used in a MQL grinding experiment of cast iron and EN24 alloy steel. Experimental results revealed that the 8% nanofluid MQL grinding achieves lower grinding force, friction coefficient, and specific grinding coefficient and higher g-ratio than the 2% nanofluid MQL grinding. Moreover, they found that the 8% nanofluid MQL grinding has lower grinding temperature and good wear resistance. Given that soybean oil is vegetable-derived and biodegradable, this nanofluid satisfies the requirements of environmental protection and reduces damages to the environment and physical health of operating personnel. Settiet et al. (Setti, Ghosh, & Rao, 2012) conducted an experiment on the waterbased Al_2O_3 nanofluid MQL grinding of Ti–6Al–4V alloy and found that the nanofluid with 1% volume fraction significantly reduces the grinding force and that the 4% nanofluid has better surface quality. Mao et al. (Mao, Tang, Zou, Huang, & Zhou, 2012) conducted an experimental study on the Al_2O_3 nanofluid MQL grinding of AISI 52100 steel. Compared with MQL, nanofluid MQL grinding exhibits better surface-processing quality because it can significantly reduce the grinding force, grinding temperature, and surface roughness. Zhang et al. (Zhang, Li, Zhang, Jia, & Zhang, 2015) prepared nanofluids with volume fractions of 1, 2, and 3% by adding carbon nanotube (CNT) nanoparticles into soybean oil and discussed the energy ratio coefficient and specific grinding energy of 45 steel MQL grinding. They found that the 2% CNT nanofluid MQL grinding has the lowest grinding temperature and energy ratio coefficient. Zhang et al. (Zhang, Li, Jia, Zhang, & Zhang, 2015) prepared nanofluids with mass fractions of 2, 4, and 6% by adding MoS_2 nanoparticles into the soybean oil and experimented on the MQL grinding of 45 steel. Experimental results showed that the 6% nanofluid contributes the lowest grinding force and surface roughness, indicating its best surface quality. Zhang et al. (Zhang et al., 2016) compared the lubrication performances of mass fraction of 6% Al_2O_3 and SiC nanofluids and Al_2O_3/SiC mixed nanofluids in the MQL grinding of Ni-based alloy, which was previously investigated by Ding et al. (Ding et al., 2013; Song, Ding, Xu, & Chen, 2012). Experimental results confirmed that the mixed nanofluids have better lubricating property than the individual nanofluids.Most of the aforementioned studies on the MQL

Table 1. The performance parameters of the workpiece material

Material	Elasticity modulus (GPa)	Density (Kg/m³)	Thermal diffusivity coefficient (10^{-5}/K)	Specific heat (J/kg·K)	Heat conductivity(W/m·K)	Poisson's ratio	Hardness
45 steel	209	7890	1.17	450	48	0.269	90±3HRB
Nickel base alloy	200	8240	1.18	435	13.7	0.3	47HRC
Cast iron	169	7090	1.08	510	47	0.305	50HRC

grinding of different workpiece materials, such as hard steel HSS, brittle hard 100Cr6 steel, Ni-based alloy, Ti–6Al–4V alloy, soft steel CK45, and cast iron, focused on the grinding force and surface roughness but disregarded the heat transfer mechanism. In this study, three typical workpiece materials were grinded using nanofluids with different concentrations of vegetable oils. The heat exchange mechanism with regard to the workpiece material property and boiling heat transfer of nanofluids was discussed using numerical simulation and experiments.

10.2 EXPERIMENTAL DESIGN

10.2.1 Workpiece Materials

Typical three workpiece materials—45 steel with 0±3HRB hardness, high-temperature nickel base alloy GH4169 with 47HRC hardness and nodular cast iron 100-70-03 with 50HRC hardness—are selected in the experiment, workpiece dimensions are 30×30×40mm and their performance parameters are shown in Table 1.

10.2.2 Nanofluids

During the grinding machining process, grinding fluid can exert lubricating, cooling, rinsing and rust-proof effects (Webster & Gruen, 2008) so as to lengthen service life of grinding wheel, guarantee workpiece machining quality and precision and improve production efficiency. Vegetable oil used in this experiment is palm oil, because Wang et al. (Wang et al., 2016) found that compared with casting-type grinding, palm oil had good lubricating property as MQL base oil, and it could significantly reduce grinding force and improve workpiece surface quality after grinding; heat transfer performances of MQL grinding of nickel base alloy with different vegetable oils taken as base oils were studied in Chapter 7, and it's found through the experiment that palm oil not only had good lubricating property but also its heat transfer performance exceeded other kinds of vegetable oils like soybean oil, castor oil and corn oil. In this experiment, CNT nanoparticles are added into palm oil to prepare nanofluids with 2% and 2.5% volume fractions, average CNT nanoparticle size is 50nm and average length is 10-30 μm; nanofluid is prepared using two-step method (Shen, 2008). In order to prevent nanoparticle agglomeration in the base fluid, 0.3% Polysorbate 80 is added as dispersing agent, and meanwhile, the prepared nanosuspension is oscillated in an ultrasonic oscillator (model: KQ3200DB), and temperature-dependent change of

Table 2. The base properties of two nanofluids

Volume fraction	Density (kg/m^3)	Specific heat capacity (J/kg·K)	Viscosity 40°C (cp)	Viscosity 90°C (cp)	Heat conductivity coefficient (W/m·K)	Surface tension (×10^{-6}) N/m
2%	909.7	1800	121.76	18.43	0.1962	1.74
2.5%	917.8	1783	123.87	18.78	0.203	3.32

nanofluid is measured using Brookfield DV2T viscosimeter and DKB-501A super thermostatic water bath; contact angle of nanofluid is measured using a contact angle measuring instrument (Radiom, Yang, & Weng, 2013). Table 2 shows basic properties of nanofluids.

10.2.3 Experimental Scheme

Grinding equipment used in this experiment are identical with those in Chapter 7. Grinding depth is set as 20 μm while other grinding parameters remain unchanged. Grinding force and grinding temperature are measured using the same method. An experimental study of heat transfer performances in NMQL grinding of different workpiece materials with different kinds of nanofluids is carried out, namely three workpiece materials—45 steel, nickel base alloy and nodular cast iron—are grinded using 2% and 2.5% CNT nanofluids, grinding force and grinding temperature in the grinding process are measured, and experimental results of lubricating property and heat transfer performance are analyzed and discussed. Experimental program is shown in Table 3.

10.3 EXPERIMENTAL RESULTS

10.3.1 Specific Grinding Force

Figure 1 and Figure 2 show experimental results of specific tangential grinding force and specific normal grinding force in the MQL grinding process of three workpiece materials with two nanofluids. It can be seen that the two nanofluids are not that different in the aspects of experimental results of specific

Table 3. Experimental program involving minimum quantity lubricant grinding of different workpiece materials

No.	Nanoparticles	Base fluid	Volume fraction	Cooling & lubricating condition	Workpiece material
1	CNT	Palm oil	2%	MQL	45 steel
2	CNT	Palm oil	2%	MQL	Nickel base alloy
3	CNT	Palm oil	2%	MQL	Nodular cast iron
4	CNT	Palm oil	2.5%	MQL	45 steel
5	CNT	Palm oil	2.5%	MQL	Nickel base alloy
6	CNT	Palm oil	2.5%	MQL	Nodular cast iron

Figure 1. Specific tangential grinding force

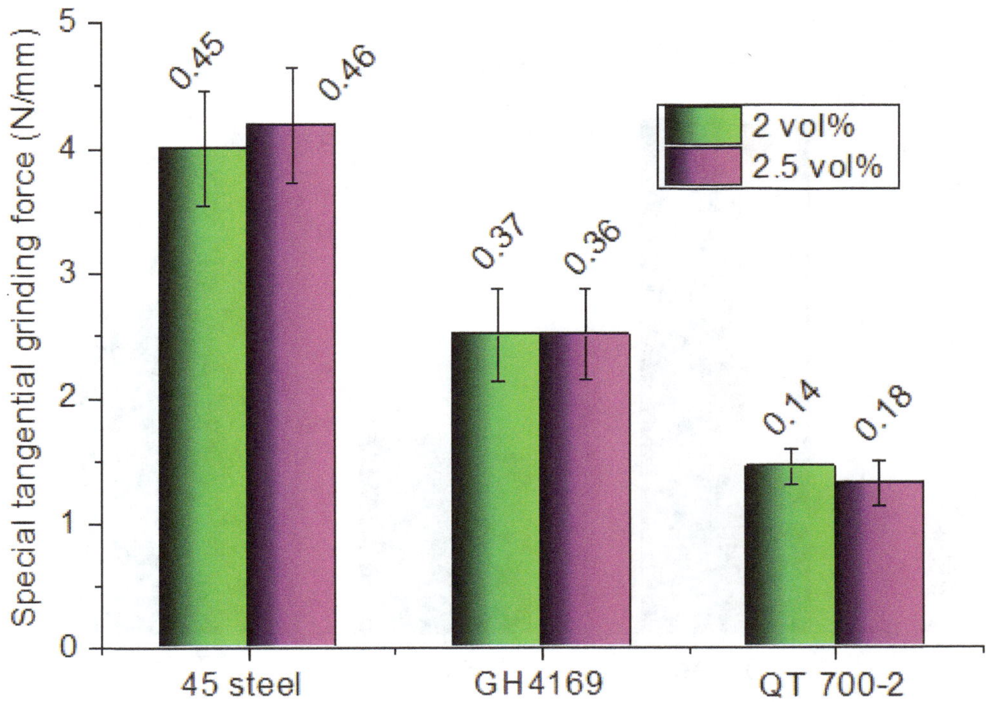

tangential grinding force and specific normal grinding force when used to grind the same workpiece material. Specific tangential grinding force and specific normal grinding force obtained through grinding of 45 steel using 2% and 2.5% nanofluids are 4 N/mm and 4.19 N/mm and 17.02 N/mm and 17.24 N/mm respectively, which may be related to properties of nanofluids. Lubricating properties of 2% and 2.5% CNT nanofluids are approximate and their lubricating effects are basically identical with they are sprayed into the grinding area, so their grinding force results are also basically identical. When the same nanofluid is used to grind different workpiece materials, they are totally different in the aspect of obtained specific tangential grinding force and specific normal grinding force. In terms of 2% NMQL grinding, specific tangential grinding forces of nickel base alloy and cast iron are 2.51 N/mm and 1.45 N/mm respectively while their specific normal grinding forces are 6.01 N/mm and 12.51 N/mm respectively, which is related to their different properties, and this will be analyzed in the following chapter in details.

10.3.2 Grinding Temperature

Figure 3 shows grinding temperatures in the NMQL grinding process of 45 steel, nickel base alloy and nodular cast iron with 2% and 2.5% nanofluids measured through the experiment. 45 steel has the highest grinding temperatures in the MQL grinding with the two nanofluids, respectively being 363.9 °C and 352.9 °C; nodular cast iron obtains the lowest grinding temperatures, being 143.2 °C and 136.8 °Crespectively; grinding temperatures of nickel base alloy are 245.7 °C and 244.5 °C respectively. Experimental results of grinding temperature keep identical with those of tangential grinding force, indicating that under the

Figure 2. Specific normal grinding force

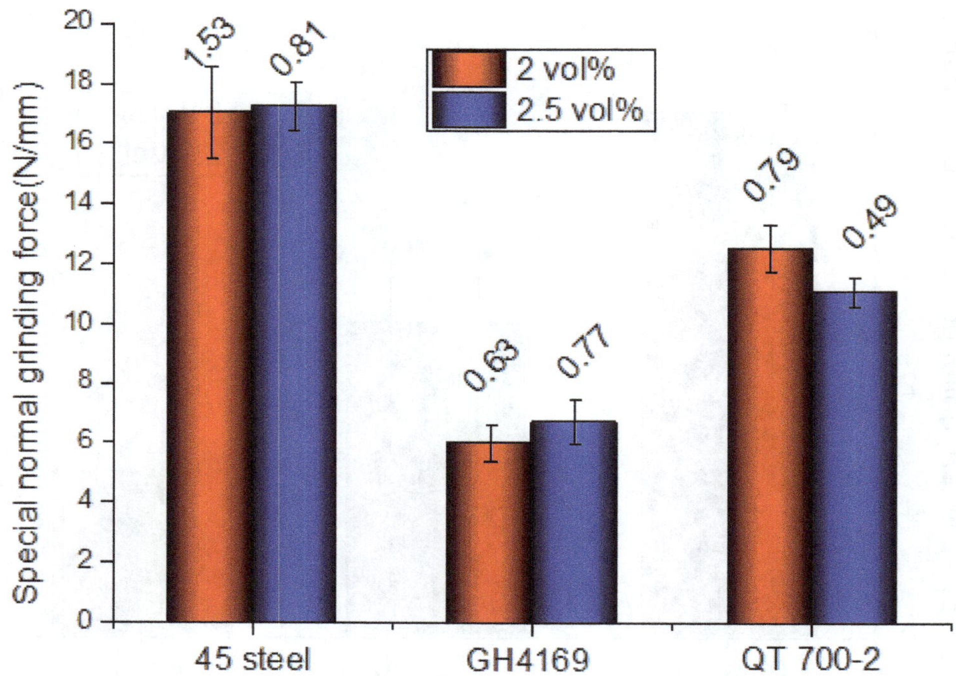

same working condition, different workpiece materials can exert a significant influence on grinding force and grinding temperature. In addition, it can be found that grinding temperature of 2% nanofluid used to grind the same workpiece material is slightly higher than that of 2.5% nanofluid, which may be related to physical characteristics of nanofluids like viscosity, surface tension, heat conductivity, etc.

10.4 SIMULATION RESULTS

According to the numerical analysis model established in Chapter 4, convective heat transfer coefficients of the two nanofluids are calculated according to the convective heat transfer coefficient model based on experimentally measured grinding force, total heat flux and proportionality coefficient of heat energy transferred to workpiece. Parameters and calculation results are input as shown in Table 4 and Table 5.

10.4.1 Temperature Field Distribution Graph

Simulation of grinding temperatures in the MQL grinding process of three different workpiece materials with two nanofluids is carried out using the finite difference method on the MATLAB platform. Figure 4 shows simulation phase diagram of grinding temperature in the 2% CNT NMQL grinding process of 45 steel. The grinding temperature is the highest in the grinding area (red area in the figure), and the farther it is from the grinding area, the lower the temperature, and grinding temperature is transferred around the grinding area. At t=1s, grinding wheel just starts grinding workpiece and grinding temperature gradually

Figure 3. Experiment grinding temperature

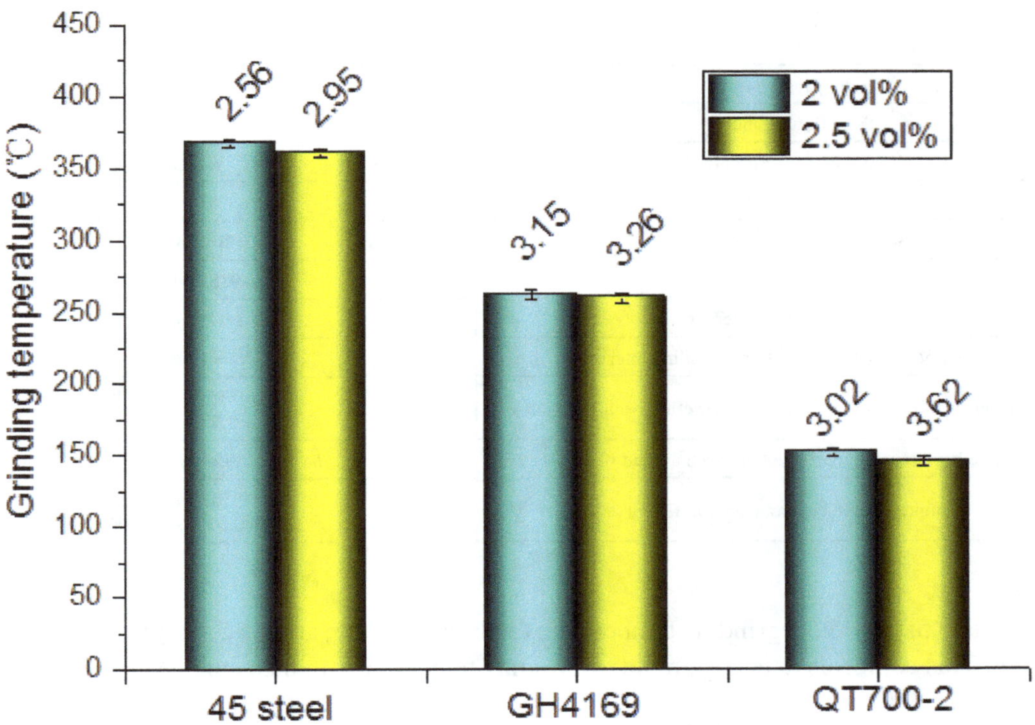

Table 4. Input parameters

Input parameter	Value	Input parameter	Value
Nozzle diameter	d_0**=1 mm**	**Grinding width**	**b=20 mm**
Internal and external pressure difference of nozzle	Δp=0.5 MPa	Fluid density	$\rho_{2\%}$=909.7 kg/m³ $\rho_{2.5\%}$=917.8 kg/m³
Air density	ρ_a=1.205 kg/m³	Fluid surface tension	$\sigma_{2\%}$=174×10⁻⁴ N/m $\sigma_{2.5\%}$=332×10⁻⁴ N/m
Dynamic viscosity of fluid	$\mu_{2\%}$=121.76 mPa·s $\mu_{2.5\%}$=123.87 mPa·s	Specific heat capacity of fluid	$c_{2\%}$=1800 J/(kg·K) $c_{2.5\%}$=1783 J/(kg·K)
Supplied fluid flow	*Q'* =5 mL/min	Droplet contact angle	$\theta_{2\%}$=51° $\theta_{2.5\%}$=53°
Workpiece length	*a*=40 mm	Heat conductivity coefficient of air	λa=0.02624 W/m·K
Workpiece feed velocity	v_w =5 mm/s	Air velocity	v_a=45 m/s
Spray angle	*β*=15°	Dynamic viscosity of air	μ_a=18.448 μPas
Nozzle height	H=12 mm	Isobaric specific heat capacity of air	C_a=1.004 kJ/(kg K)
Spray cone angle	*α*=27°		

Table 5. The calculation results

Calculation result	Value
Sauter average diameter	d_{32}=73.3 μm
Volume of nodular droplets with diameter of d_{32}	V_{32}=2.1×10^5 μm^3
Total grinding time	t=8s
Total flow quantity	Q=6.7×1011 μm^3
Total number of droplets	N_{total}=3.2×10^6
Semi-major axis of ellipse	A=30
Semi-minor axis of ellipse	B=9.82
Ellipse area	S=925.044 mm^2
Area of grinding interval	s=49 mm^2
Number of droplets falling in the grinding interval	Nz =1.7×10^5
Heat transfer coefficient of all droplets under effective heat transfer	$hn_{2\%}$=0.55 W/(mm^2·K) $hn_{2.5\%}$=0.56 W/(mm^2·K)
Convective heat transfer coefficient between air and workpiece	ha =8.1148x10^{-5} W/(mm^2·K)
Comprehensive heat transfer coefficient	$h_{2\%}$=0.56 W/(mm^2·K) $h_{2.5\%}$=0.57 W/(mm^2·K)

rises. From t=2.5s to t=5.5s, grinding temperature tends to be stable; after t>5.5 s, grinding temperature rises again. This conforms to change trend of grinding temperature in the actual grinding process.

10.4.2 Temperature Rise of Workpiece Surface

Grinding process can be divided into three areas namely cut-in area, steady area and cut-out area (Shen, Shih, & Xiao, 2011). Cut-in area is the area where grinding wheel starts grinding the workpiece; steady area is the area where grinding temperature tends to be steady; cut-out area is the area where grinding process is about to be completed. Figure 5 shows grinding temperature rise curve chart on workpiece surface of 45 steel grinded through 2% CNT MQL. In the cut-in area, as grinding wheel starts grinding the workpiece, grinding temperature rapidly rises due to the large quantity of heat used to remove material. As the grinding process proceeds, removed material volume by grinding wheel gradually tends to be stable, so is the generated heat, and at the time, grinding process enters steady area, which features balanced state between grinding heat and external heat dissipated through convective heat transfer, and from a macroscopic perspective, grinding temperature on workpiece surface is approximate to a constant value until the steady area ends; in this process, removal rate of workpiece material reaches the maximum value, but temperature on workpiece surface will not rise by a large margin as material removal rate increases. When grinding wheel is about cut out of the workpiece, workpiece surface temperature rapidly rises, because material volume participating in grinding in the cut-out area is reduced and heat quantity transferred to unit volume of workpiece material increases; on the other hand, this is possibly because as the grinding wheel cuts out of the workpiece, wedge-shaped space formed between grinding wheel and workpiece becomes small, nanofluid entering the grinding area is reduced, convective heat transfer intensity is reduced, so balance state in the steady area is broken, the generated heat quantity is far larger than that dissipated due to convective heat transfer, so temperature in the cut-out area soars up.

Figure 4. 45 steel simulation grinding phase diagram

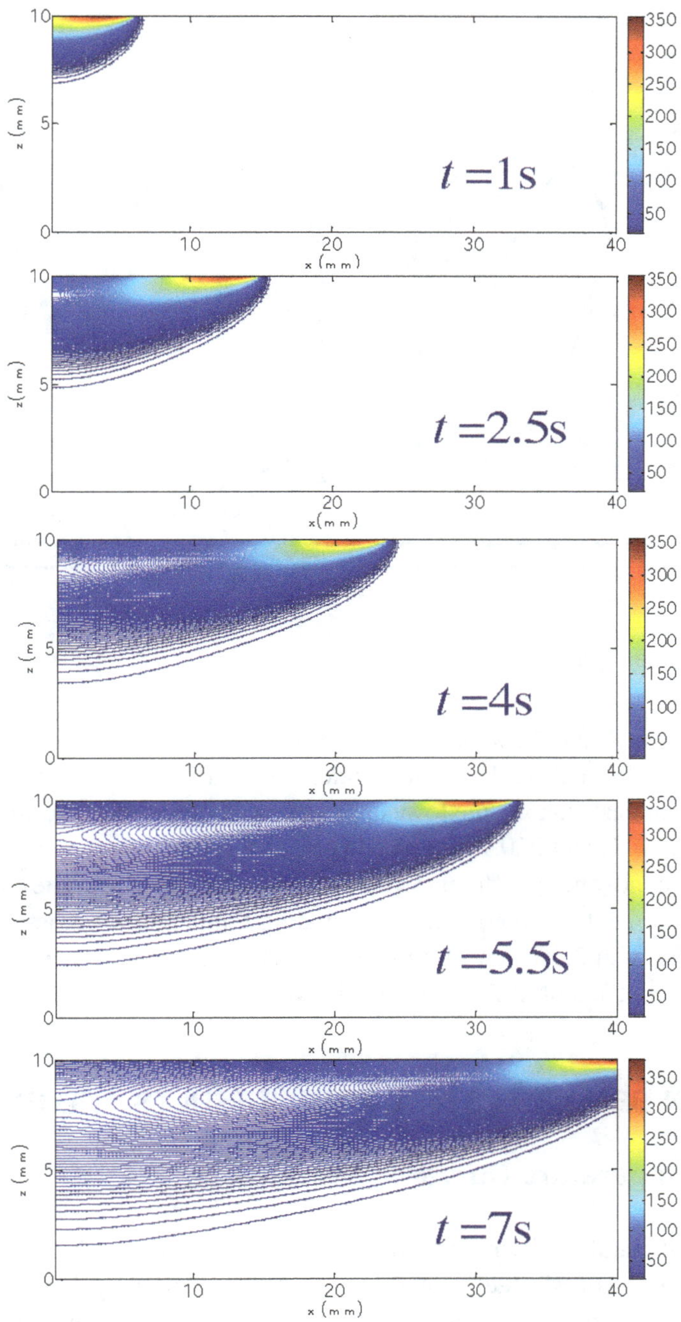

10.4.3 Simulated Grinding Temperature

Fig. 10-5 shows simulated grinding temperature results in the MQL grinding process of three different

Figure 5. Workpiece plane grinding temperature raise graph

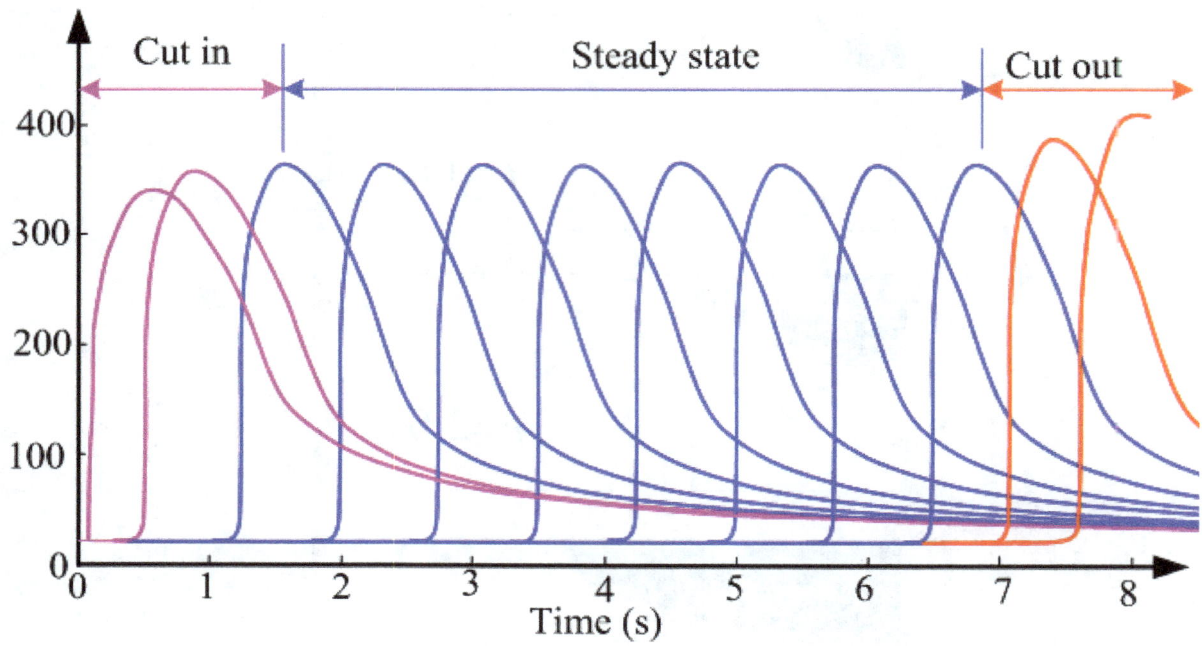

workpiece materials with 2% and 2.5% nanofluids through the finite differential heat transfer model. Simulated temperatures of 45 steel are the highest under two nanofluids, respectively being 367.1 °C and 360.4 °C, those of nodular cast iron are the lowest, respectively being 152.1 °C and 146 °C and those of nickel base alloy are 262.4 °C and 260.1 °C respectively. Therefore, grinding temperatures of the three workpiece materials are handsomely different, which may be related to properties of workpiece materials like density, specific heat, heat conductivity, etc.; grinding temperature in the grinding process of the same workpiece material with 2.5% nanofluid is lower than that with 2% nanofluid, possibly because convective heat transfer coefficient of 2.5% nanofluid is higher than that of 2% nanofluid.

10.5 COMPARISON BETWEEN SIMULATION RESULTS AND EXPERIMENT

10.5.1 Grinding Temperature Curve

Figure 7 shows curves of simulated grinding temperature and experimental grinding temperature in the 2% NMQL grinding process of 45 steel. It can be seen that simulated grinding temperature curve is basically identical with experimental temperature curve, indicating that simulation model basically accords with actual experimental status and can correctly reflect change features of grinding temperature in the experimental process.

Figure 6. Simulated grinding temperature

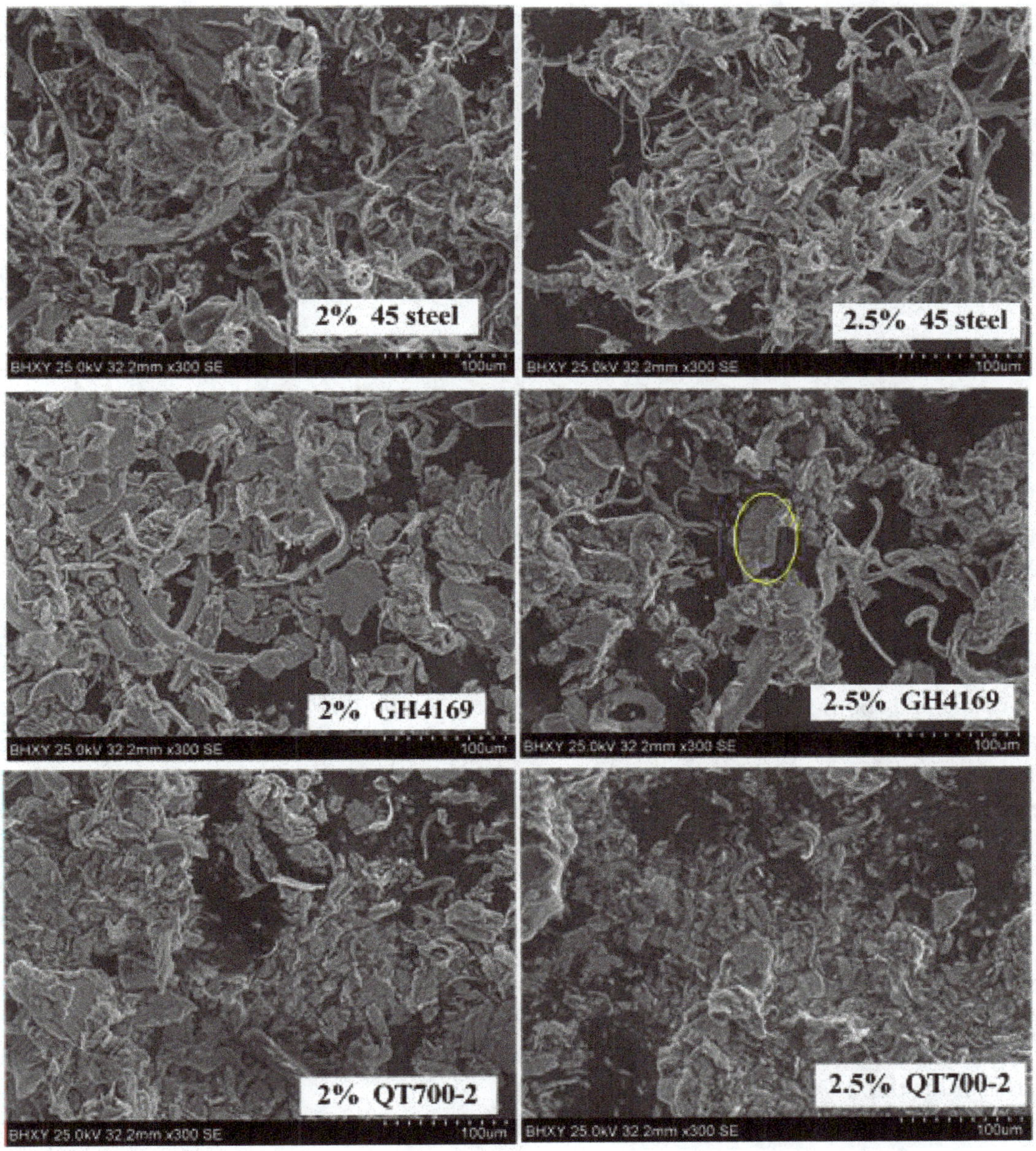

10.5.2 Grinding Temperature Comparison

Based on a comparison of experimental grinding temperature and simulated grinding temperature in Figure 3 and Figure 6, simulated grinding temperature is higher than experimental grinding temperature, possibly because only transfer of grinding heat from workpiece surface inside the workpiece is considered in this simulation process while heat transfer towards grinding wheel base and surrounding environment is neglected; this indicates that the finite differential prediction model established in Chapter 4 is reliable

Figure 7. 45 steel grinding simulation and experimental temperature curve

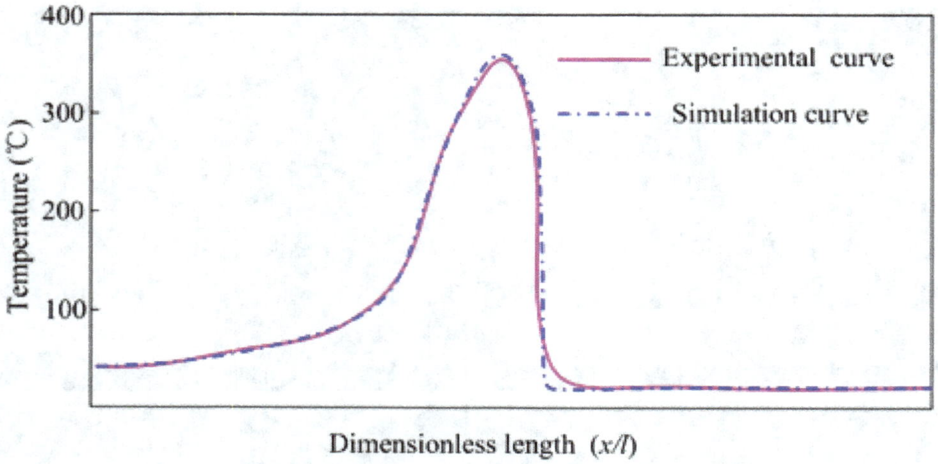

and can objectively reflect actual grinding experimental conditions. In addition, it's found that grinding temperatures when two nanofluids are used for MQL grinding of the same workpiece material show identical trend. Grinding temperature of 2% nanofluid is slightly higher than that of 2.5% nanofluid, indicating that finite differential heat transfer model is very suitable for prediction of grinding temperature under this working condition and physical characteristics of two nanofluids can have a certain influence on lubricating and cooling effect when used to grind the same workpiece material; moreover, simulated and experimental results of grinding temperature are identical with experimental results of specific grinding force. Figure 1 shows that specific tangential grinding forces in the MQL grinding process of the same workpiece material with two nanofluids are not very different, and this certifies that input energies in the grinding process are basically the same. Convective heat transfer performance of 2% nanofluid in the grinding area is slightly poorer than that of 2.5% nanofluid, so laws of obtained grinding temperatures are the same. However, difference of grinding temperatures obtained in the MQL grinding fluid of different workpiece materials with the same nanofluid is great. In terms of CNT nanofluid with 2% volume fraction, grinding temperatures of 45 steel, nickel base alloy and nodular cast iron are 363.9 °C, 245.7 °C and 143.2 °C respectively, and grinding temperatures of 45 steel and nickel base alloy are higher than that of nodular cast iron by 220.7 °C and 102.5 °C respectively possibly because of different properties of the three workpiece materials.

10.6 DISCUSSION

10.6.1 Workpiece Material Attribute Analysis

When grinding wheel grinds the workpiece, workpiece material property has a great bearing on grinding force (Malkin & Guo, 1991). Hardness values of the three workpiece materials used in the experiment are sorted as nodular cast iron >nickel base alloy >45 steel. In the sliding friction phase of workpiece grinding by grinding wheel, workpiece experiences elastic deformation under frictional and squeezing actions of abrasive grains; in the ploughing phase, as abrupt friction appears between abrasive grain and workpiece, input energy is converted into heat, yield stress of workpiece material will decline as the temperature increases. When specific normal grinding force is greater than yield stress of workpiece material, protruding abrasive grains on the grinding wheel will be pressed into the workpiece material base. Metal going through elastic deformation will move to front and two sides of abrasive grain with movement of abrasive grain so that upheaval appears on the workpiece surface. In this phase, as different workpiece materials have different properties, converted heat quantities will be different, and heat will exert a certain influence on lubricating oil film, and as a result, lubricating state between grinding wheel/workpiece interface will be changed, which will cause change of grinding force, etc. This will be discussed in details in section 10.6.2. With further driving of grinding edge of abrasive grain, after workpiece material reaches critical grinding depth, the protruding metal material ruptures and is divorced from workpiece surface to form grinding debris, and metal material not reaching critical grinding depth is stacked at two sides of furrow which grinding edge of abrasive grain slides through due to plastic deformation. During the cutting process of metal material by abrasive grain, squeezing of workpiece base by abrasive grain will cause material plastic deformation. Different properties of these workpiece materials result in different deformations, which may give rise to different grinding forces. FENG B F et al. (Feng, Zhao, & Cai, 2003) experimented on high-speed grinding of 45 steel and 20Cr steel with single abrasive grain and found that grinding force of 45 steel was higher than that of 20Cr, because sliding friction and ploughing effects of 45 steel were stronger than those of 20Cr. Therefore, according to the above analysis, sliding friction and ploughing have an important influence on elastic and plastic deformations. But compared with 45 steel, nodular cast iron has poor tenacity and plasticity. Elastic and plastic deformations of workpiece in the sliding friction phase are far smaller than those of 45 steel; plasticity and tenacity of nickel base alloy are slightly better than those of nodular cast iron, so its elasticity and plasticity are greater than those of cast iron, and then 45 steel has the maximum specific grinding force.

Furthermore, chip shape can reflect grinding force (Wang, Ye, Chen, & Bin, 2011). Chip shape is not only related to material property and sharpness of abrasive grain. Figure 8 shows chip SEM graphs obtained through MQL grinding of three workpiece materials with two nanofluids. Grinding debris formed through 45 steel grinding are banding grinding debris; grinding debris formed by nickel base alloy grinding are mainly banding grinding debris, but there are also mouldering-shaped grinding debris, and local segmental grinding debris are formed when material shear resistance is reduced due to thermoplastic deformation (Black, 1979); grinding debris formed by nodular cast iron grinding are mouldering-shaped grinding debris, mainly because hardness of cast iron is the highest with poor plasticity and tenacity. Banding grinding debris are formed because grinding wheel can't be easily blocked in the grinding process. When abrasive grain is grinding the workpiece, grinding edge is sharp. Sharp grinding edge will result in relatively high specific grinding force in the grinding process. When 45 steel is being grinded, grinding edge of abrasive grain keeps high grinding performance, which is possibly related to its maximum specific normal grinding force (Rabiei et al., 2015) as large specific normal grinding force can improve self-dressing performance of grinding wheel surface so that grinding edge of abrasive grain becomes sharper.

In addition, according to the above theoretical analysis, normal grinding forces in the grinding process should satisfy 45 steel >nickel base alloy >nodular cast iron. However, it can be found from Figure 2 that specific normal grinding force of nodular cast iron is greater than that of nickel base alloy, which may be related to grinding wheel material. Studies indicate that it's improper to use silicon carbide grinding wheel to grind high-temperature alloy (Wei, Xie, Chen, & Yang, 2009). Komandurl and Shaw

Figure 8. Abrasive dust shapes of minimum quantity lubricant grinding of three workpiece materials

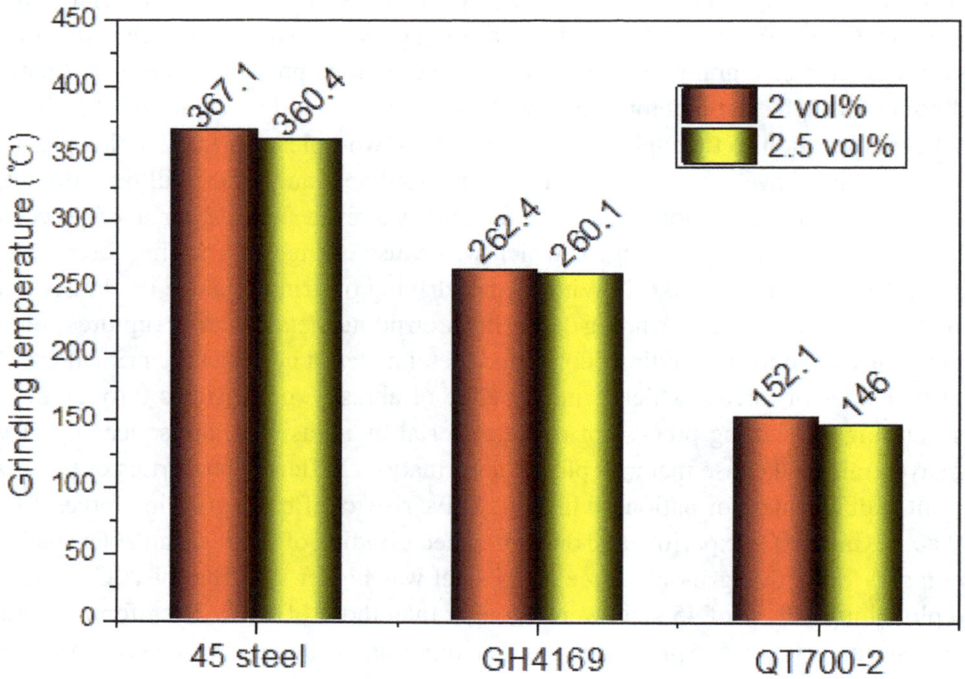

(Komanduri & Shaw, 1976, 1977) used single silicide abrasive grain to conduct grinding experiment on Co-based high-temperature alloy (including elements like Co, Ni, Cr, W, Ta, Ti, C and Zr) and found that some elements (mainly Co, Ni, Cr and W) would escape out of the surface and react with silicon to form brittle silicon compounds like $NiSi_2$, $CoSi_2$, $CrSi_2$, WSi_2, etc. These disilicides formed at the top of abrasive grain in the grinding process will be diffused on workpiece surface and experience chemical reactions with elements like Co, Ni and W in the workpiece to generate metallic carbides like Co carbide Co_3C and Ni carbide Ni_3C, and these carbides are unstable and will be decomposed into metal ions and carbon in the cooling process so as to accelerate wear of abrasive grain. In the grinding process of high-temperature nickel base alloy, brittle silicides $NiSi_2$ and $CrSi_2$ as well as Ni carbide Ni_3C which may be formed may cause wear and even falling of silicon carbide abrasive grain. As main elements of two workpiece materials—45 steel and nodular cast iron—are Fe and C, in the grinding process using silicon carbide, chemical reaction, which may cause abrasive grain consumption of grinding wheel, will not occur between abrasive grain and workpiece. In addition, hardness of nickel base alloy is far greater than those of 45 steel and cast iron. Mechanical shock and thermal shock borne by silicon carbide abrasive grain in the grinding process may be greater than grinding process of 45 steel and nodular cast iron, and abrasive grain will also generate intergranular crushing and accelerate wear, so during the grinding process of 45 steel and cast iron, abrasive grain of grinding wheel has better sharpness than that in the grinding process of nickel base alloy. After abrasive grain is worn or falls off, its depth pressed into the workpiece material base in the ploughing phase becomes small. As shown in Figure 9, actual cutting depth of abrasive grain reduces, so specific normal grinding force is reduced, so is material removal rate

Figure 9. Sharp abrasive and wear abrasive actual cutting depth comparison

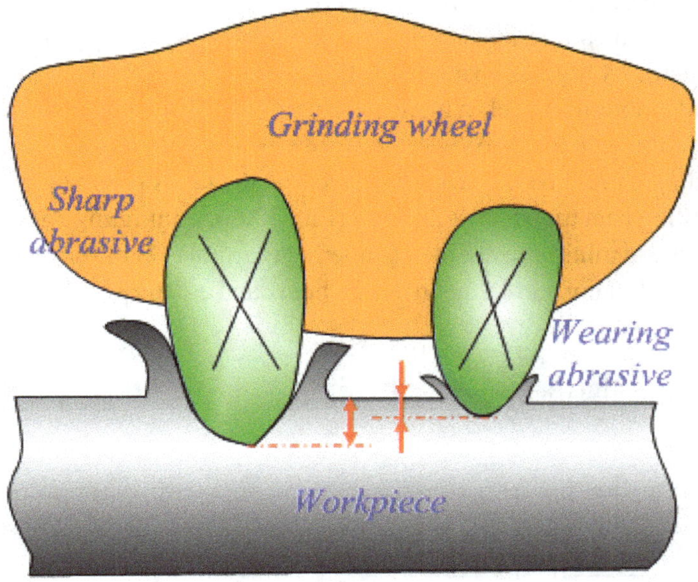

and grinding efficiency. Reduction of actual grinding depth may be because specific normal grinding force of nickel base alloy is smaller than those of nodular cast iron and 45 steel.

Under certain grinding parameters, tangential grinding force decides grinding energy used to remove unit volume of metallic material and grinding energy has an extremely great influence on grinding temperature. According to the analysis of specific tangential grinding forces in Figure 1, 45 steel has the highest grinding energy, followed by nickel base alloy and nodular cast iron in succession, so under a certain nanofluid heat transfer capability, 45 steel obtains the highest grinding temperature while nodular cast iron obtains the lowest grinding temperature, which rightly accords with experimental and simulation results of grinding temperature as shown in Figure 3 and Figure 6.

10.6.2 Analysis of Heat Transfer State of Nanofluids

In the grinding process, fierce friction occurs between abrasive grain and workpiece, a large quantity of input energy is converted into heat, but the generated grinding heat may exert nonnegligible effect on lubricating & cooling properties of MQL grinding fluid. According to different heat transfer performances of droplets, MAO C et al. (Mao, Zou, Huang, & Zhou, 2014) divided grinding area into four heat transfer areas namely non-boiling heat transfer, nucleate boiling heat transfer, transition boiling heat transfer and stable film boiling heat transfer and proposed heat transfer coefficient model of MQL grinding area, where heat transfer coefficient of non-boiling heat transfer area was a small constant value, a large quantity of heat generated in the grinding area could not be timely transferred out, which was adverse to workpiece surface quality; in the nucleate boiling area, convective heat transfer coefficient increased as the workpiece surface temperature rose, and grinding heat continuously escaped with bubbles generated by grinding fluid boiling, which effectively reduced thermal injuries on workpiece surface; heat transfer coefficients in transition boiling area and stable film boiling area rapidly declined with continuous increase of workpiece surface temperature, which was adverse to transfer of grinding heat. When workpiece surface temperature rose to a certain value, stable transition boiling area and even stable film boiling area would appear, and at the time, liquid film formed by grinding fluid on workpiece surface would be destructed by high temperature so that lubricating state of grinding fluid didn't appear between workpiece and grinding wheel, namely under dry grinding state. Therefore, that stable film boiling area appeared in the grinding area was the most adverse to grinding performance, and appearance of stable film boiling should be avoided in engineering as far as possible.

According to the boiling heat transfer mechanism (Lin, 2003), minimum degree of superheat formula (1) for bubble generation and minimum degree of superheat formula (2) of film boiling heat transfer (Lu, 2002), heat transfer state of nanofluid in the grinding area can be determined.

$$\Delta T_s = \frac{T_s^2}{B}\lg(1+\frac{2\sigma}{rp_s})\left[1+\frac{T_s}{B}\lg(1+\frac{2\sigma}{rp_s})\right] \tag{1}$$

$$\Delta T_{\min} = 0.127\frac{\rho_g'\varsigma}{K_g'}\left[\frac{g(\rho_g-\rho_l)}{\rho_l+\rho_g}\right]^{2/3}\left[\frac{\sigma}{g(\rho_g-\rho_l)}\right]^{1/2}\left[\frac{\mu_l}{g(\rho_g-\rho_l)}\right]^{1/3} \tag{2}$$

where ΔT_s is minimum degree of superheat for bubble generation (°C); σ is surface tension (N/m); T_s is liquid boiling point; r is radius of concave liquid level formed by liquid in the capillary tube (mm); B is a constant, p_s=0.1 Mpa, B=2156; ρ_l is liquid density (kg/m³); ρ_g is air density (kg/m³); μ_l is liquid density (kg/m³); ρ_g' is air film density (kg/m³); k_g' is air film heat transfer coefficient (kJ/(mKs)); ς is latent heat (kJ/kg).

According to formulas (10-1) and (10-2), ΔT_s =10 °C and ΔT_{min} =144 °C are obtained. In addition, temperature difference of nucleate boiling heat transfer is generally kept within 20 °C-50 °C (Xin, 1987). Boiling point of MQL oil—palm oil—used in the experiment is 230 °C, so when ΔT_s <10 °C, grinding temperature is not enough to cause phase change of palm oil, but only convective heat transfer of air and heat transfer through droplet heating occur, and no boiling phenomenon will appear; when 10°C<ΔT_s <60°C, palm oil will generate bubbles, which will carry away the heat, under high temperature in the grinding area, so nucleate boiling heat transfer phenomenon will occur in the grinding area; when 60°C<ΔT_s<144°C, grinding temperature continues to rise, and bubbled formed on workpiece surface basically form a whole, and a transition boiling area will appear in the grinding area; when ΔT_s >144°C, as heat flux is too high, palm oil fails to form an effective lubricating fluid film in the grinding area, and a stable film boiling area appears in the grinding area as shown in Figure 10. According to the analysis of experimental results in Figure 3, average grinding temperature in the palm oil (2% and 2.5% volume fractions)-based CNT MQL grinding process of 45 steel is 358°C, so transition boiling heat transfer phenomenon may occur in the grinding area; average grinding temperature in the grinding process of high-temperature nickel base alloy is 245°C, so nucleate boiling heat transfer will occur in the grinding area; average grinding temperature in the grinding process of nodular cast iron is 140°C, so boiling heat transfer will not occur in the grinding area. According to nanofluid heat transfer states in the grinding fluid during the MQL grinding process of three workpiece materials, nucleate boiling heat transfer hap-

Figure 10. Palm oil boiling heat transfer curve

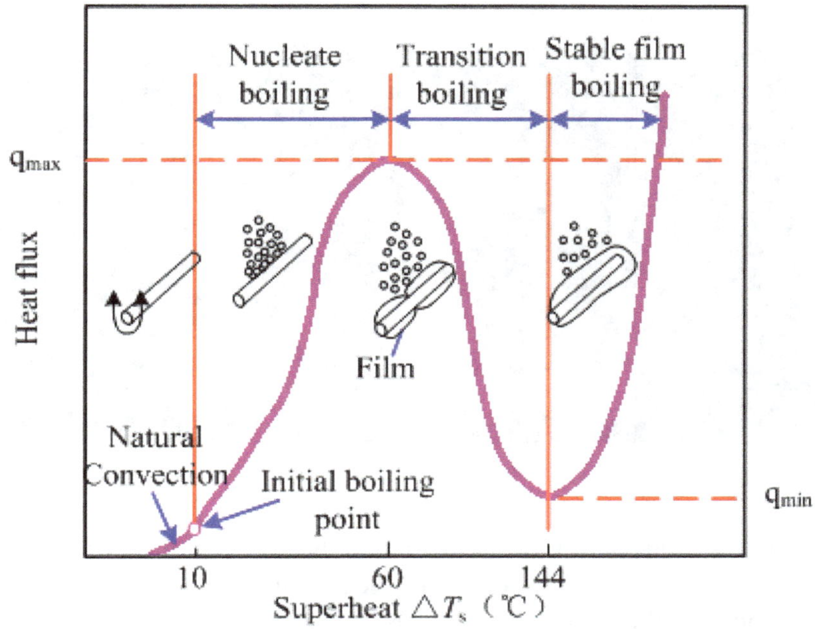

pens to two nanofluids in the grinding process of nickel base alloy, and as the grinding temperature rises, convective heat transfer coefficient gradually increases, so the heat transfer capability is the strongest. In the grinding process of cast iron, as boiling heat transfer doesn't happen to nanofluid, its heat transfer coefficient is the lowest with the poorest heat transfer capability; in the grinding process of 45 steel, as transition boiling heat transfer of nanofluid is very unstable, its heat transfer coefficient will decrease as the temperature rises, but heat transfer capability is between nucleate boiling heat transfer and non-boiling heat transfer. Based on experimental results of tangential grinding force and grinding temperature, the grinding condition used in the experiment is not suitable to grind 45 steel, which is consistent with experimental result of Barczak et al (L. Barczak & Batako, 2012; L. M. Barczak et al., 2010).

According to the above analysis, when 2 nanofluids are used to grind nickel base alloy, nucleate boiling heat transfer will occur. However, as two nanofluids have different volume fractions, their nucleate boiling heat transfer intensities are also different. On the one hand, workpiece surface morphology will exert a certain influence on boiling heat transfer of nanofluid in the grinding area. It's pointed out that bubble formation in the boiling heat transfer is related to superfine concave joints or cracks on surface (Bang & Chang, 2014; Das, Putra, & Roetzel, 2003). As these concave joints or cracks may adsorb minimal quantity of gas or steam to become the evaporation center. Figure 11 shows surface micrographs obtained through grinding of nickel base alloy using two nanofluids. It can be seen that surface morphologies obtained through MQL grinding of nickel base alloy using 2% and 2.5% nanofluids are not significantly different, so quantities of superfine concave joints or cracks formed on workpiece surface are not that different. From the analysis of the influence of surface morphology on boiling heat transfer, nucleate boiling heat transfer intensities of two nanofluids are basically identical.

On the other hand, nanofluid is a uniform suspension prepared by adding nanoparticles in the base fluid, so nanoparticles will also generate a certain influence on heat transfer effect. SHI M H et al. (Shi,

Figure 11. Surface morphology of two kinds of nanofluid minimum quantity lubricant grinding of nickel-based alloys

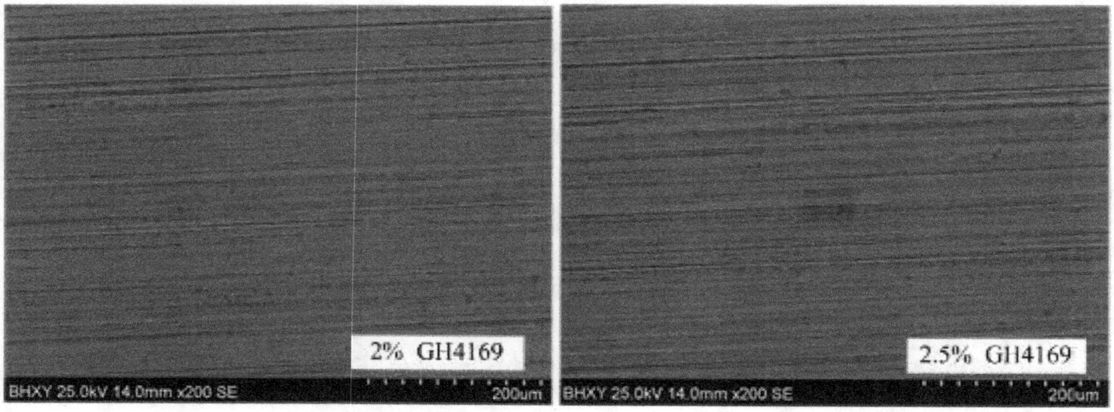

2006) studied nanofluid boiling characteristics and found that stuffing effect of nanoparticles reduced caliber and quantity of concave joints or cracks on the heated surface as shown in Figure 12, so boiling heat transfer was weakened. On the precondition that basically identical surface morphologies are obtained through MQL grinding of nickel base alloy with two nanofluids, the effect of nanoparticles on concave joints or cracks on surface becomes the main factor influencing nucleate boiling heat transfer intensity of two nanofluids. Obviously, number of nanoparticles contained in unit volume of 2.5% nanofluid is greater than that contained in the unit volume of 2% nanofluid, so more evaporation centers may be formed on workpiece surface when 2% nanofluid is used to grind nickel base alloy under the stuffing effect of nanoparticles, and the quantity of formed bubbles may be larger than that of 2.5% nanofluid, so nucleate boiling heat transfer intensity of 2% nanofluid is slightly higher than that of 2.5% nanofluid.

REFERENCES

Bang, I. C., & Chang, S. H. (2014). Boiling heat transfer performance and phenomena of Al_2O_3-water nano-fluids from a plain surface in a pool. *International Journal of Heat and Mass Transfer*, *48*(12), 2407–2419. doi:10.1016/j.ijheatmasstransfer.2004.12.047

Barczak, L., & Batako, A. (2012). Application of Minimum Quantity Lubrication in Grinding. *Materials and Manufacturing Processes*, *27*(4), 406–411. doi:10.1080/10426914.2011.577866

Barczak, L. M., Batako, A. D. L., & Morgan, M. N. (2010). A study of plane surface grinding under minimum quantity lubrication (MQL) conditions. *International Journal of Machine Tools & Manufacture*, *50*(11), 977–985. doi:10.1016/j.ijmachtools.2010.07.005

Black, J. T. (1979). Flow Stress Model in Metal Cutting. *Journal of Engineering for Industry*, *101*(4), 403–415. doi:10.1115/1.3439527

Figure 12. Nanoparticles stuffing schematic

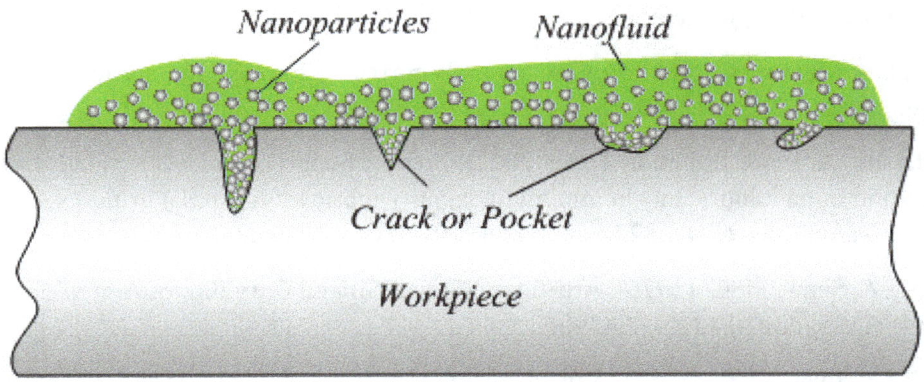

Chen, Z., Xuan, S., Lei, L., Chen, X., Fei, C., Chi, T. C., & Shung, K. (2016). 3D printing of piezoelectric element for energy focusing and ultrasonic sensing. *Nano Energy*, *27*, 78–86. doi:10.1016/j.nanoen.2016.06.048

Choi, S., Zhang, Z., Yu, W., Lockwood, F., & Grulke, E. (2001). Anomalous thermal conductivity enhancement in nanotube suspensions. *Applied Physics Letters*, *79*(14), 2252–2254. doi:10.1063/1.1408272

Das, S. K., & Choi, S. U. S. (2009). A Review of Heat Transfer in Nanofluids. *Advances in Heat Transfer*, *41*(08), 81–197. doi:10.1016/S0065-2717(08)41002-X

Das, S. K., Putra, N., & Roetzel, W. (2003). Pool boiling of nano-fluids on horizontal narrow tubes. *International Journal of Multiphase Flow*, *29*(8), 1237–1247. doi:10.1016/S0301-9322(03)00105-8

Ding, F., Xu, J. H., Chen, Z. Z., Yang, C. Y., Song, C. J., & Fu, Y. C. (2013). Fabrication and performance of porous metal-bonded CBN grinding wheels;using alumina bubble particles as pore-forming agents. *International Journal of Advanced Manufacturing Technology*, *67*(5-8), 1309–1315. doi:10.100700170-012-4567-4

Eastman, J. A., Choi, S. U. S., Li, S., Yu, W., & Thompson, L. J. (2001). Anomalously increased effective thermal conductivities of ethylene glycol-based nanofluids containing copper nanoparticles. *Applied Physics Letters*, *78*(6), 718–720. doi:10.1063/1.1341218

Fan, J., & Wang, L. Q. (2011). Review of Heat Conduction in Nanofluids. *Journal of Heat Transfer*, *133*(4), 040801. doi:10.1115/1.4002633

Feng, B., Zhao, H., & Cai, G. (2003). *Experimental study on single-grit high-speed grinding 45 and 20Cr steels*. Machinery Manufacturing Engineer. doi:10.1115/1.4003835

González-Santander, J. L., Fernández, R., Martín, G., & Arrazola, P. J. (2016). An useful analytical formula to avoid thermal damage in the adaptive control of dry surface grinding. *International Journal of Mechanical Sciences*, *117*, 152–161. doi:10.1016/j.ijmecsci.2016.08.014

Hadada, M. J., Sadeghi, M. H., & Sadeghi, B. (2012). Temperature and energy partition in minimum quantity lubrication-MQL grinding process. *International Journal of Machine Tools & Manufacture, 55*(3), 10–17. doi:10.1016/j.ijmachtools.2011.11.010

Hafenbraedl, D., & Malkin, S. (2001). Environmentally correct technology for internal cylindrical grinding. *Mach and Metals Magaz, 37*, 40–55.

Kalita, P., Malshe, A. P., & Rajurkar, K. P. (2012). Study of tribo-chemical lubricant film formation during application of nanolubricants in minimum quantity lubrication (MQL) grinding. *CIRP Annals - Manufacturing Technology, 61*(1), 327-330. doi:10.1016/j.cirp.2012.03.031

Komanduri, R., & Shaw, M. C. (1976). Attritious Wear of Silicon Carbide. *Journal of Engineering for Industry, 98*(4), 1125. doi:10.1115/1.3439065

Komanduri, R., & Shaw, M. C. (1977). A mechanism for the accelerated wear of high speed sliding surfaces in the presence of spherical particles. *Wear, 42*(1), 177–185. doi:10.1016/0043-1648(77)90178-8

Lin, Z. (2003). *Gas-Liquid Two-Phase Flow and Boiling Heat Transfer*. Xi'an Jiaotong University Press.

Lu, Z. (2002). *Two-Phase Flow and Boiling Heat Transfer*. Tsinghua University Press.

Malkin, S., & Guo, C. (1991). Grinding technology: Theory and applications of machining with abrasives. *International Journal of Machine Tools & Manufacture, 31*(3), 435–436. doi:10.1016/0890-6955(91)90088-K

Mao, C., Tang, X., Zou, H., Huang, X., & Zhou, Z. (2012). Investigation of grinding characteristic using nanofluid minimum quantity lubrication. *International Journal of Precision Engineering and Manufacturing, 13*(10), 1745–1752. doi:10.100712541-012-0229-6

Mao, C., Zou, H., Huang, Y., & Zhou, Z. (2014). Research on Heat Transfer Mechanism in Grinding Zone for MQL Surface Grinding. *Zhongguo Jixie Gongcheng, 25*(6), 826–831. doi:10.3969/j.issn.1004-132X.2014.06.024

Öğüt, E. B. (2009). Natural convection of water-based nanofluids in an inclined enclosure with a heat source. *International Journal of Thermal Sciences, 48*(11), 2063–2073. doi:10.1016/j.ijthermalsci.2009.03.014

Rabiei, F., Rahimi, A. R., Hadad, M. J., & Ashrafijou, M. (2015). Performance improvement of minimum quantity lubrication (MQL) technique in surface grinding by modeling and optimization. *Journal of Cleaner Production, 86*, 447–460. doi:10.1016/j.jclepro.2014.08.045

Radiom, M., Yang, C., & Weng, K. C. (2013). Dynamic contact angle of water-based titanium oxide nanofluid. *Nanoscale Research Letters, 8*(1), 282–282. doi:10.1186/1556-276X-8-282 PMID:23759071

Sadeghi, M. H., Hadad, M. J., Tawakoli, T., Vesali, A., & Emami, M. (2010). An investigation on surface grinding of AISI 4140 hardened steel using minimum quantity lubrication-MQL technique. *International Journal of Material Forming, 3*(4), 241–251. doi:10.100712289-009-0678-3

Setti, D., Ghosh, S., & Rao, P. V. (2012). Application of nano cutting fluid under minimum quantity lubrication (MQL) technique to improve grinding of Ti-6Al-4V alloy. *Proceedings of World Academy of Science,* (10), 493-497. doi:10.13140/2.1.3188.8009

Shen, B. (2008). Minimum Quantity Lubrication Grinding Using Nanofluids. *Dissertations & Theses - Gradworks*.

Shen, B., Shih, A. J., & Tung, S. C. (2008). Application of Nanofluids in Minimum Quantity Lubrication Grinding. *Tribology & Lubrication Technology*, *51*(6), 730–737. doi:10.1080/10402000802071277

Shen, B., Shih, A. J., & Xiao, G. (2011). A Heat Transfer Model Based on Finite Difference Method for Grinding. *Journal of Manufacturing Science and Engineering*, *133*(3), 031001. doi:10.1115/1.4003947

Shi, M. H. (2006). Expeimental Study Of Pool Boiling Heat Transfer Of Nano-Particle Suspensions OnA Plate Surface. *Journal of Engineering Thermophysics*, *14*(3), 223–231. doi:10.1615/IHTC13.p28.60

Silva, L. R., Catai, R. E., Fusse, R. Y., França, T. V., & Aguiar, P. R. (2005). Study on the behavior of the minimum quantity lubricant - MQL technique under different lubricating and cooling conditions when grinding ABNT 4340 steel. *Journal of the Brazilian Society of Mechanical Sciences and Engineering*, *27*(2), 192–199. doi:10.1590/S1678-58782005000200012

Song, C. J., Ding, W. F., Xu, J. H., & Chen, Z. Z. (2012). Grinding Performance of Metal-Bonded CBN Wheels with Regular Pores. *Applied Mechanics & Materials, 217-219*, 1857-1862.

T, T., MJ, H., & MH, S. (2010). Influence of oil mist parameters on minimum quantity lubrication-MQL grinding process. *International Journal of Machine Tools & Manufacture, 50*(6), 521-531.

Tasdelen, T., Thordenberg, H., & Olofsson, D. (2008). An experimental investigation on contact length during minimum quantity lubrication (MQL) machining. *Journal of Materials Processing Technology*, *203*(1), 221–231. doi:10.1016/j.jmatprotec.2007.10.027

Tawakoli, H., Hadad, M. J., Sadeghi, M. H., Daneshi, A., Stöckert, S., & Rasifard, A. (2009). An experimental investigation of the effects of workpiece and grinding parameters on minimum quantity lubrication—MQL grinding. *International Journal of Machine Tools & Manufacture*, *49*(12), 924–932. doi:10.1016/j.ijmachtools.2009.06.015

Tawakoli, T., Westkämper, E., & Azarhoushang, B. (2009). Effects of Vibration-Assisted Grinding on Wear Behavior of Vitrified Bond Al_2O_3 Wheel. *Advanced Materials Research, 76-78*, 21-26.

Wang, J. M., Ye, R. Z., Chen, H. P., & Bin, H. Z. (2011). Surface Grinding Force Model of Steel 55 Based on Undeformed Chip Thickness with CBN Electroplated Wheels. *Advanced Materials Research, 189-193*, 1768-1773.

Wang, Y., Li, C., Zhang, Y., Min, Y., Li, B., Jia, D., & Cong, M. (2016). Experimental evaluation of the lubrication properties of the wheel/workpiece interface in minimum quantity lubrication (MQL) grinding using different types of vegetable oils. *Tribology International*, *127*, 487–499. doi:10.1016/j.triboint.2016.03.023

Webster, J. A., & Gruen, P. (2008). Improving grinding process economics through more effective fluid application. *International Journal of Mechatronics & Manufacturing Systems*, *1*(4), 343–364. doi:10.1504/IJMMS.2008.023224

Wei, Y., Xie, H., Chen, L., & Yang, L. (2009). Investigation of thermal conductivity and viscosity of ethylene glycol based ZnO nanofluid. *Thermochimica Acta*, *491*(1), 92–96. doi:10.1016/j.tca.2009.03.007

Wu, X., Wu, H., & Cheng, P. (2009). Pressure drop and heat transfer of Al_2O_3-H_2O nanofluids through silicon microchannels. *Journal of Micromechanics and Microengineering*, *19*(19), 105020. doi:10.1088/0960-1317/19/10/105020

Xin, D. (1987). *Boiling heat transfer and strengthening*. Chongqing University Press.

Xinyu, W. U., Huiying, W. U., Jian, Q. U., & Zheng, P. (2008). Flow and heat transfer characteristics of nanofluids in silicon chip microchannels. *Journal of Chemical Industry & Engineering*.

Yang, Y., Chen, Z., Song, X., Zhang, Z., & Chen, Y. (2017). Biomimetics: Biomimetic Anisotropic Reinforcement Architectures by Electrically Assisted Nanocomposite 3D Printing (Adv. Mater. 11/2017). *Advanced Materials*, *29*(11). doi:10.1002/adma.201605750 PMID:28185341

Yang, Y., Chen, Z., Song, X., Zhu, B., Hsiai, T., Wu, P. I., & Zhou, Q. (2016). Three dimensional printing of high dielectric capacitor using projection based stereolithography method. *Nano Energy*, *22*, 414–421. doi:10.1016/j.nanoen.2016.02.045

Zhang, D., Li, C., Zhang, Y., Jia, D., & Zhang, X. (2015). Experimental research on the energy ratio coefficient and specific grinding energy in nanoparticle jet MQL grinding. *International Journal of Advanced Manufacturing Technology*, *78*(5-8), 1275–1288. doi:10.100700170-014-6722-6

Zhang, X., Li, C., Zhang, Y., Jia, D., Li, B., Wang, Y., & Zhang, X. (2016). Performances of Al_2O_3 / SiC hybrid nanofluids in minimum-quantity lubrication grinding. *International Journal of Advanced Manufacturing Technology*, *86*(9-12), 1–15. doi:10.100700170-016-8453-3

Zhang, Y., Li, C., Jia, D., Zhang, D., & Zhang, X. (2015). Experimental evaluation of MoS_2 nanoparticles in jet MQL grinding with different types of vegetable oil as base oil. *Journal of Cleaner Production*, *87*(1), 930–940. doi:10.1016/j.jclepro.2014.10.027

Chapter 11
Experimental Evaluation of the Lubrication Properties of the Wheel/Workpiece Interface in MQL Grinding Using Vegetable Oils

ABSTRACT

Given the increasing attention to environmental and health problems caused by machining, the development of an environmentally friendly grinding fluid has become an urgent task. In this study, seven typical vegetable oils (i.e., soybean, peanut, maize, rapeseed, palm, castor, and sunflower oil) were used as the minimum quantity lubrication (MQL) base oil to conduct an experimental evaluation of the friction properties of the grinding wheel/workpiece interface. With nickel-based alloy GH4169 as workpiece material, the flood grinding and MQL grinding were selected. Experimental results indicated that castor oil MQL grinding had a friction coefficient and specific grinding energy of 0.30 and 73.47 J/mm3, which decreased by 50.1% and 49.4%, respectively, compared with flood grinding. Moreover, maize oil had the highest G-ratio of 29.15. Peanut, sunflower, and soybean oil with more saturated fatty acids, castor oil with more castor acids, and palm oil with numerous palmitic acids were suitable as lubricating fluids.

11.1 INTRODUCTION

To obtain a higher precision and surface quality of the workpiece, the grinding process is indispensable during machining. The grinding process is a random integration of scratching, ploughing, and cutting using abundant irregular abrasive particles that are scattered on the grinding wheel/workpiece interface (Chen, Rowe, & W, 1996; Xun & Rowe, 1996). Grinding employs a higher unit grinding force and grinding speed than other cutting processes and therefore involves a significantly higher grinding energy. Most of this energy is converted into thermal energy. This thermal energy has different distribution patterns from

DOI: 10.4018/978-1-7998-1546-4.ch011

other cuttings. Most grinding heat (approximately 80%) is transferred into the workpiece (Komanduri & Hou, 2001). Aerospace materials are widely used, especially nickel-based alloy. However, their hardness is great, and their thermal conductivity is poor, so that more heat is consumed during the machining process of these hard materials; these shortcomings severely influence the surface quality, machining precision, and machining efficiency of the workpiece (Guy Albert, 2010). A high temperature in the grinding process also has significant influences on the grinding property of abrasive particles, which may directly decrease the service life of the grinding wheel.

To decrease the temperature in the grinding zone and to improve the machining quality of the workpiece and the life of the cutter, different cooling lubrication technologies are applied in the grinding process. In machining, flood cooling technology is the most common cooling lubrication method, which requires about 60 L/h of grinding fluid flow. The grinding fluid can cool the lubrication grinding wheel and workpiece and remove debris. Flood cooling grinding results in better workpiece surface quality than dry grinding; however, the grinding wheel at a high speed forms a layer of gas barrier, which makes entry into the grinding wheel and workpiece interface difficult for numerous grinding fluids. Thus, an effective flow rate between the grinding wheel and workpiece interface is only 5%–40%. Moreover, in the machining process, the volatilization, leakage, or overflow of the grinding fluid may cause significant damage to the health of workers, and may induce water pollution primarily because of the grinding fluid. Direct contact between the smog caused by the heat and volatilization of the grinding fluid and the human body poses threat to human health and induces various diseases of the skin, respiratory tract, and lung. Mineral oil is also used as a cooling lubricating liquid. It has excellent lubrication properties, but its low cooling property and high cost limit its application in machining. To protect the environment and decrease costs, dry grinding is adopted because it is significantly advantageous in environment protection. However, given that removing materials per unit volume through the grinding process consumes more energy than other processes, more heat gathers in the grinding zone (Malkin & Guo, 2007). Only 10% heat is removed by the grindings in the grinding process (Mao et al., 2013). The gathered heat leads to high temperature, and without the lubrication and cooling effect of the grinding fluid, the grinding wheel wears seriously, and the workpiece precision and surface integrity deteriorate. Furthermore, the process of dry grinding has relatively high requirements in terms of the grinding wheel, workpiece materials, and machine tool. The grinding wheel should have high hardness, toughness, and wear resistance. These specific conditions required by dry grinding limit its wide application in machining.

In 1996, scholars W. D. Hewson and G. K. Gerow in the USA Michigan Tech University proposed the concept of minimum quantity lubrication (MQL) (Hewson & Gerow, 1999). MQL technology is another environment-friendly processing technology. After the mixture and atomization of high-pressure gas and minimum quantity liquid, grinding liquid fogdrop at the micron scale is sprayed to the grinding zone. Among them, high-pressure gas causes cooling and debris removal. The lubricating fluid adheres to the surface of the workpiece and forms a layer of protective film for lubrication. MQL technology can precisely control the supply quantity of the grinding fluid, with a flow of as high as 30–100 ml/h, which is far lower than the 60 L/h fluid flow of flood cooling grinding. This technology can realize a residue-free complete processed workpiece. Sadeghi et al. (2010) reported that MQL improves surface quality and reduces the tangential grinding force and friction coefficient in hardened steel technique compared with other coolant lubricant environments. Mao et al. (2013) analyzed the heat transfer coefficient of the workpiece surface during MQL grinding. Their analysis shows that the theory of surface heat transfer coefficient during MQL grinding is creditable. Meanwhile, numerous studies and experiments have indicated that MQL technology can improve grinding properties and can reach and even exceed the flood lubricating effects under certain experimental conditions.

The grinding lubricating fluid is usually composed of base oils and additives, and base oils comprise the majority, with approximately 95%. Traditional lubricating fluids mostly use mineral oil as base oil. Mineral oil can reduce frictional wear, save energy, lengthen the service life of a machine, and satisfy lubrication requirements under rigorous working conditions. With its low cost, this oil has been widely used. However, mineral base oils consist of hydrocarbons with paraffinic, naphthenic, or aromatic structures. The biodegradability of mineral oil is very poor, which can cause long-term pollution to the environment. Moreover, mineral oil is classified under non-renewable resources. Synthetic and semi-synthetic base oils composed of chemicals with additives and diluted in water are free from mineral oil. Synthetic base oils form transparent solutions and provide the cutting operation good visibility. The addition of organic and inorganic chemical solutions in synthetic base oils provides water softening, corrosion resistance, lubrication, reduction of surface tension, and blending. Synthetic base oils are good coolant, but provide insufficient lubrication compared with other cutting fluids because of the lack of oiliness (Debnath, Reddy, & Yi, 2014). Semi-synthetic base oils are chemical emulsions, which contain mineral oil diluted in water with some additives to reduce the size of oil particles; hence, this type of lubricant fluid is more effective (Kuram, Ozcelik, & Demirbas, 2013).

With the development of technology and the realization of the importance of environmental protection, environment-friendly, resource-saving, and energy-efficient sustainable manufacturing has gained a more prominent role. To reduce the environmental pollution and processing cost of the grinding fluid, some scholars have attempted to use vegetable oil as MQL base oil. In 1999, Lambert and Johnson (1997) used vegetable oil with high oleic acid, vegetable oil additive, and antioxidant additive to develop internal combustion oil. This kind of lubricating oil can be mainly used as internal combustion engine oil, rail oil, and chain oil and provides excellent lubrication, heat conduction, and biodegradability.-In machining, vegetable oil is an ideal option as MQL base oil. Vegetable oil has unique advantages over mineral oil. (1) Vegetable oil is a renewable resource and has no or very light toxicity. It has clean and abundant special types, with wide planting scope. The refining and processing technology of vegetable oil is mature, and its price is relatively low (Cetin, Ozcelik, Kuram, & Demirbas, 2011). (2) Vegetable oil has excellent biodegradability, because its glycerine ester group is easy to hydrolyze and the unsaturated double bond in the ester group chain is extremely easy to be attacked by microorganisms, thereby causing further β-oxidation. Moreover, the natural fatty acid in vegetable oil can promote the biodegradation process (Lawal, Choudhury, & Nukman, 2013). (3) Vegetable oil has high lubrication property, which is determined by the basic structure of vegetable oil molecules and chemical components (Asadauskas, Perez, & Duda, 1996). Vegetable oil molecules can form an adsorption film on the metallic surface, and the fatty acid in this oil can react with the metal surface, forming a monofilm of metallic soap. Both can influence anti-friction and anti-wear. (4) Vegetable oils have a higher flash point, which reduces smoke formation and fire hazard. A higher flash point allows the use of cutting fluid under high temperature conditions (Kuram, Ozcelik, Bayramoglu, Demirbas, & Simsek, 2013). In conclusion, vegetable oil-based grinding fluid not only conforms to the environmental and health requirements of environment-friendly manufacturing but also satisfies the requirements of lubrication properties (Shashidhara & Jayaram, 2010).

Adopting vegetable oil as MQL base oil, scholars at home and abroad have made numerous relevant experimental studies. Emami et al. (2014) investigated the performance of four types of lubricants (i.e., mineral, hydrocracked, synthetic, and vegetable oils) and evaluated them with regard to reduction in cutting force, specific energy, and surface roughness during the MQL grinding of Al_2O_3 engineering ceramics. The results showed that MQL considerably affects the cutting force, specific energy, and surface roughness and can also decrease the environmental hazards of cutting fluids based on vegetable oil. Khan et al. (2009) studied the effects of MQL by vegetable oil-based cutting fluids (VBCFs) on the turning performance of low alloy steel AISI 9310. They indicated that VBCFs reduce the tool wear and enhances the tool life and surface finish, compared with dry and wet machining. Furthermore, MQL is environment friendly and improves the machinability characteristics. Zhang et al. (2015) used soybean, palm, and rapeseed oil as base oil in an experimental evaluation of the grinding and lubrication property of MQL with nanoparticles. The following four grinding working conditions were considered: dry grinding, flood grinding, MQL (using three vegetable oils and liquid paraffin), and nanoparticle jet MQL (using nanoparticles of different concentrations). The experimental results indicated that palm oil-based nanofluid with MoS_2 has the best lubrication property under the nanoparticle jet MQL condition. This finding is due to the high fatty acid content and high film-forming property of carboxyl in palm oil. In a series of experimental studies, Belluco and Chiffre (2004) evaluated the performance of six cutting fluids (i.e., a commercial mineral-based oil and five vegetable-based cutting fluids) in drilling AISI 316L stainless steel using conventional HSS-Co tools. Their evaluation was conducted by measuring the tool life, tool wear, cutting forces, and chip formation. The results indicated that the VBCFs perform better than the mineral, with the best performance resulting in 177% increase in tool life and 7% decrease in thrust force vis-à-vis commercial mineral oil. Lawal et al. (2014) utilized the Taguchi method to evaluate vegetable and mineral oil cutting fluids in turning AISI 4340 steel with coated carbide tools. The results showed that both palm oil- and cottonseed oil-based cutting fluids have better heat conductivity and environment-friendly properties and that they are better alternatives for machining AISI 4340 steel with coated carbide. Cetin et al. (2011) used four different VBCFs (i.e., sunflower and canola oils) and two commercial types of semi-synthetic and mineral and evaluated them in terms of reducing surface roughness. They also used cutting and feed forces during the turning of AISI 304L austenitic stainless steel with a carbide insert tool. The results showed that sunflower and canola oils perform better than others as MQL base oil. Ozcelik et al. (2011) studied the performance of both VBCFs (i.e., refined sunflower and canola oils), including the different percentages of extreme pressure (EP) additive and two commercial cutting fluids (i.e., semi-synthetic and mineral oil), during the turning of AISI 304L. Their results indicated that a canola oil-based cutting fluid with 8% of EP has better performance than the others. Ozcelik et al. (2013) also investigated the performance of different VBCFs (i.e., crude sunflower, refined sunflower, and refined canola oil) and commercial semi-synthetic cutting fluid during the drilling of AISI 304 austenitic stainless steel with an HSSE tool. Their experimental results showed that canola oil-based cutting fluid gives the best performance among all the cutting fluids because of its good lubricant properties under constant cutting conditions. Silva et al. (2013) conducted a series of experimental research and used vegetable oil as MQL base oil in a grinding experiment to compare the conventional cooling method for the cylindrical plunge grinding process of hardened AISI 4340 steel. Sharma and Sidhu (2014) investigated the effects of dry and near dry machining on AISI D2 steel using VBCFs. They found that VBCFs can improve the surface integrity and decrease the cutting temperature by nearly 50%. Kuram et al. (2013) evaluated the performance of four different types of VBCFs and a commercial mineral cutting fluid with an EP additive in the turning of Al 7075-T6. They found that the performance of VBCFs during the turning of Al 7075-T6 was better than that of the commercial mineral cutting fluid. They also claimed that MQL using vegetable oil protects the environment from pollution because vegetable oil discharges significantly less fumes, mist, and oxides than mineral oil, which is non-biodegradable. Xavior and Adithan (2009) studied the performance of coconut oil in the machining of AISI 304 material with carbide a tool and found that coconut oil reduces the tool wear and improves the surface finish more significantly than mineral oil.

Numerous studies on vegetable oil-based lubricating fluid and the previous studies of relevant scholars have proven that vegetable oil can be used as MQL base oil in turning, drilling, and other metal machining to good cooling lubrication effects(Guo et al., 2017; Li et al., 2016; Mao et al., 2014; Rahim & Sasahara, 2011a, 2011b; Wang et al., 2016). However, in such studies, only one to two vegetable oils are used as base oil, and no systematic comparative performance evaluation is conducted on different vegetable oils. This is crucial because vegetable oils of different varieties with various ingredients differ in lubricating and cooling effects. In this context, the present study investigates the grinding of GH4169 with seven different types of vegetable oil-based lubricating fluids and a commercial mineral oil using the MQL technique. Experimental evaluations are conducted on the friction lubrication property of the grinding wheel/workpiece interface in terms of friction coefficient, specific grinding energy, grinding ratio (G-ratio) of the grinding wheel, surface morphology, and surface roughness under different base oil lubricating conditions and flood lubricating conditions.

11.2 EXPERIMENTAL DESIGN

11.2.1 Experimental Equipment and Grinding Parameters

Selecting proper experimental and analysis equipment is an important precondition to successfully complete the experiment. Experimental equipment can be divided into two parts: experimental equipment and analysis equipment, where the former includes K-P36 NC surface grinder, dynamometer, grinding wheel and MQL device and the latter includes roughness measuring instrument, digital viscosimeter, contact angle measuring instrument and SEM.

1. K-P36 NC Surface Grinder

K-P36 NC surface grinder produced by German SCHLEIFRING is used in this experiment. Main technical parameters are as follows. Maximum output power of spindle: 4.5 KW; range of spindle speed: 45~4800 r/min; size of magnetic workbench: 950 mm×1000 mm; working range: 600mm; transverse feed velocity: 30~30000 mm/min; longitudinal feed velocity: 4~4000 mm/min. The experimental equipment is shown in Figure 1.

2. Dynamometer and Grinding Wheel

Three-way grinding dynamometer YDM-III99developed by Dalian University of Technology is used to measure grinding force, and charge amplifier YE5850D is used to measure data and record grinding tangential force, normal force and axial force, where sampling frequency for measurement of grinding force is 1 kHz. As shown in Figure 1, grinding force signal after sampling is input into "dynamic grinding force test system" for smoothing processing, and grinding force image and grinding force data are finally output. In the end, 100 force data points are selected in the steady areas of grinding force in all directions to solve average grinding forces.

Figure 1. Surface grinding measuring experiment equipment: 1-grinding wheel, 2 –magnetic workbench, 3-workpiece, 4-YDM-III99 grinding dynamometer, 5-data acquisition unit, 6-charge amplifier, 7-MQL device, 8-operation panel of NC grinder

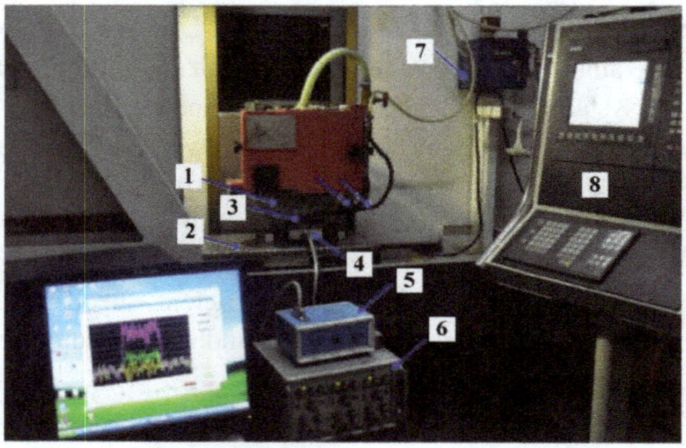

White corundum ceramic-bonded grinding wheel is used in this experiment, its model is WA80MV12P, dimensions are 300 mm×20 mm×76.2 mm, granularity is 80# and the highest linear velocity is 50 m/s with large pores and medium-soft hardness.

3. MQL Device

MQL device in the experiment is KS-2106 MQL system produced by Shanghai KINS Energy-saving Technology Co., Ltd as shown in Figure 2. This system uses an impulsator to regulate flow quantities of lubricating oil and compressed air of the MQL system, proper gas-liquid ratio can be selected for processing, and compressed air and MQL lubricating oil are sprayed from the nozzle into machining area through high-pressure gas after mixing and atomization.

4. Viscosimeter and Contact Angle Measuring Instrument

Viscosity and contact angle characteristics of lubricating fluid used in the experiment are measured using DV2TLV digital viscosimeter produced by Sinoinstrument Co., Ltd (Figure 3 (a)) and JC2000C1B contact angle measuring instrument produced by Shanghai Zhongcheng Digital Technic Apparatus Equipment Co., Ltd (Figure 3 (b)) respectively.

5. Roughness Measuring Instrument and SEM

TIME3220 surface roughness measuring instrument produced by Time Group is used to measure roughness of the machined workpiece as shown in Figure 4 (a). Five different points with 0.8mm measuring length in vertical grinding direction are averagely selected on the workpiece surface to measure roughness values R_a and R_z. SEM is S-3400N produced by Hitachi as shown in Figure 4 (b), which is used to observe morphological structure of workpiece surface.

Figure 2. Minimum quantity oil supply system: 1- MQL fluid reservoir, 2- Air mass flow regulating valve, 3-impulsator, 4-control valve with pressure gage

Grinding parameters used in the experiment are shown in Table. 1. In order to ensure identical sharpness of grinding wheel measured each time, grinding wheel is dressed before each group of experiment

Figure 3. The digital viscometer and contact angle measuring instrument

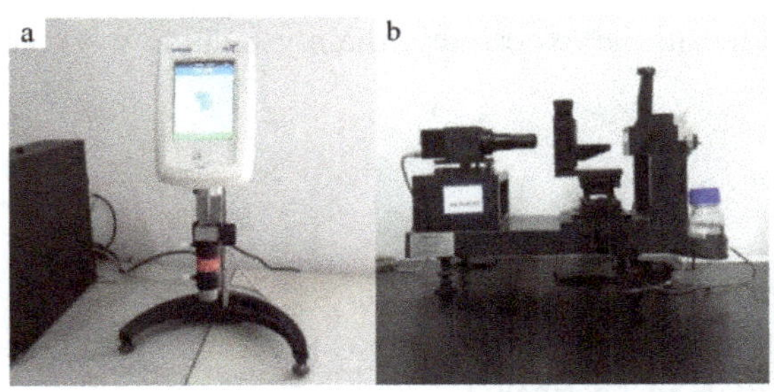

Figure 4. TIME3220 roughness measuring instrument and S-3400N scanning electron microscope

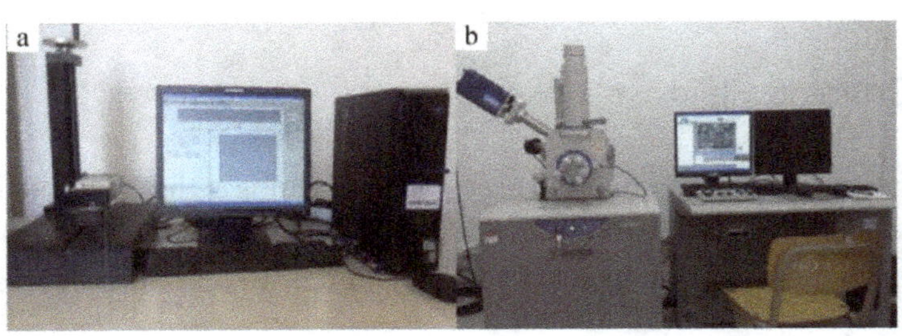

Table 1. Grinding parameters

Grinding type	Surface grinding, down grinding
Grinding wheel	Al_2O_3:WA80MV12P (ceramic bond)
Lubricating conditions	Flood grinding, MQL
Flow rate under Flood grinding	60 L/h
Flow rate under MQL grinding	50 ml/h
Gas-liquid ratio	0.3
Linear velocity of grinding wheel (V_s)	30 m/s
Workpiece feed velocity (V_w)	3000 mm/min
Grinding depth (a_p)	10 μm
MQL nozzle distance (d)	12 mm
MQL nozzle angle (α)	15°
MQL air pressure (P)	0.6 MPa
Dresser	Single-point diamond
Total dressing depth (a_d)	40 μm
Dressing velocity (v_d)	300 mm/min

and dressing parameters are listed in Table. 1. Down grinding condition is adopted on the 30×20mm workpiece surface in each grinding experiment.

11.2.2 Experimental Materials

Experimental materials selected in the experiment mainly include workpiece material, grinding fluid and MQL oil.

1. Workpiece Material

Workpiece material machined in the experiment is nickel base alloy 718. Workpiece dimensions are 40×30×30 mm as shown in Figure 5. Chemical compositions and mechanical properties of the material are shown in Table 2 and Table 3 respectively.

Nickel base alloy has its unique advantages as the experiment material. Its high strength and large hardness will give rise to high grinding force and grinding temperature. Besides, obvious wear of grinding wheel, surface adhesion and pore blocking will occur so that experimental phenomenon can be more obvious.

2. Water-Based Grinding Fluid

Water-based grinding fluid is one of commonly used grinding fluids in the traditional grinding machining process with favorable cooling & lubricating and rinsing effects(Shen & Shih, 2009). Water-based grinding fluid in this experiment is Syntilo 9930 water-soluble grinding fluid produced by Talent Chemicals Co.,

Figure 5. The sample piece of workpiece

Table 2. Chemical compositions of nickel base alloy

Element	C	Ni	Fe	Cu	Mn	Mo	Cr	Al	Co	Si
Composition /%	0.08	50-55	trace	0.3	0.35	2.8-3.3	17-21	0.95	1.0	0.35

Table 3. Mechanical properties of nickel base alloy

Density g/cm³	Heat conductivity W/(m·K)⁻¹	Hardness (HBS)	Elasticity modulus (GPa)	Tensile strength (GPa)	Shear modulus (GPa)	Yield stress (MPa)	Poisson's ratio
8.24	14.7 (100°C)	435	199.9	965	77.2	1110	0.3

Ltd. Ratio of pure grinding fluid to total grinding fluid after preparation is about 3~5% (mass fraction), and concentration of water-based grinding fluid used in the experiment is 5%.

3. MQL Fluid

Liquid paraffin and seven different kinds of vegetable oils (soybean oil, peanut oil, corn oil, rapeseed oil, palm oil, castor oil and sunflower oil) are respectively used in the experiment as MQL oils. In order to certify effects of polar molecules in vegetable oils and then select a nonpolar oil, liquid paraffin is taken as the oil in the control group. Liquid paraffin presents white liquid state, and colorless and ordourless mixture is obtained through distillation of crude oil, mainly being saturated cycloparaffin and alkane mixture.

Table 4. Fatty acid composition of seven kinds of vegetable oil

	Palmitic acid %	Stearic acid %	Oleic acid %	Linoleic acid %	Linolenic acid %
Soybean oil	7-10	3-5	22-31	49-55	6-11
Corn oil	9-19	1-3	26-40	40-55	<1
Peanut oil	6-9	3-6	55-71	13-25	0.5
Rapeseed oil	2-4	1-2	40-60	19-20	7-8
Sunflower oil	4-19	3-6	14-35	50-75	0.1
Castor oil	-	-	3-9	3-5	Minimal quantity
Palm oil	35-48	4-6	38-41	8-12	Minimal quantity

In order to study lubricating properties of different vegetable oils, typical vegetable oils are selected as lubricating oils, where most are vegetable oil varieties with large yield in China and relatively low prices. Table 4 lists fatty acid compositions of seven vegetable oils.

Lubricating fluid contains vegetable oil, surfactant and additives. As vegetable oil molecules contain many C=C double bonds, especially linoleic acid or linolenic acid containing two or three double bonds can be rapidly oxidized in the initial oxidation phase so that vegetable oil has very poor oxidative stability. Antioxidant Vitamin E is added into the base oil namely vegetable oil in the experiment, and concentration of added Vitamin E is volume fraction 0.2%.

11.2.3 Experimental Scheme

The research goal of this experiment is to discuss about lubricating properties at grinding wheel/workpiece interface in the MQL grinding process with different kinds of vegetable oil and formulate a set of experimental programs under different lubricating conditions so as to select the vegetable oil with the optimal lubricating property according to experimental results. By studying friction coefficient, specific grinding energy and grinding wheel G-ratio under lubrication conditions with different base oils and flood grinding, an experimental study of lubricating property at grinding wheel/workpiece interface is carried out. Univariate method is used in this experiment, and only lubricating fluid type is changed in each group of experiment while other factors remain unchanged.

Lubricating fluids include: water-based grinding fluid with 5% concentration, liquid paraffin and seven vegetable oils. Three lubrication conditions are: Flood grinding, paraffin oil MQL and vegetable oil MQL. Lubricating properties of water-based grinding fluid, liquid paraffin and seven vegetable oils are comparatively studied. Experimental program is shown in Table 5.

11.3 EXPERIMENTAL RESULTS

Figure 6 shows grinding force signal images under four working conditions. Figure 6 (a): water-soluble grinding fluid has high grinding force. Figure 6 (b): castor oil is used as MQL grinding fluid. Figure 6 (c): peanut oil is used as MQL grinding fluid. Figure 6 (d): paraffin oil is used as MQL grinding fluid. Grinding force is obviously small when MQL technology is used.

Table 5. Experimental program

Experiment No.	Lubricating fluid	Lubrication condition
3-1	Water-based grinding fluid (5 vol%)	Flood grinding
3-2	Liquid paraffin oil	MQL
3-3	Soybean oil	MQL
3-4	Peanut oil	MQL
3-5	Corn oil	MQL
3-6	Rapeseed oil	MQL
3-7	Palm oil	MQL
3-8	Castor oil	MQL
3-9	Sunflower oil	MQL

Figure 6. Grinding force measuring signal images of four different lubrication

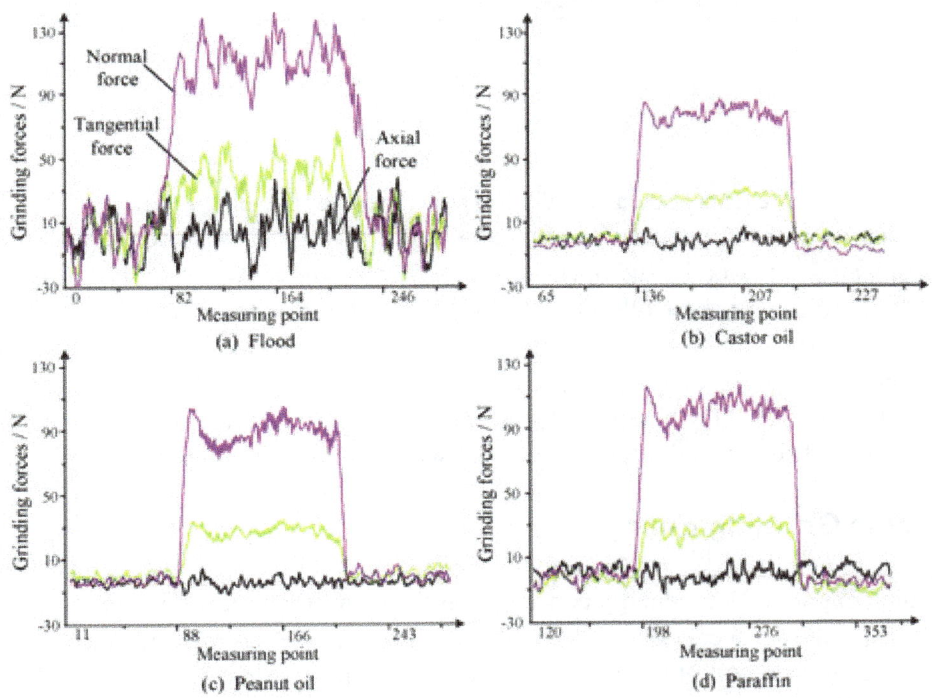

11.3.1 Friction Coefficient

After Surface grinding experiment, the relationship between different lubricating fluids and friction coefficient is shown in Figure 7. Each drawn data point is average value obtained after 100 grinding

Figure 7. Friction coefficient of different lubrication conditions

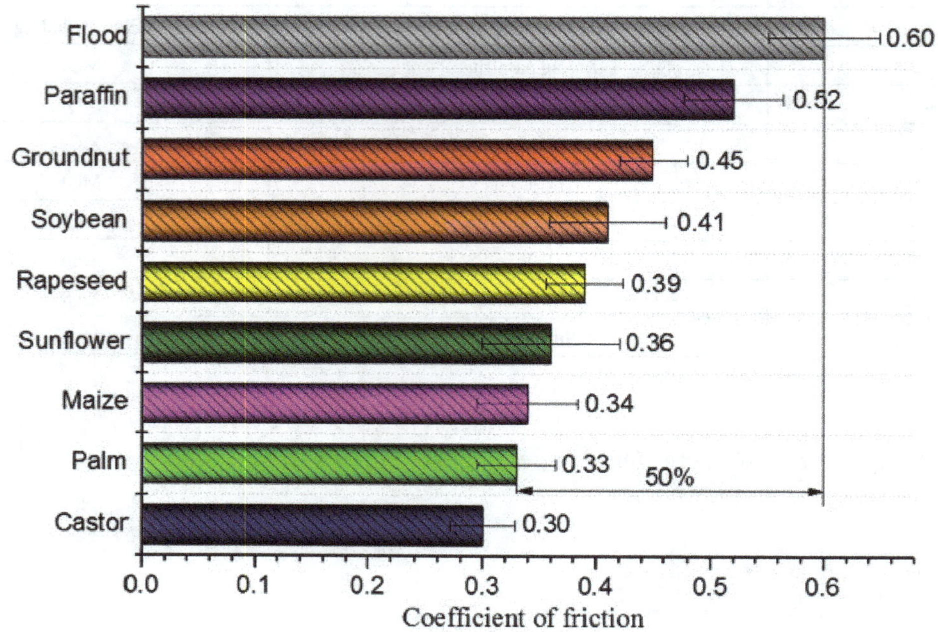

travels, and error bar represents standard deviation of friction coefficient.

It can be seen from the figure that friction coefficient is the highest under flood grinding, being μ_{flood}=0.60. Compared with flood grinding, friction coefficients under MQL grinding with both liquid oil and seven vegetable oils being base oils are obviously reduced, where friction coefficient under paraffin oil condition is μ_{para}=0.52, and those of peanut oil, soybean oil and rapeseed oil are lower than that of paraffin oil, reduced by μ_{mai}=0.45, μ_{sun}=0.41 and μ_{palm}=0.39 respectively. Friction coefficients of sunflower oil, corn oil and palm oil are further educed, and their differences are not large. Castor oil obtains the minimum friction coefficient, and it reduces by 50% when compared with that under flood grinding.

11.3.2 Specific Grinding Energy

Specific grinding energy reflects energy consumption in the grinding machining process. Low specific grinding energy means that energy consumed in the machining process is low and this working condition is more of green, energy conservation and environmentally friendly properties (Jia et al., 2017; Masoumi, Safavi, & Salehi, 2014; Ren, Zhang, & Zhou, 2009). Figure 8 shows specific grinding energy values under lubrication conditions with different grinding fluids.

Figure 8 indicates that specific grinding energy presents basically similar change trend with friction coefficient under different working conditions. Under flood grinding, lubricating state is poor in the grinding area and the energy consumed in the grinding process, reaching 145.23 (J/mm^3). Lubricating effect of paraffin oil MQL is better than that of flood grinding, its specific grinding energy is 108.63 (J/mm^3), so lubricating effect of vegetable oil-based cutting fluid is the most ideal. When the seven kinds

Figure 8. Specific grinding energies under different lubrication conditions

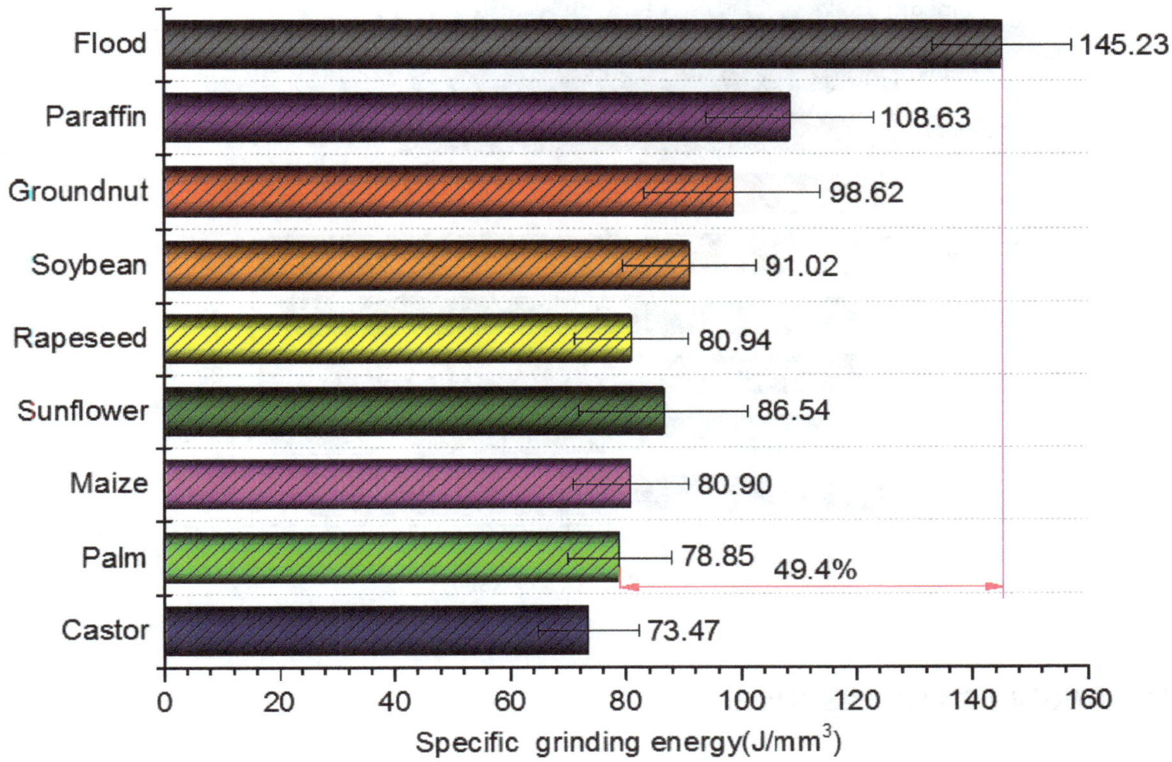

of vegetable oil are used as lubricating fluids, specific grinding energies decline to different degrees. Compared with flood grinding, specific grinding energies of peanut oil, soybean oil and rapeseed oil decline by 32.1%, 37.3% and 44.3% respectively. Although specific grinding energies of sunflower oil, corn oil and palm oil are reduced by 40.4%, 44.3% and 45.7% respectively relative to flood grinding, their numerical values are not that different. In the experiment under lubrication conditions with different grinding fluids, castor oil obtains the minimum specific grinding energy is 73.47 (J/mm^3), which reduces by 49.4% when compared with that under flood grinding condition.

11.3.3 Grinding Ratio

G ratios under different lubrication conditions calculated after 100 grinding travels are shown in Figure 9. It can be known that MQL grinding obtains higher G ratio than flood grinding, where G ratios of paraffin oil, peanut oil and soybean oil are 22.14, 22.92 and 26.50 respectively. Sunflower oil and palm oil are not that different in G ratio, and their G ratios are 28.19 and 28.63 respectively, G ratios of rapeseed oil and corn oil are very large, being 29.13 and 29.15 respectively. Among vegetable oil-based cutting fluids, that with the most obvious change of friction coefficient relative to G ratio is castor oil, and its G ratio only increases by 39.6% when compared with that (only being 19.25%) under flood grinding.

Figure 9. G-ratios under different lubrication conditions

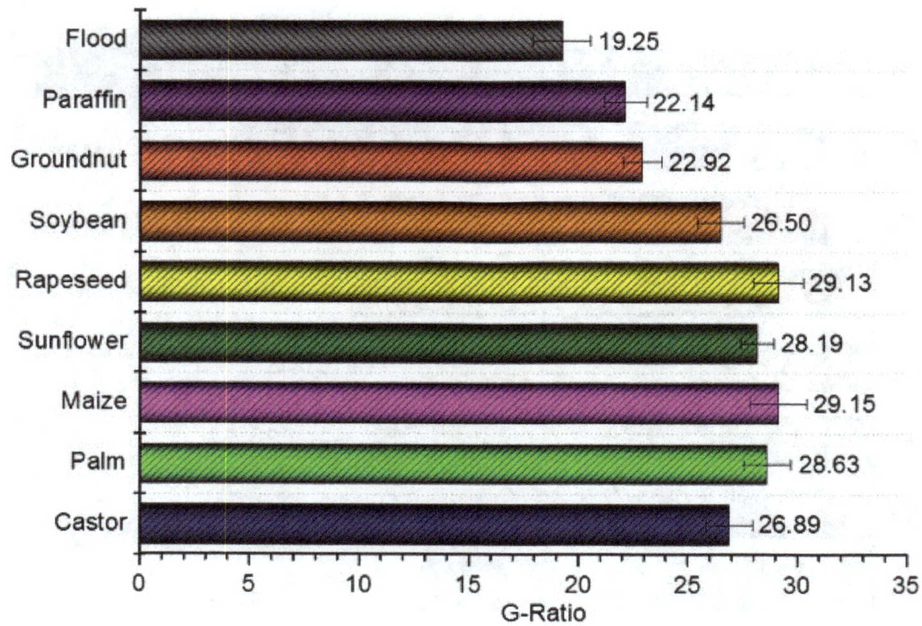

11.4 RESULTS AND DISCUSSION

11.4.1 Comparison of the Lubrication Property of Vegetable Oil-Based MQL and Flood Lubrication

According to friction coefficient, specific grinding energy and G ratio obtained under flood grinding and vegetable oil MQL conditions, flood grinding obtains the highest friction coefficient and specific grinding energy, indicating that flood grinding has not ideal lubricating property at grinding wheel/ workpiece interface. The main cause for this phenomenon is that high hydrodynamic pressure around grinding wheel prevents grinding oil from entering the grinding area, grinding fluid mainly exerts the local cooling effect on workpiece, so lubricating property in the grinding area is poor and friction coefficient increases. Lubricating properties of vegetable oil MQL at grinding wheel/workpiece interface is better than that of flood grinding. Causes for this phenomenon are as follows:

Under MQL grinding condition, although use level of lubricating fluid is only 1/1000 of flood grinding, using high-pressure gas to carry away grinding fluid can break through air-bond layer generated by revolution of grinding wheel so that the grinding area is well lubricated(Li, Li, Wang, & Zhang, 2013). Moreover, high-pressure gas also exerts effects of cooling workpiece and carrying away grinding debris, so it can also exert a certain lubricating effect in the grinding area.

Vegetable oil mainly consists of fatty acids and triglyceride. Both carboxyl (-COOH) in fatty acid molecules and ester groups in triglyceride are polar groups which have superior lubricating properties(Alves, Barros, Trajano, Ribeiro, & Moura, 2013; Debnath et al., 2014; Khan, Mithu, & Dhar, 2009; Su, Gong, & Chen, 2015). Fig. 11-10 shows typical polar molecules in the vegetable oil. They will be closely adsorbed

Figure 10. Polar molecule in vegetable oil

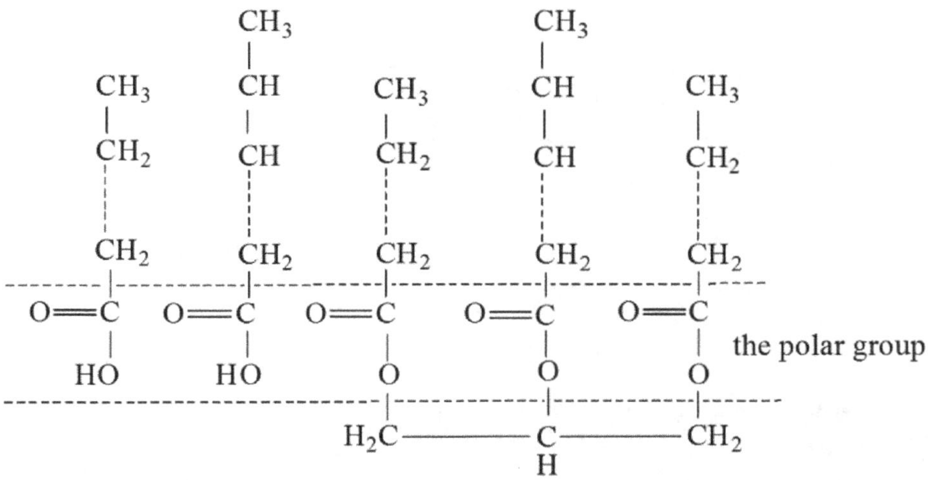

onto molecular films presenting directional arrangement on metallic surface like tiny magnets to exert anti-friction and anti-wear effects. Under high temperature, polar molecules can also react with metal to generate metallic soap film, which then forms boundary film through chemical adsorption(Wang & Zhang, 2005), and schematic diagram of lubricating effect of metallic soap film on workpiece surface in the grinding process is shown in Fig. 11-11. This characteristic of vegetable oil endows it very good boundary lubricating property and then decides that it has extremely good lubricating effect when added into the grinding fluid (Salete Martins; Alves & de Oliveira, 2008).

All in all, vegetable oil, which is used as MQL base oil, has superior characteristics like good lubricating property, low volatility, nontoxicity, high biodegradation rate and recyclability and it can realize goals of reducing environmental pollution and production cost and recycling resources. Therefore, it is the first choice as base oil in the environmentally friendly and green cutting process.

When paraffin oil is used as MQL grinding fluid, its lubricating property is superior to that under flood grinding condition but its lubricating property is poorer than vegetable oil. MQL technology reduces frictional wear at grinding wheel/workpiece interface and improves lubricating property which is better than lubricating property under flood grinding condition. As a mineral oil, paraffin oil consists of C_{16}-C_{20} n-alkanes which are pure hydrocarbons and don't contain other elements, functional groups and saturated hydrocarbons of carbon branched chain, and their molecules consist of several CH_2 atomic groups and 2 H atoms (Zhang et al., 2015). Paraffin oil contains polar molecules, so it doesn't contain polar atoms and polar groups like vegetable oils do, but instead, its molecules are randomly arranged on metal surface and then it's impossible to form physical adsorption film with strong adsorption capacity.

11.4.2 Influences of the Molecular Structure of Vegetable Oil on Lubrication Property

Lubricating property of vegetable oil is obviously improved when compared with both water-based grinding fluid and paraffin oil, but lubricating properties of different vegetable oils are also different. Through a comprehensive analysis of evaluation parameters of lubricating properties, castor oil has the

Figure 11. Schematic of lubrication in the grinding interface

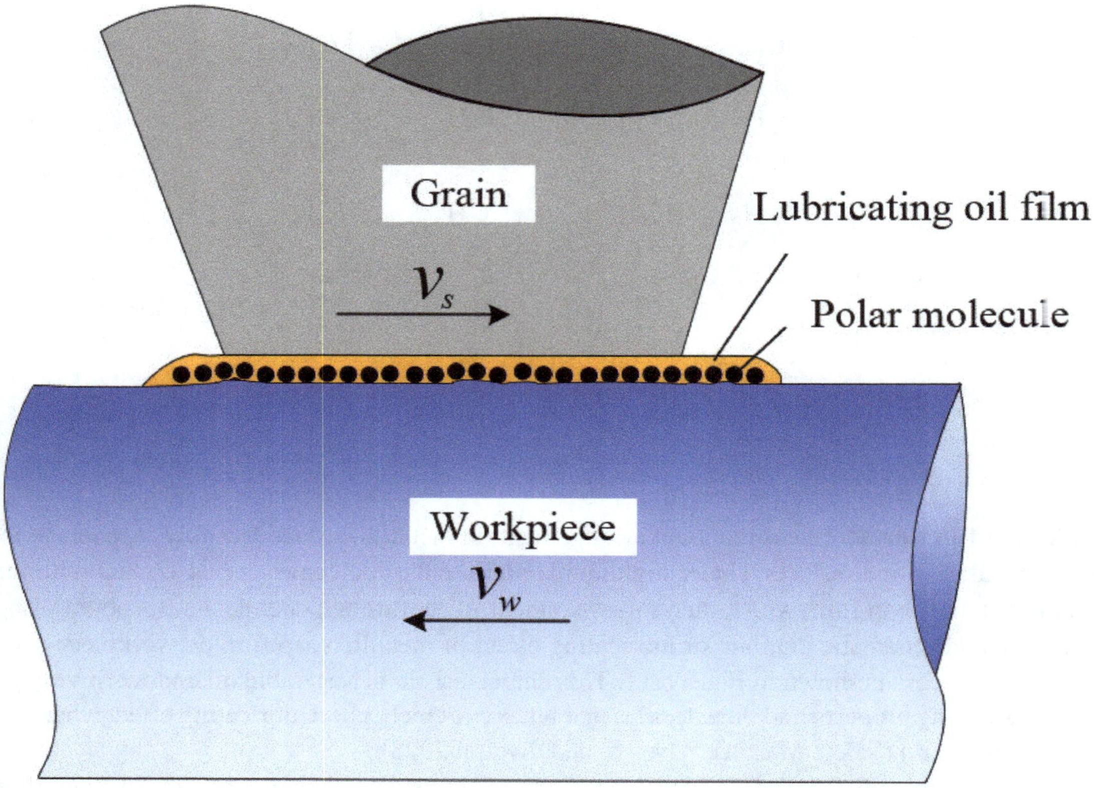

best lubricating property, followed by palm oil. In addition, peanut oil, sunflower oil, soybean oil and rapeseed oil also show superior lubricating properties. However, lubricating property of corn oil is not that ideal. There are many factors influencing the difference of lubricating properties, including vegetable oil composition, molecular structure, viscosity, density, flash point, etc..

The main composition of vegetable oil is triglyceride formed by single-molecule glyceride and three-molecule straight-chain fatty acids as shown in Figure 12, where R_1, R_2 and R_3 are carbon chains of fatty acid which are connected to glycerin molecules through ester bond. Chain length of fatty acid is generally 14-22 carbon atoms with unsaturated bonds of different degrees. For saturated fatty acid, when its carbon chain length is greater than 16, friction property is not influenced by carbon number; on the contrary, for unsaturated fatty acid, as adsorption film is not compact due to the effect of double bonds, molecular cohesion of fatty acid is very important and then fatty acid with greater carbon number has good lubricating property.

Typical fatty acids include saturated fatty acids (palmitic acid and stearic acid $C_{17}H_{35}COOH$) and unsaturated fatty acids (oleic acid, linoleic acid $C_{17}H_{31}COOH$ and linolenic acid $C_{17}H_{29}COOH$). Table 4 lists basic compositions in fatty acids of seven kinds of vegetable oil.

Figure 12. Molecular structure of triglyceride

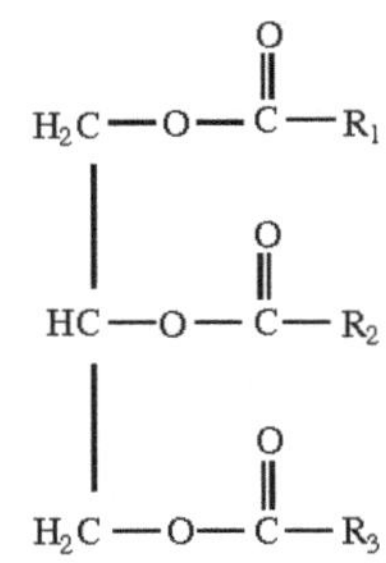

Lubricating property of saturated fatty acid is better than that of unsaturated fatty acid(Abdalla; & Patel, 2006). Saturated fatty acids have significant influences on reducing friction and wear, especially stearic acid provides a more powerful protection for oil film by strengthening intermolecular acting force. Differences of saturated fatty acids in content and structure will result in different cohesions between molecules adsorbed on metal surface, different compactness degrees of the adsorption film, different adsorption energies and different strengths of the adsorption film. Palm oil contains many palmitic acids, so does sunflower oil, and the main fatty acid contained in rapeseed oil is erucic acid (C_{22}) containing one double bond, and it belongs to unsaturated fatty acid. Therefore, research results are analyzed through the saponification film-forming mechanism. As the main fatty acid types in palm oil are palmitic acid (accounting for about 30%~50%) and stearic acid, both of which are saturated fatty acids, and carbon number is 16, so a lubricating film with strong adsorption force can be formed so as to reach good lubricating property. Compared with rapeseed oil, sunflower oil contains many palmitic acids (saturated fatty acids) and few oleic acids (unsaturated acids), so lubricating property of sunflower oil is better than that of rapeseed oil. According to the experiment, their friction coefficients can be sorted as palm oil <sunflower oil <rapeseed oil <soybean oil.

Double bonds existing in unsaturated fatty acids are relatively unstable and can easily experience chemical reactions like oxidization. Moreover, double bonds of unsaturated fatty acids will weaken intermolecular acting force, which results in relatively poor lubricating property, as double bonds endow molecules a certain rigidity, but no rigidity exists in saturated fatty acids. Double bonds elevate rigidity level of fatty acid chains and prevent close packing of fatty acid chains so as to result in weak molecular force. On the contrary, saturated fatty acids don't contain double bonds so they are not rigid, and these characteristics make it possible for them to be closely packed (Baumgart, Canzi, Hanashiro, Doezema, & Siniawski, 2010). For peanut oil, although its viscosity is about 36.31cp under 40 °C, which is higher than those of soybean oil and sunflower oil, peanut oil contains about 50%~60% of oleic acid and 15%~20% of linoleic acid, both of which belong to unsaturated fatty acids with poor lubricating properties, and this is why friction coefficient and specific grinding energy of peanut oil are high.

11.4.3 Influences of Vegetable Oil Components on Lubrication Property

Molecular structure of fatty acids in vegetable oil decides its physiochemical properties. HU Z M (1999) used fatty acids in vegetable oil as lubricant additive to investigate frictional characteristics of fatty acids. Through a comparison of friction coefficients and wear conditions among several kinds of fatty acids, for saturated fatty acids, when their carbon chain length is greater than 16, friction property is not influenced by carbon number; for unsaturated fatty acids, adsorption film is not compact due to the effect of double bonds, so molecular cohesion of fatty acids is very critical and the longer the carbon chain length, the better the strength and lubricating property of adsorption film of unsaturated fatty acids. A great number of nonpolar methyl groups exist in carbon chain of fatty acids, and intermolecular acting force is mainly dispersion force. Any molecule is of instantaneous dipole characteristic, which will induce adjacent molecules to generate instantaneous dipole, so two molecules can be mutually attracted through the instantaneous dipole. Dispersion force is related to interactive molecular deformability: the greater the deformability (the greater the molecular weight, the greater the deformability), the greater the dispersion force. Dispersion force increases with carbon number, and schematic diagram of its acting force is shown in Figure 13. Fatty acids of long carbon chain show better lubricating properties than fatty acids with short carbon chain. According to this phenomenon, main fatty acid of rapeseed oil namely erucic acid (C_{22}) molecule contains 1 double bond and it belongs to unsaturated acid. Carbon number of the main fatty acid of soybean oil namely linoleic acid is 18, which contains two unsaturated bonds. Rapeseed oil has longer carbon chain (C_{22}) than soybean oil and it only contains one double bond with lower unsaturation degree than soybean oil. Therefore, friction coefficient of rapeseed oil is lower than that of soybean oil. Ricinoleic acid content in castor oil exceeds 80% and its carbon chain length is longer than other fatty acids. Its viscosity is the maximum, and moreover, ricinoleic acid belongs to single-hydroxyl unsaturated acid which causes strong hydrogen bond. Due to energy effect of exoelectron in the friction process, frictional polymerization reaction occurs, and then a frictional polymerization film is formed on friction surface to exert anti-wear effect and improve bearing capacity of oil film. This is the reason why friction coefficient and specific grinding energy of castor oil are the minimum. Figure 14 shows molecular structures of several kinds of fatty acids.

11.4.4 Influences of the Viscosity of Vegetable Oil on Lubrication Property

Vegetable oil viscosity plays a significant role in the machining industry(Guo et al., 2018; Quinchia, Delgado, Valencia, Franco, & Gallegos, 2010; Zhang et al., 2018). During the grinding process, it mainly influences lubricating property and heat transfer performance in the grinding area. Vegetable oil has very high natural viscosity. As the machining temperature rises, decline rate of viscosity of vegetable oil is lower than that of mineral oil. Vegetable oil can keep better mobility as the temperature declines so that grinding debris can be removed from workpiece more rapidly. Within the range of working temperature, high viscosity index of vegetable oil ensures that it can provide more stable lubricating effect. In addition, high-viscosity lubricating oil is of uneasy flow. When vegetable oil-based cutting fluid is sprayed through the nozzle into grinding wheel/workpiece interface, grinding fluid can stay in the grinding area for a longer time so that it can exert better lubricating effect and prevent rapid tool wear (Krahenbuhl, 2002). Figure 15 shows viscosity differences between all kinds of base oils, where viscosity range of all kinds of vegetable oils under 25 °C is 41.8-535.3 cp. Viscosity of soybean oil is the minimum while that of castor oil is the highest, reaching 535.3 cp. Viscosity values of peanut oil, rapeseed oil, sunflower oil, corn oil and palm oil increase successively, but their differences are not that large. According to experimental results, lubricating properties of seven vegetable oils are sorted as corn oil <rapeseed oil <soybean oil <sunflower oil< peanut oil< palm oil <castor oil, so as the viscosity increases, friction coefficients of corresponding base oils basically decline. This is more obvious for castor oil, as its viscosity is far higher than those of other base oils, and then its friction coefficient and specific grinding energy are the minimum. This phenomenon indicates that high-viscosity fluid has better anti-friction

Figure 13. Schematic of lubrication in the grinding interface

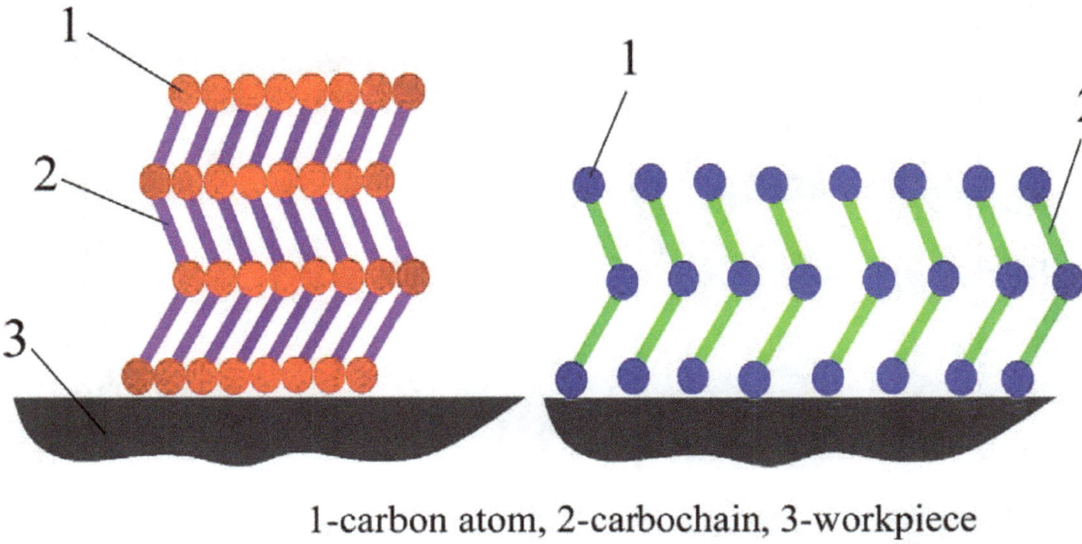

and anti-wear effects. Based on the heat exchange angle analysis, high-viscosity base oil has strong molecular adhesive force and weak Brownian movement, which reduces heat transfer capability of base oil.

According to G ratios under different working conditions in Figure 9, G ratios under MQL grinding conditions are all higher than that under flood grinding, mainly because the film oil formed by vegetable oil-based cutting fluid at grinding wheel/workpiece interface is of extremely good lubricating property, which reduces friction and wear as well as tangential grinding force. It can be obtained that all kinds of

Figure 14. Structures of some fatty acids

Figure 15. Viscosity values of eight base oils

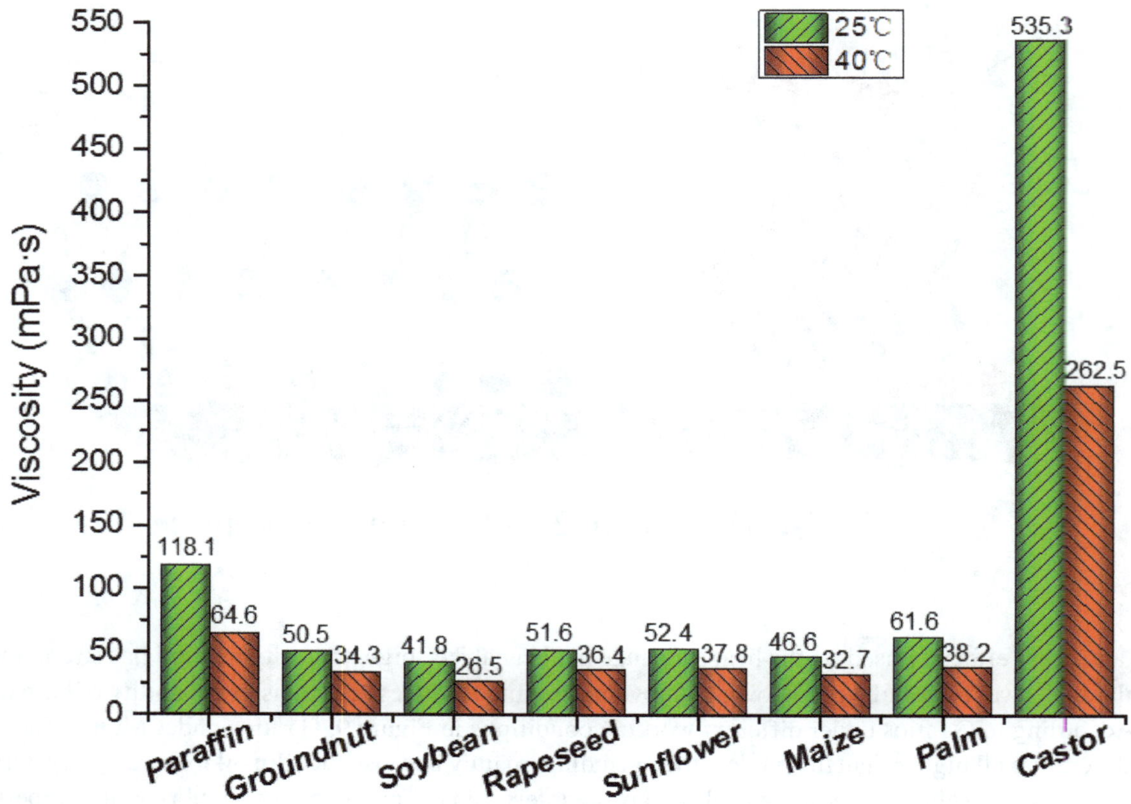

lubricating fluid change tangential grinding force by influencing lubricating property so as to change G ratio. G ratios of seven vegetable oils obtained through the experiment are not identical with change trends of corresponding friction coefficient and specific grinding energy, indicating that lubricating property is only one factor influencing G ratio. Through the analysis of typical castor oil, it's obvious that although castor oil has low friction coefficient and specific grinding energy with good lubricating property, its G ratio is not the best. The possible reason is that the grinding area temperature is high when castor oil is used as MQL grinding fluid, which causes serious abrasive wear of grinding wheel. Jindal (2012) used low-viscosity coconut oil to mix with other oils for drilling experiment. When coconut oil-based cutting fluid was used, the temperature at tool/workpiece interface was low, thus reducing tool wear. Therefore, temperature is also a factor influencing G ratio.

11.4.5 Surface Morphology and Surface Roughness

Based on the above evaluation parameters of lubricating property, workpiece surface micromorphology (SEM) and surface roughness are combined to further study lubricating properties of different grinding fluids. Workpiece surface quality is an important parameter evaluating grinding performance. Studying micromorphology is the most effective research method of evaluating grinding surface quality. In a similar way, surface roughness is also a key factor of workpiece surface quality. Small surface roughness means better surface smoothness. Figure 16 and Figure 17 show surface micrographs and roughness values obtained through flood grinding, liquid paraffin oil MQL grinding and MQL grinding using different vegetable oils.

It can be obviously observed from Figure 16 and Figure 17 that workpiece surface morphology and roughness obtained under different lubricating environments are obviously different. For instance, the workpiece with good surface morphology has low roughness value, so surface morphology keeps identical with surface roughness. According to comparison results, the following can be observed:

The analysis of surface morphology and roughness indicates that the best and poorest workpiece surface qualities are obtained respectively under vegetable oil MQL and flood grinding. There are many large and wide furrows on workpiece surface under flood grinding as shown in Figure 16 (a) with the maximum roughness value as shown in Figure 17. When vegetable oil is used for MQL grinding, furrows on workpiece surface are obviously reduced and narrowed with several small speckles as shown in Figure 16 (d). This also certifies that MQL has better lubricating effect than flood grinding.

Workpiece surface qualities obtained under MQL grinding conditions using seven different kinds of vegetable oils are sorted as corn oil <rapeseed oil <soybean oil <sunflower oil< peanut oil <palm

Figure 16. Surface morphology of nickel base alloy under different lubrication conditions

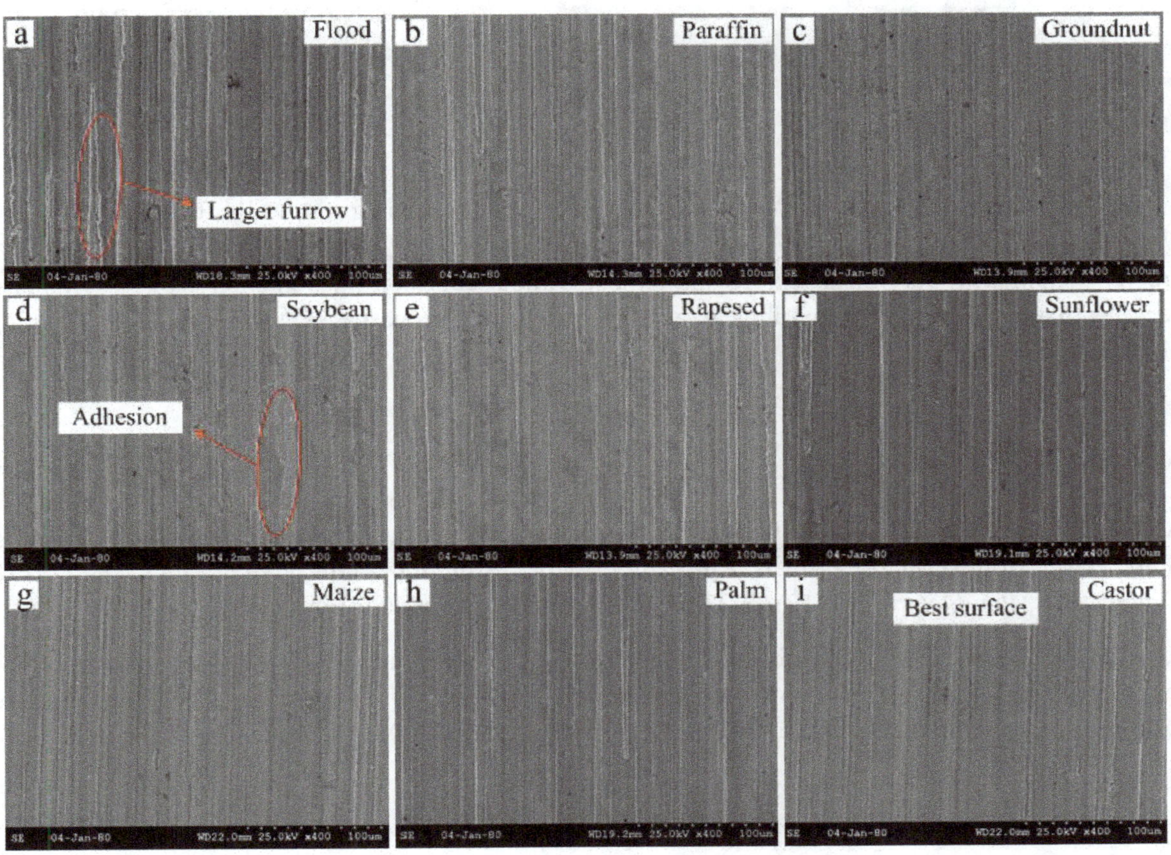

Figure 17. Surface roughness under different working conditions

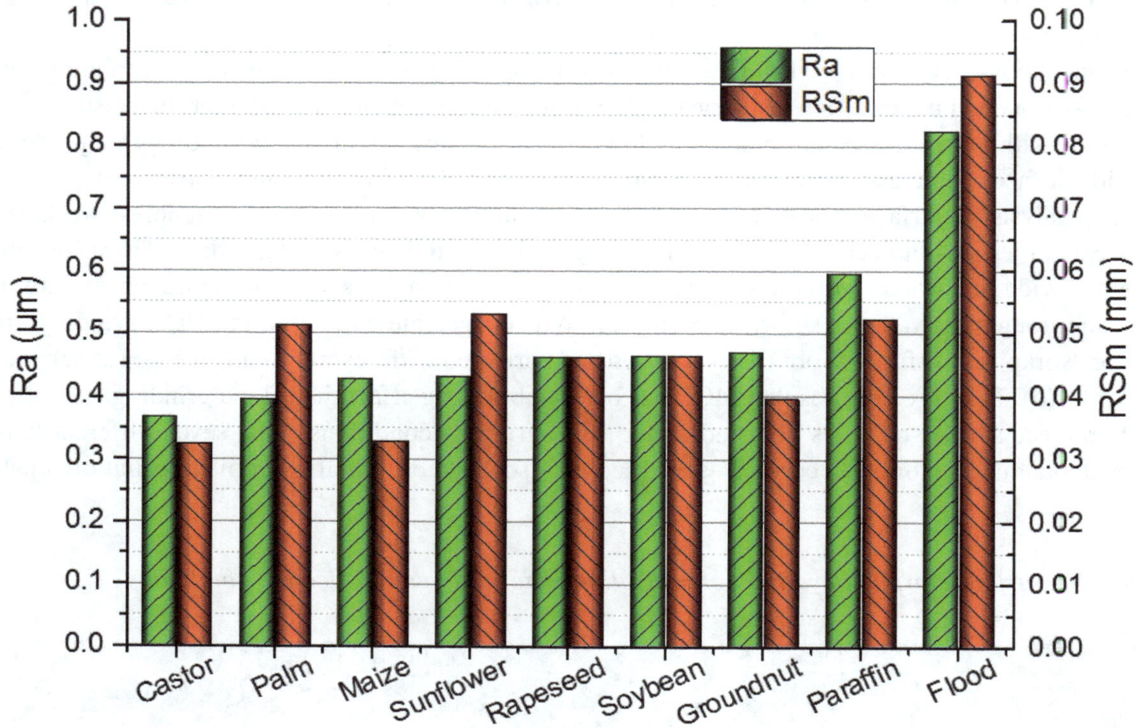

oil <castor oil. Especially, the workpiece surface quality obtained using castor oil is better with smooth workpiece surface morphology. When surface morphology is amplified by 400 times, surface furrows are very small with very clear texture and almost without adhesion phenomenon (16 (i)). This also certifies that castor oil and palm oil have favorable lubricating properties. On the contrary, lubricating oil of corn oil is not ideal.

REFERENCES

Abdalla, H. S., & Patel, S. (2006). The performance and oxidation stability of sustainable metalworking fluid derived from vegetable extracts. *Proceedings of the Institution of Mechanical Engineers. Part B: Engineering Manufacture*, *220*(12), 2027–2040. doi:10.1243/09544054JEM357

Alves, S. M., Barros, B. S., Trajano, M. F., Ribeiro, K. S. B., & Moura, E. (2013). Tribological behavior of vegetable oil-based lubricants with nanoparticles of oxides in boundary lubrication conditions. *Tribology International*, *65*, 28–36. doi:10.1016/j.triboint.2013.03.027

Alves, S. M., & de Oliveira, J. F. G. (2008). Vegetable Based Cutting Fluid: An Environmental Alternative to Grinding Process. *LCE 2008: 15th CIRP International Conference on Life Cycle Engineering: Conference Proceedings.*

Asadauskas, S., Perez, J., & Duda, J. (1996). *Lubrication properties of castor oil - Potential basestock for biodegradable lubricants.* Paper presented at the Society-of-Tribologists-and-Lubrication-Engineers 51st Annual Meeting, Cincinnati, OH.

Baumgart, P., Canzi, G., Hanashiro, T., Doezema, L. A., & Siniawski, M. T. J. L. S. (2010). *Influence of fatty acid additives on the tribological performance of sunflower oil.* Academic Press.

Belluco, W., & Chiffre, L. D. (2004). Performance evaluation of vegetable-based oils in drilling austenitic stainless steel. *Journal of Materials Processing Technology, 148*(2), 171–176. doi:10.1016/S0924-0136(03)00679-4

Cetin, M. H., Ozcelik, B., Kuram, E., & Demirbas, E. (2011). Evaluation of vegetable based cutting fluids with extreme pressure and cutting parameters in turning of AISI 304L by Taguchi method. *Journal of Cleaner Production, 19*(17), 2049–2056. doi:10.1016/j.jclepro.2011.07.013

Chen, R., & Rowe, W. B. (1996). Analysis and simulation of the grinding process. Part I: Generation of the grinding wheel surface. *International Journal of Machine Tools & Manufacture, 36*(8), 871–882. doi:10.1016/0890-6955(96)00116-2

Debnath, S., Reddy, M. M., & Yi, Q. S. (2014). Environmental friendly cutting fluids and cooling techniques in machining: A review. *Journal of Cleaner Production, 83*, 33–47. doi:10.1016/j.jclepro.2014.07.071

Emami, M., Sadeghi, M. H., Sarhan, A. A. D., & Hasani, F. (2014). Investigating the Minimum Quantity Lubrication in grinding of Al_2O_3 engineering ceramic. *Journal of Cleaner Production, 66*(3), 632–643. doi:10.1016/j.jclepro.2013.11.018

Guo, S. M., Li, C. H., Zhang, Y. B., Wang, Y. G., Li, B. K., Yang, M., ... Liu, G. T. (2017). Experimental evaluation of the lubrication performance of mixtures of castor oil with other vegetable oils in MQL grinding of nickel-based alloy. *Journal of Cleaner Production, 140*, 1060–1076. doi:10.1016/j.jclepro.2016.10.073

Guo, S. M., Li, C. H., Zhang, Y. B., Yang, M., Jia, D. Z., Zhang, X. P., ... Ji, H. J. (2018). Analysis of volume ratio of castor/soybean oil mixture on minimum quantity lubrication grinding performance and microstructure evaluation by fractal dimension. *Industrial Crops and Products, 111*, 494–505. doi:10.1016/j.indcrop.2017.11.024

Guy Albert, G. (2010). *Cutting of Nickel-Based High Temperature Alloy.* Aeronautical Manufacturing Technology.

Hewson, W. D., & Gerow, G. K. (1999). *U.S. Patent No. 08/778530.* Washington, DC: US Patent Office.

Hu, Z. (1999). *Study on tribology of hydroxy and sulfurized vegetable oil fatty acids.* Shanghai University.

Jia, D. Z., Li, C. H., Zhang, Y. B., Yang, M., Wang, Y. G., Guo, S. M., & Cao, H. J. (2017). Specific energy and surface roughness of minimum quantity lubrication grinding Ni-based alloy with mixed vegetable oil-based nanofluids. *Precision Engineering-Journal of the International Societies for Precision Engineering and Nanotechnology*, *50*, 248–262. doi:10.1016/j.precisioneng.2017.05.012

Jindal, A. (2012). Analysis of tool wear rate in drilling operation using scanning electron microscope (SEM). *Journal of Minerals & Materials Characterization & Engineering*, *11*(1), 43–54. doi:10.4236/jmmce.2012.111004

Khan, M., Mithu, M. A. H., & Dhar, N. R. (2009). Effects of minimum quantity lubrication on turning AISI 9310 alloy steel using vegetable oil-based cutting fluid. *Journal of Materials Processing Technology*, *209*(15), 5573–5583. doi:10.1016/j.jmatprotec.2009.05.014

Khan, M. M. A., Mithu, M. A. H., & Dhar, N. R. (2009). Effects of minimum quantity lubrication on turning AISI 9310 alloy steel using vegetable oil-based cutting fluid. *Journal of Materials Processing Technology*, *209*(15-16), 5573–5583. doi:10.1016/j.jmatprotec.2009.05.014

Komanduri, R., & Hou, Z. B. (2001). A review of the experimental techniques for the measurement of heat and temperatures generated in some manufacturing processes and tribology. *Tribology International*, *34*(10), 653–682. doi:10.1016/S0301-679X(01)00068-8

Krahenbuhl, U. (2002). Vegetable oil-based coolants improve cutting performance. (Cutting Fluids). *Tooling & Production*, *68*, 34–35.

Kuram, E., Ozcelik, B., Bayramoglu, M., Demirbas, E., & Simsek, B. T. (2013). Optimization of cutting fluids and cutting parameters during end milling by using D-optimal design of experiments. *Journal of Cleaner Production*, *42*(3), 159–166. doi:10.1016/j.jclepro.2012.11.003

Kuram, E., Ozcelik, B., Cetin, M. H., Demirbas, E., & Askin, S. J. L. S. (2013). *Effects of blended vegetable-based cutting fluids with extreme pressure on tool wear and force components in turning of Al 7075-T6*. Academic Press.

Kuram, E., Ozcelik, B., & Demirbas, E. (2013). *Environmentally Friendly Machining: Vegetable Based Cutting Fluids*. Academic Press.

Lambert, J. W., & Johnson, D. L. (1997). *US91213097A*. Academic Press.

Lawal, S. A., Choudhury, I. A., & Nukman, Y. (2013). A critical assessment of lubrication techniques in machining processes: A case for minimum quantity lubrication using vegetable oil-based lubricant. *Journal of Cleaner Production*, *41*, 210–221. doi:10.1016/j.jclepro.2012.10.016

Lawal, S. A., Choudhury, I. A., & Nukman, Y. (2014). Evaluation of vegetable and mineral oil-in-water emulsion cutting fluids in turning AISI 4340 steel with coated carbide tools. *Journal of Cleaner Production*, *66*(3), 610–618. doi:10.1016/j.jclepro.2013.11.066

Li, B. K., Li, C. H., Zhang, Y. B., Wang, Y. G., Jia, D. Z., & Min, Y. (2016). Grinding temperature and energy ratio coefficient in MQL grinding of high-temperature nickel-base alloy by using different vegetable oils as base oil. *Chinese Journal of Aeronautics*, *29*(4), 1084–1095. doi:10.1016/j.cja.2015.10.012

Li, C., Li, J., Wang, S., & Zhang, Q. (2013). Modeling and numerical simulation of the grinding temperature field with nanoparticle jet of MQL. *Advances in Mechanical Engineering, 7*(2), 137-181. doi:10.1155/2013/986984

Malkin, S., & Guo, C. (2007). Thermal Analysis of Grinding. *CIRP Annals - Manufacturing Technology, 56*(2), 760-782.

Mao, C., Zou, H., Huang, Y., Li, Y., & Zhou, Z. (2013). Analysis of heat transfer coefficient on workpiece surface during;minimum quantity lubricant grinding. *International Journal of Advanced Manufacturing Technology*, *66*(1-4), 363–370. doi:10.100700170-012-4330-x

Mao, C., Zou, H. F., Huang, Y., Li, Y. F., & Zhou, Z. X. (2013). Analysis of heat transfer coefficient on workpiece surface during minimum quantity lubricant grinding. *International Journal of Advanced Manufacturing Technology*, *66*(1-4), 363–370. doi:10.100700170-012-4330-x

Mao, C., Zou, H. F., Zhou, X., Huang, Y., Gan, H. Y., & Zhou, Z. X. (2014). Analysis of suspension stability for nanofluid applied in minimum quantity lubricant grinding. *International Journal of Advanced Manufacturing Technology*, *71*(9-12), 2073–2081. doi:10.100700170-014-5642-9

Masoumi, H., Safavi, S. M., & Salehi, M. (2014). Grinding Force, Specific Energy and Material Removal Mechanism in Grinding of HVOF-Sprayed WC-Co-Cr Coating. *Materials and Manufacturing Processes*, *29*(3), 321–330. doi:10.1080/10426914.2013.872261

Ozcelik, B., Demirbas, E., & Şik, E. (2013). Effects of vegetable-based cutting fluids on the wear in drilling. *Sadhana*, *38*(4), 687–706. doi:10.100712046-013-0179-4

Ozcelik, B., Kuram, E., Cetin, M. H., & Demirbas, E. (2011). Experimental investigations of vegetable based cutting fluids with extreme pressure during turning of AISI 304L. *Tribology International*, *44*(12), 1864–1871. doi:10.1016/j.triboint.2011.07.012

Quinchia, L. A., Delgado, M. A., Valencia, C., Franco, J. M., & Gallegos, C. (2010). Viscosity modification of different vegetable oils with EVA copolymer for lubricant applications. *Industrial Crops and Products*, *32*(3), 607–612. doi:10.1016/j.indcrop.2010.07.011

Rahim, E. A., & Sasahara, H. (2011a). An Analysis of Surface Integrity When Drilling Inconel 718 Using Palm Oil and Synthetic Ester under Mql Condition. *Machining Science and Technology*, *15*(1), 76–90. doi:10.1080/10910344.2011.557967

Rahim, E. A., & Sasahara, H. (2011b). Investigation of tool wear and surface integrity on MQL machining of Ti-6AL-4V using biodegradable oil. *Proceedings of the Institution of Mechanical Engineers. Part B, Journal of Engineering Manufacture*, *225*(9B9), 1505–1511. doi:10.1177/0954405411402554

Ren, Y. H., Zhang, B., & Zhou, Z. X. (2009). Specific energy in grinding of tungsten carbides of various grain sizes. *Cirp Annals-Manufacturing Technology*, *58*(1), 299–302. doi:10.1016/j.cirp.2009.03.026

Sadeghi, M. H., Hadad, M. J., Tawakoli, T., Vesali, A., & Emami, M. (2010). An investigation on surface grinding of AISI 4140 hardened steel using minimum quantity lubrication-MQL technique. *International Journal of Material Forming*, *3*(4), 241–251. doi:10.100712289-009-0678-3

Sharma, J., & Sidhu, B. S. (2014). Investigation of effects of dry and near dry machining on AISI D2 steel using vegetable oil. *Journal of Cleaner Production*, *66*(2), 619–623. doi:10.1016/j.jclepro.2013.11.042

Shashidhara, Y. M., & Jayaram, S. R. (2010). Vegetable oils as a potential cutting fluid—An evolution. *Tribology International*, *43*(5), 1073–1081. doi:10.1016/j.triboint.2009.12.065

Shen, B., & Shih, A. J. (2009). Minimum quantity lubrication (MQL) grinding using vitrified CBN wheels. *Transactions of NAMRI/SME, 37*, 129-136.

Silva, L. R., Corrêa, E. C. S., Brandão, J. R., & Ávila, R. F. D. (2013). Environmentally friendly manufacturing: Behavior analysis of minimum quantity of lubricant - MQL in grinding process. *Journal of Cleaner Production*. doi:10.1016/j.jclepro.2013.01.033

Su, Y., Gong, L., & Chen, D. D. (2015). An Investigation on Tribological Properties and Lubrication Mechanism of Graphite Nanoparticles as Vegetable Based Oil Additive. *Journal of Nanomaterials*, *2015*, 1–7. doi:10.1155/2015/276753

Wang, J., & Zhang, J. (2005). On Formation and Breakup of Boundary Lubricating Layer. *Lubrication Engineering, 2005*(6), 4-8.

Wang, Y. G., Li, C. H., Zhang, Y. B., Yang, M., Li, B. K., Jia, D. Z., ... Mao, C. (2016). Experimental evaluation of the lubrication properties of the wheel/workpiece interface in minimum quantity lubrication (MQL) grinding using different types of vegetable oils. *Journal of Cleaner Production*, *127*, 487–499. doi:10.1016/j.jclepro.2016.03.121

Xavior, M. A., & Adithan, M. J. J. o. M. P. T. (2009). *Determining the influence of cutting fluids on tool wear and surface roughness during turning of AISI 304 austenitic stainless steel*. Academic Press.

Xun, C., & Rowe, W. B. (1996). Analysis and simulation of the grinding process. Part II: Mechanics of grinding. *International Journal of Machine Tools & Manufacture*, *36*(8), 883–896. doi:10.1016/0890-6955(96)00117-4

Zhang, J. C., Li, C. H., Zhang, Y. B., Yang, M., Jia, D. Z., Liu, G. T., ... Cao, H. J. (2018). Experimental assessment of an environmentally friendly grinding process using nanofluid minimum quantity lubrication with cryogenic air. *Journal of Cleaner Production*, *193*, 236–248. doi:10.1016/j.jclepro.2018.05.009

Zhang, Y., Li, C., Jia, D., Zhang, D., & Zhang, X. (2015). Experimental evaluation of MoS_2 nanoparticles in jet MQL grinding with different types of vegetable oil as base oil. *Journal of Cleaner Production*, *87*, 930–940. doi:10.1016/j.jclepro.2014.10.027

Chapter 12
Experimental Study of Lubricating Property at Grinding Wheel/Workpiece Interface Under NMQL Grinding

ABSTRACT

Nanofluid is the suspension formed by lubricating oil and nanoparticles with particles sizes of 1~100 nm, and common nanoparticles include metal nanoparticles (Cu, Ag, etc.), oxide nanoparticles (Al2O3, SiO2, ZrO2, etc.), carbides (CNT, diamond), and MoS2 nanoparticles, etc. Different nanoparticles exhibit various physicochemical properties (e.g., structure and shape), which can influence their tribological characteristics. In this work, six nanofluids, namely, MoS2, SiO2, diamond, carbon nanotubes (CNTs), Al2O3, and ZrO2, were used as minimum quantity lubrication grinding fluids to select the kind of nanoparticles with optimum lubrication performance in grinding nickel alloy GH4169. Experimental results concluded the following: 1) Nanoparticles with spherical or sphere-like molecular structure and nanofluids with high viscosity demonstrate superior lubrication performances. 2) The polishing effect of nanodiamond particles enhances their surface morphology. 3) The lubricating property of the six nanofluids is described in the following order: ZrO2 < CNTs < ND < MoS2 < SiO2 < Al2O3.

12.1 INTRODUCTION

Grinding is widely used in machining and the process is a random integration of sliding, ploughing, and cutting by abundant irregular abrasive grains scattered on a grinding wheel/workpiece interface viewed from microstructure (Malkin & Guo, 1991). The abovementioned process exerts higher units of grinding force and grinding speed than other machining processes, thus utilizing significantly higher grinding energies. In particular, hard materials in the grinding process will generate a lot of additional heat (Ding et al., 2015). Flood technology is the most common cooling–lubrication method used to improve the machining quality of workpieces and cutters. However, the surface processing quality of workpieces

DOI: 10.4018/978-1-7998-1546-4.ch012

is not satisfactory because the grinding wheel at high-speed forms a layer of gas-barrier, which makes entry into the grinding wheel and workpiece interface difficult for numerous grinding fluids (Li, Wang, & Zhang, 2013). Minimum quantity lubrication (MQL) technique is a green processing method (Baheti, & Guo, 1998). The MQL technology uses a minimum quantity of lubricants to achieve satisfying lubrication performance. This technique mixes a minimum quantity of lubricants in high-pressure gases, which can break through the air-barrier layer of the grinding wheel and arrive at the grinding area after mixing atomization. In particular, high-pressure gas plays a role in cooling and debris removal. Li et al. (Li, Hou, Xiu, & Cai, 2008) performed experiments to evaluate the performance of MQL technology compared with that of conventional flood cooling. Mao et al. (Mao et al., 2013) analyzed the heat-transfer coefficient on workpiece surface during MQL grinding and concluded that the theory of surface heat-transfer coefficient during MQL grinding is creditable. Numerous studies and experiments have indicated that the grinding process is improved by MQL technology. More importantly, the dosage of the grinding fluid significantly decreases, thereby reducing hazards to workers and the environment (Zhang, Li, Jia, Zhang& Zhang, 2015).

To increase the machining performance and disposal of grinding fluids, as well as to address the related environmental and health problems, some scholars attempted to use vegetable oils as MQL base oil. Emami et al. (Emami, Sadeghi, Sarhan, & Hasani, 2014) investigated the performances of four types of lubricants, namely, mineral, hydrocracked, synthetic and vegetable oils; their results showed that MQL not only considerably affects the cutting force, specific energy, and surface roughness, but the vegetable oils can also diminish environmental hazards of cutting fluids. Rahim et al. (Rahim & Sasahara, 2011; Rahim & Sasahara, 2011a, 2011b) conducted a series of experimental studies in which palm oil and synthetic ester were used as MQL base oils. To develop a more advanced processing method using lubricants with less energy consumption and less environmental pollution, nanoparticles are employed as additives of nanofluids to significantly improve the lubrication effect and heat-transfer. Nanofluids are colloidal mixtures of nanometer-sized particles (1 nm to 100 nm) in a base fluid. Such nanoparticles can be metallic (Fe, Cu, Ag), non-metallic (diamond), oxide (Al_2O_3, SiO_2, ZrO_2), carbide (carbon nanotubes), sulfide (MoS_2), ceramic, and hybrid. Nanofluids demonstrate many advantages, including superior lubrication performance and thermal conductivity (Hu, Bai, Lv, Wang, & Li, 2014). Given that different nanoparticles exhibit various physicochemical properties like molecular structure and shape, the corresponding nanofluids will demonstrate diverse lubrication properties and heat-transfer capabilities. Shen et al. (Shen, Shih, & Tung, 2008) investigated the wheel wear and tribological characteristics in wet, dry, and MQL grinding with water-based Al_2O_3 and diamond nanofluids. Their results showed that a dense and hard slurry layer was formed on the wheel surface, thereby reducing the grinding force. The authors claimed that MQL grinding can also significantly reduce the grinding temperature. Ge et al. (Ge & Cao, 2015) studied the surface morphology of the workpiece after tribological tests with oleophilic nanometer-TiO_2 and nanometer-SiO_2 as additives. In their experiment, a layer of protective oil film was generated by the deposited Ti and Si or metallic oxide, which can reduce friction and improve the tribological properties. Kalita et al. (Kalita, Malshe, Kumar, Yoganath, & Gurumurthy, 2012) per formed nanofluid MQL grinding experiments by using soybean oil and paraffin oil as base oils; their results revealed good tribological property of MoS_2 nanoparticles. Lee et al. (Lee, Kim, Nam, & Lee, 2012) added ND and nanoAl_2O_3 particles into paraffin oil with nanofluids for MQL microgrinding; their findings showed that nanofluid MQL can effectively reduce the grinding forces and enhance the surface quality. Mao et al. (Mao, Tang, Zou, Huang, & Zhou, 2012) used four different lubrication conditions (i.e., dry, flood,pure MQL, and Al_2O_3 nanofluid MQL) for grinding experiments; their results indicated that water-based Al_2O_3 nanofluid MQL grinding presented the optimal workpiece surface quality. Jia et al. (Jia, Li, Zhang, Zhang, & Zhang, 2014) used ZrO_2, diamond, and MoS_2 nanoparticles with MQL grinding of AISI 1045 steel. The authors reported that MoS_2 nanoparticles achieved the optimum lubrication performance, followed by diamond nanoparticles. Zhang et al. (Guo et al., 2017) performed an experimental evaluation on the lubrication performance of mixed nanofluid MQL grinding. Their results showed that the MoS_2/CNT hybrid nanoparticles achieved better lubrication effect than single nanoparticles. Alves et al. (Alves, Barros, Trajano, Ribeiro, & Moura, 2013) studied the development of vegetable-oil-based lubricants and the addition of oxide nanoparticles (ZnO and CuO) as additives. The authors found that with the addition of nanoparticles to conventional lubricants, the lubrication properties can be significantly improved.

According to experimental study results in Chapter 11, castor oil and palm oil have good lubricating characteristics, so they are suitable to be MQL base oils. However, application of castor oil has been restricted due to its poor mobility. Therefore, on basis of research contents in the last chapter, palm oil is used as base oil in this chapter to prepare different nanofluids and investigate lubricating properties of different nanoparticles.

In practical production,nanoparticles can exert "bearing-like effect", "oil film protection effect", "stuffing effect", etc. on the grinding wheel/workpiece interface. Meanwhile, nanoparticles have superior anti-friction and anti-wear effects, so nanoparticles will further improve lubricating property in the grinding area (Zhou, Wu, Zhang, Liu, & Dang, 2001) . In predecessors' researches, only 1~2 kinds of nanofluids are used lubricating fluids without systematic comparison and experimental evaluation of properties between multiple kinds of nanofluids. Different nanoparticles have different physiochemical characteristics like molecular structure and shape, and the formed nanofluids have different lubricating properties and heat transfer performances. Therefore, lubricating properties of several kinds of typical common nanoparticles (ZrO_2, CNT, ND, MoS_2, SiO_2 and Al_2O_3) are investigated in this experiment so as to obtain influence laws of different nanofluids on lubricating properties.

12.2 EXPERIMENTAL DESIGN

12.2.1 Experimental Equipment and Grinding Parameters

Equipment used in this experiment and grinding parameters are totally identical with those in section 11.2.1, namely: K-P36 NC surface grinder, dynamometer, grinding wheel and MQL device, roughness measuring instrument, digital viscosimeter and SEM.

12.2.2 Experimental Materials

(1) Workpiece Material and Water-Based Grinding Fluid

In order to keep consistency with the experiment in Chapter 11, workpiece material used in this experiment is still high-temperature nickel base alloy and grinding fluid is water-soluble grinding fluid. The related introduction is identical with that in section 11.2.2

Table 1. Physical and chemical properties of the palm oil

Density	Flash point	Viscosity (40°C)	Acid value	Iodine value	Saponification value	Melting point
0.91 g/cm^3	240 °C	39.2 cp	0.4	53	193 mg/g	57.6 °C

(2) MQL Base Oil

According to experimental conclusions in Chapter 11, among selected seven vegetable oils, palm oil has favorable comprehensive lubricating property, so palm oil is used as base oil in this experiment. The reasons why palm oil is taken as MQL base oil are: 1. Palm oil contains a large quantity of polar triglycerides, which can form a physical adsorption film which is good for lubrication; 2. Palm oil has high viscosity and can stay in the grinding area for a longer time to exert better lubricating effect; 3. Main fatty acid types contained in palm oil are palmitic acid (30%~50%) and stearic acid, both of which are saturated fatty acids with carbon number of 16, so a lubricating film with strong adsorption force can be formed. This can realize better lubricating property in the grinding area. Table 1 introduces related physiochemical properties of palm oil.

(3) Nanofluid

Nanofluid is formed mainly mixing base oil and nanoparticles (Saeedinia, Akhavanbehabadi, & Nasr, 2012). Base oil is the above selected palm oil. Six kinds of nanoparticles are used namely MoS_2, SiO_2, ND, CNT, Al_2O_3 and ZrO_2, where average length of CNT is 10-30 μm and average diameter is 50nm. Particle sizes of other nanoparticles are all 50nm. Selected nanoparticles belong to typical particles, where MoS_2 nanoparticles have stratified structure. CNT presents coaxial circular tube constituted by several layers to dozens of layers of carbon atoms presenting hexagonal arrangement. SiO_2 nanoparticles are ellipsoid-shaped and other particles are spherical. ND, SiO_2 and ZrO_2 nanoparticles are of high hardness while MoS_2 nanoparticles are soft. These provide distinct features for studying their lubricating properties.

Nanofluids are prepared in this experiment using two-step method, and all kinds of nanoparticles are respectively dispersed into the base oil. According to the previous experimental study (Zhang, Li, Jia, Zhang, & Zhang, 2015), nanofluid is prepared by mass fraction of 6%, as nanofluid with this mass fraction has good lubricating property. As nanoparticles are not stable in the base oil and can easily go through agglomeration phenomenon, adding a small quantity of surfactant in the nanofluid can improve nanofluid stability. Surfactant lauryl sodium sulfate (SDS) is respectively added into nanofluids. Study results of Demas (Demas, Routbort, & Fenske, 2012) indicate that: adding a small quantity of SDS almost has no influence on friction property, but can exert a great effect on dispersion and stability of nanoparticles. NC ultrasonic oscillator (KQ3200DB) is used for 2h nanofluid oscillation and stirring in order to ensure uniform dispersion and stable suspension property. Figure 1 shows prepared different 100ml nanofluids.

Figure 1. The digital camera photos of different nanofluids

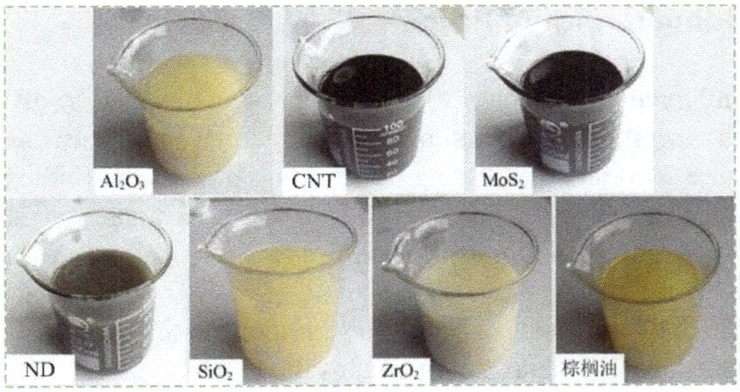

12.2.3 Experimental Scheme

This experimental study aims at discussing about influences of different nanofluids on lubricating properties at grinding wheel/workpiece interface and workpiece surface qualities in the MQL grinding process. According to experimental results, the nanofluid with the optimal lubricating property is selected. Based on the study of sliding grinding forces, specific sliding grinding energies and grinding wheel G ratios as well as grinding debris and workpiece surface micromorphology under different lubrication conditions, an experimental evaluation of lubricating properties at grinding wheel/workpiece interface is carried out. Univariate method is used in the experiment, other factors remain unchanged, and only lubricating fluid type is changed in each group of experiments.

Grinding fluids include water-based grinding fluid, pure palm oil and six nanofluids. Three lubrication conditions are casting type, pure palm oil MQL and NMQL. Water-based grinding fluid and pure palm oil are taken as the comparison group to study lubricating properties of six nanofluids. Experimental program is shown in Table 2.

Table 2. Experimental program

Experiment No.	Lubricating fluid	Lubrication condition
4-1	Water-based grinding fluid (5 vol%)	Flood
4-2	Al_2O_3 nanofluid	MQL
4-3	CNT nanofluid	MQL
4-4	MoS_2 nanofluid	MQL
4-5	ND nanofluid	MQL
4-6	SiO_2 nanofluid	MQL
4-7	ZrO_2 nanofluid	MQL
4-8	Pure palm oil	MQL

12.3 EXPERIMENTAL RESULTS

12.3.1 Specific Sliding Grinding Force

Specific sliding grinding force (Wang et al., 2016) is the product obtained by dividing sliding grinding force by grinding width. Specific tangential sliding grinding force and specific normal sliding grinding force are shown in Figure 2 (a) and (b). During the grinding process, a grinding force data point will be drawn every 5 grinding travels, and value of each data point is average value of measured results in each grinding travel of grinding wheel on workpiece.

It can be seen from Figure 2 that general change trend of force is as such: except for ND nanofluid, all tangential sliding grinding forces and normal sliding grinding forces increase with the grinding travel. Under all working conditions, flood lubrication obtains the maximum force. Compared with flood lubrication, pure palm oil MQL effectively improves lubricating property in the grinding area, which then reduces sliding grinding force to a certain degree. NMQL is the most effective in the aspect of improving lubricating effect, and compared with pure palm oil MQL grinding, grinding force is reduced, especially changes of sliding grinding forces of Al_2O_3 and MoS_2 nanofluids are the most obvious.

From the beginning of grinding till the 30th travel, sliding grinding forces under all working conditions are not very different, but with accumulation of grinding travel, force curve basically presents a rising trend, but the dispersion range is broadened. Change trend of ND nanofluid curve is the most abnormal: when grinding proceeds to the 30th travel, sliding grinding force starts declining and then the curve tends to be gentle.

Figure 2. Specific sliding grinding forces under different lubrication conditions: a-Specific tangential sliding forces, b-specific normal sliding forces

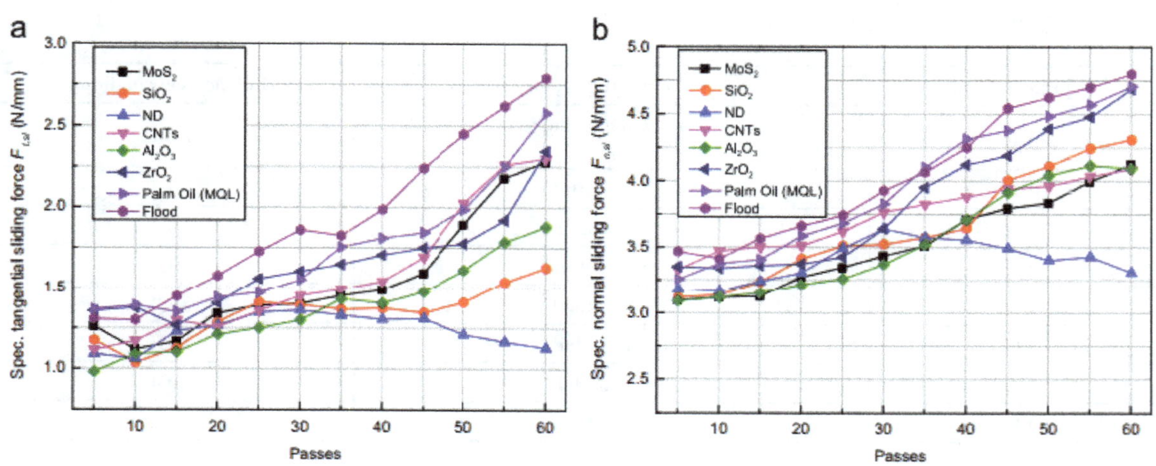

Figure 3. Sliding friction coefficients under different working conditions

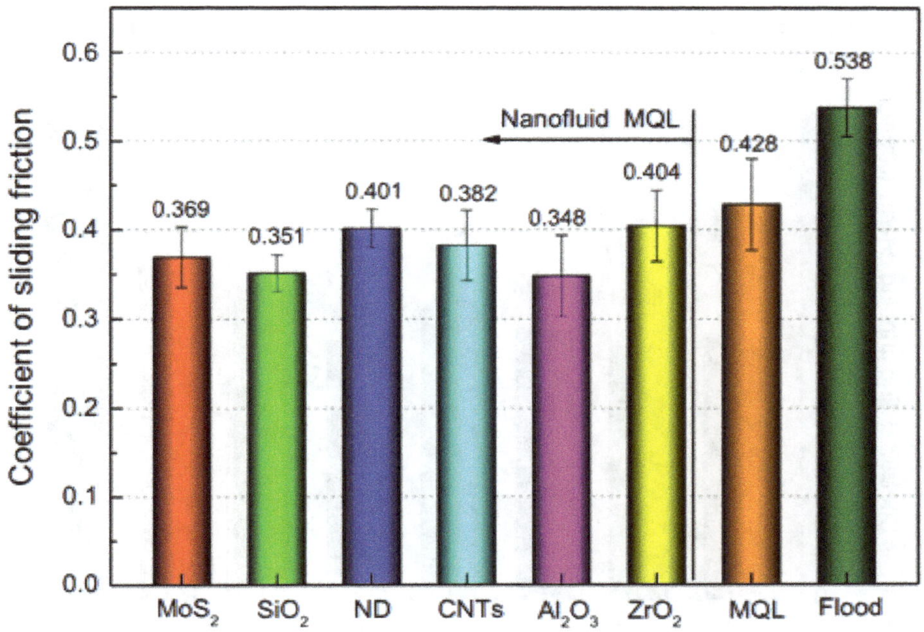

12.3.2 Sliding Friction Coefficient

Sliding friction coefficient μ_{sl} is the ratio of tangential sliding friction force $F_{t,\,sl}$ to normal sliding friction force $F_{n,\,sl}$. The relationship between different grinding conditions and sliding friction coefficient is obtained as shown in Figure 3. Each drawn data point is average value solved through 60 grinding travels, and error bar represents standard deviation of sliding friction coefficient.

Under the same experimental conditions, sliding friction coefficient under flood grinding condition is the maximum, being $\mu_{\text{sl-flood}}$=0.538. Compared with flood lubrication, sliding friction coefficients under pure palm oil MQL condition and different NMQL conditions decline to a certain degree, where pure palm oil MQL significantly improves lubricating property and sliding friction coefficient decreases to $\mu_{\text{sl-MQL}}$=0.428. After nanoparticles are added into the base oil, lubricating property between grinding wheel and workpiece is further improved. Compared with pure palm oil MQL, lubricating effect is slightly improved after nanoparticles ZrO_2, ND and CNT are added, and it is significantly improved after nanoparticles MoS_2, SiO_2 and Al_2O_3 are added.

12.3.3 Specific Sliding Grinding Energy

Macroscopic specific grinding energy reflects energy proportion consumed in the grinding process, and low specific grinding energy represents that the grinding process is more of green and energy conservation properties. According to the formula, calculation method of specific sliding grinding force can be known (Ren & Hua, 2011). Specific sliding grinding energies under different working conditions are shown in Figure 4, and error bar represents standard deviation of specific sliding grinding energy.

Figure 4. Specific sliding grinding energies under different lubrication conditions

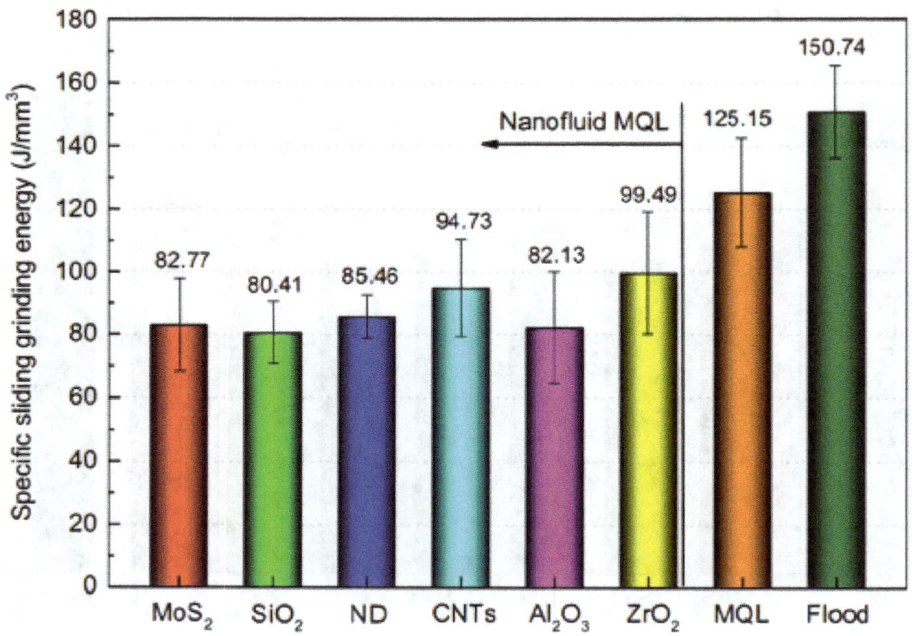

Figure 4 shows that change laws of specific sliding grinding energy and sliding friction coefficient under different working conditions are largely the same. Under the flood condition, lubricating state in the grinding area is not good enough and the energy consumed is the maximum. Compared with flood lubrication, specific sliding grinding energy is reduced by 16.9% under palm oil MQL condition. NMQL displays better lubricating effect, and this effect can be more significant when MoS_2 and Al_2O_3 nanoparticles are used with their specific grinding energies reduced by 33.8% and 34.5% respectively compared with palm oil MQL. Jia et al. (Jia et al., 2014) studied lubricating effects of MoS_2, ZrO_2 and ND nanofluids, and results also showed that MoS_2 had the optimal lubricating property, followed by ND.

12.3.4 Grinding Ratio

Figure 5 shows G ratios calculated after 60 grinding travels.

Flood lubrication obtains the lowest G ratio, compared with which, G ratio under pure palm oil MQL grinding is slightly elevated, being 25.83. G ratio significantly increases under NMQL conditions. Especially, G ratios under SiO_2, ND and MoS_2 NMQL grinding conditions increase by 41.9%, 40.2% and 39.8% respectively when compared with that (19.65%) under flood lubrication condition. When Al_2O_3 nanofluid is used for MQL grinding, wear extent of grinding wheel is the minimum, thus reaching the best anti-friction and anti-wear effects. This study result is similar to that of Shen et al. (Shen, 2008; Shen et al., 2008).

Figure 5. G-ratios under different lubrication conditions

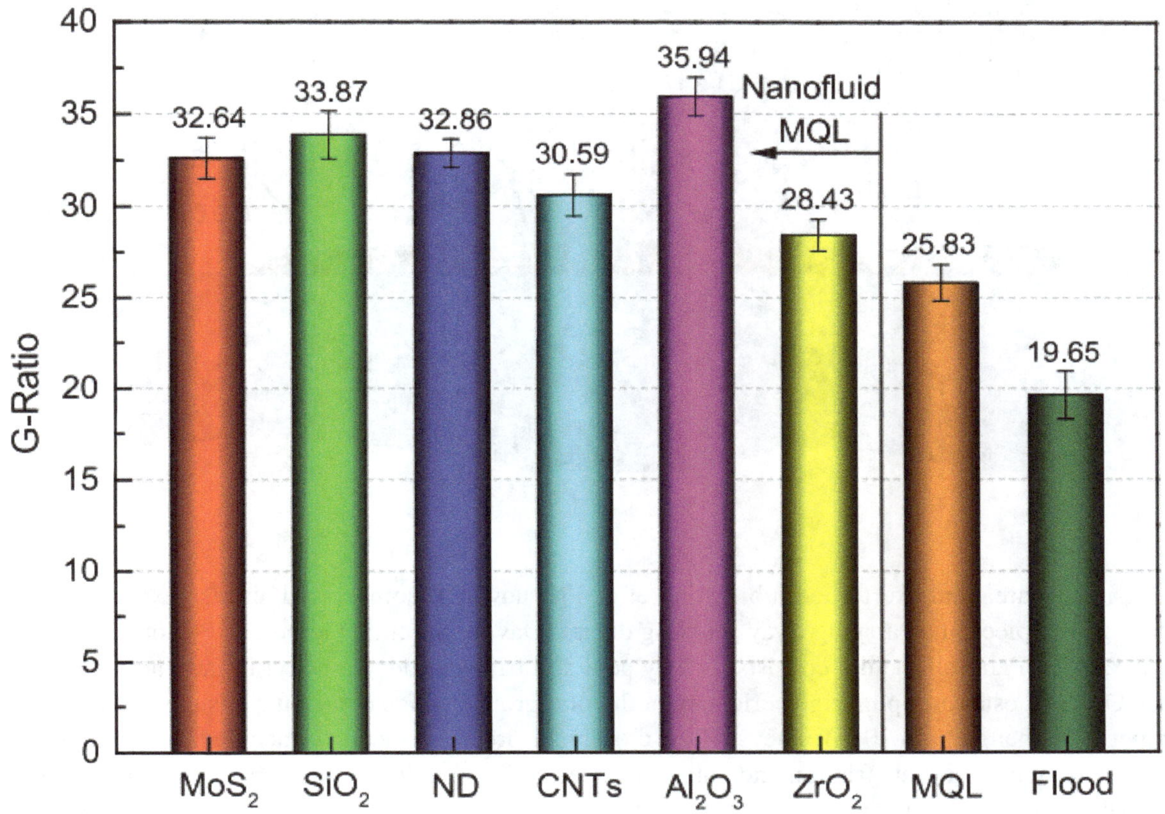

12.4 RESULTS AND DISCUSSION

12.4.1 Comparison of Lubrication Properties of the Three Working Conditions

Based on evaluation parameters obtained under flood lubrication, palm oil MQL and NMQL conditions, flood lubrication obtains the highest sliding grinding force, specific sliding grinding energy and the lowest G ratio, indicating that lubricating property in the grinding area is the poorest under flood lubrication condition. As an air-bond layer will be formed around grinding wheel under high-speed revolution of grinding wheel, high hydrodynamic pressure at air-bond layer hinders grinding fluid from entering the grinding area, it will be difficult for a large quantity of grinding fluid to enter the grinding fluid, which then results in declining lubricating property and very poor workpiece surface quality (Mao et al., 2012; Ebbrell, Woolley, Tridimas, Allanson, & Rowe, 2000; Morgan et al., 2008; Z. X. Zhou, Zhou, Mao, & Yang, 2009).

Lubricating effect of pure palm oil MQL on grinding wheel/workpiece interface is better than flood lubrication. Under MQL condition, as high-pressure gas carries away minimal quantity of grinding fluid, the gas can break through air-bond layer generated by revolution of grinding wheel so as to enter

Figure 6. Schematic diagram of lubricating film at the grinding interface

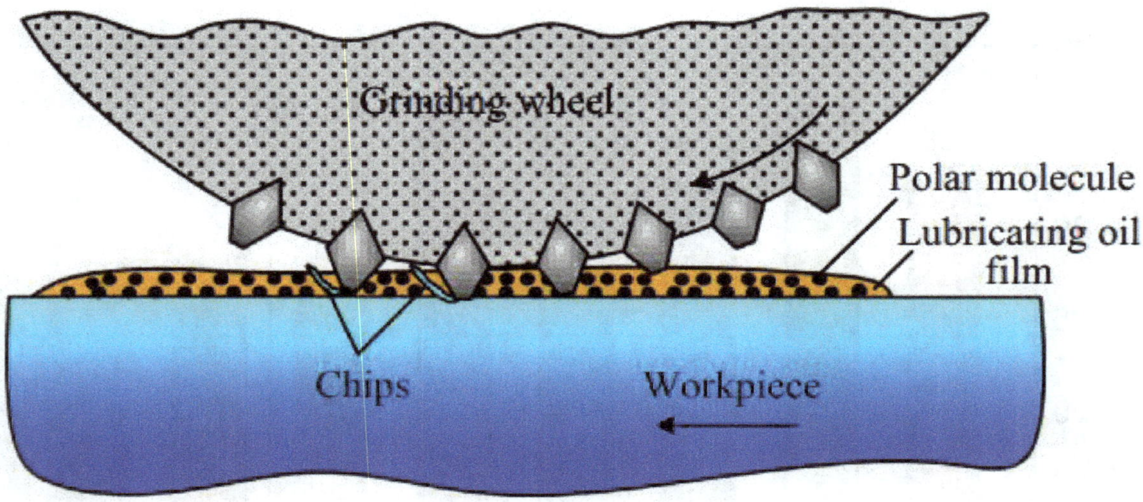

the grinding area and exert good lubricating effect. Meanwhile, compressed air also exerts effects of cooling workpiece and carrying away grinding debris (Davim, Sreejith, Gomes, & Peixoto, 2006; Mao et al., 2013). Palm oil mainly consists of fatty acid and triglyceride. Carboxyl in fatty acid molecules (-COOH) and ester group in triglyceride are both polar groups which contribute to superior lubricating properties (Zhang et al., 2015). They will be closely adsorbed on metal surface and form a molecular film, which can exert anti-friction and anti-wear effects, through directional arrangement just like tiny magnets. Under high temperature, polar molecules can react with metal to generate metal soap film as shown in Figure 6. This characteristic of vegetable oil endows it very good boundary lubricating property, so it decides superior lubricating effect of vegetable oil in the grinding fluid and further improves workpiece surface quality (Salete Martins Alves, 2008).

The reasons why NMQL had superior lubricating effect than pure palm oil MQL are as follows:

Under NMQL grinding condition, nanofluid is carried by high-pressure gas to pass through the nozzle and form tiny droplets. High-polarity nanoparticles are adsorbed on the surfaces of grinding wheel and workpiece through precipitation or adsorption effect, thus forming an oil film, where nanoparticles are embedded into the base oil film. Nanofluid can better infiltrate into the grinding area through high pressure so that nanofluid can exert anti-friction and anti-wear effects and greatly improve lubricating effect in the grinding area (Kalita, Malshe, & Rajurkar, 2012).

Most nanoparticles are spherical, near spherical or filiform (Kelly, Coronado, Zhao, & Schatz, 2003). In the sliding friction phase, nanoparticles enter abrasive grain/workpiece interface to exert "bearing-like" effect, the original sliding friction is turned into sliding-rolling complex friction, so lubricating property is improved.

During the workpiece machining process, micro-scratches, grooves or cracks will appear on the surface. Nanoparticles will be precipitated or adsorbed to these valleys on workpiece surface to exert stuffing and repair effects on the workpiece subsurface, reduce workpiece wear and improve lubricating

Figure 7. Anti-wear mechanism of nanoparticles

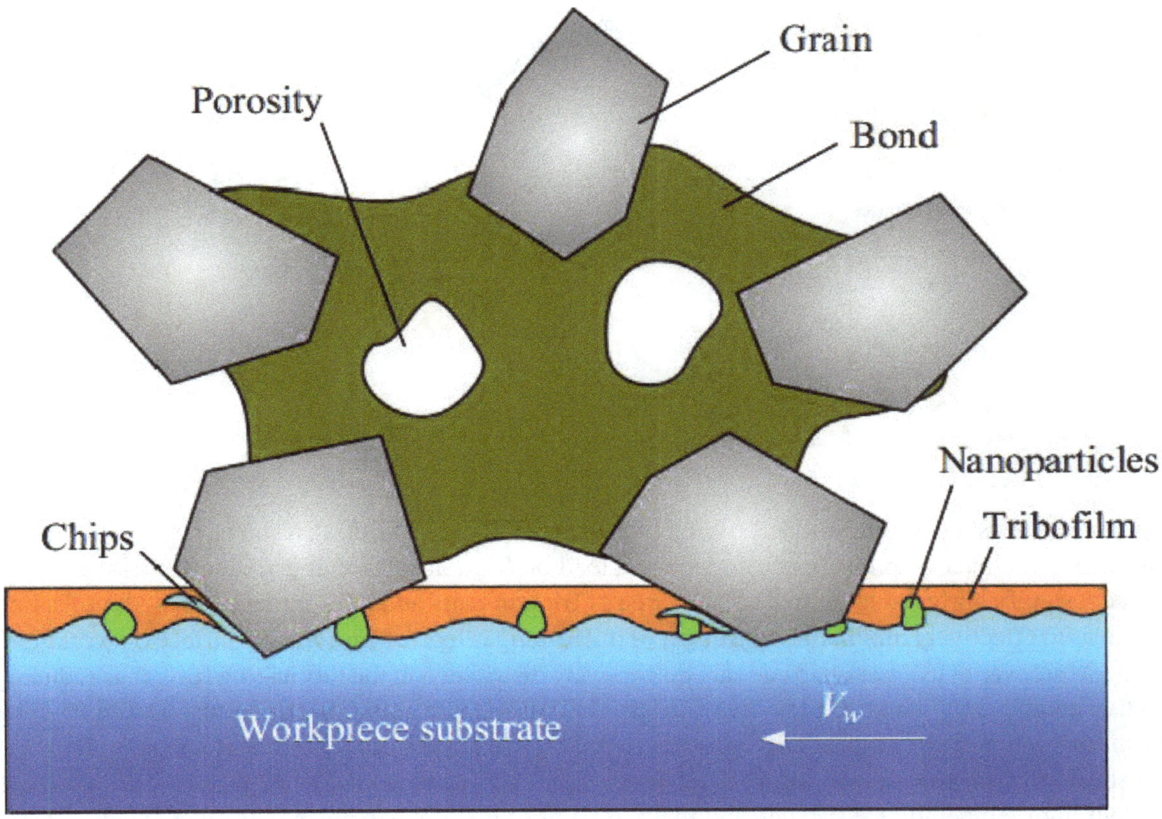

effect on grinding wheel/workpiece interface (Ebbrell et al., 2000). Schematic diagram of its anti-wear mechanism is shown in Figure 7.

12.4.2 Effect of Physical Properties of Nanoparticles on Lubrication Properties

Nanoparticles with different structures, shapes and sizes will generate different physical effects, so there will be different lubricating mechanisms (NEHME, MOURHATCH, ASWATH, & Pranesh, 2010).

According to nanoparticle surface effect (Choi, Jung, Choi, Lee, & Oh, 2011; Hwang, Park, Lee, & Jung, 2006) (Figure 8), specific surface area and surface energy are both very high with many surface atoms due to their small particle sizes. There are few adjacent atoms around surface atoms with many dangling bonds. With unsaturation property and extreme instability, they can be easily bonded with polar atoms in palm oil so that the nanofluid can generate higher surface energy and more powerfully adsorbed on workpiece/grinding wheel interface to exert better lubricating effect.

That Al_2O_3 nanofluid has good lubricating property is related to its structural behaviors (Zhang et al., 2017). Being of spherical structure, Al_2O_3 has superior characteristics like high strength, high hardness

Figure 8. Schematic diagram of nanoparticle surface effect

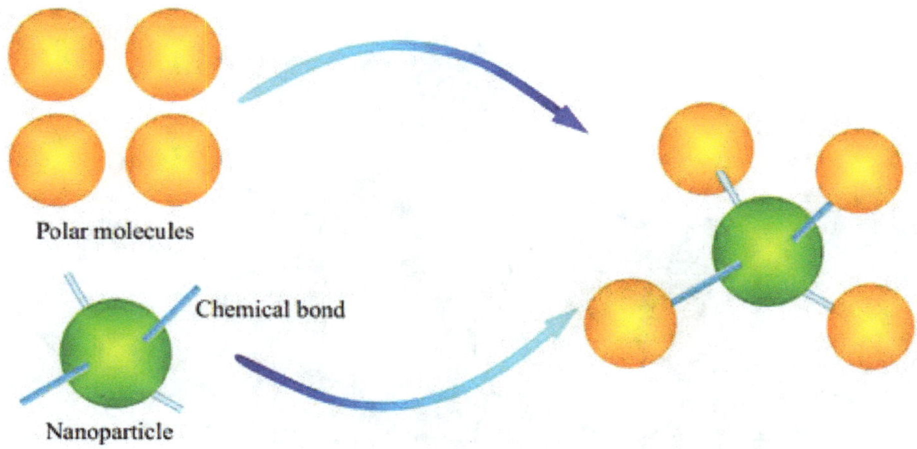

and high heat resistance. As the added Al_2O_3 is hard phase (HR=2700-3000), it shows favorable abrasive resistance in the friction process, can support the load on friction surface at grinding wheel/workpiece interface, significantly reduce actual contact area of friction pairs and reduce tangential sliding friction force so as to effectively inhibit plowing effect of micro-bulges located on abrasive surface on workpiece surface. Moreover, Al_2O_3 nanoparticles are dispersively distributed in the lubricating film, thus reducing effective contact area which experiences adhesive effect. Al_2O_3 nanoparticles can turn sliding friction into sliding and rolling complex friction so as to reduce sliding friction coefficient and effectively reduce shear stress in the wear process (BUSSE et al., 2011). Furthermore, as Al_2O_3 nanoparticles have excellent high-temperature resistance and melting point of oil film can reach 2200°C, they can exert a certain hindering effect on dislocation sliding and growth of crystal particles, and as a result, lubricating film is strengthened, its hardness increases and high-temperature resistance is improved (Sun, Yang, & Lib, 2008). Moreover, Al_2O_3 nanoparticles sprayed into the grinding area exert a certain mechanical polishing effect on workpiece surface, which will contribute to formation of a good workpiece surface morphology (Figure 15c).

SiO_2 nanoparticles are spherical, and surface molecules are of 3D network structure. As there are many unsaturated dangling bonds on the surface with very high surface energy and surface activity, they can be easily precipitated on friction surface of workpiece. SiO_2 nanoparticles, which are uniformly dispersed in the base oil, have very high diffusivity and self-diffusion capacity. When the grinding just begins, SiO_2 nanoparticles with high surface activity can go through intense chemical adsorption on friction surface between grinding wheel and workpiece through hydroxyls under the effects of compressive stress and shear stress in the grinding process, nanoparticles can be easily adsorbed on surface of friction pairs, and then a layer of physical adsorption film is formed. As the frictional motion further proceeds, under boundary lubrication conditions and local high temperature and external high load on the friction surface, SiO_2 nanoparticles located at grinding wheel/workpiece interface may be under molten, semi-molten or sintering state due to their extremely small particle size and low melting point (Yu et al., 2013), and meanwhile, some elements in nanoparticles may infiltrate into workpiece surface or subsurface due to dispersion and precipitation effects so as to form a solid solution with base components, and then a layer of so-called "third body" ceramic-like nanofilm is formed (Yu et al., 2013), which is very different from ordinary oil films, as its tenacity and compressive strength are obviously super to those of other oil films. Existence of this film isolates direct contact of abrasive grain-workpiece friction pairs so that frictional wear on grinding wheel/workpiece interface is significantly reduced, and a good workpiece surface morphology is obtained (Figure 15d).

MoS_2 nanoparticles are ellipsoid-shaped and connected through stratified structure. MoS_2 stratified structure is a hexagonal structure formed by bonding Mo and S through covalent bonds. Each crystal consists of a large quantity of MoS_2 molecules, each MoS_2 molecular layer is divided into 3 atomic layers, where upper and lower layers are S atomic layers and the middle layer is Mo atomic layer, and each Mo atom is surrounded by 6 S atoms as shown in Fig. 12-9a. Good lubricating property of MoS_2 is decided by its crystal structure. As bonding force between S atoms and Mo atoms is very strong at each molecular layer while bonding force between S atoms and Mo atoms between molecules is very weak, a surface with low shear force is generated. When small shear force exists between molecules, this molecular layer can easily rupture and then glide surfaces are formed. Due to these numerous glide surfaces, direct contact between friction pairs under relative friction is turned into relative sliding between MoS_2 molecular layers (Kalin, Kogovšek, & Remškar, 2012). As MoS_2 nanoparticles are of certain looseness, flexibility and ductility, thin physical MoS_2 film which falls off in the sliding process can be rapidly supplemented and updated through follow-up adsorption to keep this sliding characteristic and exert good lubricating effect (Figure 15e).

Molecular structure of CNT can be simplified as several layers to dozens of layers of coaxial circular tube constituted by carbon atoms presenting hexagonal arrangement as shown in Fig. 12-10b. As CNT has special tubular structure and extremely high tenacity, it has good lubricating effect and wear-resisting property, etc. SP2 hybridization is adopted for carbon atoms in CNT. Compared with SP3 hybridization, S orbital composition in SP2 hybridization is large so that CNT has high modulus and high strength. When CNT content in the nanofluid is small, nanoparticles exist in the form of "independent microtube" in the grinding area. With high strength and high hardness, CNT will not be grinded into hard film under a large load, so good sliding friction effect is achieved, sliding friction coefficient in the grinding area is reduced, and lubricating property of nanofluid is improved. However, due to geometric features like CNT shape and easy twining, it can't effectively exert rolling "micro-bearing" effect just like spherical nanoparticles, so its anti-wear effect is limited (Guo et al., 2017) (Figure 15g).

ZrO_2 nanoparticles are mainly spherical. When a certain load is applied to grinding wheel/workpiece interface and a certain temperature is reached and if any defect exists on workpiece surface, at the defect is the high-energy aggregation area, and at the time, ZrO_2 nanoparticles with high surface energy and surface activity (Wenjun et al., 2009) can be easily adsorbed onto workpiece surface to form chelate, and as a result, nanoparticles are infiltrated into the defect on workpiece surface, a layer of lubricating film is built on workpiece surface and the lubricating property is improved. However, as there are many hydrogen-bonded hydroxyls on surface of ZrO_2 nanoparticles, they become hydrophilic and oleophobic, and then agglomeration can be easily generated.

12.4.3 Effect of the Viscosity of Nanofluids on Lubrication Performance

Viscosity is variable exchange formed due to irregular movement between molecules and adhesive force between adjacent molecules. In the grinding area, relative motion between grinding wheel and workpiece will generate shear force on lubricating fluid so that internal friction force is generated in the internal

Figure 9. Atomic structures of nanoparticles: (a)- MoS_2 structure and lubrication mechanism, (b)-CNT structure, c-ND structure

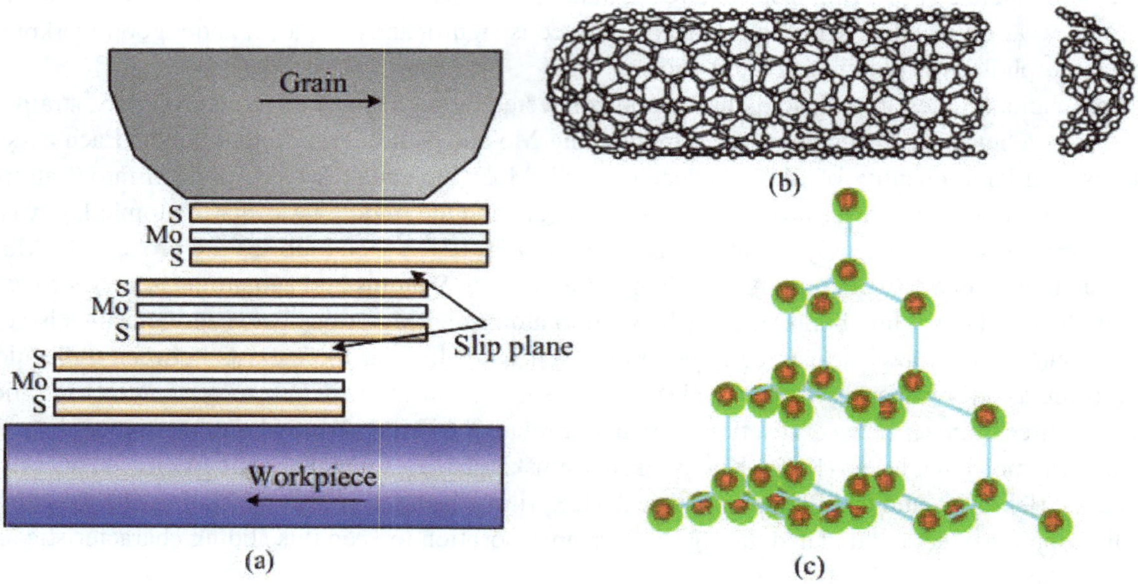

fluid layer of lubricating fluid, and the property influencing internal friction force is namely viscosity. Nanoparticles are added into the base oil to form suspension and viscosity will be further enhanced relative to viscosity of pure palm oil. In the grinding process, viscosity of lubricating fluid (Serrato, Maru, & Padovese, 2007) is another important factor influencing lubricating property. Figure 10 shows change curves of nanofluid viscosity and temperature.

According to the above figure, nanofluid viscosity declines with temperature (Alvi, Latif, Hussain, & Asghar, 2017). Before 70°C, viscosity value decreases very obviously, where viscosity values of SiO_2, Al_2O_3 and CNT nanofluids are high. On the one hand, high-viscosity lubricating fluid can stay in the grinding area for a longer time so as to improve lubricating property. On the other hand, Brownian movement and viscous force between lubricating fluid molecules increase with viscosity, thus enlarging thickness and strength of adsorption film. As sliding grinding force and specific sliding grinding energy under both SiO_2 and Al_2O_3 nanofluid lubrication conditions are relatively low, high G ratios are obtained. Viscosity values of ZrO_2 and ND nanofluids are relatively low, so their lubricating properties are relatively poor. This phenomenon indicates that high-viscosity nanofluid has better lubricating effect.

12.4.4 Analysis of Lubrication Properties of Diamond Nanofluids

From Figure 2 to 5, when ND nanofluid is used as lubricating fluid, grinding force starts declining at the 30th grinding travel and then it tends to be gentle. The obtained sliding friction coefficient is high and specific sliding grinding energy is low. Moreover, Figure 15f shows that workpiece surface morphology is very good, and small textures on the surface may be formed under the effect of ND nanoparticles. Through the research and analysis, this phenomenon may be generated for the following reasons.

Figure 10. Relationship between nanofluid viscosity and temperature

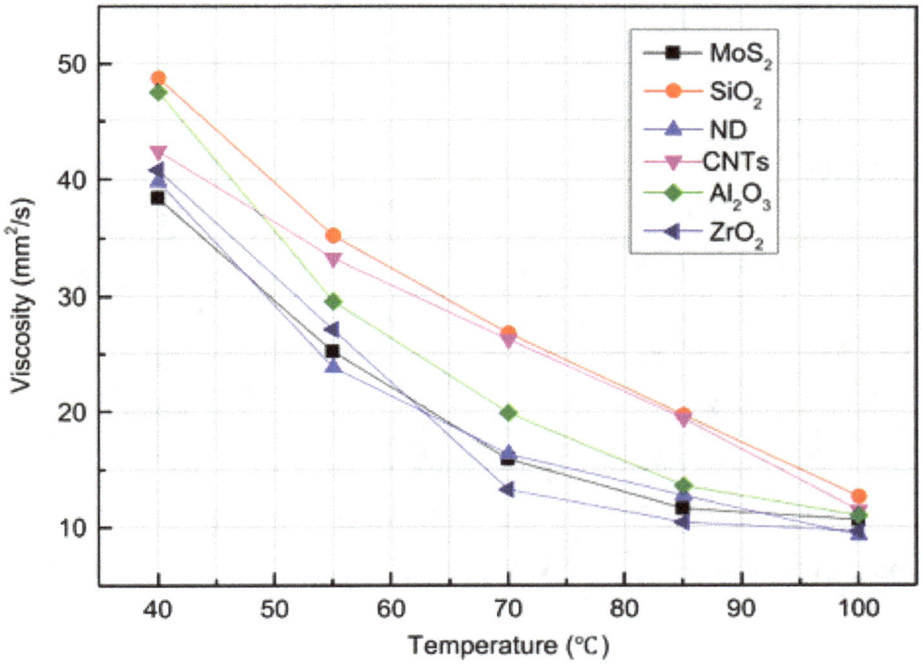

ND is cubic structure and its shape is spherical or ellipsoid-shaped as shown in Figure 9 (c). ND hardness is extremely high (HV=98 GPa), and wear resistance and mill capacity are the optimal among all grinding materials (xiao & Liu, 2000). Especially, ND has extremely high elasticity modulus (980 GPa) and compressive strength is about 13GPa, and meanwhile, friction coefficient of ND is very small in air. These high performances will certainly have an influence on lubricating property (Hu, 2005). After grinding wheel conducts grinding to a certain degree, grinding wheel can easily cause wear of abrasive grains and pore blocking. When ND nanofluid is used for lubrication, high-hardness ND nanoparticles are instantaneously sprayed onto grinding wheel and workpiece under high-pressure effect. At the time, ND nanoparticles with high kinetic energy impact the surface of grinding wheel and break through the bonding agent so that abrasive grain is protruded out to exert dressing effect on grinding wheel.

Grinding wheel consists of abrasive grain, bonding agent and pore. When ND nanoparticles with high kinetic energy are sprayed into grinding wheel/workpiece interface, some ND nanoparticles may be embedded into pores of grinding wheel under high-temperature and heavy-load environment, and others will be adhered around cutting edge of abrasive grains. Then high-hardness ND nanoparticles may temporarily exert the effect of abrasive grains and cut the workpiece surface. The schematic diagram is shown in Figure 11.

In the grinding machining process, gap h_0 (tiny gap formed due to protrusion of abrasive grains and existence of chip space on grinding wheel surface) at grinding wheel/workpiece interface is very small, ND nanoparticles entering the grinding wheel/workpiece interface are forcefully pressed and embedded into recession on grinding wheel surface, and then ND micropowder grinding wheel is formed for grinding and polishing (Li, 2005) as shown in Figure 12. As grinding wheel revolves at high speed, when

Figure 11. Schematic diagram of ND nanoparticles serving as abrasive grains

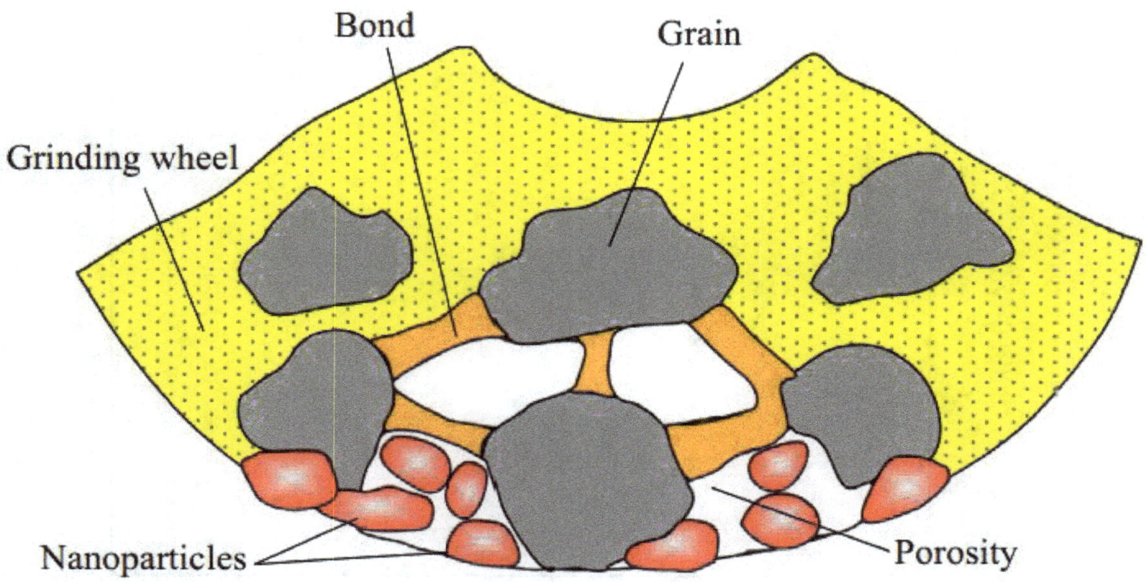

ND nanofluid is sprayed into the wedge-shaped gap between grinding wheel and workpiece, oil wedge pressing phenomenon by dynamic fluid pressure (Min, Nelson, & Hirasaki, 1983) is formed, 3D velocity field and pressure field are formed, and ND nanofluid obtains energy under high-speed revolution of grinding wheel as well as effects of centrifugal force, fluid pressure field and velocity field. Under boundary constraint conditions of grinding wheel, workpiece micro-removal machining is completed on grinding wheel/workpiece interface so as to exert grinding and polishing effects.

12.4.5 Surface Roughness and Surface Morphology

(1) Surface Roughness

As shown in Figure 13, Al_2O_3 NMQL grinding and flood grinding obtain the best and poorest surface roughness respectively. Workpiece surface grinded by pure palm oil MQL grinding is better than that by flood grinding, but NMQL grinding improves surface quality more significantly. Good surface quality is obtained through MQL technology is mainly because it can realize more effective lubrication at the grinding wheel/workpiece interface. Thanks to better lubrication, grinding debris can slide on surface of abrasive grain more easily so as to obtain a better surface roughness value.

(2) Graph Analysis of Grinding Debris and Workpiece Surface

In order to explain the above phenomenon, SEM graphs of grinding debris and workpiece surface are used to further demonstrate lubricating properties of different lubricating fluids. Micromorphology

Figure 12. Schematic diagram of ND nanoparticles grinding and polishing

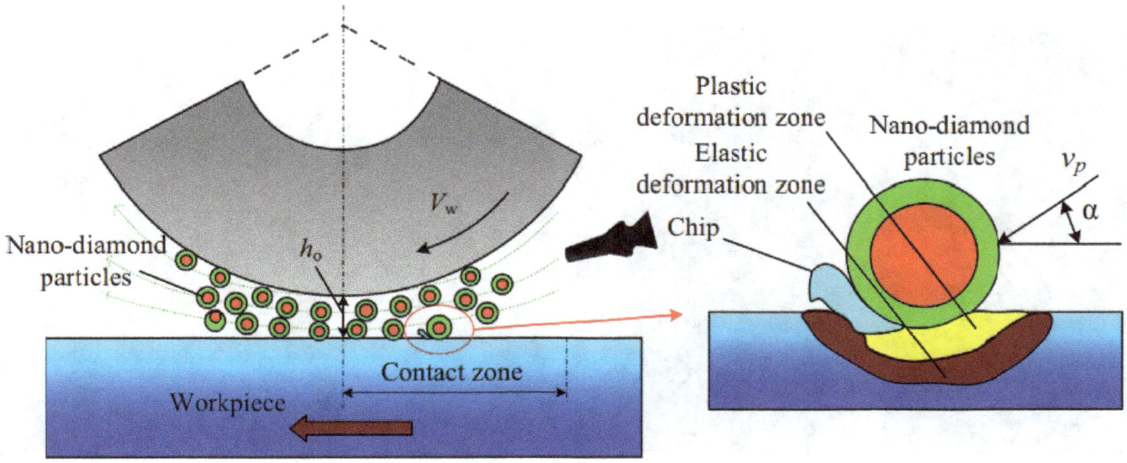

(Stoops & Gérard, 2007), an important index used to study workpiece surface quality, can not only reflect interaction state between abrasive grain and workpiece but also can reflect metal material removal

The following can be mainly observed from the above micrographs:

Flood lubrication obtains poor grinding quality with rough grinding surface, large chip thickness, and there are irregular slide marks on the surface. Moreover, furrows, scratches and material stacking

Figure 13. Workpiece surface roughness values under different lubrication conditions

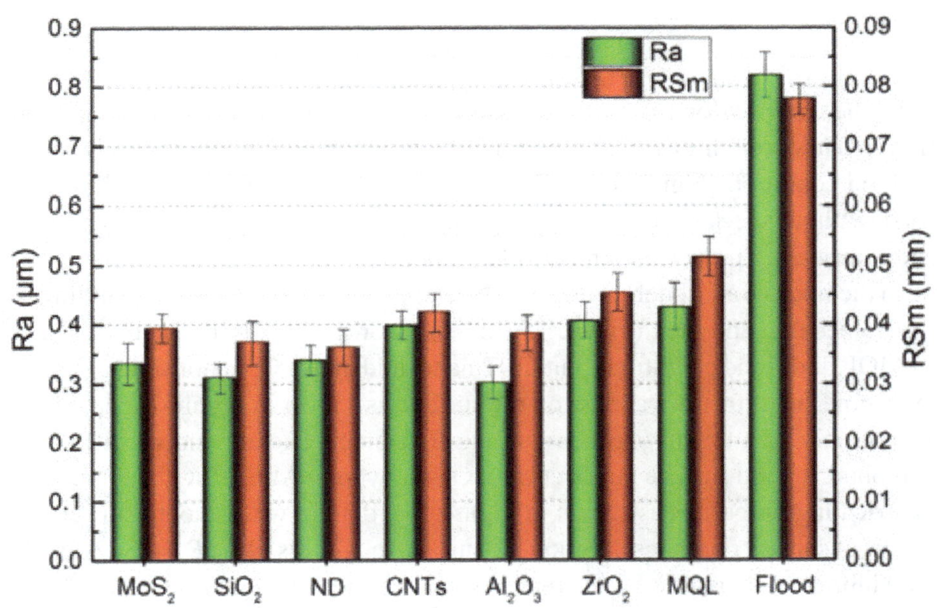

Figure 14. SEM micrographs of grinding debris under different conditions

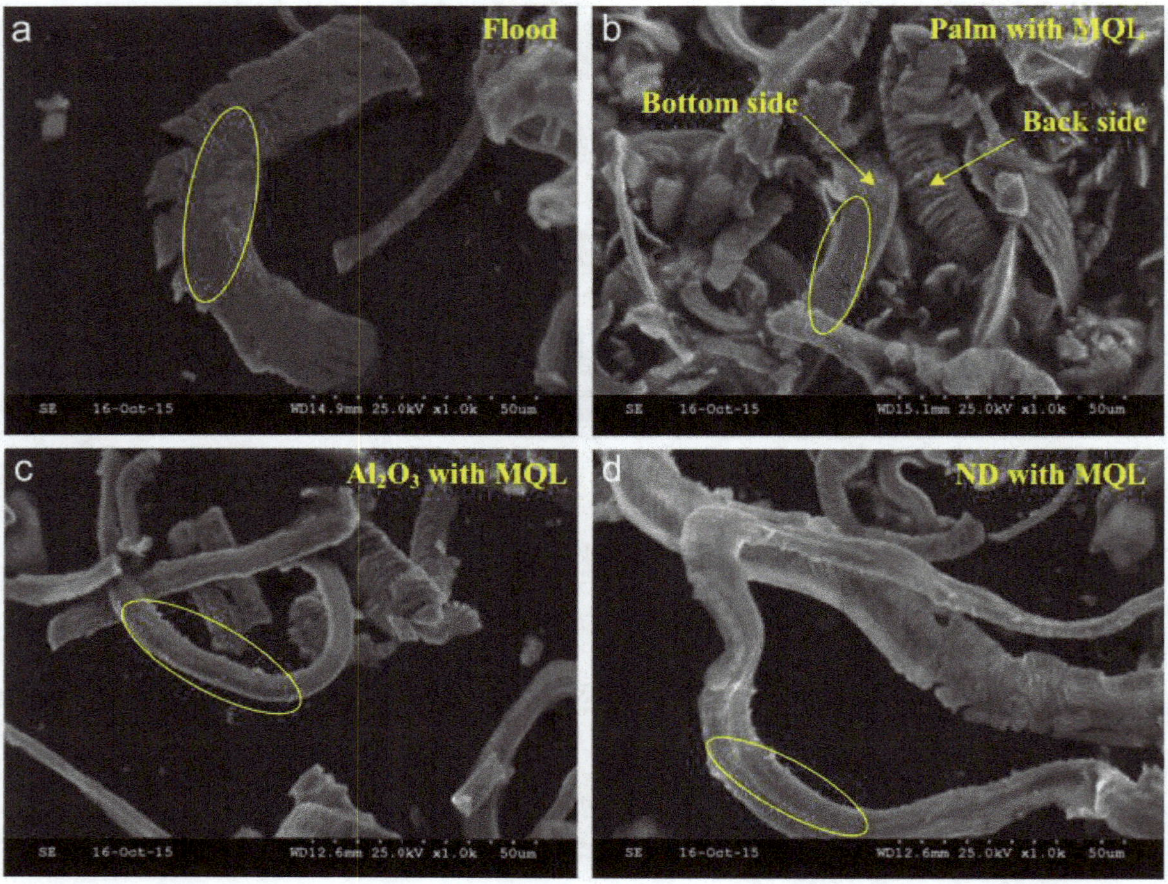

phenomenon can be observed on the workpiece surface (Figure 15a). Grinding debris under palm oil MQL condition also have rough surface, but undersurface morphology is significantly improved relative to that under flood condition (Figure 14b). Furrows on workpiece surface are obviously reduced and narrowed almost without speckles, but slight plastic plowing (Figure 15b) occurs. This also proves that MQL has better lubricating property than flood lubrication.

According to the micrographs, debris undersurface obtained through NMQL grinding is smooth and debris shape becomes thin and long (Figure 15c and d). Workpiece surface quality is better than those under pure oil MQL and flood conditions, but CNT nanofluid and ZrO_2 nanofluid have not shown favorable effect, and some small furrows appear on the surface as shown in Figure 15g and h. Nevertheless, Al_2O_3 NMQL not only significantly reduces sliding friction coefficient and specific sliding grinding energy but also obtains good surface quality (15c). On the whole, Al_2O_3, SiO_2 and ND nanofluids have shown good lubricating properties as shown in Figure 15c, d and f with smooth surface morphologies. However, in terms of surface roughness, workpiece surface roughness value R_a=0.302 μm obtained under Al_2O_3 nanofluid lubrication condition is the minimum. Roughness values of SiO_2 and ND nanofluids are 0.311 μm and 0.340 μm respectively. Therefore, analysis results indicate that Al_2O_3 nanofluid has the

Figure 15. SEM micrographs of grinding debris under different conditions

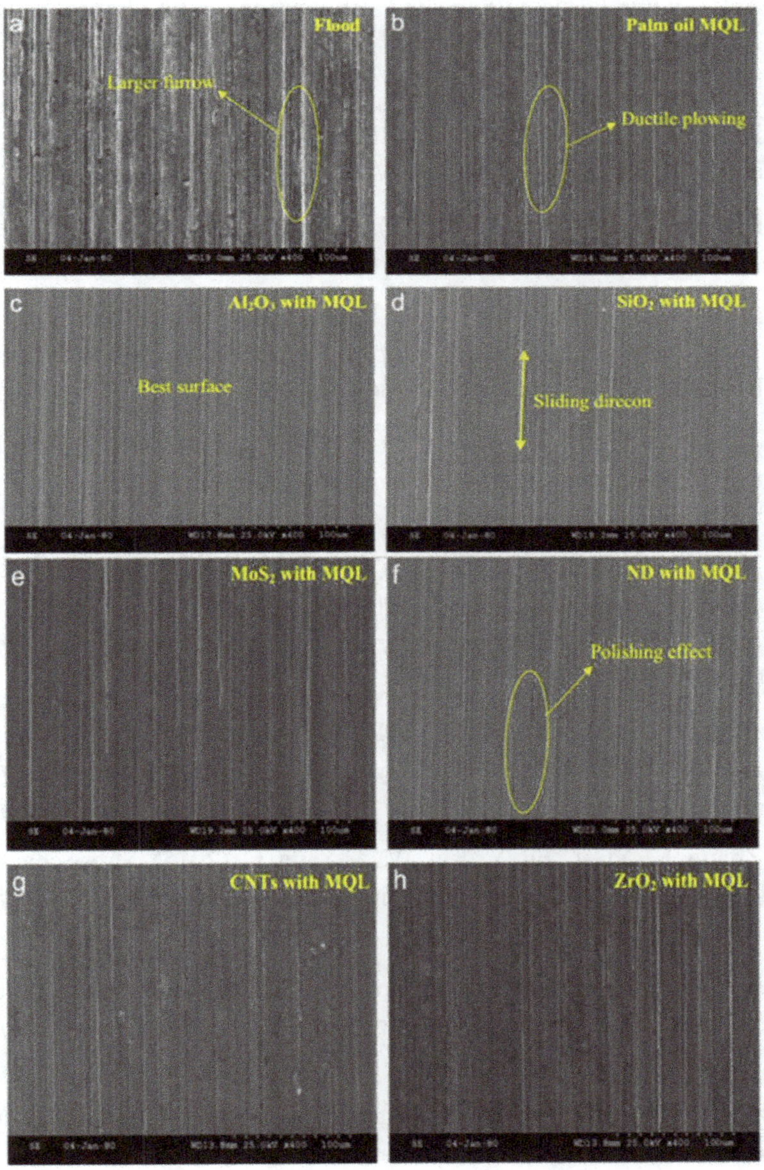

optimal lubricating effect, followed by SiO_2 and ND nanofluids. Workpiece surface morphology under NMQL condition is better than surface quality under pure palm oil MQL condition.

REFERENCES

Alves, S. M. (2008). *Vegetable based cutting fluid - an environmental alternative to grinding process.* Academic Press.

Alves, S. M., Barros, B. S., Trajano, M. F., Ribeiro, K. S. B., & Moura, E. (2013). Tribological behavior of vegetable oil-based lubricants with nanoparticles of oxides in boundary lubrication conditions. *Tribology International, 65*(3), 28–36. doi:10.1016/j.triboint.2013.03.027

Alvi, N., Latif, T., Hussain, Q., & Asghar, S. (2017). Peristalsis of Nanofluid with Temperature Dependent Viscosity. *Journal of Computational and Theoretical Nanoscience, 14*(3), 1417-1430. doi:10.1166/jctn.2017.6465

Busse, P., Peter, K., Karl, C. W., Geisler, H., & Klüppel, M. (2011). Reducing friction with Al_2O_3/SiO_2-nanoparticles in NBR. *Wear, 271*(7), 1066–1071. doi:10.1016/j.wear.2011.05.017

Choi, C., Jung, M., Choi, Y., Lee, J., & Oh, J. (2011). Tribological properties of lubricating oil-based nanofluids with metal/carbon nanoparticles. *Journal of Nanoscience and Nanotechnology, 11*(1), 368–371. doi:10.1166/jnn.2011.3227 PMID:21446457

Davim, J. P., Sreejith, P. S., Gomes, R., & Peixoto, C. (2006). Experimental studies on drilling of aluminium (AA1050) under dry, minimum quantity of lubricant, and flood-lubricated conditions. *Proceedings of the Institution of Mechanical Engineers. Part B, Journal of Engineering Manufacture, 220*(10), 1605–1611. doi:10.1243/09544054JEM557

Demas, N. G., Routbort, J. L., & Fenske, G. R. (2012). Tribological Effects of BN and MoS_2 Nanoparticles Added to Polyalphaolefin Oil in Piston Skirt/Cylinder Liner Tests. *Tribology Letters, 47*(1), 91–102. doi:10.100711249-012-9965-0

Ding, W., Zhu, Y., Zhang, L., Xu, J., Fu, Y., Liu, W., & Yang, C. (2015). Stress characteristics and fracture wear of brazed CBN grains in monolayer grinding wheels. *Wear, 332-333*, 800–809. doi:10.1016/j.wear.2014.12.008

Ebbrell, S., Woolley, N. H., Tridimas, Y. D., Allanson, D. R., & Rowe, W. B. (2000). The effects of cutting fluid application methods on the grinding process. *International Journal of Machine Tools & Manufacture, 40*(2), 209–223. doi:10.1016/S0890-6955(99)00060-7

Emami, M., Sadeghi, M. H., Sarhan, A. A. D., & Hasani, F. (2014). Investigating the Minimum Quantity Lubrication in grinding of Al_2O_3 engineering ceramic. *Journal of Cleaner Production, 66*(3), 632–643. doi:10.1016/j.jclepro.2013.11.018

Ge, X., & Cao, Z. (2015). Tribological properties and insulation effect of nanometer TiO_2 and nanometer SiO_2 as additives in grease. *Tribology International, 92*, 454–461. doi:10.1016/j.triboint.2015.07.031

Guo, S., Li, C., Zhang, Y., Wang, Y., Li, B., Min, Y., & Liu, G. (2017). Experimental evaluation of the lubrication performance of mixtures of castor oil with other vegetable oils in MQL grinding of nickel-based alloy. *Journal of Cleaner Production, 140*, 1060–1076. doi:10.1016/j.jclepro.2016.10.073

Hu, C., Bai, M., Lv, J., Peng, W., & Li, X. (2014). Molecular dynamics simulation on the friction properties of nanofluids confined by idealized surfaces. *Tribology International*, *78*(4), 152–159. doi:10.1016/j.triboint.2014.05.018

Hu, Z. (2005). The Tribological Characteristics of Nano-scale Diamond in Polyoxyethylene Glycol. *Lubrication Engineering,* (5), 7-10+13.

Hwang, Y., Park, H. S., Lee, J. K., & Jung, W. H. (2006). Thermal conductivity and lubrication characteristics of nanofluids. *Current Applied Physics*, *6*(1), 67–71. doi:10.1016/j.cap.2006.01.014

Jia, D., Li, C., Zhang, D., Zhang, Y., & Zhang, X. (2014). Experimental verification of nanoparticle jet minimum quantity lubrication effectiveness in grinding. *Journal of Nanoparticle Research*, *16*(12), 1–15. doi:10.100711051-014-2758-7

Kalin, M., Kogovšek, J., & Remškar, M. (2012). Mechanisms and improvements in the friction and wear behavior using MoS_2 nanotubes as potential oil additives. *Wear, 280–281*(11), 36–45. doi:10.1016/j.wear.2012.01.011

Kalita, P., Malshe, A. P., Kumar, S. A., Yoganath, V. G., & Gurumurthy, T. (2012). Study of specific energy and friction coefficient in minimum quantity lubrication grinding using oil-based nanolubricants. *Journal of Manufacturing Processes*, *14*(2), 160–166. doi:10.1016/j.jmapro.2012.01.001

Kalita, P., Malshe, A. P., & Rajurkar, K. P. (2012). Study of tribo-chemical lubricant film formation during application of nanolubricants in minimum quantity lubrication (MQL) grinding. *CIRP Annals - Manufacturing Technology, 61*(1), 327-330. doi:10.1016/j.cirp.2012.03.031

Kelly, K. L., Coronado, E., Lin Zhao, L., & Schatz, G. C. (2003). The Optical Properties of Metal Nanoparticles: The Influence of Size, Shape, and Dielectric Environment. *ChemInform*, *34*(16), 668–677. doi:10.1002/chin.200316243

Lee, P.-H., Kim, J. S., Nam, J., & Lee, S. W. (2012). Experimental Study on Micro-Grinding Process of Titanium Alloy Using Electro-Hydro-Dynamic Spray With Nanofluid and Compressed Air. *Manuf Processes*, *14*, 160–166. doi:10.1115/MSEC20159296

Li, C. (2005). Investigation on Material Removal Mechanism for Abrasive Jet Precision Finishing with Grinding Wheel as Restraint. *Zhongguo Jixie Gongcheng*.

Li, C., Hou, Y., Xiu, S., & Cai, G. (2008). Application of lubrication theory to near-dry green grinding-feasibility analysis. *Advanced Materials Research*, *44*, 135–142.

Li, C. H., Li, J. Y., Wang, S., & Zhang, Q. (2013). Modeling and Numerical Simulation of the Grinding Temperature Field with Nanoparticle Jet of MQL. *Advances in Mechanical Engineering, 2013*, 761-776. doi:0.1155/2013/986984

Malkin, S., & Guo, C. (1991). Grinding technology: Theory and applications of machining with abrasives. *International Journal of Machine Tools & Manufacture*, *31*(3), 435–436. doi:10.1016/0890-6955(91)90088-K

Mao, C., Tang, X., Zou, H., Huang, X., & Zhou, Z. (2012). Investigation of grinding characteristic using nanofluid minimum quantity lubrication. *International Journal of Precision Engineering and Manufacturing, 13*(10), 1745–1752. doi:10.100712541-012-0229-6

Mao, C., Zou, H. F., Huang, Y. Z. X., Li, Y., & Zhou, Z. (2013). Analysis of heat transfer coefficient on workpiece surface during;minimum quantity lubricant grinding. *International Journal of Advanced Manufacturing Technology*, *66*(1-4), 363–370. doi:10.100700170-012-4330-x

Min, J. T., Nelson, R. C., & Hirasaki, G. J. (1983). Study of the Oil Wedge Phenomenon Through the Use of a Chemical Flood Simulator. *Society of Petroleum Engineers Journal*, *23*(5), 746–758. doi:10.2118/10729-PA

Morgan, M. N., Jackson, A. R., Wu, H., Baines-Jones, V., Batako, A., & Rowe, W. B. (2008). Optimisation of fluid application in grinding. *CIRP Annals - Manufacturing Technology, 57*(1), 363-366.

Nehme, Mourhatch, Aswath, & Pranesh. (2010). Effect of contact load and lubricant volume on the properties of tribofilms formed under boundary lubrication in a fully formulated oil under extreme load conditions. *Wear*, *268*(9), 1129–1147. doi:10.1016/j.wear.2010.01.001

Rahim, E. A., & Sasahara, H. (2011). An analysis of surface integrity when drilling inconel 718 using palm oil and synthetic ester under mql condition. *Machining Science and Technology*, *15*(1), 76–90. do i:10.1080/10910344.2011.557967

Rahim, E. A., & Sasahara, H. (2011a). Investigation of tool wear and surface integrity on MQL machining of Ti-6AL-4V using biodegradable oil. *Proceedings of the Institution of Mechanical Engineers. Part B, Journal of Engineering Manufacture*, *225*(9), 1505–1511. doi:10.1177/0954405411402554

Rahim, E. A., & Sasahara, H. (2011b). A study of the effect of palm oil as MQL lubricant on high speed drilling of titanium alloys. *Tribology International*, *44*(3), 309–317. doi:10.1016/j.triboint.2010.10.032

Ren, J., & Hua, D. (2011). Grinding principle. Beijing: Electronic Industry Press.

Saeedinia, M., Akhavanbehabadi, M. A., & Nasr, M. (2012). Experimental study on heat transfer and pressure drop of nanofluid flow in a horizontal coiled wire inserted tube under constant heat flux. *Experimental Thermal & Fluid Science, 36*, 158-168. doi:10.1016/j.expthermflusci.2011.09.009

Serrato, R., Maru, M. M., & Padovese, L. R. (2007). Effect of lubricant viscosity grade on mechanical vibration of roller bearings. *Tribology International*, *40*(8), 1270–1275. doi:10.1016/j.triboint.2007.01.025

Shen, B. (2008). Performance of novel MoS2 nanoparticles based grinding fluids in minimum quantity lubrication grinding. *Transactions of NAMRI/SME, 36*, 357-364.

Shen, B., Shih, A. J., & Tung, S. C. (2008). Application of Nanofluids in Minimum Quantity Lubrication Grinding. *Tribology & Lubrication Technology*, *51*(6), 730–737. doi:10.1080/10402000802071277

Stoops, G., & Gérard, M. (2007). *Micromorphology. Soils of Volcanic Regions in Europe*. Springer Berlin Heidelberg.

Sun, L. H., Yang, Z. G., & Li, X. H. (2008). Study on the friction and wear behavior of POM/Al_2O_3 nanocomposites. *Wear*, *264*(7), 693–700. doi:10.1016/j.wear.2007.06.005

U, B., S, M., & Guo, C. (1998). Environmentally conscious cooling and lubrication for grinding. *Proc Int Semin Improv Mach Tool Perform, 2*, 643-654.

Wang, Y., Li, C., Zhang, Y., Li, B., Yang, M., Zhang, X., & Liu, G. (2016). Experimental evaluation of the lubrication properties of the wheel/workpiece interface in MQL grinding with different nanofluids. *Tribology International, 99*, 198–210. doi:10.1016/j.triboint.2016.03.023

Wenjun, M., Luqi, L., Zhong, Z., Rong, Y., Gang, L., Taihua, Z., & Zhiqiang, N. (2009). High-strength composite fibers: Realizing true potential of carbon nanotubes in polymer matrix through continuous reticulate architecture and molecular level couplings. *Nano Letters, 9*(8), 2855–2861. doi:10.1021/nl901035v PMID:19586030

Xiao, H. U., & Liu, J. (2000). Matrix wear resistance of diamond saw blade. *Chinese Journal of Nonferrous Metals*.

Yu, H., Yi, X., Shi, P., Wang, H., Min, W., Zhao, K., & Wang, H. (2013). Microstructure, mechanical properties and tribological behavior of tribofilm generated from natural serpentine mineral powders as lubricant additive. *Wear, 297*(1-2), 802–810. doi:10.1016/j.wear.2012.10.013

Zhang, D. K., Li, C., Jia, D., Zhang, Y., & Zhang, X. (2015). Specific grinding energy and surface roughness of nanoparticle jet minimum quantity lubrication in grinding. *Chinese Journal of Aeronautics, 28*(2), 570–581. doi:10.1016/j.cja.2014.12.035

Zhang, X., Li, C., Zhang, Y., Wang, Y., Li, B., Min, Y., & Zhang, N. (2017). Lubricating property of MQL grinding of Al_2O_3 /SiC mixed nanofluid with different particle sizes and microtopography analysis by cross-correlation. *Precision Engineering, 47*, 532–545. doi:10.1016/j.precisioneng.2016.09.016

Zhang, Y., Li, C., Jia, D., Zhang, D., & Zhang, X. (2015). Experimental evaluation of MoS 2 nanoparticles in jet MQL grinding with different types of vegetable oil as base oil. *Journal of Cleaner Production, 87*(1), 930–940. doi:10.1016/j.jclepro.2014.10.027

Zhou, J., Wu, Z., Zhang, Z., Liu, W., & Dang, H. (2001). Study on an antiwear and extreme pressure additive of surface coated LaF3 nanoparticles in liquid paraffin. *Wear, 249*(5), 333–337. doi:10.1016/S0043-1648(00)00547-0

Zhou, Z. X., Zhou, D. W., Mao, C., & Yang, J. (2009). Experimental Study of the Contact Length in Surface Grinding. *Journal of Hunan University, 36*(2), 27–31. doi:10.1007/978-0-387-74660-9_12

Chapter 13
Comparative Study on Tribological Properties of Nanofluids in Friction–Wear Experiments and Grinding Processing

ABSTRACT

This chapter presents the lubricating properties of different vegetable-oil-based nanofluids through a comparative evaluation between frictional test and grinding experiment. The first experiment aimed to prejudge the lubricating properties of different nanofluids with a frictional test, which simulated the interface state of grinding between the abrasive grains and the workpiece. The second aimed to test and verify the lubricating properties of the same nanofluids through a grinding experiment. The mechanism of oil-film formation of nanofluids in the grinding zone was analyzed by morphology and element analysis of the worn surface. The experimental results show that Al2O3 nanofluids have the best tribological properties. Compared with pure base oil, the friction coefficient is reduced by 20%, and the optimal friction surface morphology is obtained. The good anti-friction and anti-wear properties of nanofluids are attributed to the formation of the protective oil film formed by chemical reaction on the surface.

13.1 INTRODUCTION

Influence laws of different nanofluids on lubricating property are obtained through the grinding experiment in Chapter Twelve. Much research work has been done regarding both application of tribological properties of nanoparticles to machining process and their verification in the frictional wear experiment (Kulkarni, Nadakatti, Patil, & Kulkarni, 2017; Sharma, Tiwari, & Dixit, 2015; Virdi, Chatha, & Singh, 2019). However, as different nanoparticles have different physical characteristics, their tribological properties may also be different (Li, Zhang, Jia, Wang, & Hou, 2015; Zhang, Li, Jia, Zhang, & Zhang,

DOI: 10.4018/978-1-7998-1546-4.ch013

2015a). Zhang et al. (2015b) found that MoS_2/CNT hybrid nanoparticles had better lubrication effect than single nanoparticles. The best mixture ratio of MoS_2/CNT and the concentration of nanofluids were 2:1 and 6 wt.%, respectively. The research of Mao (2012) showed that compared with dry grinding, MQL grinding can significantly improve the grinding performance in improving the quality of grinding workpiece, reducing the grinding temperature and grinding force. Zhang et al. (2016) found that the hybrid nanoparticles composed of alumina and silicon carbide nanoparticles have lower *R*a value and surface roughness than the pure nanoparticles due to their "physical synergistic effect". Zhao et al. (2017) studied the stability of particle dispersion by Zeta potential method. The particle size of additive conglomerate in nanofluids was measured by dynamic light scattering. The results show that the dispersion of nanoparticles has an important influence on the friction properties of nanofluids. The results of EDS and XPS in Battez et al. (2010) study show that the antifriction behavior of nano lubricating oil on the wear surface can be attributed to the third body and friction sintering mechanism. Gara and Zou (2012) studied the effects of the type, concentration and surface roughness of nanoparticles on the friction and wear characteristics. The friction and wear tests were carried out with UMT-2 Micro Friction and wear tester. It is found that nanoparticles have a great influence on the friction and wear characteristics of lubricating oil. Kao and Lin (2009) found that the friction coefficient of nanofluid was always lower than that of no nanoparticles, although the viscosity loss would be caused by the increase of temperature. Nanoparticles provide rolling function, surface repair and increased viscosity as a lubricant. Yu et al. (2008) studied the effect of temperature on the tribological properties of copper nanoparticles on a four ball friction and wear tester. The results show that a copper protective film with low elastic modulus and low hardness is formed on the worn surface, which makes the copper nanoparticles have good tribological properties, especially at high oil temperature. Jiao et al. (2011) studied the tribological properties of modified alumina / silica composite nanoparticles as lubricant additives. The results show that their wear resistance and friction properties are better than those of pure alumina or silica nanoparticles. Due to the double effects of grinding and heating, the modifier is completely dispersed on the surface of particles, and combines with Si to form Si-O-C, and with Al to form double tooth coordination acyl alumina (Li, Ren, Jia, Liu, & Zhu, 2010). Peng et al. (2009) studied the tribological properties of diamond and silica nanoparticles prepared by oleic acid surface modification method in liquid paraffin. As liquid paraffin additives, the two nanoparticles have better antiwear and antifriction properties than pure paraffin oil at a very small concentration. Liu et al. (2017) studied the effect of nano particle size on the tribological properties of SiO_2/Polyalkylene glycol nanofluids under different lubrication conditions. It is found that the mixed nano additive has better anti-wear and anti friction performance than the nano additive with uniform particle size. Gara and Zou (2013) studied the effects of surfactant, ultrasonic time, particle concentration, applied load, sliding speed and surface roughness on the friction and wear properties of nanofluids. The results show that under certain conditions, the oil-based nanofluids containing nano zinc oxide reduce the friction and wear, the application of oleic acid as dispersant reduces the friction and wear to a certain extent, and improves the dispersion and stability of the nanofluids. Han et al. (2019) pointed out that diamond nanofluids produce higher ball wear than zinc oxide nanofluids. Ran et al. (2017)'s research found that oil-based nanofluids containing nano zinc oxide can significantly reduce friction and wear. Zin et al. (2016) found that the dispersed nanostructures play an important role in protecting the surface from wear and improving the friction performance and bearing capacity of crude oil. Pavan et al. (2019) studied the properties of graphene nanosheets cutting fluid in Inconel 718 Alloy MQL grinding. The experimental results show that the grinding force, temperature, specific grinding energy and surface roughness are reduced significantly. Minh et al. (2017)'s research shows that due to its better tribological performance and cooling and lubricating effect, when using alumina nanofluids for minimum lubrication, the tool life increases by nearly 177% - 230% (depending on the type of nanofluids), and the surface roughness and cutting force decrease by nearly 35% - 60%. Kumar et al. (2017) found that the use of nanofluids greatly improved the surface finish and reduced sub surface damage. Madanchi et al. (2019) studied the effect of nanoparticles and particles on the properties of cutting fluid and tribology. The effects of two different metal oxides, Al_2O_3 and ZrO_2, as well as the size, concentration and base solution of silica were studied. Although the lubricity of most nanoparticles has been improved, the opposite has been found in alumina. Dambatta et al. (2019) found that with the increase of nanofluid concentration, the grinding force of engineering ceramics decreased significantly, and the surface quality was improved. Lv et al. (2018) used graphene oxide/SiO_2 hybrid nanoparticles water-based lubricant as cutting fluid, and compared the tribological and processing characteristics of four ball milling test devices. The results showed that the friction coefficient and wear scar diameter of hybrid nanoparticles water-based MQL are significantly lower than those of single nanoparticles water-based MQL. Furthermore, in order to improve the processing performance and reduce the oil mist concentration in the processing workshop, Lv et al. (2018) proposed a near dry processing technique called electrostatic minimum quantity lubrication. The tribology and deposition performance of nanoparticles lubricant electrostatic minimum quantity lubrication and ordinary minimum lubrication were compared processability.

Few researches have compared tribological properties between different kinds of nanoparticles, and it's difficult to determine which kind of nanoparticles has the optimal tribological properties. In addition, using friction experimental apparatus to study tribological properties of lubricating fluid is a very conventional method, but judging tribological properties of lubricating fluid through the grinding experiment has problems of difficult capturing of experimental phenomena, high difficulty of experimental implementation, etc. Friction experiment is different from grinding experiment, but grinding experimental conditions can be simulated and tribological characteristics of grinding fluid can be obtained in accordance with principles of friction experiment.

Main objectives of this experiment are: firstly, an experimental study of tribological properties of six different kinds of nanoparticles is conducted through the friction experimental apparatus. Lubricating properties of nanofluids in the grinding process are studied based on the comparison with friction experiment, thus providing theoretical support and experimental verification for investigating lubricating properties of nanofluids in the grinding area.

13.2 EXPERIMENTAL DESIGN

13.2.1 Experimental Equipment and Grinding Parameters

The experiment is carried out using ball-disk reciprocating sliding friction experimental apparatus (UMT-3 Instruments, US CETR). S-3400N SEM, equipped with energy spectrum analysis (EDS) equipment, is used to observe and analyze micromorphology and chemical elements on surfaces of steel ball grinding cracks and grinding wheel. In order to reduce measuring errors, worn samples are put under ultrasonic cleaning using acetone for 10min after the experiment, and then they are blown dry and weighed. Wear extents of two frictional pairs are measured using AR224CN electronic scale (precision 0.1mg). Furthermore, DV2TLV digital viscosimeter is used to measure temperature-dependent change of dynamic

Figure 1. Diagram of the test equipments and the experimental sequence

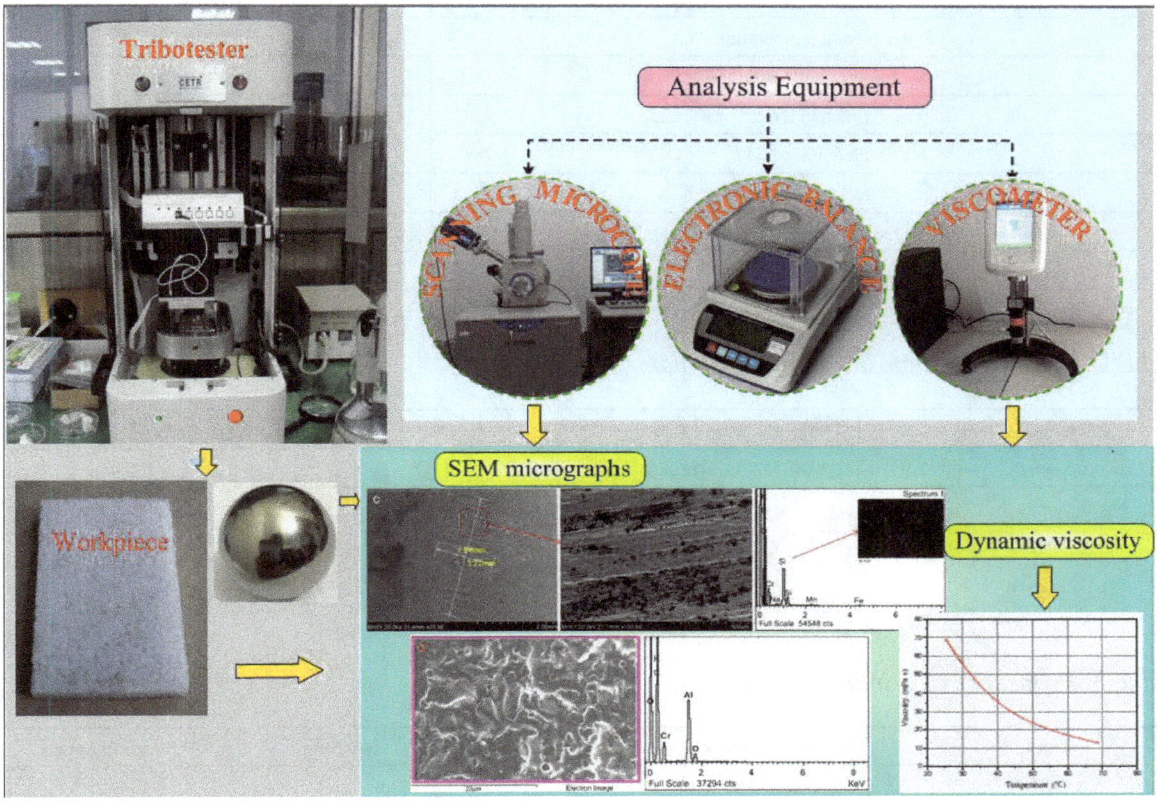

viscosity of different nanofluids (20~100°C). Figure 1 shows experimental and analytical equipment and their use order. Friction experimental parameters are seen in Table 1.

13.2.2 Experimental Materials

1.Workpiece Material

In order to keep consistency with the experiment in Chapter 12 and reach approximate grinding conditions, contact friction pairs in the experiment are 440C steel ball and grinding wheel block (grinding wheel is processed into cuboid block). Sliding between steel ball and abrasive grain and approximate experimental parameters are used to simulate lubricating state at the abrasive grain/workpiece interface in the grinding experiment. Steel and chemical compositions of steel ball and performance parameters of grinding wheel block are summarized in Table 2 and Table 3.

Table 1. Tribotesting parameters

Parameter	Value
Experimental temperature (°C)	25±5
Experimental time (s)	1200
Load (N)	10
Frequency (Hz)	2
Steel ball material	440C
Steel ball diameter (mm)	9.5
Steel ball hardness (HRC)	58-60

Table 2. Chemical compositions of 440C material

Element	C	Si	S	Cu	Mn	Mo	Cr	P	Ni
Component	0.95	1.00	remainder	0.3	1.00	0.75	16-18	0.035	0.60

Table 3. Properties of grinding wheel

Model	Dimensions (mm)	Granularity	Hardness	Bonding agent
WA80H12V	40×25×5	80 #	Soft	Ceramic

2.MQL Base Oil and Nanofluid

Selected MQL base oil and nanofluid are identical with those in Chapter 12, as introduced in section 12.2.2 in details.

13.2.3 Experimental Scheme

The experiment is an experimental study of tribological properties of six nanofluids based on the comparative grinding experiment using the friction experimental apparatus.

Seven different kinds of lubricating fluids (pure palm oil + six nanofluids) are used in the friction experiment. Experimental method is control variable method. Only category of lubricating fluid is used in each experiment, and the overall experimental design is listed in Table 4. In order to reduce experimental measurement errors, each group of experiment is conducted for three times. After the experiment ends, impurities like lubricating fluid and debris staying on the sample are removed, which is convenient for sample weighing and chemical component analysis of the surface.

Table 4. The conditions of different lubricants

Experiment No.	Lubricant
1	Pure palm oil
2	Al_2O_3 nanofluid
3	CNT nanofluid
4	MoS_2 nanofluid
5	ND nanofluid
6	SiO_2 nanofluid
7	ZrO_2 nanofluid

13.3 RESULTS AND DISCUSSION

13.3.1 Tribological Experimental Results

Each kind of nanoparticles can reduce friction coefficient between friction pairs. However, different nanoparticles cause different friction coefficients, so it's necessary to discuss about the effectiveness of anti-friction and anti-wear effects of nanoparticles.

In the grinding process, grinding wheel G ratio is a very important parameter characterizing tribological properties of lubricating fluid (Rascalha, Ribeiro, Lauro, & Brandao, 2017; Wang et al., 2017). In the friction experiment, for the sake of comparison of two friction pairs (steel ball regarded as tool and grinding wheel as workpiece), G ratio is used to express lubricating properties of lubricating fluid. According to the weighing method, wear extents of two friction pairs are measured before and after the experiment. Ratio of wear mass of grinding wheel block to that of steel ball is defined as G ratio, and then the greater the wear G ratio, the larger the relative wear extent of grinding wheel block, the smaller the wear extent of steel ball, and the better the lubricating property of the lubricating fluid. Its formula can be defined as:

$$G = \frac{m_s}{m_b} \quad (13\text{-}1)$$

where m_s is wear mass of grinding wheel block (mg); m_b is wear mass of steel ball (mg).

The relationship between friction coefficient and wear G ratio under different lubricants is shown in Figure 2. All friction coefficients present fluctuating reduction rend with time, the machining area becomes smooth after 500s, friction coefficient changes little and even declines somehow and then tends to be stable. Figure 2b shows average friction coefficients under different lubricants. When pure palm oil (PPO) is used, high friction coefficient and low wear G ratio are obtained. After nanoparticles are added, anti-friction and anti-wear property of base oil is significantly improved. In the grinding process, nanoparticles exert good anti-friction and anti-wear effects. Especially Al_2O_3 nanoparticles present the most significant effect on improving tribological properties of base oil with low friction coefficient. Compared with pure base oil, reduction of average friction coefficient reaches 19.2% and G ratio increases by 65.8%. This also indicates that Al_2O_3 nanoparticles have more significant anti-friction properties than

Figure 2. The (a) friction coefficient, (b) average friction coefficient, (c) G-ratio under different lubricants

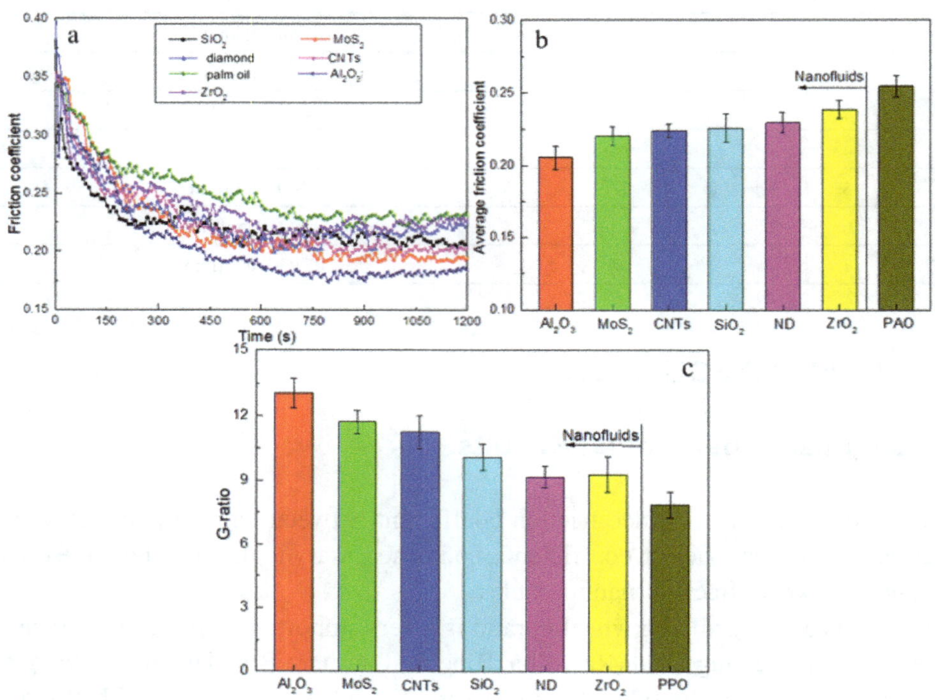

other nanoparticles and obtain the best tribological properties. On the contrary, SiO_2 nanoparticles have a high friction coefficient. In addition, friction coefficient and G ratio are approximate with small difference value under SiO_2, CNT and ND nanofluid lubrication conditions. Moreover, friction coefficient of CNT nanofluid fluctuates a lot with time, possibly because geometric characteristics of CNT itself like tubular shape (Fig. 13-8a) and easy intertwining result in incomplete oil film, and major intertwining and stacking occur on the surface of friction pairs (Yao & Yao, 2005). CNT can't have the "bearing-like" effect like spherical nanoparticles, so its anti-friction property is limited. In addition, friction coefficient of MoS_2 nanofluid is low, indicating that stratified MoS_2 nanoparticles can significantly influence anti-friction property (Hu, Xu, Hu, & Hu, 2017; Zhang et al., 2016). Low friction coefficient and high G ratio indicate that nanofluid lubricant has favorable lubricating property in the friction experiment. All nanoparticles have improved lubricating properties of base oil, not only because friction film is formed on the friction surface but also because nanofluid viscosity exerts a certain influence. Figure 3 shows viscosity-temperature curve of nanofluids. Al_2O_3 and SiO_2 nanofluids have high viscosity, thus having low friction coefficient. As the viscosity increases, intermolecular colloidal force, Brownian movement and viscous force correspondingly increase, which contribute to formation of oil film and enhancement of thickness and strength of adsorption film so as to improve lubricating properties (Sia, Bassyony, & Sarhan, 2014). It's not easy for lubricant with small viscosity to form oil film with enough thickness on the high-temperature friction surface, and oil film has small bearing capacity and can easily be dam-

Figure 3. Effective viscosities of different nanofluids with as a function of temperature

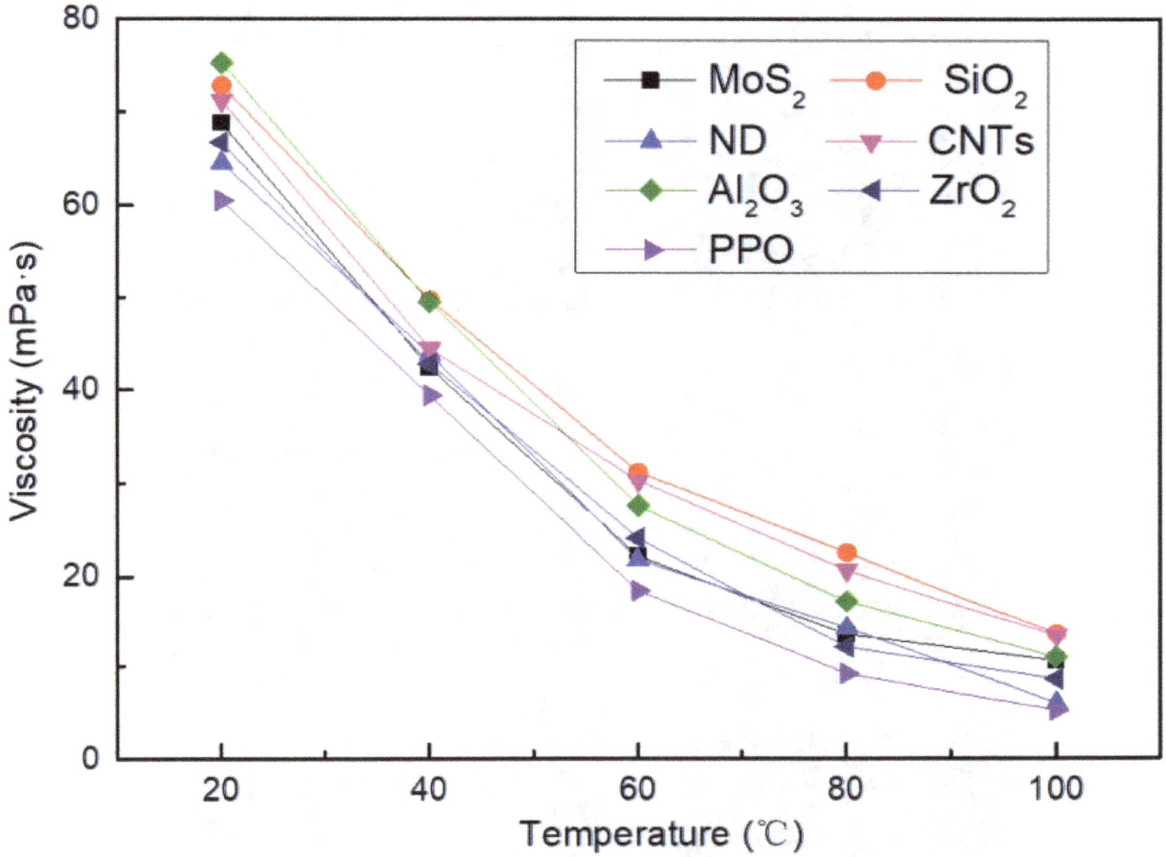

aged under load effect, which causes insufficient lubrication of friction pairs and enlargement of friction coefficient.

13.3.2 Analysis of Wear Surface Morphology and Formation of Lubrication Film

In order to study lubricating mechanisms of different nanoparticles, wear surface is analyzed through SEM-EDS. Figure 4 shows SEM images of steel ball friction surface under four different nanofluid lubrication conditions, from which different grinding crack lengths and surface morphology amplified by 100 times can be observed. Grinding crack is approximately elliptic, and grinding crack length is divided into major-axis length 2a and minor-axis length 2b.

Under nanofluid lubrication condition, MoS_2 nanofluid lubrication obtains the shortest grinding crack length, and surface of grinding crack is relatively smooth without obvious furrows and scratches (Figure 4a). Compared with other nanofluids, under ND nanofluid lubrication, the longest grinding crack length is obtained (Figure 4d), and friction surface becomes rough with large furrows and wear scars. Length of grinding crack length and quality of surface morphology obtained through SiO_2 nanofluid and CNT nanofluid are between the two. Figure 5 displays morphologies of worn surfaces of steel ball under other

Figure 4. SEM images of worn ball surfaces lubricated with different nanofluids

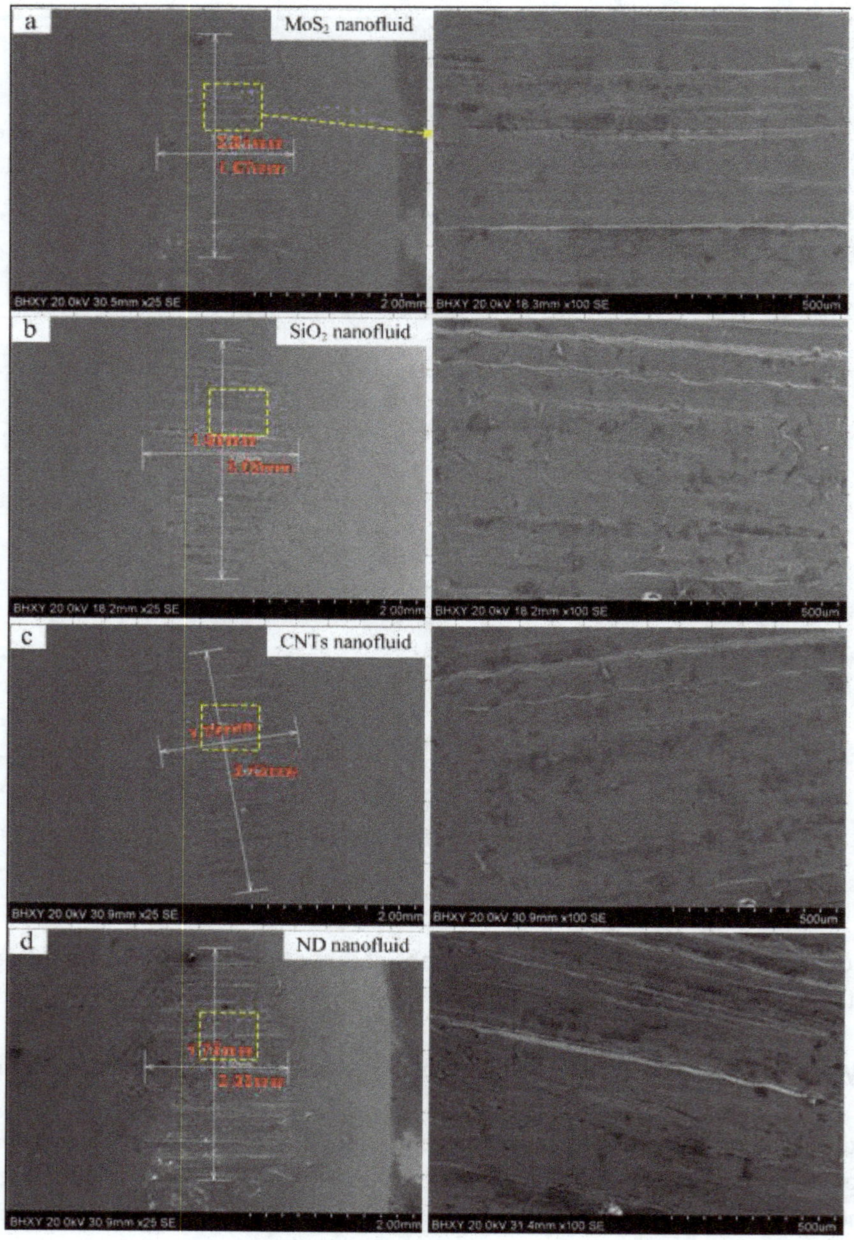

lubrication conditions. Different lubricating fluids have different lengths of grinding cracks. Furrows and pits due to plastic deformation are generated on the friction surface under pure oil lubrication, and materials fall off in large area, indicating that serious sliding injury occurs on the friction surface (Figure 5a). As base oil doesn't contain any nanoparticle, it can't form friction film on the friction surface under other lubrication conditions. Although the friction surface obtained through ZrO_2 nanofluid lubrica-

Figure 5. SEM morphology and EDS analysis of worn surface under different lubrication conditions

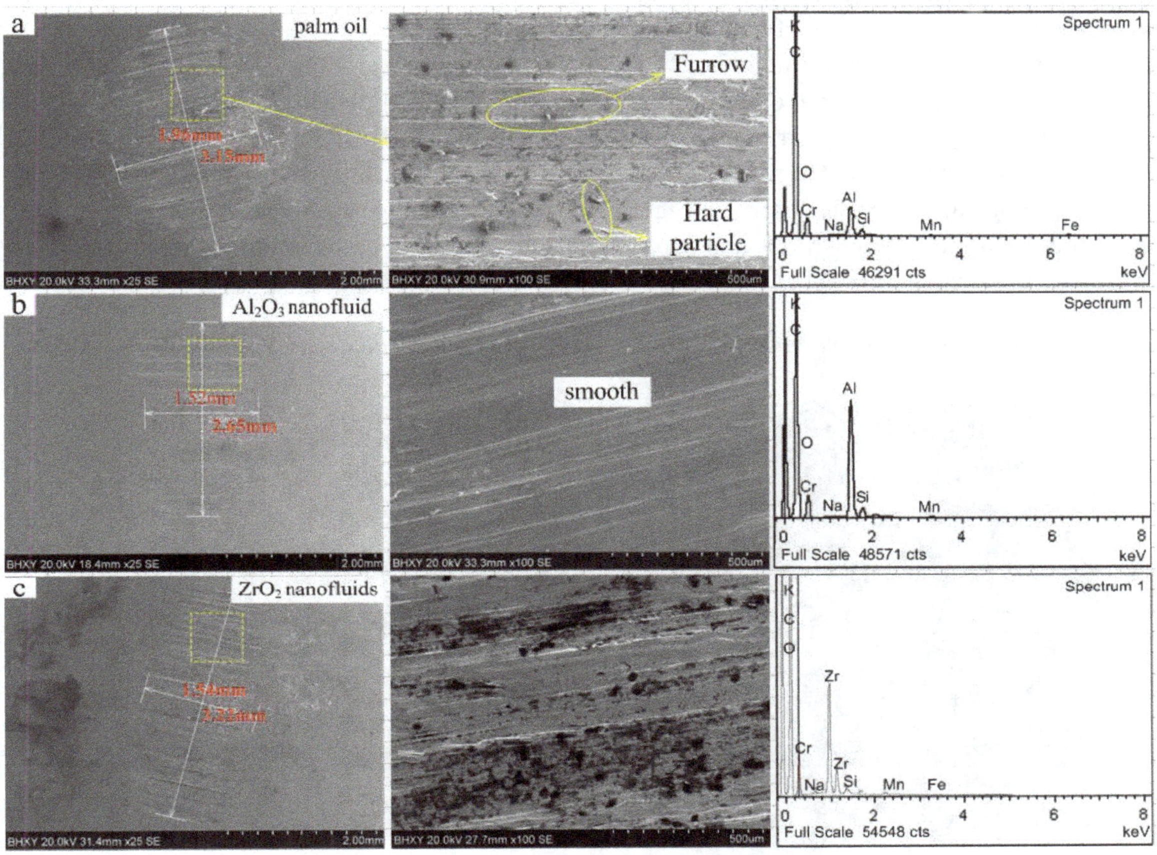

tion is better than that obtained through pure oil, but there are some furrows, freckles and other defects on the surface (Figure 5c). However, when Al_2O_3 nanoparticles are contained in the base oil, grinding crack length is smaller than those under other conditions, friction surface is the smoothest, brightest and cleanest, and there are only minor wear and small furrows (Figure 5b). This is consistent with stable lubricating effect and excellent anti-wear property of Al_2O_3 nanoparticles. Therefore, it can be obtained that all additives can significantly improve tribological properties of base oil. Moreover, Al_2O_3 and MoS_2 nanoparticles have better lubricating effects than other additives.

In order to evaluate formation of friction film on the worn surface, EDS elements on friction surface under base oil, Al_2O_3 nanofluid and ZrO_2 nanofluid lubrication conditions are analyzed as shown in Figure 5. It can be found that there are minimal quantities of Al and Si elements on friction surface under base oil lubrication condition. Comparatively speaking, through the comparison of EDS results, high contents of Al, Zr and O elements are detected from friction surface under Al_2O_3 and ZrO_2 nanofluid lubrication conditions. Nanoparticles are precipitated and spread on the friction surface to form physical friction film in the grinding process. Meanwhile, these Al, Zr and O elements constitute elements of friction film on friction surface. As the base oil doesn't contain any anti-friction additive, reducing friction only by relying on its own lubricating property can only exert the effect of reducing friction to a certain degree. As for Al_2O_3 nanofluid lubrication, Al_2O_3 nanoparticles contribute to formation of lubricating film in the friction process (Ge, Xia, & Cao, 2015; Mao et al., 2014; Yin et al., 2018). Lubricating film improves tribological properties of nanofluids.

In order to further explain formation of friction film and friction and lubrication mechanism, SEM-EDS analysis is carried out for another friction pair namely grinding wheel block surface as shown in Figure 6. Abrasive grains are protruding and sharp on the surface of grinding wheel before the experiment (Figure 6a), which can effectively reduce contact area between two friction pairs. After abrasive grain morphology after wear is observed (Figure 6b), rubdown and wear phenomena of abrasive grains exist in the worn area. When Al_2O_3 and MoS_2 nanofluids are used as lubricating fluids respectively,

Figure 6. SEM morphology and element analysis of wheel block surface using different nanofluids

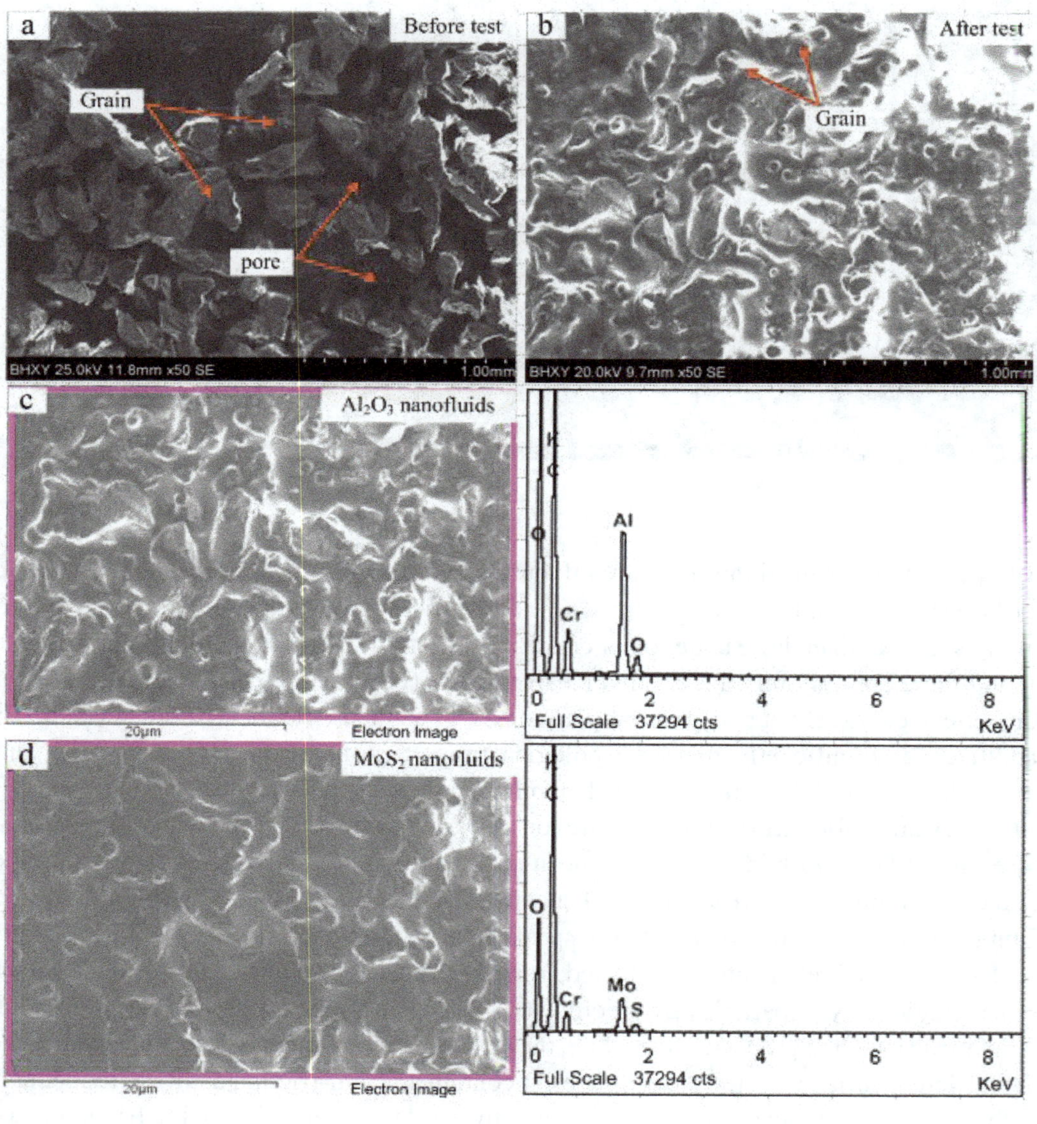

EDS results in Figure 6c and d certify that friction surface contains a large quantity of Al, Mo and S elements, because nanoparticles are transferred and precipitated onto the friction surface in the sliding friction surface, and this is good for formation of friction film. In addition, precipitated nanoparticles on the worn surface can reduce shear stress so as to reduce friction and wear. It's well known that friction film exerts an important effect on improving lubricating property between friction pairs.

Under blunting condition of abrasive grain, actual contact area between steel ball and abrasive grain is enlarged, friction occurs between friction pairs and material is removed, and at the time, nanoparticles not only exert anti-friction and anti-wear effect but also exert "bearing effect" and "stuffing effect" (Choi, Jung, Choi, Lee, & Oh, 2011; Hwang, Park, Lee, & Jung, 2006). However, when abrasive grain conducts ploughing and removes the material, friction film will be damaged. It can be observed from Figure 2a that friction coefficient change with time also indicates reduction of friction coefficient due to existence of friction film. In the fluctuation phase of friction coefficient, viscosity of lubricating fluid reduces, so is oil film thickness as the temperature increases. If oil fil thickness between friction surfaces continues to reduce, it will rupture, thus forming local dry friction and aggravating the wear phenomenon. However, with re-infiltration of nanofluid, friction film is formed again. In other words, nanoparticles can be more easily adsorbed on surface of friction pair. Adsorption film, which falls off due to wear in the friction process, can be supplemented and updated through follow-up adsorption. Additive containing nanoparticles and polar end of palm oil molecules will be physically adsorbed on surface of steel ball to form an adsorption layer (Figure 7) (Rahim & Sasahara, 2011). Afterwards, under pressure and temperature conditions, some nanoparticles experience chemical reaction with steel ball to rapidly generate chemical compounds and form chemical reaction film. Effective friction film is formed under joint action of adsorption layer and chemical reaction layer so as to reduce wear on steel ball surface and exert anti-friction and anti-wear effect.

13.3.3 Comparison of Friction-Wear and Grinding Experiment

Based on the above experimental results and analysis and through a comparative analysis between frictional wear experiment and grinding experiment, they are not only different in machining condition but

Figure 7. Schematic diagram of the formation of vegetable oil adsorption film

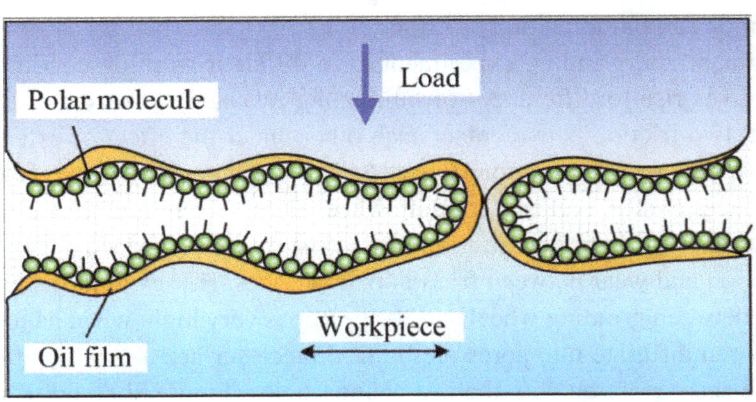

also are significantly different in lubrication mechanism. However, they are also associated, lubricating properties and film-forming mechanisms of different nanofluids are explored through the frictional wear experiment, and this can provide a theoretical analysis of lubricating properties of nanofluids in the grinding experiment.

1.Comparison of Working Conditions

In the friction experiment, tribological properties of different nanofluids are evaluated through the ball-disk friction experimental apparatus. First of all, steel ball 440C and grinding wheel block represent grinding wheel and workpiece material in the grinding experiment respectively. Secondly, pressure intensity on contact surface between steel ball and abrasive grain in the friction experiment is identical with the pressure intensity on contact surface between abrasive grain and workpiece in the grinding experiment, so friction experiment accords with grinding experiment in the aspect of working conditions.

Velocity of relative motion between friction pairs in the friction experiment is low, and the workpiece wear extent is small. More importantly, lubricating fluid only purely exerts its tribological properties while not being influenced by external factors. Therefore, it can more easily characterize tribological properties of this lubricating fluid. However, the essence of grinding machining is a random process of scratching, ploughing and cutting effects completed by abrasive grain, which are irregularly and discretely distributed on grinding wheel in quantity. The prominent feature is that grinding wheel has high circular velocity and high energy consumption. Grinding machining is usually negative rake cutting of abrasive grain, the contact time between each abrasive grain and workpiece is extremely short, so the energy consumed to remove unit volume of material in the grinding process is far greater than other machining conditions (Zhao & Wang, 2007). Especially, due to the air-bond layer formed under high-speed revolution of grinding wheel (circular velocity of grinding wheel is 30 m/s), it's difficult for a large quantity of grinding fluid to enter the grinding area. Rightly because of this, grinding experiment can't exert the effect of lubricating fluid just like friction experiment. In other words, friction experiment can better evaluate tribological properties of a lubricating fluid than grinding experiment.

2.Comparison of Lubrication Mechanisms

Even if the same lubricating fluid is used, lubricating properties will be different in different lubricating environments. Friction experiment and grinding experiment have their respective features in aspects of experimental conditions and lubrication conditions.

Grinding experiment and grinding experiment have different working conditions and lubrication conditions, which gives rise to differences of lubricating mechanisms of the same nanofluid. In the friction experiment, two friction pairs contact each other under the effect of external force and make conventional relative reciprocating motion (Figure 8a). According to the above analysis and SEM and EDS images on surfaces of steel ball and grinding wheel under different lubricants as described in Figure 4 and Figure 5, nanofluid can form friction film at the friction interface under the frictional effect so as to reduce friction and wear between friction pair surfaces. However, in the grinding experiment, as relative velocity between grinding wheel and workpiece is very high, when nanoparticles are added, some nanoparticles can infiltrate into pores on the workpiece surface under the effect of external load and then experience shear reaction with the rest nanoparticles (Figure 8b), and then a shear oil film is formed (Yan, Zhang, & Kuriyagawa, 2011). This phenomenon will not occur in the friction experiment,

Figure 8. Schematic diagram of the lubrication mechanisms in (a) grinding experiment and (b) frictional wear experiment

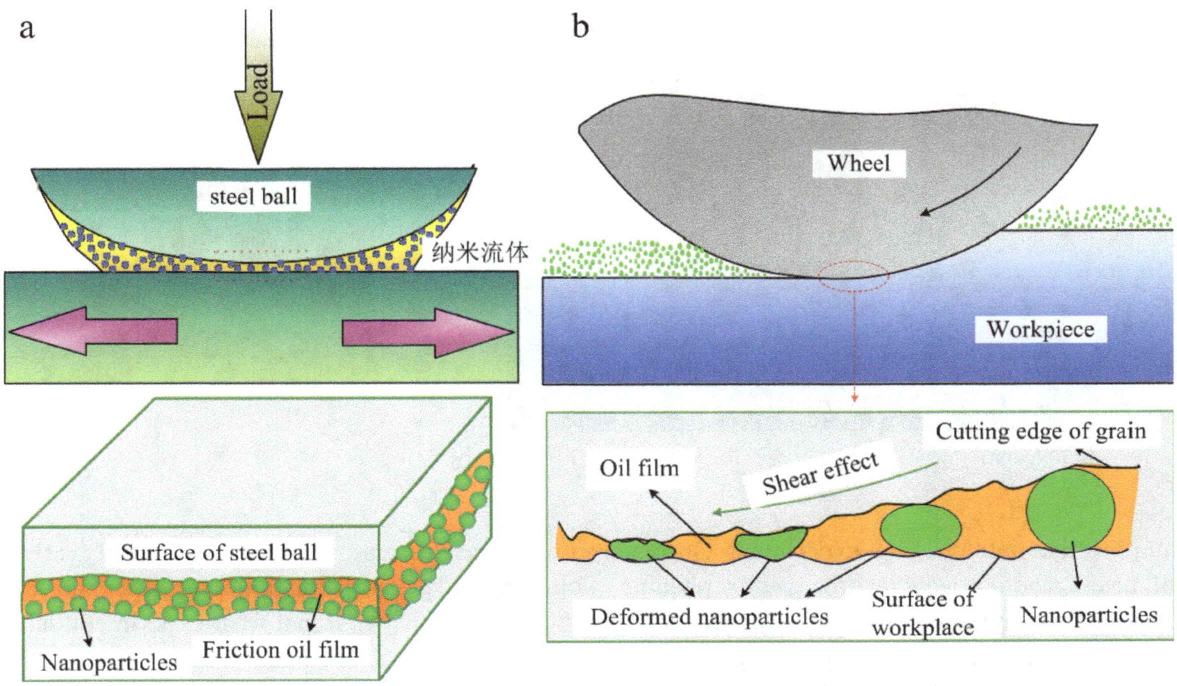

however. In short, nanoparticles can exert bearing effect and stuffing effect in the lubrication area in both frictional wear experiment and grinding experiment.

Through the analysis of working conditions and lubrication conditions of friction experiment and grinding experiment, even if the same nanofluid is used, it will present different lubricating properties in the two experiments. ND nanofluid is of cubic structure and it is spherical or ellipsoidal with extremely high hardness, and its wear resistance and mill capacity exceed all grinding materials. In the high-speed revolution of grinding wheel, ND nanofluid is brought into the wedge-shaped gap between grinding wheel and workpiece, oil wedge pressing phenomenon by dynamic fluid pressure will occur, 3D velocity filed and pressure field will be formed, and ND nanofluid obtains energy under effects of grinding wheel revolution as well as centrifugal force, fluid pressure field and velocity field. Under boundary constraining effect of grinding wheel, micro-removal machining of workpiece is completed nearby the contact area between grinding wheel and workpiece, and grinding and polishing effect is formed (Figure 9) (Li, Cai, & Li, 2005). Figure 10a shows friction surface obtained under surface grinding. On the contrary, Figure 10b shows friction surface obtained through the friction experimental apparatus. It can be seen that polishing effect exists in the grinding experiment.

In a word, friction experiment is reciprocating grinding of friction pairs using lubricating fluid for a long time, where material removal rate is low, most materials exist in the scratching form, and a layer of chemical reaction and friction oil film is formed between friction pairs through experimental verification (Figure 5 and Figure 6). For comparison, grinding experiment is a single-travel cutting process of workpiece by grinding wheel, and surfaces of friction pairs can't be always exposed in the lubricating

Figure 9. Schematic diagram of grinding and polishing effects of ND nanoparticles

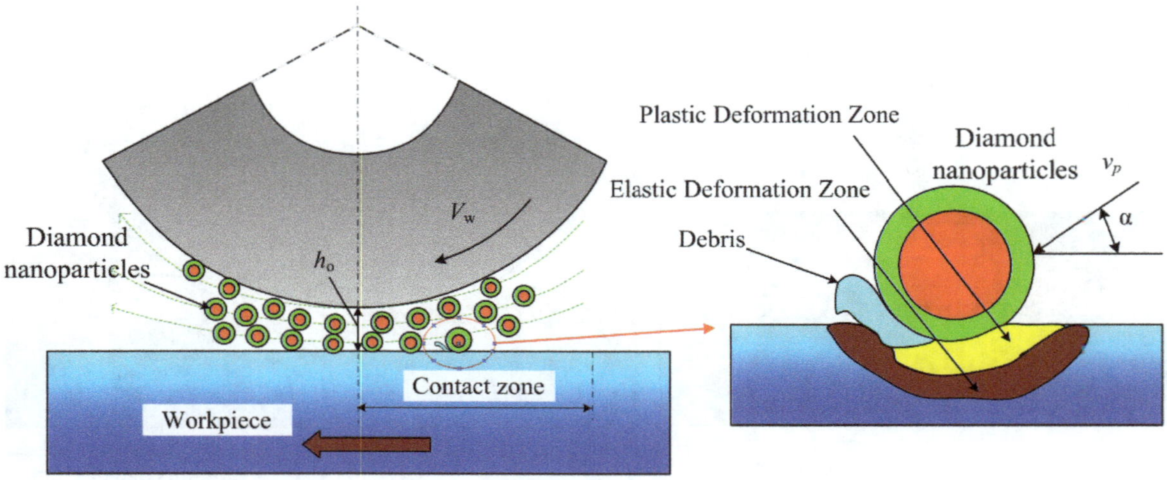

fluid just like in the friction experiment. Once a layer of material is removed, the next layer of material will be cut, and it is a cyclic process of scratching, ploughing and debris formation. If chemical reaction oil film is formed in the grinding experiment like that formed in the frictional wear experiment, it can only be formed on surface of grinding wheel (Jia, Li, Zhang, Zhang, & Zhang, 2014; Shen, Malshe, Kalita, & Shih, 2008; Shen, Shih, & Tung, 2008). During the grinding process, film-forming mechanism of nanofluid in the grinding area should be further studied according to this experiment.

Figure 10. SEM morphologies of the worn surfaces lubricated by the diamond nanofluid using surface grinding and friction experiment

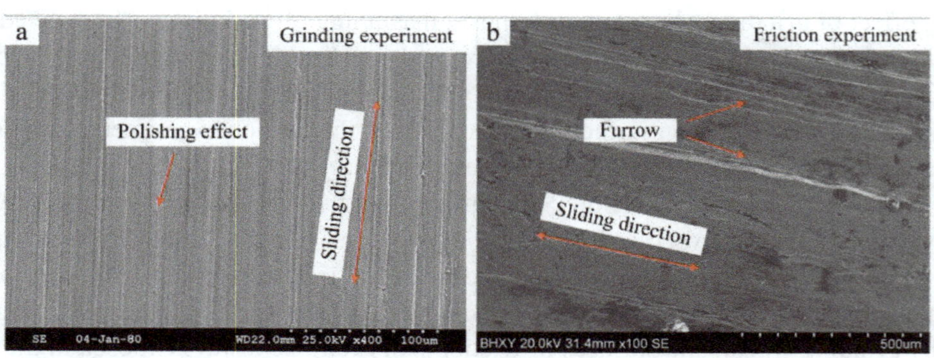

REFERENCES

Battez, A. H., Viesca, J. L., Gonzalez, R., Blanco, D., Asedegbega, E., & Osorio, A. (2010). Friction reduction properties of a CuO nanolubricant used as lubricant for a NiCrBSi coating. *Wear*, *268*(1-2), 325–328. doi:10.1016/j.wear.2009.08.018

Choi, C., Jung, M., Choi, Y., Lee, J., & Oh, J. (2011). Tribological Properties of Lubricating Oil-Based Nanofluids with Metal/Carbon Nanoparticles. *Journal of Nanoscience and Nanotechnology*, *11*(1), 368–371. doi:10.1166/jnn.2011.3227 PMID:21446457

Dambatta, Y. S., Sayuti, M., Sarhan, A. A. D., Hamdi, M., Manladad, S. M., & Reddy, M. (2019). Tribological performance of SiO_2-based nanofluids in minimum quantity lubrication grinding of Si3N4 ceramic. *Journal of Manufacturing Processes*, *41*, 135–147. doi:10.1016/j.jmapro.2019.03.024

Gara, L., & Zou, Q. (2012). Friction and Wear Characteristics of Water-Based ZnO and Al_2O_3 Nanofluids. *Tribology Transactions*, *55*(3), 345–350. doi:10.1080/10402004.2012.656879

Gara, L., & Zou, Q. (2013). Friction and Wear Characteristics of Oil-Based ZnO Nanofluids. *Tribology Transactions*, *56*(2), 236–244. doi:10.1080/10402004.2012.740148

Ge, X. Y., Xia, Y. Q., & Cao, Z. F. (2015). Tribological properties and insulation effect of nanometer TiO_2 and nanometer SiO_2 as additives in grease. *Tribology International*, *92*, 454–461. doi:10.1016/j.triboint.2015.07.031

Han, X., Barber, G. C., Zhang, Z. P., Thrush, S., Schall, J. D., Li, Z. G., & Wang, B. X. (2019). Tribological performance of oil-based ZnO and diamond nanofluids. *Lubrication Science*, *31*(3), 73–84. doi:10.1002/ls.1449

Hu, E. Z., Xu, Y., Hu, K. H., & Hu, X. G. (2017). Tribological properties of 3 types of MoS_2 additives in different base greases. *Lubrication Science*, *29*(8), 541–555. doi:10.1002/ls.1387

Hwang, Y., Park, H. S., Lee, J. K., & Jung, W. H. (2006). Thermal conductivity and lubrication characteristics of nanofluids. *Current Applied Physics*, *6*, e67–e71. doi:10.1016/j.cap.2006.01.014

Jia, D. Z., Li, C. H., Zhang, D. K., Zhang, Y. B., & Zhang, X. W. (2014). Experimental verification of nanoparticle jet minimum quantity lubrication effectiveness in grinding. *Journal of Nanoparticle Research*, *16*(12), 2758. doi:10.100711051-014-2758-7

Jiao, D., Zheng, S. H., Wang, Y. Z., Guan, R. F., & Cao, B. Q. (2011). The tribology properties of alumina/silica composite nanoparticles as lubricant additives. *Applied Surface Science*, *257*(13), 5720–5725. doi:10.1016/j.apsusc.2011.01.084

Kao, M. J., & Lin, C. R. (2009). Evaluating the role of spherical titanium oxide nanoparticles in reducing friction between two pieces of cast iron. *Journal of Alloys and Compounds*, *483*(1-2), 456–459. doi:10.1016/j.jallcom.2008.07.223

Kulkarni, H. B., Nadakatti, M. M., Patil, M. S., & Kulkarni, R. M. (2017). A Review on Nanofluids for Machining. *Current Nanoscience*, *13*(6), 634–653. doi:10.2174/1573413713666170623094121

Kumar, A., Ghosh, S., & Aravindan, S. (2017). Grinding performance improvement of silicon nitride ceramics by utilizing nanofluids. *Ceramics International*, *43*(16), 13411–13421. doi:10.1016/j.ceramint.2017.07.044

Li, C., Zhang, D., Jia, D., Wang, S., & Hou, Y. (2015). Experimental evaluation on tribological properties of nano–particle jet MQL grinding. *International Journal of Surface Science and Engineering*, *9*(2-3), 159–175. doi:10.1504/IJSURFSE.2015.068243

Li, C. H., Cai, G. Q., & Li, Q. (2005). Investigation on material removal mechanism for abrasive jet precision finishing with grinding wheel as restraint. *Zhongguo Jixie Gongcheng*, *16*(23), 2116–2119.

Li, L. S., Ren, X., Jia, W. X., Liu, J. B., & Zhu, J. (2010). Surface hydrophobic modification of ultra-fine aluminum silicate by mechanically grinding and heating. *Applied Surface Science*, *256*(22), 6824–6828. doi:10.1016/j.apsusc.2010.04.095

Liu, X. Y., Xu, N., Li, W. M., Zhang, M., Chen, L. F., Lou, W. J., & Wang, X. B. (2017). Exploring the effect of nanoparticle size on the tribological properties of SiO_2 / polyalkylene glycol nanofluid under different lubrication conditions. *Tribology International*, *109*, 467–472. doi:10.1016/j.triboint.2017.01.007

Lv, T., Huang, S. Q., Hu, X. D., Ma, Y. L., & Xu, X. F. (2018). Tribological and machining characteristics of a minimum quantity lubrication (MQL) technology using GO/SiO_2 hybrid nanoparticle water-based lubricants as cutting fluids. *International Journal of Advanced Manufacturing Technology*, *96*(5-8), 2931–2942. doi:10.100700170-018-1725-3

Lv, T., Huang, S. Q., Liu, E. T., Ma, Y. L., & Xu, X. F. (2018). Tribological and machining characteristics of an electrostatic minimum quantity lubrication (EMQL) technology using graphene nano-lubricants as cutting fluids. *Journal of Manufacturing Processes*, *34*, 225–237. doi:10.1016/j.jmapro.2018.06.016

Madanchi, N., Zellmer, S., Winter, M., Flach, F., Garnweitner, G., & Herrmann, C. (2019). Investigation on the Effects of Nanoparticles on Cutting Fluid Properties and Tribological Characteristics. *International Journal of Precision Engineering and Manufacturing-Green Technology*, *6*(3), 433–447. doi:10.100740684-019-00053-0

Mao, C., Huang, Y., Zhou, X., Gan, H., Zhang, J., & Zhou, Z. X. (2014). The tribological properties of nanofluid used in minimum quantity lubrication grinding. *International Journal of Advanced Manufacturing Technology*, *71*(5-8), 1221–1228. doi:10.100700170-013-5576-7

Mao, C., Tang, X., Zou, H., Zhou, Z., & Yin, W. (2012). Experimental investigation of surface quality for minimum quantity oil–water lubrication grinding. *International Journal of Advanced Manufacturing Technology*, *59*(1-4), 93–100. doi:10.100700170-011-3491-3

Minh, D. T., The, L. T., & Bao, N. T. (2017). Performance of Al_2O_3 nanofluids in minimum quantity lubrication in hard milling of 60Si2Mn steel using cemented carbide tools. *Advances in Mechanical Engineering*, *9*(7), 1687814017710618. doi:10.1177/1687814017710618

Pavan, R. B., Gopal, A. V., Amrita, M., & Goriparthi, B. K. (2019). Experimental investigation of graphene nanoplatelets-based minimum quantity lubrication in grinding Inconel 718. *Proceedings of the Institution of Mechanical Engineers. Part B, Journal of Engineering Manufacture*, *233*(2), 400–410. doi:10.1177/0954405417728311

Peng, D. X., Kang, Y., Hwang, R. M., Shyr, S. S., & Chang, Y. P. (2009). Tribological properties of diamond and SiO_2 nanoparticles added in paraffin. *Tribology International*, *42*(6), 911–917. doi:10.1016/j.triboint.2008.12.015

Rahim, E. A., & Sasahara, H. (2011). A study of the effect of palm oil as MQL lubricant on high speed drilling of titanium alloys. *Tribology International*, *44*(3), 309–317. doi:10.1016/j.triboint.2010.10.032

Ran, X., Yu, X. Y., & Zou, Q. (2017). Effect of Particle Concentration on Tribological Properties of ZnO Nanofluids. *Tribology & Lubrication Technology*, *73*(11), 90–95.

Rascalha, A., Ribeiro, S. L. M., Lauro, C. H., & Brandao, L. C. (2017). Effects of different cutting strategies on G-ratio in through-feed center-less grinding. *Acta Scientiarum. Technology*, *39*(1), 45–49. doi:10.4025/actascitechnol.v39i1.29224

Sharma, A. K., Tiwari, A. K., & Dixit, A. R. (2015). Progress of Nanofluid Application in Machining: A Review. *Materials and Manufacturing Processes*, *30*(7), 813–828. doi:10.1080/10426914.2014.973583

Shen, B., Malshe, A. P., Kalita, P., & Shih, A. (2008). Performance of novel MoS_2 nanoparticles based grinding fluids in minimum quantity lubrication grinding. *Trans. NAMRI/SME, 36*(357), 357-364.

Shen, B., Shih, A. J., & Tung, S. C. (2008). Application of Nanofluids in Minimum Quantity Lubrication Grinding. *Tribology Transactions*, *51*(6), 730–737. doi:10.1080/10402000802071277

Sia, S. Y., Bassyony, E. Z., & Sarhan, A. A. D. (2014). Development of SiO2 nanolubrication system to be used in sliding bearings. *International Journal of Advanced Manufacturing Technology*, *71*(5-8), 1277–1284. doi:10.100700170-013-5566-9

Virdi, R. L., Chatha, S. S., & Singh, H. (2019). Experiment evaluation of grinding properties under Al_2O_3 nanofluids in minimum quantity lubrication. *Materials Research Express, 6*(9). doi:10.1088/2053-1591/ab301f

Wang, Y. G., Li, C. H., Zhang, Y. B., Yang, M., Zhang, X. P., Zhang, N. Q., & Dai, J. J. (2017). Experimental evaluation on tribological performance of the wheel/workpiece interface in minimum quantity lubrication grinding with different concentrations of Al_2O_3 nanofluids. *Journal of Cleaner Production*, *142*, 3571–3583. doi:10.1016/j.jclepro.2016.10.110

Yan, J. W., Zhang, Z. Y., & Kuriyagawa, T. (2011). Effect of Nanoparticle Lubrication in Diamond Turning of Reaction-Bonded SiC. *International Journal of Automotive Technology*, *5*(3), 307–312.

Yao, P., & Yao, K. (2005). Friction and Wear Properties of Carbon Nanotubes as Lubricating Oil Additives. *Tribology & Lubrication Technology*, *25*(5), 394–397.

Yin, Q. A., Li, C. H., Zhang, Y. B., Yang, M., Jia, D. Z., Hou, Y. L., ... Dong, L. (2018). Spectral analysis and power spectral density evaluation in Al2O3 nanofluid minimum quantity lubrication milling of 45 steel. *International Journal of Advanced Manufacturing Technology*, *97*(1-4), 129–145. doi:10.100700170-018-1942-9

Yu, H. L., Xu, Y., Shi, P. J., Xu, B. S., Wang, X. L., & Liu, Q. (2008). Tribological properties and lubricating mechanisms of Cu nanoparticles in lubricant. *Transactions of Nonferrous Metals Society of China*, *18*(3), 636–641. doi:10.1016/S1003-6326(08)60111-9

Zhang, X. P., Li, C. H., Zhang, Y. B., Jia, D. Z., Li, B. K., Wang, Y. G., ... Zhang, X. W. (2016). Performances of Al_2O_3/SiC hybrid nanofluids in minimum-quantity lubrication grinding. *International Journal of Advanced Manufacturing Technology*, *86*(9-12), 3427–3441. doi:10.100700170-016-8453-3

Zhang, Y. B., Li, C. H., Jia, D. Z., Li, B. K., Wang, Y. G., Yang, M., ... Zhang, X. W. (2016). Experimental study on the effect of nanoparticle concentration on the lubricating property of nanofluids for MQL grinding of Ni-based alloy. *Journal of Materials Processing Technology*, *232*, 100–115. doi:10.1016/j.jmatprotec.2016.01.031

Zhang, Y. B., Li, C. H., Jia, D. Z., Zhang, D. K., & Zhang, X. W. (2015a). Experimental evaluation of MoS_2 nanoparticles in jet MQL grinding with different types of vegetable oil as base oil. *Journal of Cleaner Production*, *87*, 930–940. doi:10.1016/j.jclepro.2014.10.027

Zhang, Y. B., Li, C. H., Jia, D. Z., Zhang, D. K., & Zhang, X. W. (2015b). Experimental evaluation of the lubrication performance of MoS_2/CNT nanofluid for minimal quantity lubrication in Ni-based alloy grinding. *International Journal of Machine Tools & Manufacture*, *99*, 19–33. doi:10.1016/j.ijmachtools.2015.09.003

Zhao, C. L., Chen, Y. K., & Ren, G. (2017). A Study of Tribological Properties of Water-Based Ceria Nanofluids. *Tribology & Lubrication Technology*, *73*(3), 56–66.

Zhao, H. H., & Wang, Y. (2007). Development and Actuality of Grinding Process. *Machine Building Automation, 2*.

Zin, V., Barison, S., Agresti, F., Colla, L., Pagura, C., & Fabrizio, M. (2016). Improved tribological and thermal properties of lubricants by graphene based nano-additives. *RSC Advances*, *6*(64), 59477–59486. doi:10.1039/C6RA12029F

Chapter 14
Experimental Evaluation on Tribological Performance of the Wheel/Workpiece Interface in NMQL Grinding With Different Concentrations of Al2o3 Nanofluids

ABSTRACT

As a result of the growing need for environmental protection and the increasing number of health problems faced by workers, traditional lubricants are gradually being replaced. Nanofluids, which contain nanoparticles in the proper base fluid, can serve as a low carbon, "green" lubricant. Nanofluids show improved heat transfer capability and lubricating properties. Therefore, increasing lubricating effects is an effective way to improve machining performance. The tribological properties of grinding wheel/workpiece interface with different concentration of Al2O3 nanofluid micro-lubrication grinding were studied. The influences of the force ratio, viscosity and contact angle of Al2O3 nanofluids with different concentrations on the grinding force and the surface quality of workpieces are discussed. The best concentration of Al2O3 nanofluid with good lubrication performance in grinding zone was obtained.

14.1 INTRODUCTION

Based on grinding experimental study in Chapter Twelve and friction experimental verification in Chapter Thirteen, compared with other nanofluids, Al_2O_3 nanofluid has very good tribological properties (Zhang et al., 2016; Zhang et al., 2017). An optimized study of tribological properties under the optimal nanofluid concentration on the above basis is carried out in this chapter.

DOI: 10.4018/978-1-7998-1546-4.ch014

Shen et al. (2008, 2009) applied water-based alumina and diamond nanofluids to the grinding process of MQL, and compared the results with that of pure water grinding. In the process of nanofluid MQL grinding, a dense and hard slurry layer is formed on the surface of grinding wheel, which is conducive to improving the grinding performance. Minh et al. (2017) discussed the potential performance of Al_2O_3 nanoparticle based cutting fluid in cemented carbide milling under different minimum lubrication conditions. Hard milling with 0.5% volume fraction of nanofluid alumina under the condition of minimum lubrication has achieved good results. Yin et al. (2018) studied the spectrum analysis of force spectrum and power spectral density (PSD) of the surface microstructure. The milling force, friction coefficient, specific energy, surface roughness and the microstructure of wear debris of 45 steel under different lubrication conditions were studied. Mao et al. (2013) studied surface grinding experiments with different nanofluid parameters. The results show that the lubrication and cooling properties of the grinding zone are improved with the increase of the concentration of nanoparticles. Lee et al. (2012) used small grinding wheel to effectively achieve the micro grinding of nano fluid MQL. Nano diamond and nano alumina particles were selected, and paraffin oil was used as the base fluid. The experimental results show that the nano fluid MQL has a good effect on reducing grinding force and improving surface quality. Wang et al. (2018) evaluated the properties of three lubricants (pure palm oil, MoS_2 Nanofluids and Al_2O_3 nanofluids) and three materials (Inconel 718, nodular cast iron and AISI 1045 steel) with good lubricating properties from three aspects of force ratio, specific grinding energy and G ratio. The optimum process combination of lubricant and workpiece material under the same experimental conditions was determined by orthogonal test. Amrita et al. (2014) prepared a new type of nanofluid and evaluated its lubricity. Compared with the traditional cutting fluid, the new nano fluid has a great improvement in the lubrication performance, and showed excellent performance. Zhang et al. (2016) found that MoS_2/CNTs nanofluids MQL can improve the machining accuracy and surface quality. The high frequency continuous oscillation of autocorrelation curve of 8% MoS_2/CNTs in different concentrations of nanofluids showed that the surface quality of the workpiece is improved. Zhang et al. (2016) comprehensively analyzed the influence of the physical properties (viscosity, surface tension) of vegetable oil on the cooling effect. The mechanism of the influence of viscosity and surface tension of nanofluids on cooling performance was discussed. Low viscosity and surface tension vegetable oil has good cooling performance, high viscosity and surface tension vegetable oil has good lubrication performance. Lee et al. (2012) added Al_2O_3 nanoparticles and diamond nanoparticles to paraffin oil to prepare nanofluids for MQL grinding. It is found that MQL can effectively reduce the grinding force and improve the surface quality, and different volume fractions of nanoparticles have a great impact on the grinding performance. Setti et al. (2015) used different volume fractions of Al_2O_3 and CuO nanofluids MQL to grind Ti-6Al-4V. By analyzing the surface quality of workpiece and the shape of grinding wheel. It is found that MQL can significantly reduce the tangential grinding force and grinding temperature, and the grinding performance of Al_2O_3 nanofluid is better than that of CuO nanofluid. Zhang et al. (2018) proposed a new green processing technology, namely the Cryogenic air nanofluid minimum quantity lubrication (CNMQL) in low temperature air. Compared with traditional processing methods and other green technologies, low-temperature air nanofluids have many advantages, such as good economy, low carbon utilization rate, good energy saving, good cooling and lubrication performance. Chatha et al. (2016) found that the excellent performance of nmql can be attributed to the rolling effect of nanoparticles and excellent cooling performance, which results in low friction at the tool / chip and tool workpiece interface. Seyedzavvar et al. (2019) studied the lubrication and cooling performance of environmental friendly graphite nanofluids in MQL grinding, and evaluated the performance of the synthesized nanofluids as lubricants under different grinding parameters. Kim et al. (2017) found that the nanofluid MQL cooled by cold CO_2 gas was effective in reducing milling force, friction coefficient, tool wear and surface roughness. The research also shows that the lower concentration (0.1 wt.%) of nano diamond fluid is more effective to reduce milling force, friction coefficient and tool wear. By mixing alumina based nanofluids with graphene nanoflakes (GNP), Singh et al. (2017) developed mixed nanofluids (NF) with better thermal and tribological properties at volume concentrations of 0.25, 0.75 and 1.25% by volume. The results show that the thermal conductivity and viscosity of composite nanofluids increase with the increase of nanoparticle concentration, but the thermal conductivity is lower than the component and the viscosity is between the components. Dambatta et al. (2019) found that with the increase of nanofluid concentration, grinding force decreased significantly and surface quality improved.

A proper quantity of nanoparticles is added into the base oil to prepare nanofluids, and it's already proved that they can more effectively improve anti-friction and anti-wear properties of lubricating fluid compared with traditional solid lubricant additives (Mao et al., 2014). Nanofluids have been extensively applied to fields like biotechnology, chemical engineering, heat exchanger and lubricating oil and become the research emphasis in the disciplines of materials, physical chemistry, heat transfer theory and tribology. Scholars have not conducted systematic studies of tribological property parameters of different concentrations of nanofluids previously especially under high-speed revolution of grinding wheel in the grinding process. Therefore, related analysis is carried out in this chapter.

14.2 EXPERIMENTAL DESIGN

14.2.1 Experimental Equipment and Grinding Parameters

Equipment and grinding parameters used in this experiment are identical with those in Chapter Eleven, namely: K-P36 NC plane grinder, dynamometer, grinding wheel, MQL device, roughness measuring instrument, digital viscosimeter and SEM as well as contact angle measuring instrument.

14.2.2 Experimental Materials

Experimental materials selected in this experiment mainly include: workpiece material, MQL base oil and nanofluid.

1.Workpiece Material and MQL Base Oil

In order to keep consistency with the experiment in Chapter 13, workpiece material used in the experiment is still nickel base alloy 718 and base oil is palm oil.

2.Nanofluid

Al_2O_3 nanoparticles are selected as additives in this experiment (Mao et al., 2014). According to experimental studies in Chapter 12 and Chapter 13, Al_2O_3 nanoparticles are nontoxic to human body with spherical shape and high hardness, which can improve lubricating properties. Physical properties of

Table 1. Physical properties of α-Al_2O_3 nanoparticles

Property	Value
Shape	Spherical
Purity (%)	99.9
Particle size (nm)	50
Apparent density (g/cm^3)	0.33
Specific surface area (m^2/g)	30.21
Heat conductivity (W/m K)	36

Al_2O_3 nanoparticles are listed in Table 1. Figure 1a shows SEM image of Al_2O_3 nanoparticles. In the experiment, Al_2O_3 nanoparticles with different volume fractions are added into the base oil to prepare Al_2O_3 nanofluids with different volume concentrations, where nanofluid preparation method refers to contents introduced in section 12.2.2.

In the preparation process of Al_2O_3 nanofluid, when volume fraction is smaller than 2.0%, Al_2O_3 nanoparticles can stably exist in the base oil even 48h later. When volume fraction exceeds 2.0%, some nanoparticles are precipitated 2h later, but they are enough to meet experimental time. When volume concentration exceeds 3.0%, nanoparticles will go through serious agglomeration phenomenon. Figure 1b shows nanofluid pictures when concentration of Al_2O_3 nanoparticles is 2.0%.

14.2.3 Experimental Scheme

This chapter is an experimental study of tribological properties at the grinding wheel/workpiece interface under Al_2O_3 NMQL grinding conditions with different concentrations, and the nanofluid concentration with the optimal tribological properties is selected according to experimental results. Univariate method is used in this experiment, other factors are consistent, and only volume concentration of nanofluid is changed. Different concentrations of Al_2O_3 nanofluid are used for MQL grinding of nickel base alloy workpiece. Through an analysis of macroscopic grinding property parameters (force ratio, specific grinding energy and G ratio) and microscopic grinding property parameters (abrasive grain, abrasive

Figure 1. (a) The SEM micrographs of Al_2O_3 nanoparticles and (b) CNTs nanofluid

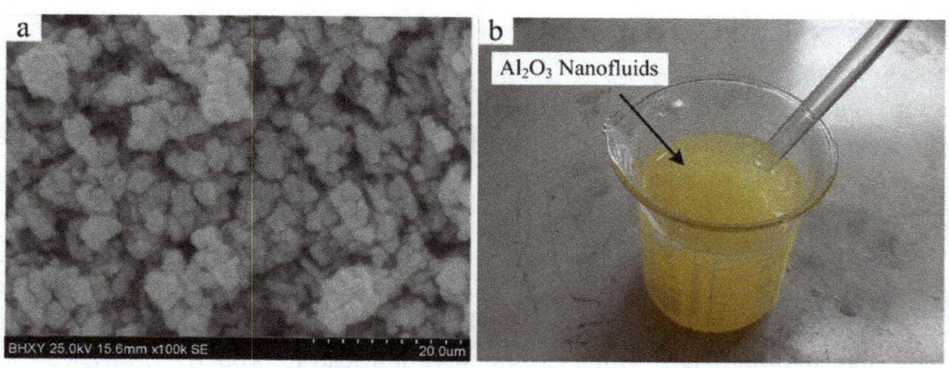

debris and workpiece surface morphologies and nanofluid viscosity and contact angle), a systematic experimental study of tribological properties at the grinding wheel/workpiece interface is carried out. The optimal additive amount of Al_2O_3 nanoparticles is selected.

Volume concentrations of nanofluid are 0%, 0.5%, 1.0%, 1.5%, 2.0%, 2.5%, 3.0%, 3.5% and 4.0% respectively, where pure palm oil has 0% volume fraction, and overall experimental program is listed in Table 2.

14.3 RESULTS AND DISCUSSION

14.3.1 Macro Parameters of Grinding Performance

Figure 2a, b and Figure 3 show force ratio, specific grinding energy and G ratio, all of which are macroscopic parameters used to evaluate tribological properties at the grinding wheel/workpiece interface. Each drawn data point is average value obtained after 100 grinding travels, and error bar represents standard deviation of this group of values.

Low force ratio reflects good lubricating state in the grinding area. Under different volume concentrations of nanoparticles (Figure 2a), pure oil MQL has the highest force ratio. With addition of nanoparticles, force ratio is obviously reduced. As nanoparticles have good anti-friction and anti-wear effects, they can significantly improve lubricating property in the grinding area so as to reduce grinding force. In addition, grinding force can be reduced by enhancing anti-pressure ability of lubricating oil film and increasing heat conductivity of Al_2O_3 nanoparticles. When nanoparticle concentration continuously increases, these nanoparticles can infiltrate into pores on workpiece surface (Figure 14a) and go through shear reaction with the rest nanoparticles, thus forming shear oil film. After excessive nanoparticles are added, agglomerated nanoparticles will impact the rough workpiece surface and thus affect its surface quality and generate high grinding force (Gao et al., 2019; Zhang, Li, Jia, Zhang, & Zhang, 2015). Therefore, this is why improper nanoparticle concentration will degrade grinding performance.

Table 2. Experimental program

Experiment No.	Lubricating fluid	Volume concentration (%)	Nanoparticle weight in 100 *ml* nanofluid (g)	SDS weight (g)	Lubrication condition
1	Pure palm oil	0	0	0	Pure oil MQL
2	Al_2O_3 nanofluid	0.5	0.165	0.0165	NMQL
3		1.0	0.33	0.033	
4		1.5	0.495	0.0495	
5		2.0	0.66	0.066	
6		2.5	0.825	0.0825	
7		3.0	0.99	0.099	
8		3.5	1.155	0. 1155	
9		4.0	1.32	0. 132	

Figure 2. Effect of different nanoparticle concentrations on (a) force ratio and (b) specific energy

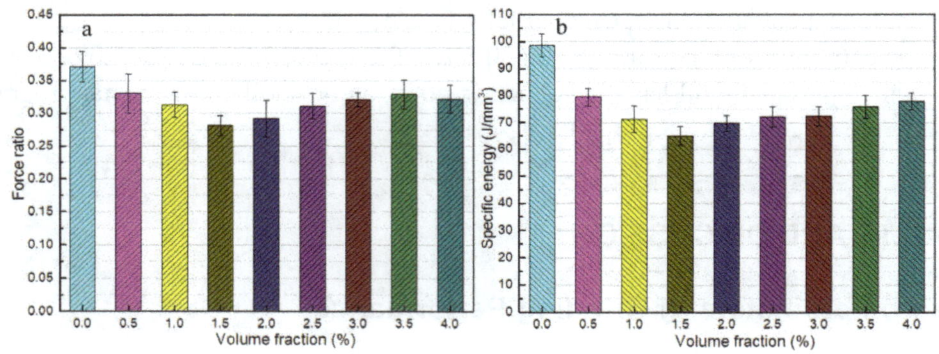

Figure 3. Effect of different nanoparticle concentrations on G-ratio

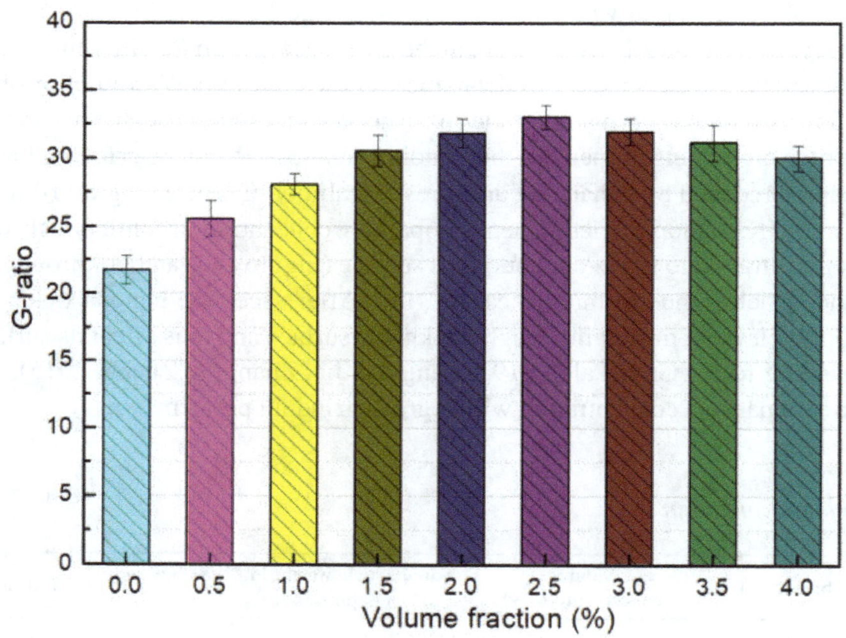

Figure 2b shows that specific grinding energy presents similar change trend to force ratio. The highest specific energy consumed by pure oil MQL is 98.61 J/mm^3, and specific energy is significantly reduced when NMQL condition is adopted. Especially for nanofluid with 1.5 vol% concentration, energy consumption is reduced by 34.1% when compared with pure oil MQL condition. As when Al_2O_3 nanofluid is sprayed onto grinding wheel/workpiece interface, one part will be spread on workpiece surface and another part is adhered to grinding wheel surface to exert the lubricating effect, and sliding friction force is reduced by lubricating already blunted abrasive grain. However, lubrication condition can't be completely reflected only by relying on force ratio and specific grinding energy. Figure 3 shows G ratio, and high G ratio means low wear rate of grinding wheel. Pure oil MQL obtains the lowest value, being 27.9 with serious wear of grinding wheel. Comparatively speaking, wear rate of grinding wear is significantly reduced under NMQL condition. Nanofluid exerts the significant lubricating effect, and furthermore, abrasive grains infiltrated into the nanofluid (Figure 4b) can effectively remit blunting.

Low force ratio and specific grinding energy and high G ratio indicate that NMQL has high machining efficiency. Tribological properties are not improved as the nanoparticle concentration increases. According to Figure 2 and Figure 3, lubricating property of nanofluid firstly increases and then declines as the nanoparticle concentration rises, and the optimal values are reached under 1.5 vol% nanoparticle concentration (force ratio: 0.28; specific energy: 64.95 J/mm^3; G ratio: 30.56). This indicates that proper nanoparticle concentration can improve tribological properties of the nanofluid, but too high nanoparticle concentration is not good for improving tribological properties on the contrary. Because if nanoparticle concentration is too high, nanoparticles in the suspension will collide and agglomerate in the motion process thus forming particle clusters and causing agglomeration. 3.5 vol% and 4.0 vol% Al_2O_3 nanofluids have high force ratio and specific energy and low G ratio, possibly because high-concentration Al_2O_3 nanoparticles cause blocking of grinding wheel, reduce lubricating properties of nanofluid and then degrades cutting performance of abrasive grains.

14.3.2 Microstructure Characterization and Morphology Analysis

1. Grinding Wheel Morphology and Debris Formation

Figure 4 shows surface morphologies of grinding wheel before and after NMQL grinding. Abrasive grains on the dressed grinding wheel (Figure 4a) are more protruding and sharper, which can effectively reduce contact area between abrasive grain and workpiece in the grinding area. Large pores can not be easily blocked and their durability and cutting performance are good. Figure 4b shows the nanofluid retained on the surface of grinding wheel. Pores or pore capillaries on grinding wheel are just like reservoirs which can temporarily store the nanofluid. This ensures that pores passing through the surface of grinding wheel can carry Al2O3 nanoparticles into the grinding area (Kalita, Malshe, & Rajurkar, 2012). In addition, high-porosity grinding wheel can carry more nanofluid so that there will be more lubricant sources in the grinding area. Moreover, nanofluid coated on the surface of abrasive grain form a layer of Al2O3 nanofluid lubricating film through physical or chemical reaction, which can effectively hinder abrasive grain from being rubbed down and fall off (Malkin & Guo, 2008). More importantly, spherical Al2O3 nanoparticles exert “bearing-like” effect in the grinding area (Figure 13) and hinder direct contact between two friction pairs. Original sliding friction between abrasive grain and workpiece is turned into rolling friction, rubdown of abrasive grain is remitted by reducing the contact area between abrasive grain and workpiece so as to reduce force ratio, and this contributes to low energy consumption and low wear rate of grinding wheel. SEM-EDS energy spectrum analysis (Figure 4b) shows that there are Al and O elements on the surface of abrasive grains, proving that there are Al2O3 nanoparticles on pores and abrasive grains, and a lubricating oil film is formed under external temperature and pressure. High-pressure gas MQL technology can not only clean the surface of grinding wheel and prevent blocking of grinding wheel but also nanoparticles can help to remove impurities retained on workpiece surface texture like debris, and therefore, benign working state of grinding wheel can be kept (Wenfeng, Ze, Jiuhua, Ding Shubao, & Yang, 2011).

Figure 4. SEM-EDS surface analysis of abrasive grains before grinding and after grinding with nanofluid

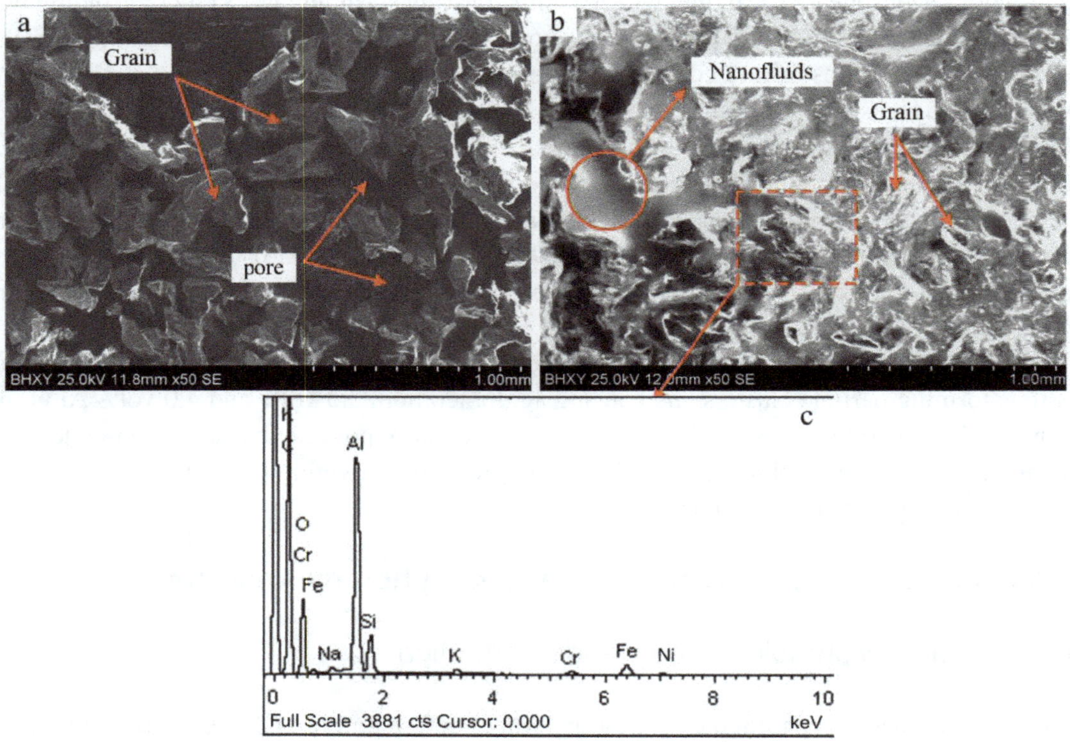

Figure 5 displays grinding debris collected under NMQL conditions with different concentrations. Relative to pure oil MQL, grinding debris obtained under NMQL condition is not that different, debris are thin and long under disperse distribution and their mutual intertwining phenomenon is not serious as shown in Figure 5 (b-i). Short and small grinding debris shown in Figure 5 (g-i) are formed mainly because nanoparticle concentration is too high, which causes blocking of grinding wheel and reduces sharpness of abrasive grain.

The contact surface where grinding debris is divorced from abrasive grains is front face of grinding debris and the other surface is rear face of grinding debris. Relative to rear face, front face of grinding debris is smooth, and rear face is ripple-shaped due to extruding shear force in the cutting process. Figure 6a shows grinding debris micrograph under pure oil MQL condition. Compared with other grinding debris, on the one hand, the grinding debris are mutually intertwined and present a cluster shape. On the other hand, grinding debris is coarse and short. Front face of grinding debris is coarse with obvious scratches by abrasive grains. Figure 6 shows grinding debris obtained through 2.0 vol% Al_2O_3 NMQL condition have better shape and surface morphology than those obtained with other volume concentrations. This means that under 2.0 vol % Al_2O_3 NMQL condition, intertwining phenomenon between grinding debris is less and they are long, thin and banding shaped with smooth appearance, indicating that deformation force is small in the grinding process, and abrasive material fits in well with workpiece material. Grinding is typical negative rake machining. As grinding velocity is high, deformation of grinding debris is more approximate to adiabatic shear (Oliveira et al., 2012). As Al_2O_3 nanofluid has considerable lubricating

Figure 5. SEM images of grinding debris obtained through Al_2O_3 NMQL with different concentrations

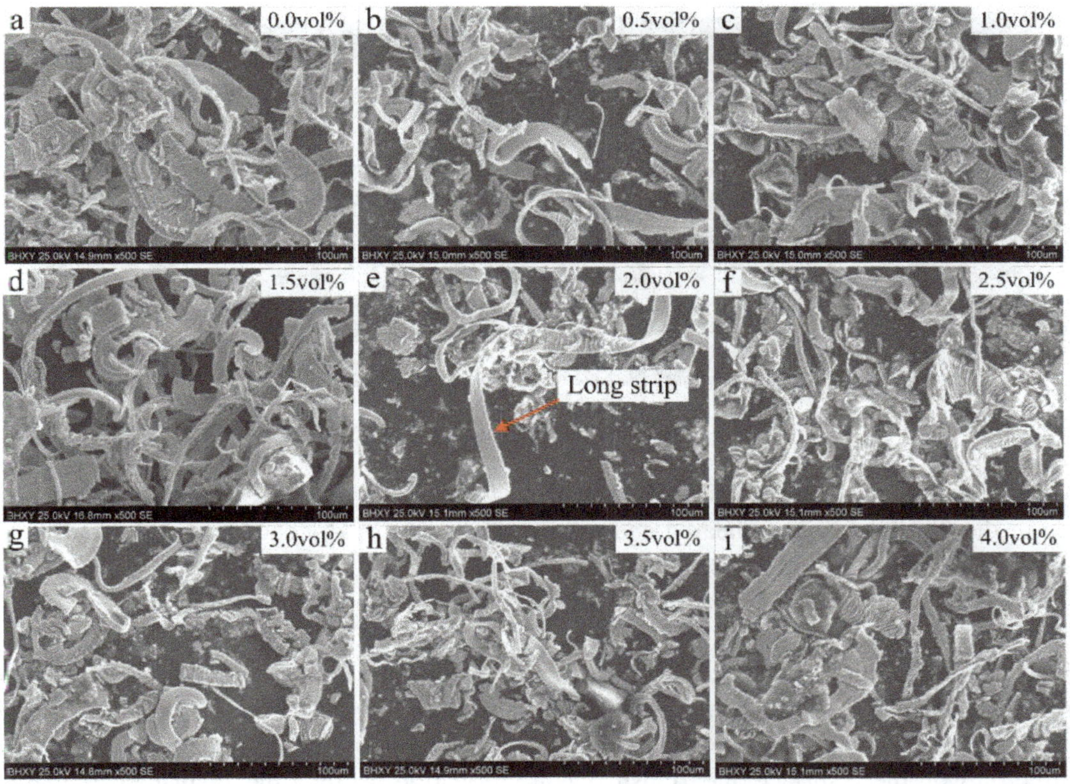

effect and nanoparticles have favorable anti-friction and anti-wear effects, nanofluid infiltrating into the abrasive grain/chip interface can exert a good lubricating effect in the abrasive grain cutting process. Under the effect of grinding force in the machining process, grinding debris has already crimped. Force ratio and specific energy consumption are both small under good lubricating state.

2. Analysis of Surface Roughness and Surface Morphology

Figure 7 shows typical SEM graph and roughness value of workpiece under MQL grinding conditions with different nanofluid concentrations. From the integral analysis, micromorphology of workpiece surface is corresponding to its roughness value. Different roughness values exist under different nanofluid concentrations: roughness value under pure oil MQL condition is the highest, and correspondingly, the surface morphology is the poorest. Under NMQL condition (Figure 7b-i), there are almost no large furrows or plastic deformation layer on the workpiece surface, and surface roughness value is small, indicating that nanoparticles can effectively improve workpiece surface quality by reducing adhesion between abrasive grain and workpiece by virtue of their lubricating effect. Obviously, surface roughness value is the minimum under 2.0 vol% Al_2O_3 nanoparticles, meaning that the best workpiece surface quality is obtained under this working condition. However, surface quality declines as the concentration increases. High-concentration nanoparticles are precipitated into incomplete part of workpiece surface, and these

Figure 6. SEM micrographs of grinding debris under pure oil and 2.0 vol% Al_2O_3 nanofluids MQL

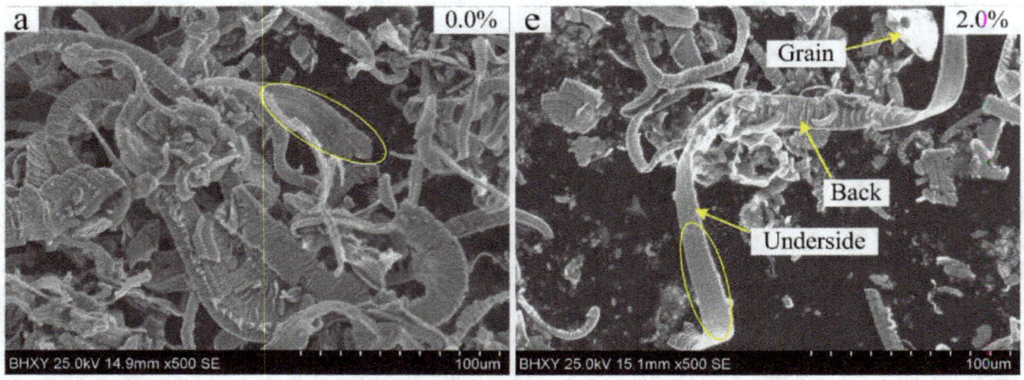

nanoparticles experience intense shear effect with newly sprayed nanoparticles so that nanoparticles retained in the thin oil film are sheared and fall off. Therefore, high-concentration nanoparticles are adverse to formation of a good surface on the contrary (Rahmati, Sarhan, & Sayuti, 2014).

Figure 8a and e show SEM-EDX images of workpiece surface under pure oil MQL and 2.0 vol% NMQL conditions. It can be found that under pure oil MQL condition, surface morphology is more complicated, and plastic deformation layer, furrow and material precipitation are serious on the surface. Under a large magnification factor, these surface defects are clearer. When abrasive grains at the bottom of the surface of grinding wheel is embedded into workpiece surface to conduct cutting, furrows are caused on workpiece surface due to insufficient lubrication. At the time, displacement and plastic ploughing occur between metallic crystals of workpiece under high temperature and high pressure, and metallic particles are re-precipitated to two furrow sides of workpiece by adhering to abrasive grains, thus forming a plastic deformation layer. Under 2.0 vol% NMQL condition, surface texture is smooth and clean with the best surface morphology and the lowest surface roughness value. Nanoparticles are precipitated onto workpiece surface to form a very thin lubricating oil film which exerts anti-friction and anti-wear effect. Through EDX component analysis, after NMQL is conducted, compared with pure oil MQL, O and Al contents obviously increase, indicating that Al_2O_3 nanoparticles are transported and precipitated onto workpiece surface, further proving existence of oil film.

14.3.3 Influence of Dynamic Viscosity and Contact Angle

Nanofluid viscosity is generated mainly because irregular movement of nanoparticles and base oil molecules generate interactive force between molecules. As nanoparticles are very small and their Brownian movement is fierce, internal energy exchange in the nanofluid flow process significantly increases, which enlarges nanofluid viscosity (Lee et al., 2008; Ni, Yang, & Wu, 2019). Relative motion between grinding wheel and workpiece will generate shear force of grinding fluid in the grinding area, and internal frictional force is generated between internal fluid layers. On the one hand, microscopic increase of viscosity will increase internal friction force between internal fluid layers, thus enlarging tangential force and decreasing force ratio in the abrasive grain grinding process. From the macroscopic scale, nanofluid is sprayed into the grinding area in the form of droplets. High-viscosity grinding fluid

Figure 7. SEM micrographs of workpiece surface under different concentrations of Al_2O_3 nanofluids

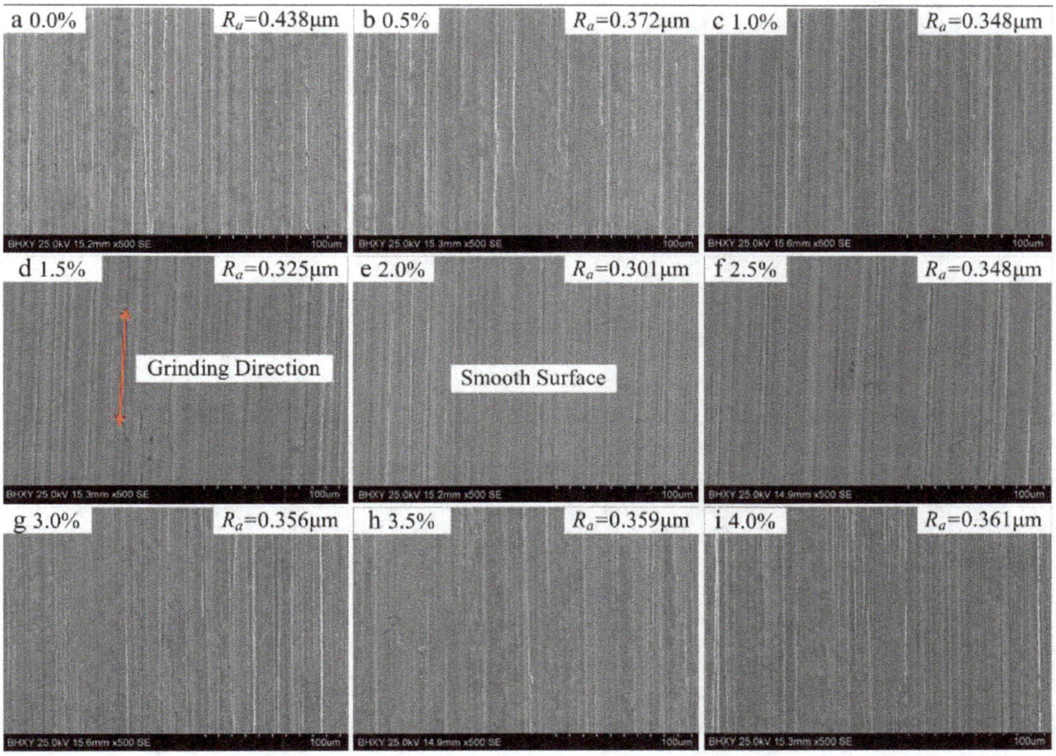

has greater ability in hindering its own flow trend, thus lengthening lubrication time between tool and workpiece and effectively preventing rapid tool wear.

Figure 9 shows that nanofluid viscosity presents nonlinear growth as volume fraction of nanoparticles increase. Viscosity reaches 67.04 mPa·s when volume concentration is about 2.5%, and afterwards, increase amplitude of concentration is significantly reduced, and viscosity presents a fluctuating trend within a small scope, because too high nanoparticle concentration results in agglomeration phenomenon of nanoparticles, nanoparticles in the suspension collide and aggregate in the random movement process, and particle clusters are formed. Some nanoparticles lose dynamic stability and are precipitated (Demas, Timofeeva, Routbort, & Fenske, 2012). Therefore, viscosity presents a fluctuating trend after nanoparticles reach saturation degree in the nanofluid, and at the time, viscosity increase is not good for improving lubricating properties of the nanofluid on the contrary.

Wettability is used to describe abilities of the fluid in spreading, infiltrating and covering the tool and workpiece. The smaller the contact angle of droplets, the better the wettability, and the larger the infiltration area of droplets with the same volume on the surface (Zhang, Cheng, Xing, Li, & Zhu, 2011). Larger infiltration area means that grinding fluid can effectively cover a larger area so as to generate better lubricating effect. Figure 10 shows basic transformation relations among droplet radius, contact angle and wetting area.

Figure 8. SEM-EDS analysis of workpiece surface under pure oil MQL condition and 0 vol% and 2.0 vol% Al_2O_3 NMQL conditions

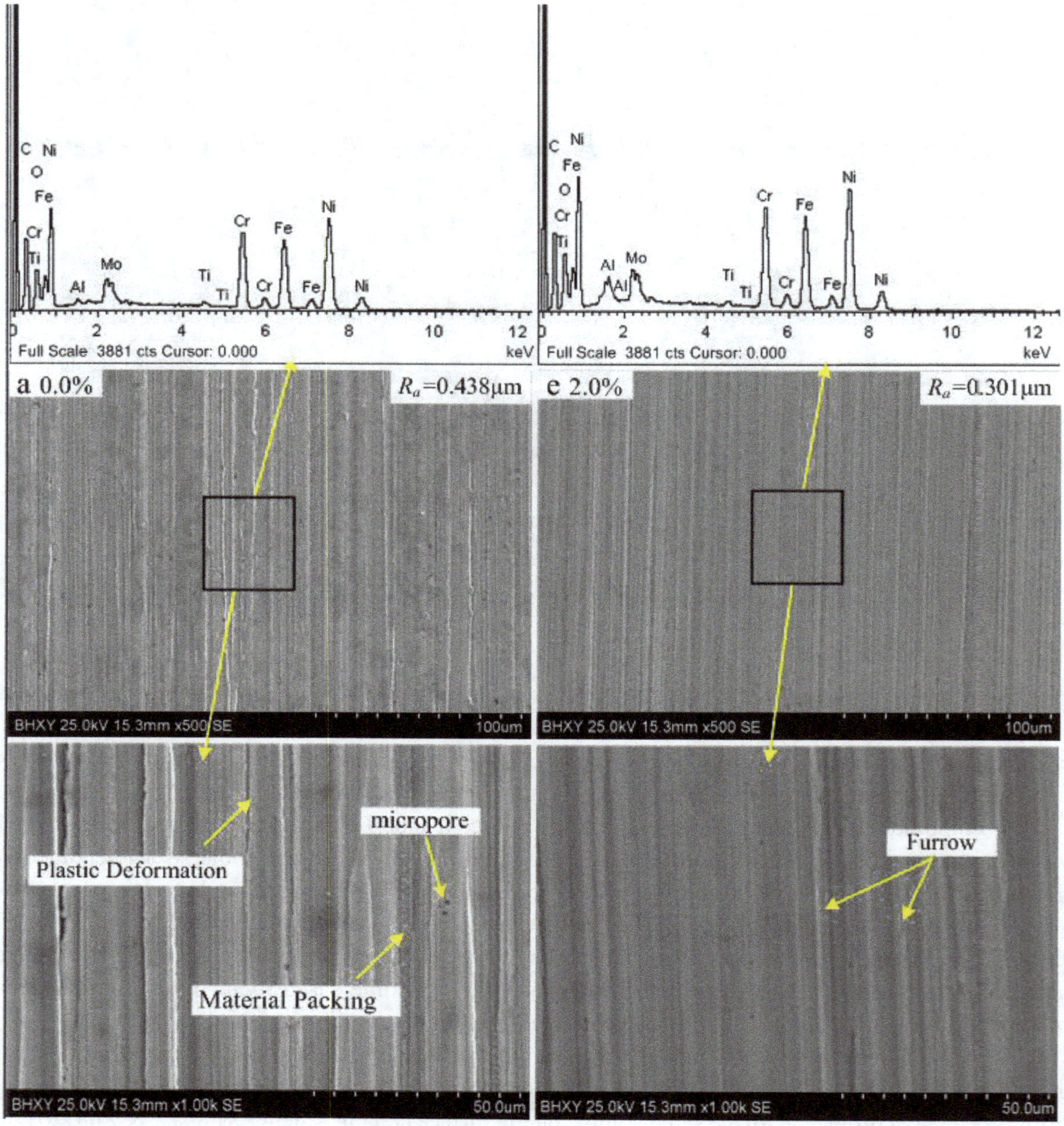

Figure 9. Variation of dynamic viscosity with different volume fractions of nanoparticles (25°C)

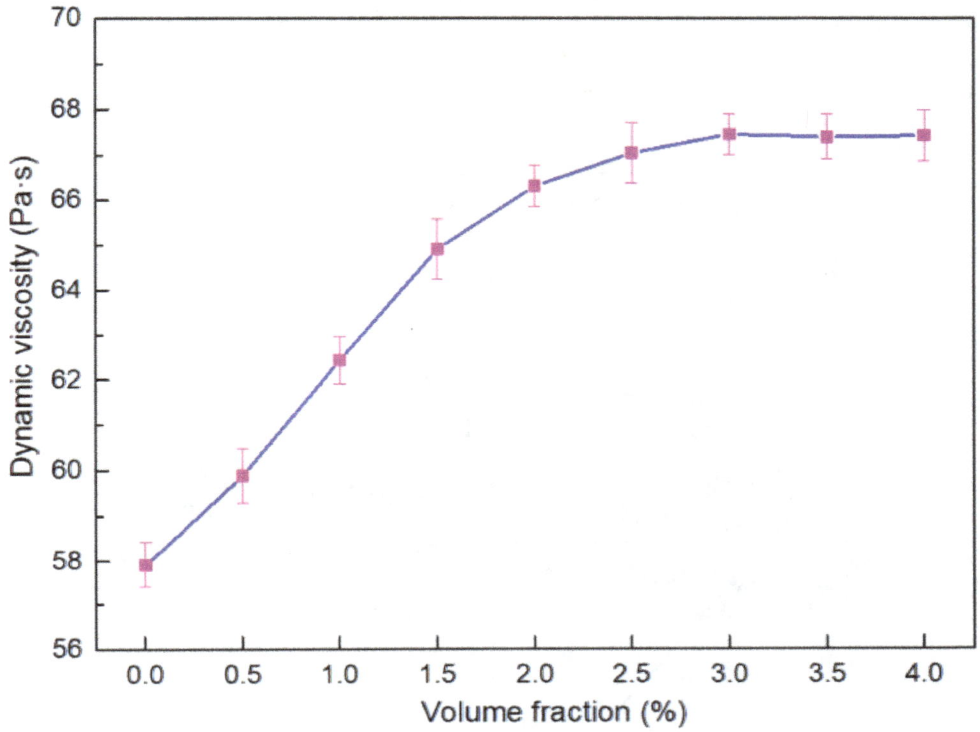

In the analysis of droplet diameter change of nanofluids with different concentrations, assume volumes of droplets with radius of r_{drop} in the sprayed nanofluid are all equal to:

$$V_{drop} = \frac{4}{3}\pi r_{drop}^{3} \tag{1}$$

Volumes V_{drop} of droplets keep unchanged before and after impacting the workpiece surface. Diameter of contact area between deformed liquid and surface is r, and the following can be obtained according to the law of conservation of mass (Sai, ManojKumar, & Ghosh, 2015):

$$r = \sqrt[3]{\frac{V_{drop}}{\frac{\pi}{2}\frac{1}{tg\theta}\left(\frac{1}{\cos\theta}-1\right)\left[1+\frac{1}{3}\frac{1}{tg^{2}\theta}\left(\frac{1}{\cos\theta}-1\right)^{2}\right]}} \tag{2}$$

where V_{drop} is volume of droplets before impacting; θ is contact angle; t is surface tension; g is gravitational acceleration.

Figure 10. Changes of droplet shape and wetting area from the air to plane

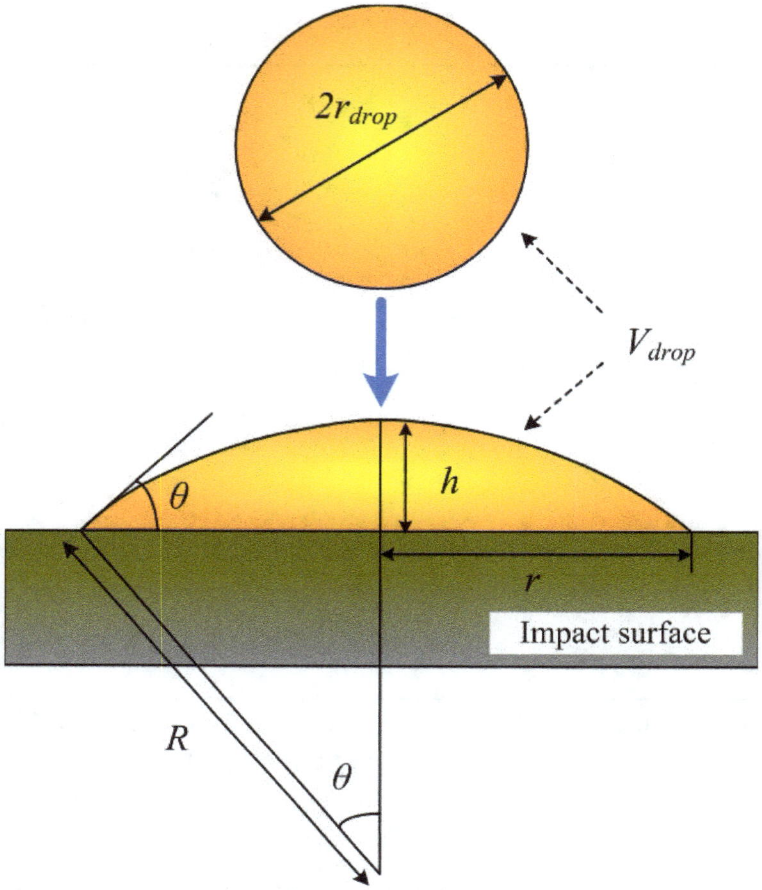

Wetting area A of droplets can be calculated through the following formula:

$$A = \pi r^2 \tag{3}$$

Figure 11 shows contact values of nanofluids with different concentrations sprayed onto workpiece surface. Contact angle is reduced with addition of nanoparticles, possibly because spherical nanoparticles in base oil as shown in Figure 12 are 50nm in particle size, which is larger than palm oil molecules by many times. These particles have higher orders of magnitudes than oil molecules in size and density, they will apply additional downward extruding force oil molecules towards contact surface, and then contact angle of droplets is reduced and wettability is improved. As a proper number of nanoparticles are continuously added in the base oil, more particles will participate in this effect and further reduce fluid contact angle (Chinnam, Das, Vajjha, & Satti, 2015). Correspondingly, addition of nanoparticle also improves strength of lubricating oil film. However, when volume concentration is higher than 2.0%, contact angle is enlarged on the contrary, because the added high-concentration nanoparticles can be dispersed in the base oil, serious agglomeration phenomenon occurs, they lose dynamic stability and then precipitated, and this reduces fluid wettability and workpiece surface quality becomes poor as shown in Figure 7 (g-i).

Figure 13 shows maximum and minimum contact angles under two nanofluid concentrations. Large wetting area enlarges the lubricating area and reduces grinding force and workpiece surface roughness. It can be seen from Figure 7a and e that roughness value corresponding to minimum contact angle is smaller than that corresponding to maximum contact angle by 31%, and therefore, workpiece surface quality is improved.

14.3.4 Anti-Wear and Anti-Friction Mechanism of Al_2O_3 Nanoparticles as Lubricant Additives

Adding a proper number of nanoparticles in the base oil can improve lubricating properties and moreover, this has significant effects of reducing grinding force and wear of grinding wheel and improving workpiece surface quality.

Al_2O_3 is a hexagonal closely stacked crystal material with high hardness, heat resistance and wear resistance. In the crystal lattice of α-Al_2O_3, oxygen ions are closely stacked in hexagonal form, Al_3+ ions are symmetrically distributed in the octahedral coordination center enclosed by oxygen ions with great lattice energy, so melting point and boiling point are very high (Jia, 2014). Therefore, compared with other materials, Al_2O_3 shows very good anti-friction and anti-wear properties. Moreover, when nanoparticles are added into the base oil, 50 nm Al_2O_3 nanoparticles can enter grinding wheel/workpiece interface more easily as carried by MQL fluid under the external stress due to small size effect and high surface

Figure 11. Transformation relation between nanoparticle concentration and contact angle (25°C)

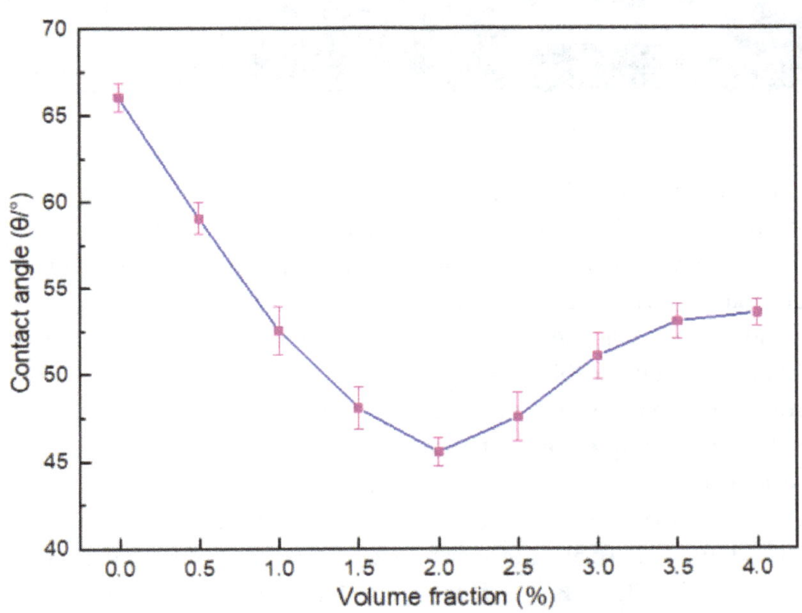

Figure 12. Change las of contact angle after different concentrations of nanoparticles are added

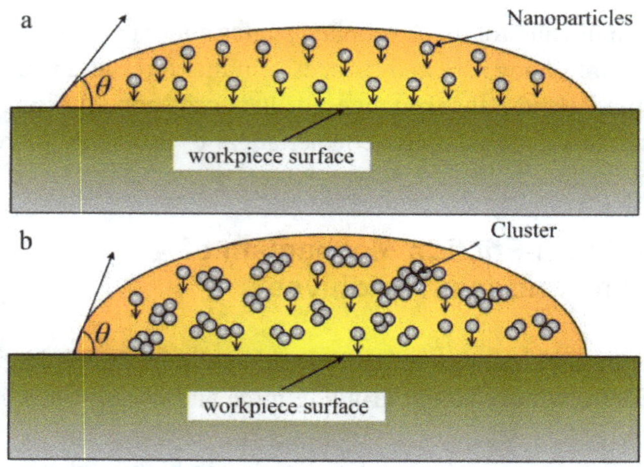

Figure 13. Contact angles of pure palm oil and 2.0 vol% Al_2O_3 nanofluid

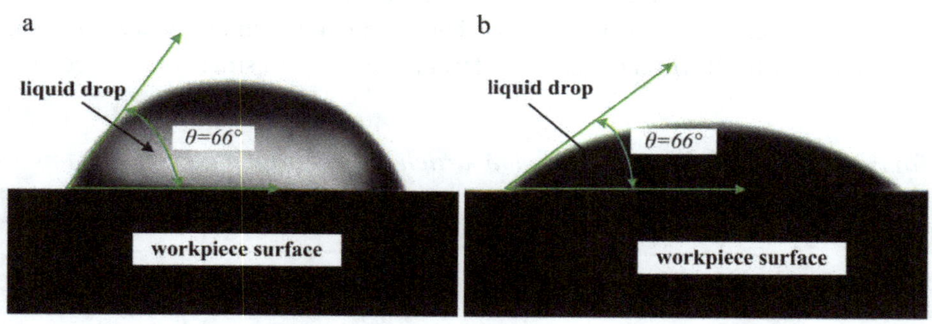

energy. A stacked oil film is formed, where nanoparticles can change the friction condition between abrasive grain and workpiece. Sliding friction is turned into sliding and rolling complex friction, thus reducing sliding friction force, force ratio and energy consumption. Hence, they exert "ball bearing-like effect" and "effect protective film", etc. (Radice & Mischler, 2006) (Figure 14a).

Furthermore, nanoparticles can stuff cavity and incomplete parts on workpiece surface to exert "wear repair effect" (Radice & Mischler, 2006) (Figure 14b) and improve lubricating effect. Spherical Al_2O_3 nanoparticles are of high porosity, so their elasticity is improved within a certain load scope and the gap between grinding wheel and workpiece is further enlarged. Therefore, compression-resisting nanoparticles in the base oil and gaps between abrasive grain and workpiece have high contact resistance, which is good for formation of oil film on workpiece surface. Finaly, Al_2O_3 nanoparticles can improve tribological properties of nanofluid just as described in section 12.4.2 by virtue of high hardness, high-temperature resistance and polishing effect (Cao, Bai, & Huang, 2010; Han, Guan, & Song, 2006; Wang et al., 2016; D. K. Zhang, Li, Zhang, Jia, & Zhang, 2015).

Figure 14. Schematic diagram of anti-friction and anti-wear effects of nanoparticles in the grinding area

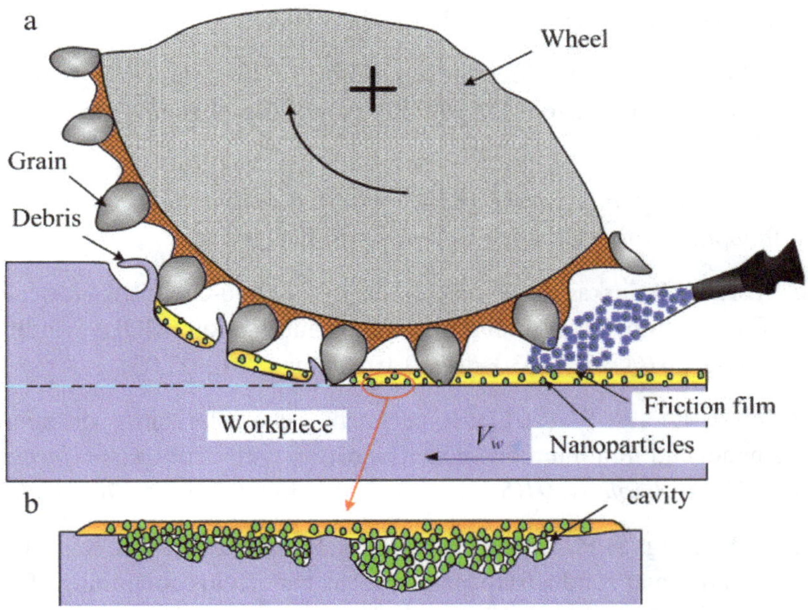

REFERENCES

Amrita, M., Srikant, R. R., & Sitaramaraju, A. V. (2014). Performance Evaluation of Nanographite-Based Cutting Fluid in Machining Process. *Materials and Manufacturing Processes*, *29*(5), 600–605. doi:10.1080/10426914.2014.893060

Cao, P., Bai, Y., & Huang, D. (2010). Tribological Characteristics of MoS_2,Film under Different Lubrication Conditions. *Lubrication Engineering*, *35*(10), 18–21.

Chatha, S. S., Pal, A., & Singh, T. (2016). Performance evaluation of aluminium 6063 drilling under the influence of nanofluid minimum quantity lubrication. *Journal of Cleaner Production*, *137*, 537–545. doi:10.1016/j.jclepro.2016.07.139

Chinnam, J., Das, D., Vajjha, R., & Satti, J. (2015). Measurements of the contact angle of nanofluids and development of a new correlation. *International Communications in Heat and Mass Transfer*, *62*, 1–12. doi:10.1016/j.icheatmasstransfer.2014.12.009

Dambatta, Y. S., Sayuti, M., Sarhan, A. A. D., Hamdi, M., Manladad, S. M., & Reddy, M. (2019). Tribological performance of SiO_2-based nanofluids in minimum quantity lubrication grinding of Si3N4 ceramic. *Journal of Manufacturing Processes*, *41*, 135–147. doi:10.1016/j.jmapro.2019.03.024

Demas, N. G., Timofeeva, E. V., Routbort, J. L., & Fenske, G. R. (2012). Tribological Effects of BN and MoS_2 Nanoparticles Added to Polyalphaolefin Oil in Piston Skirt/Cylinder Liner Tests. *Tribology Letters*, *47*(1), 91–102. doi:10.100711249-012-9965-0

Gao, T., Li, C. H., Zhang, Y. B., Yang, M., Jia, D. Z., Jin, T., ... Li, R. Z. (2019). Dispersing mechanism and tribological performance of vegetable oil-based CNT nanofluids with different surfactants. *Tribology International*, *131*, 51–63. doi:10.1016/j.triboint.2018.10.025

Han, D., Guan, D., & Song, X. (2006). Effect of Nano-Al_2O_3 in lubricating oil on the characteristics of wear and friction of cylinder liner/Piston ring rubbing pair of diesel engine. *Materials for Mechanical Engineering*, *30*(3), 44–47.

Jia, D. (2014). *Lubrication Mechanism and experimental study of fluid field characteristics under grinding using MQL by nanoparticle jet*. Qingdao University of Technology.

Kalita, P., Malshe, A. P., & Rajurkar, K. P. (2012). Study of tribo-chemical lubricant film formation during application of nanolubricants in minimum quantity lubrication (MQL) grinding. *CIRP Annals-Manufacturing Technology*, *61*(1), 327–330. doi:10.1016/j.cirp.2012.03.031

Kim, J. S., Kim, J. W., & Lee, S. W. (2017). Experimental characterization on micro-end milling of titanium alloy using nanofluid minimum quantity lubrication with chilly gas. *International Journal of Advanced Manufacturing Technology*, *91*(5-8), 2741–2749. doi:10.100700170-016-9965-6

Lee, J. H., Hwang, K. S., Jang, S. P., Lee, B. H., Kim, J. H., Choi, S. O. U. S., & Choi, C. J. (2008). Effective viscosities and thermal conductivities of aqueous nanofluids containing of Al_2O_3 low volume concentrations nanoparticles. *International Journal of Heat and Mass Transfer*, *51*(11-12), 2651–2656. doi:10.1016/j.ijheatmasstransfer.2007.10.026

Lee, P. H., Nam, J. S., Li, C., & Lee, S. W. (2012). An Experimental Study on Micro-Grinding Process with Nanofluid Minimum Quantity Lubrication (MQL). *International Journal of Precision Engineering and Manufacturing*, *13*(3), 331–338. doi:10.100712541-012-0042-2

Malkin, S., & Guo, C. (2008). *Grinding technology: Theory and application of machining with abrasives*. Industrial Press Inc.

Mao, C., Huang, Y., Zhou, X., Gan, H., Zhang, J., & Zhou, Z. X. (2014). The tribological properties of nanofluid used in minimum quantity lubrication grinding. *International Journal of Advanced Manufacturing Technology*, *71*(5-8), 1221–1228. doi:10.100700170-013-5576-7

Mao, C., Zhang, J., Huang, Y., Zou, H. F., Huang, X. M., & Zhou, Z. X. (2013). Investigation on the Effect of Nanofluid Parameters on MQL Grinding. *Materials and Manufacturing Processes*, *28*(4), 436–442. doi:10.1080/10426914.2013.763970

Minh, D. T., The, L. T., & Bao, N. T. (2017). Performance of Al_2O_3 nanofluids in minimum quantity lubrication in hard milling of 60Si(2)Mn steel using cemented carbide tools. *Advances in Mechanical Engineering*, *9*(7). doi:10.1177/1687814017710618

Ni, J., Yang, Y. F., & Wu, C. (2019). Assessment of water-based fluids with additives in grinding disc cutting process. *Journal of Cleaner Production*, *212*, 593–601. doi:10.1016/j.jclepro.2018.12.066

Oliveira, D. D., Guermandi, L. G., Bianchi, E. C., Diniz, A. E., de Aguiar, P. R., & Canarim, R. C. (2012). Improving minimum quantity lubrication in CBN grinding using compressed air wheel cleaning. *Journal of Materials Processing Technology*, *212*(12), 2559–2568. doi:10.1016/j.jmatprotec.2012.05.019

Radice, S., & Mischler, S. (2006). Effect of electrochemical and mechanical parameters on the lubrication behaviour of Al_2O_3 nanoparticles in aqueous suspensions. *Wear*, *261*(9), 1032–1041. doi:10.1016/j.wear.2006.03.034

Rahmati, B., Sarhan, A. A. D., & Sayuti, M. (2014). Morphology of surface generated by end milling AL6061-T6 using molybdenum disulfide(MoS_2) nanolubrication in end milling machining. *Journal of Cleaner Production*, *66*, 685–691. doi:10.1016/j.jclepro.2013.10.048

Sai, S. S. (2015). Assessment of spray quality from an external mix nozzle and its impact on SQL grinding performance. *International Journal of Machine Tools & Manufacture*, *89*, 132–141. doi:10.1016/j.ijmachtools.2014.10.004

Setti, D., Sinha, M. K., Ghosh, S., & Rao, P. V. (2015). Performance evaluation of Ti-6Al-4V grinding using chip formation and coefficient of friction under the influence of nanofluids. *International Journal of Machine Tools & Manufacture*, *88*, 237–248. doi:10.1016/j.ijmachtools.2014.10.005

Seyedzavvar, M., Shabgard, M., & Mohammadpourfard, M. (2019). Investigation into the performance of eco-friendly graphite nanofluid as lubricant in MQL grinding. *Machining Science and Technology*, *23*(4), 569–594. doi:10.1080/10910344.2019.1575403

Shen, B., Shih, A. J., & Tung, S. C. (2008). Application of Nanofluids in Minimum Quantity Lubrication Grinding. *Tribology Transactions*, *51*(6), 730–737. doi:10.1080/10402000802071277

Shen, B., Shih, A. J., & Tung, S. C. (2009). Application of Nanofluids in Minimum Quantity Lubrication Grinding. *Tribology & Lubrication Technology*, *65*(3), 73–80.

Singh, R. K., Sharma, A. K., Dixit, A. R., Tiwari, A. K., Pramanik, A., & Mandal, A. (2017). Performance evaluation of alumina-graphene hybrid nano-cutting fluid in hard turning. *Journal of Cleaner Production*, *162*, 830–845. doi:10.1016/j.jclepro.2017.06.104

Wang, Y. G., Li, C. H., Zhang, Y. B., Li, B. K., Yang, M., Zhang, X. P., ... Liu, G. T. (2016). Experimental evaluation of the lubrication properties of the wheel/workpiece interface in MQL grinding with different nanofluids. *Tribology International*, *99*, 198–210. doi:10.1016/j.triboint.2016.03.023

Wang, Y. G., Li, C. H., Zhang, Y. B., Yang, M., Li, B. K., Dong, L., & Wang, J. (2018). Processing Characteristics of Vegetable Oil-based Nanofluid MQL for Grinding Different Workpiece Materials. *International Journal of Precision Engineering and Manufacturing-Green Technology*, *5*(2), 327–339. doi:10.100740684-018-0035-4

Wenfeng, Z. C., Jiuhua, X., Shubao, D., & Yang, T. L. (2011). Simulation of chip formation in grinding titanium alloy TC4 with single abrasive grit. *Diamond & Abrasives Engineering, 2.*

Yin, Q. A., Li, C. H., Zhang, Y. B., Yang, M., Jia, D. Z., Hou, Y. L., ... Dong, L. (2018). Spectral analysis and power spectral density evaluation in Al_2O_3 nanofluid minimum quantity lubrication milling of 45 steel. *International Journal of Advanced Manufacturing Technology*, *97*(1-4), 129–145. doi:10.100700170-018-1942-9

Zhang, D. K., Li, C. H., Zhang, Y. B., Jia, D. Z., & Zhang, X. W. (2015). Experimental research on the energy ratio coefficient and specific grinding energy in nanoparticle jet MQL grinding. *International Journal of Advanced Manufacturing Technology*, *78*(5-8), 1275–1288. doi:10.100700170-014-6722-6

Zhang, J. C., Li, C. H., Zhang, Y. B., Yang, M., Jia, D. Z., Liu, G. T., ... Cao, H. J. (2018). Experimental assessment of an environmentally friendly grinding process using nanofluid minimum quantity lubrication with cryogenic air. *Journal of Cleaner Production*, *193*, 236–248. doi:10.1016/j.jclepro.2018.05.009

Zhang, S., Cheng, Y., Xing, F., Li, M., & Zhu, Z. (2011). Study on contact angle and surface free energy current situation and prospect. *J Meikuang Jixie*, *32*(10), 8–10.

Zhang, X. P., Li, C. H., Zhang, Y. B., Jia, D. Z., Li, B. K., Wang, Y. G., ... Zhang, X. W. (2016). Performances of Al2O3/SiC hybrid nanofluids in minimum-quantity lubrication grinding. *International Journal of Advanced Manufacturing Technology*, *86*(9-12), 3427–3441. doi:10.100700170-016-8453-3

Zhang, X. P., Li, C. H., Zhang, Y. B., Wang, Y. G., Li, B. K., Yang, M., ... Zhang, N. Q. (2017). Lubricating property of MQL grinding of Al2O3/SiC mixed nanofluid with different particle sizes and microtopography analysis by cross-correlation. *Precision Engineering-Journal of the International Societies for Precision Engineering and Nanotechnology*, *47*, 532–545. doi:10.1016/j.precisioneng.2016.09.016

Zhang, Y. B., Li, C. H., Jia, D. Z., Li, B. K., Wang, Y. G., Yang, M., ... Zhang, X. W. (2016). Experimental study on the effect of nanoparticle concentration on the lubricating property of nanofluids for MQL grinding of Ni-based alloy. *Journal of Materials Processing Technology*, *232*, 100–115. doi:10.1016/j.jmatprotec.2016.01.031

Zhang, Y. B., Li, C. H., Jia, D. Z., Zhang, D. K., & Zhang, X. W. (2015). Experimental evaluation of MoS_2 nanoparticles in jet MQL grinding with different types of vegetable oil as base oil. *Journal of Cleaner Production*, *87*, 930–940. doi:10.1016/j.jclepro.2014.10.027

Zhang, Y. B., Li, C. H., Yang, M., Jia, D. Z., Wang, Y. G., Li, B. K., ... Wu, Q. D. (2016). Experimental evaluation of cooling performance by friction coefficient and specific friction energy in nanofluid minimum quantity lubrication grinding with different types of vegetable oil. *Journal of Cleaner Production*, *139*, 685–705. doi:10.1016/j.jclepro.2016.08.073

Chapter 15
Optimization Design of Process Parameters for Different Workpiece Materials in NMQL Grinding With Different Vegetable Oils

ABSTRACT

This research shows with superior lubricating, heat-conducting properties, and proper market price, Al2O3 and MoS2 nanoparticles have broad application prospects as lubricant additives. This work has been researched, and most researches are restricted to using one kind of lubricants to machine specific workpiece or multiple workpieces. There has been no systematic and detailed analysis of grinding performances from cutting mechanisms and debris formation mechanisms of different workpieces. Therefore, analysis of signal to noise ratio(S/N), variance, microstructure, and morphology analysis were used to study the influence of different typical nanofluid lubricants on the grinding performance of different materials in this chapter. Results showed that the optimal grinding parameters are using nodular cast iron and MoS2 nanofluid. Compared with other lubricants, MoS2 nanofluid can exert a significant effect on reducing wear of grinding wheel. In order to reduce surface roughness, MQL lubricating oils of the following sequence can be used: Al2O3 nanofluid, MoS2 nanofluid, and pure palm oil.

15.1 INTRODUCTION

In order to improve machining performance, scholars have conducted many studies on different nanoparticles (Kalita, Malshe, & Rajurkar, 2012; Li, Zhang, Jia, Wang, & Hou, 2015; Mao et al., 2014; Mao, Tang, Zou, Huang, & Zhou, 2012; Mao et al., 2013; B. Shen, Malshe A P, Kalita P, 2008; B. Shen, Shih, & Tung, 2008; Zhang, Li, Jia, & Zhang, 2015; Zhang, Li, Jia, Zhang, & Zhang, 2015a, 2015b). From the previous chapters, it can be seen that, with superior lubricating, heat-conducting properties

DOI: 10.4018/978-1-7998-1546-4.ch015

and proper market price, Al_2O_3 and MoS_2 nanoparticles have broad application prospects as lubricant additives (Ali et al., 2016; Dong, Duc, & Long, 2019; Emami, Sadeghi, Sarhan, & Hasani, 2014; Han, 2006; Jamil et al., 2019; Kumar & Patel, 2018; Ping & Gao, 2010; Sen et al., 2019; Wang et al., 2016; Yang et al., 2017; Yin, Li, Dong, et al., 2018; Zhang et al., 2016; Zhang et al., 2017). Meanwhile, for different workpieces: (1) nickel base alloy (Shi, Wang, Xu, Yu, & Sun, 2019), it is extensively applied to aerospace and chemical engineering departments. The grinding of nickel base alloy is accompanied by large cutting force, high cutting temperature and serious hardening phenomenon due to its high strength and large hardness, so it belongs to typical difficult-to-machine material. (2) nodular cast iron (Fernandes et al., 2018), a new-type high-strength cast iron material with low cost, good casting performance, favorable shock absorption and good wear resistance, and it has been successfully used to cast some parts with complicated stress-bearing condition and high requirements for strength, tenacity and wear resistance, e.g. connecting rod and bent axle in auto parts, etc. (3) 45 steel, it is a kind of typical medium carbon steel with low hardness and good plasticity and favorable mechanical properties, and it has been extensively applied to multiple kinds of important structural parts, especially connecting rod, bolt, gear and bearing working under alternating load. Therefore, it's very necessary to study machining performance of the above materials.

Different nanofluids have obtained different machining performances when applied to MQL grinding of different workpiece materials (Jia, Li, Zhang, Zhang, & Zhang, 2014; Lee et al., 2009; Lee, Kim, & Lee, 2018; Li, Li, Zhang, Wang, Jia, et al., 2017; Li, Li, Zhang, Wang, Yang, Jia, Zhang, & Wu, 2017; Li, Li, Zhang, Wang, Yang, Jia, Zhang, Wu, et al., 2017; Li et al., 2018; Liu et al., 2018; Wang et al., 2017; Zhang & Xu, 2019). Even though so far, this work has been researched frequently, most researches are restricted to using one kind of lubricants to machine specific workpiece or multiple workpieces. In addition, there has been no systematic and detailed analysis of grinding performances from aspects of cutting mechanisms and debris formation mechanisms of different workpieces. Therefore, a fundamental study of the influence of different typical nanofluid lubricants on grinding performances of different workpiece materials will be carried out in this chapter.

15.2 EXPERIMENTAL PART

15.2.1 Experimental Equipment and Grinding Parameters

Equipment used in this experiment are identical with those in Chapter 11, namely K-P36 NC surface grinder, IDM-III99 three-phase grinding dynamometer, surface roughness coarseness profiling instrument, digital viscosimeter and SEM.

Silicon carbide ceramics-bonded grinding wheel is used, and its concrete model is GC80K12V. Performance parameters of grinding wheel are as follows: dimensions: 300 mm × 20 mm × 76.2 mm; granularity: 80 # with medium soft hardness, and the highest linear velocity is 50m/s.

Grinding parameters and grinding wheel dressing parameters used in the experiment are shown in Table 1. In order to realize controlled grinding process and ensure consistency of grinding wheel sharpness in each experiment, grinding wheel was dressed before experiment, where down-grinding condition was adopted on 30×20 mm workpiece surface in each grinding experiment.

Table 1. Grinding experimental parameters

Grinding type	Surface grinding, down-grinding
Grinding wheel type	SiC: GC80K12VCeramic bond
Lubricating condition	flood, MQL
Flow rate under flood grinding condition	60 L/h
Flow rate under MQL condition	50 ml/h
Gas-liquid ratio	0.3
Linear velocity of grinding wheel (V_s)	30 m/s
Workpiece feed velocity (V_w)	3000 mm/min
Grinding depth (a_p)	20 μm
MQL nozzle distance (d)	12 mm
MQL nozzle angle (α)	15°
MQL air pressure (P)	0.6 MPa
Dresser	Single-point diamond
Total dressing depth (a_d)	40 μm
Dressing velocity (v_d)	300 mm/min

15.2.2 Experimental Materials

Experimental materials are mainly workpiece material and nanofluid lubricant. Different lubricants were used to machine different workpiece materials to observe their respective grinding performances.

1.Workpiece Materials

Three different kinds of typical workpiece materials were selected in the experiment: ①nickel base alloy 718, ②nodular cast iron 100-70-03, ③45 steel. Dimensions of workpiece specimen are 40 mm×30 mm ×30 mm. Chemical compositions and mechanical properties of the selected workpiece are shown in Table 2 and Table 3.

Table 2. Chemical compositions of workpiece materials (wt%)

Workpiece	C	Ni	Fe	Cu	Mn	Mo	Cr	Al	Co	Si	P	S
Nickel base alloy	0.08	50~55	trace	0.3	0.35	2.8~3.3	17~21	0.95	1.0	0.35	~	~
Nodular cast iron	3.5	~	> 95	~	0.15	~	0.35	~	~	2.25	0.05	0.025
45 steel	0.45	≤0.3	> 95	≤0.25	0.6	≤0.1	≤0.3	~	~	0.25	≤0.04	≤0.035

Table 3. Mechanical properties of workpiece materials

Workpiece	Density (g/cm^3)	Elasticity modulus (GPa)	Heat conductivity (W/m K)	Hardness	Yield strength (MPa)	Tensile strength (MPa)	Poisson's ratio
Nickel base alloy	8.24	199.9	14.7	≤363 (HB)	550	965	0.30
Nodular cast iron	7.32	154	35.2	50 (HRC)	≥370	≥600	0.26
45 steel	7.85	210	50.2	25~30(HRC)	≥355	≥600	0.31

2.MQL Base Oil and Nanofluids

Palm oil was used as MQL base oil in this experiment. Three different lubricants were used, namely pure palm oil, MoS_2 nanofluid and Al_2O_3 nanofluid, and the mechanical properties of nanoparticles are shown in Table 4. The used nanofluids are identical with those used in Chapter 12. According to previous researches, two nanofluids with 2% volume fraction have good anti-friction and anti-wear properties (Wang et al., 2016; Zhang, Li, Zhang, Jia, & Zhang, 2015), and they have typical different anti-friction and anti-wear mechanisms.

Table 4. The mechanical properties of nanoparticles

Properties	MoS_2	Al_2O_3
Crystal structure	Stratified	Closely stacked in hexagonal form
Melting point (°C)	1185	2050
Apparent density (g/cm^3)	0.39	0.33
Heat conductivity (W/m K)	138	36
Color	Lead grey/black	White
Moh's hardness	1.0-1.5	8.8-9.0

15.2.3 Experimental Scheme

Each lubricant and different workpiece materials were treated and optimized in the experiment. The research aims at optimizing matching between different MQL lubricants and workpieces. Two-factor three-level comprehensive orthogonal test was adopted in this experiment as shown in Table 5. Lubricant type and workpiece type were variable quantities of grinding parameters. In this experiment, force ratio, specific grinding energy, G ratio and surface morphology were used to characterize grinding performance. Taguchi orthogonal experimental design (L_9) and signal-to-noise ratio (S/N) objective function were used to optimize grinding performance parameters. The experiment was designed with two factors (MQL lubricant and workpiece) and three levels (palm oil, MoS_2 nanofluid and Al_2O_3 nanofluid; nickel base alloy, nodular cast iron and 45 steel) as shown in Table 6.

Table 5. Grinding factors and their levels

Symbol	Factor	Level 1	Level 2	Level 3
W	Workpiece	Nickel base alloy 718	Nodular cast iron	45 steel
M	Lubricant	Pure palm oil	MoS_2 nanofluid	Al_2O_3 nanofluid

In addition, workpiece surface roughness values (R_a and R_z) obtained through the measurement are correspondingly recorded in Table 6. Minitab 17 software was used to analyze experimental results.

Table 6. L_9 orthogonal array and measurements

Experiment No.	Workpiece	Lubricant	F_t/F_n	U (J/mm³)	G ratio	R_a (μm)	R_z (μm)
1	Nickel base alloy 718	Pure palm oil	0.45	63.76	29.54	0.34	2.34
2		MoS_2 nanofluid	0.40	75.05	36.62	0.30	2.29
3		Al_2O_3 nanofluid	0.38	62.20	39.27	0.26	2.16
4	Nodular cast iron	Pure palm oil	0.16	35.17	34.19	0.27	1.69
5		MoS_2 nanofluid	0.13	36.77	44.68	0.18	1.54
6		Al_2O_3 nanofluid	0.11	42.63	42.44	0.24	1.85
7	45 steel	Pure palm oil	0.23	66.71	36.68	0.18	1.30
8		MoS_2 nanofluid	0.20	108.94	43.16	0.14	1.82
9		Al_2O_3 nanofluid	0.15	100.57	40.25	0.20	1.11

15.3 ANALYSIS AND RESULTS

Orthogonal test is an optimal selection method of grinding parameters using principles of mathematical statistics. Experimental arrangement through orthogonal array is representative and can comprehensively reflect influences of factors and levels on the objective. S/N, which is an index measuring quality stability of target product, finds the optimal parameter-level combination through a statistical analysis of all kinds of experimental programs. Therefore, S/N analysis is very crucial for selecting the optimal point. Furthermore, variance analysis aims at studying which factors can exert significant influences on force ratio, specific grinding energy, G ratio and surface roughness. S/N analysis, variance analysis and verification experiment were used for a statistical analysis of experimental results.

15.3.1 Analysis of Signal to Noise Ratio (S/N)

S/N is ratio of average value to standard deviation and an index measuring deviation degree from the expected value. In the optimization process, performance characteristics can be divided into smaller-the-better (STB) characteristic, larger-the-better (LTB) characteristic and nominal-the-best (NTB) characteristic (Yang & Tarng, 1998). S/N expresses disturbance degree borne by product from noise factor. The experiment aims at optimizing grinding performance parameters to obtain better (lower value) force

Table 7. Corresponding S/N (dB) ratios for the observations

Experiment No.	F_t/F_n	U (J/mm³)	G ratio	R_a (μm)	R_z (μm)	S/N (F_t/F)	S/N (U)	S/N (G)	S/N (R_a)	S/N (R_z)
7-1	0.45	63.76	29.54	0.34	2.34	6.74	-36.18	31.03	9.49	-7.38
7-2	0.40	75.05	36.62	0.30	2.29	7.49	-37.53	28.82	9.36	-8.03
7-3	0.38	62.20	39.27	0.26	2.16	8.42	-35.98	32.11	11.67	-6.80
7-4	0.16	35.17	34.19	0.27	1.69	15.65	-30.97	31.41	11.32	-4.56
7-5	0.13	36.77	44.68	0.18	1.54	15.79	-31.42	33.08	14.85	-3.79
7-6	0.11	42.63	42.44	0.24	1.85	19.71	-32.63	31.54	12.37	-5.38
7-7	0.23	66.71	36.68	0.18	1.30	12.71	-36.50	29.33	14.28	-2.75
7-8	0.20	108.94	43.16	0.14	1.82	13.70	-40.99	34.30	11.29	-5.20
7-9	0.15	100.57	40.25	0.20	1.11	17.39	-40.26	31.81	7.12	-9.2

ratio, specific energy and roughness value as well as (larger value) G ratio. Therefore, it's necessary to study STB and LTB characteristics respectively, and their calculation formulas are:

$$S / N = -10\log\left[\frac{1}{n}\left(\sum_{i=1}^{n} y_i^2\right)\right], \tag{1}$$

$$S / N = -10\log\left[\frac{1}{n}\left(\sum_{i=1}^{n} \frac{1}{y_i^2}\right)\right] \tag{2}$$

where y_i is obtained experimental data; n is number of experimental data.

No matter which type S/N performance characteristic belongs to, technological parameter setting with the highest S/N ratio can always generate the optimal quality with the minimum variance. Therefore, optimal levels of machining parameters should be the highest levels corresponding to S/N value. Table 7 shows S/N values calculated corresponding to force ratio, specific energy, G ratio and surface roughness.

Main effect plot for S/N ratios for different grinding parameters is shown in Figure 1. According to S/N analysis in Figure 1, optimal grinding parameters for force ratio are nodular cast iron (level 2) and Al_2O_3 nanofluid (level 3). Force ratio generated under cast iron grinding process presents a minimum value in all workpieces, which is ascribed to physiochemical characteristics of the nodular cast iron itself. MQL lubricating oil has the potential of reducing force ratio, and force ratios under different conditions are sorted in a descending order as Al_2O_3 nanofluid, MoS_2 nanofluid and pure palm oil. Compared with other lubricating oils, force ratio of Al_2O_3 nanofluid has obviously greater influence. During the MQL grinding process, influences of different lubricating oils on grinding force ratio can be explained through lubricating properties of the oil itself. In the machining process, abrasive grain completes sliding friction, ploughing and cutting effects on workpiece surface, so lubricating properties play a significant role (Dambatta, Sarhan, Sayuti, & Bin Abd Shukor, 2019; Dambatta, Sayuti, Sarhan, Shukor, et al., 2019; Dambatta, Sayuti, Sarhan, Hamdi, et al., 2019; Dong & Zhang, 2019; Jia et al., 2019; Kumar, Ghosh, &

Figure 1. Main effects plot of S/N ratios for force ratio F_t/F_n

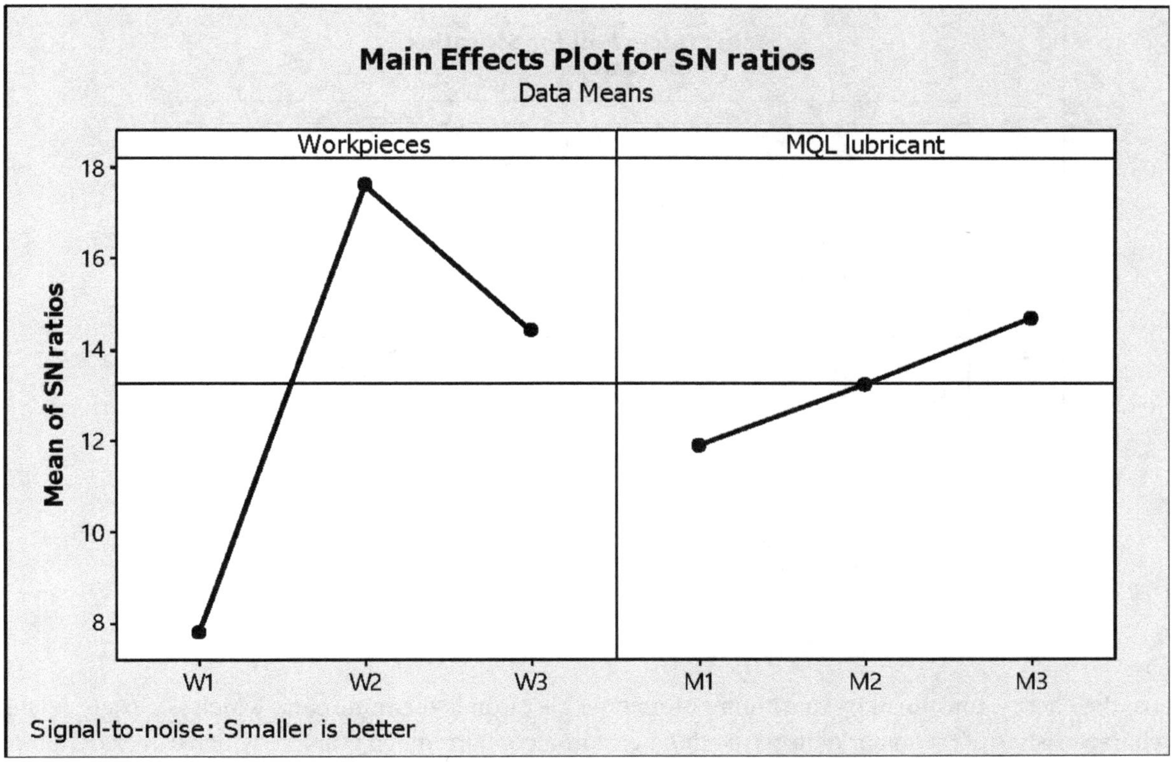

Aravindan, 2019; Pashmforoush & Bagherinia, 2018; Paul & Ghosh, 2019; Seyedzavvar, Shabgard, & Mohammadpourfard, 2019). Firstly, nanoparticles have superior anti-friction and anti-wear properties. Secondly, nanoparticles can exert "bearing-like" and "oil film protection effect" in the grinding area, which greatly enhances lubricating properties of lubricating oil (Bai, Li, Dong, & Yin, 2019; Ganesan, Babu, Santhanakumar, & Muthukrishnan, 2018; Jia, 2014; Khan et al., 2019; Nam & Lee, 2018; Nguyen, Lee, Guo, Park, & Kwon, 2019; Sharma, Singh, Dixit, & Tiwari, 2017; Uysal, 2018; Yildirim, Sarikaya, Kivak, & Sirin, 2019; Yin, Li, Zhang, et al., 2018; Yuan, Hou, Wang, & Chen, 2018; Zhang et al., 2018). In addition, physical properties of Al_2O_3 nanofluid described in section 14.3.4 are conducive to improving lubricating properties, especially the lubricating film formed on friction surface. In order to verify this fact, EDS graph of workpiece surface after Al_2O_3 nanofluid MQL grinding was used as shown in Figure 6. Al and O elements exist in the energy spectrum, which verifies formation of friction film on workpiece surface. In addition, nanoparticles can be stuffed in cavity and incomplete parts on workpiece surface to exert "wear repair effect" and improve lubricating effect. For MoS_2 nanoparticles, their structural features analyzed in section 12.4.2 can significantly improve grinding performance.

Main effect plot of S/N ratios for specific grinding energy is shown in Figure 2. According to the S/N analysis, optimal grinding parameters for specific grinding energy are nodular cast iron (level 2) and pure palm oil (level 1). The optimal parameter combination was selected to reduce machining energy consumption so as to improve grinding efficiency. That's to say, the lower the specific grinding energy in the material removal process, the higher the workpiece machining efficiency. Under the same condi-

Figure 2. Main effects plot of S/N ratios for specific grinding energy U

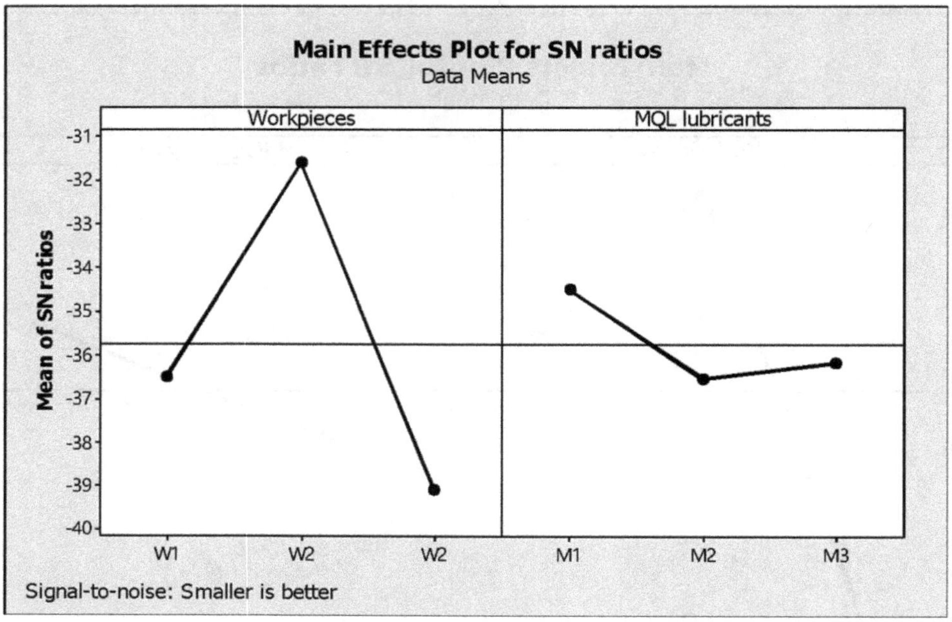

tions, the energy consumed by machining of nodular cast iron is the minimum, which is closely related to characteristics of the material itself. Nodular cast iron contains much graphite which can exert cutting and lubricating effects. Energy needed to machine nickel base alloy with high hardness is lower than that of 45 steel, possibly because the selected nanofluid is more suitable to machine nickel base alloy, thus reducing energy consumption.

For G ratio, optimal grinding parameters are nodular cast iron (level 2) and MoS_2 nanofluid (level 2) as shown in Figure 3. In order to elevate G ratio, lubricants of the following sequence can be used: MoS_2 nanofluid, Al_2O_3 nanofluid and pure palm oil. Therefore, compared with other lubricants, MoS_2 nanofluid can exert a significant effect on reducing wear extent of grinding wheel.

Figure 4a and b show main effect plots of S/N ratios for surface roughness values R_a and R_z. Optimal grinding parameters of surface roughness values R_a and R_z are 45 steel (level 3) and Al_2O_3 nanofluid (level 3). In order to reduce surface roughness, MQL lubricating oils of the following sequence can be used: Al_2O_3 nanofluid, MoS_2 nanofluid and pure palm oil. The influence of MQL lubricating oil on surface roughness is mainly ascribed to performance of lubricant additive and characteristics of the workpiece itself. Nanoparticles can not only exert the above effects, especially additive can react with workpiece surface, which is good for forming a good morphology. In the material removal process under high pressure and high temperature, nanoscale additive becomes more active and will react with workpiece surface. Nanoparticles are precipitated and adsorbed on workpiece surface to form a friction film so as to influence cutting form and chip shape (section 15.3.4), thus reducing surface roughness. In order to prove the hypothesis for friction film, EDS analysis of the workpiece surface was conducted after the experiment. Rightly because of formation of oil film, it can be observed from Figure 5a that Al and O contents are high on workpiece surface and high Mo and S elements can be observed in Figure 5b. Meanwhile, lubricant viscosity also affects formation of friction film, and viscosity values of pure palm

Figure 3. Main effects plot of S/N ratios for G-ratios

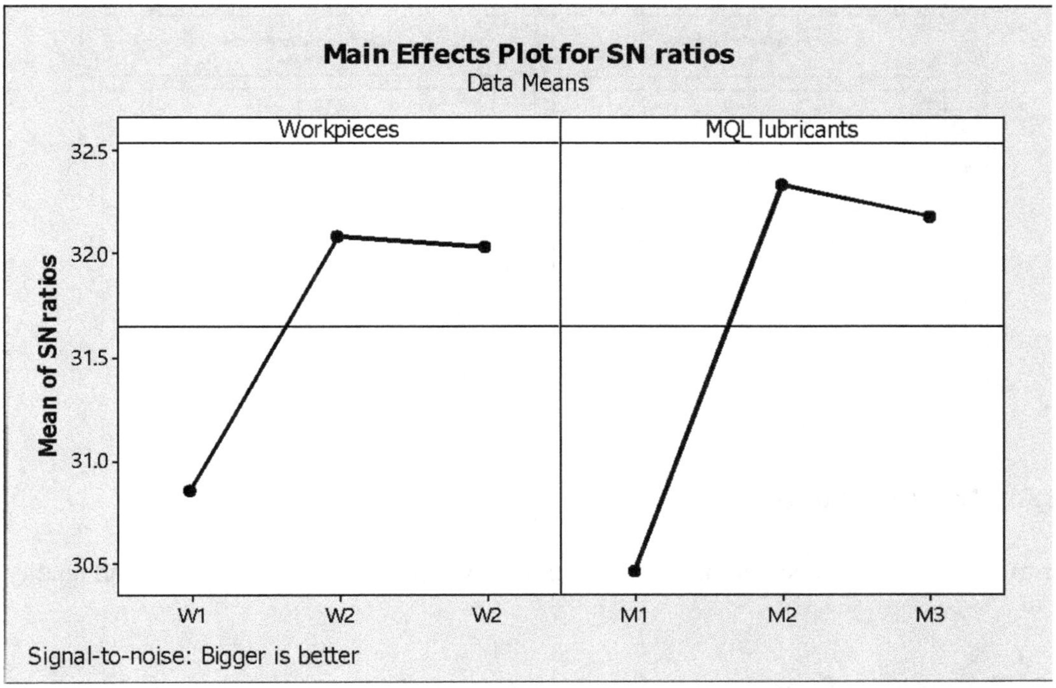

oil, MoS_2 nanofluid and Al_2O_3 nanofluid (under 40°C) are 39.4, 42.3 and 49.5 mm^2/s respectively. As the viscosity increases, colloidal force, Brownian movement and viscous force between grinding fluid molecules correspondingly increase, which contributes to formation of oil film on friction surface and enhancement of thickness and strength of adsorption film. In addition, researches by related scholars have proved that additive in the grinding process forms friction film on workpiece surface (Luo, Wei, Huang, Huang, & Yang, 2014; Sayuti, Sarhan, Tanaka, Hamdi, & Saito, 2013). As shown in Figure 4, Al_2O_3 nanofluid causes the lowest surface roughness in the experimental results, namely nanoparticles exert anti-friction and anti-wear effects. On the other hand, palm moil generates the highest surface roughness, which is mainly ascribed to its chemical compositions. In the MQL grinding process, palm oil contains a large quantity of fatty acids and doesn't contain any additive. Relative to dry grinding, palm oil has its own advantages in the lubrication process under flood lubrication condition. Palm oil contains a large quantity of fatty acids and polar groups, which can be easily adsorbed on metal surface to form boundary lubricating film, thus generating metallic soap of fatty acids (Fox & Stachowiak, 2007). However, thermal stability and bearing capacity of this type of lubricating film are limited. Therefore, palm oil obtains the highest surface roughness possibly because lubricating film with poor stability can easily lose effect under severe environment (high temperature, high pressure and high cutting velocity) at the grinding wheel/workpiece interface.

Figure 4. Main effects plot of S/N ratios for surface roughness values R_a and R_z

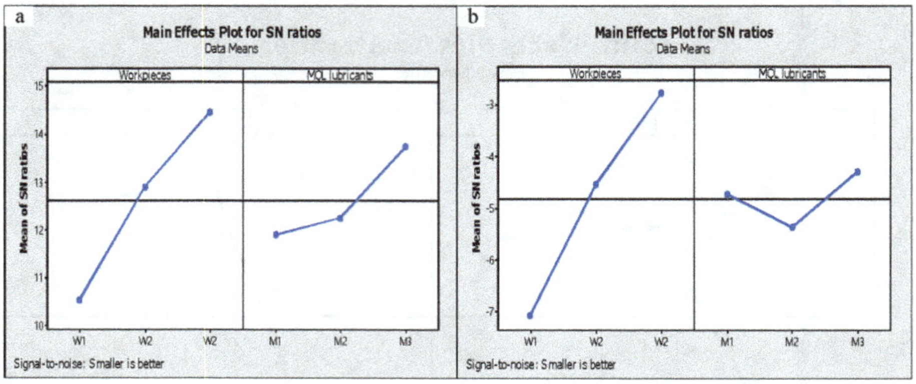

15.3.2 Analysis of Variance

In order to obtain the influence of binary interaction between machining parameters and quality char-

Figure 5. EDS analysis of workpiece surface: a- Al_2O_3 nanofluid MQL grinding of nickel base alloy, b- MoS_2 nanofluid MQL grinding of nodular cast iron

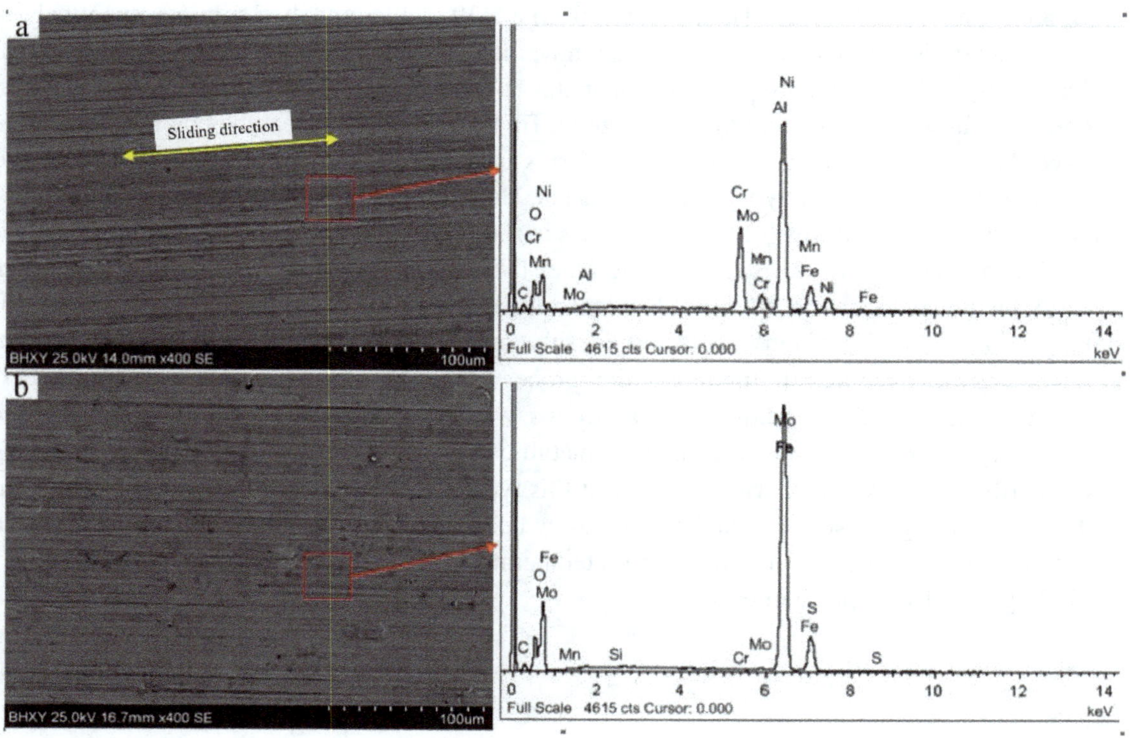

Table 8. Results of the ANOVA for force ratio, specific energy and G-ratio

Variance source	Degree of freedom	Quadratic sum	Mean square	F value	Contribution rate *P* %
***Force ratio* (F_t/F_n)**					
Workpiece	2	0.127089	0.063544	571.90	94.69
Lubricant	2	0.006689	0.003344	30.10	4.98
Error	4	0.000444	0.000111		0.33
Total	8	0.134222			
Specific grinding energy (U)					
Workpiece	2	4362.13	2181.06	14.79	79.44
Lubricant	2	539.44	269.72	1.83	9.82
Error	4	589.72	147.43		10.74
Total	8	5491.29			
G ratio					
Workpiece	2	52.064	26.0321	6.45	28.18
Lubricant	2	116.562	58.2808	14.43	63.08
Error	4	16.153	4.0384		8.74
Total	8	184.779			

acteristics, variance (ANOVA) analys was conducted. ANOVA aims at analyzing which machining parameters can exert a significant influence on grinding performance characteristics. Table 8 and Table 9 list ANOVA results of force ratio, specific grinding energy, G ratio and surface roughness respectively. According to Table 8, workpiece material has a great effect on force ratio and its contribution rate is about 94.69%. ANOVA results indicate that contribution rates of workpiece material and lubricant to specific grinding energy are 79.44% and 9.82% respectively. For G ratio, lubricant has a greater effect than workpiece material with its contribution rate reaching 63.08%. In order to lengthen service life of grinding wheel and improve machining efficiency, nanofluid lubricant should be selected.

Similarly, Table 9 describes ANOVA results of surface roughness values R_a and R_z. Contribution rates of workpiece material and lubricant to surface roughness R_a value are 50.65% and 12.20% respectively and those to roughness R_z value are 77.16% and 3.23% respectively. According to ANOVA analysis results in Table 9, workpiece material can exert a significant effect on surface roughness.

15.3.3 Verification Experiment

In order to very reasonability of the optimal grinding parameter combination in the parameter design, a verification experiment should be conducted. A confirmatory experiment was conducted for optimal-random points through the Taguchi method in order to predict response characteristics of set factors. Confirmatory experiment was used to verify prediction results of Taguchi. Table 10 lists confirmatory experimental results of predicted values and experimental values of force ratio, specific grinding energy, G ratio and surface roughness. Base on the analysis of data in each group, predicted values accord with

Table 9. Results of the ANOVA for surface roughness

Variance source	Degree of freedom	Quadratic sum	Mean square	F value	Contribution rate P %
Surface roughness R_a					
Workpiece	2	0.016689	0.008344	2.73	50.65
Lubricant	2	0.004022	0.002011	0.66	12.20
Error	4	0.012244	0.003061		37.15
Total	8	0.032956			
Surface roughness R_z					
Workpiece	2	1.13336	0.566678	7.87	77.16
Lubricant	2	0.04749	0.023744	0.33	3.23
Error	4	0.28804	0.072011		19.6
Total	8	1.46889			

experimental values very well. For the reliable statistical analysis, error value must be smaller than 20%. Error value of G ratio is relatively larger, but on the whole, this error value is within the acceptable range.

Table 10. Confirmation test results by Taguchi method

Experiment	Selected point	Orthogonal experimental result		
		Experimental value	Predicted value	Error (%)
Force ratio (F_t/F_n)				
1c	A_2 B_3 Optimal point A_2 B_3	0.11	0.10	9.10
2c	A_1 B_2 Random point A_1 B_2	0.40	0.37	7.50
Specific grinding energy(J/mm³)				
3c	A_2 B_1 Optimal point A_2 B_1	35.17	38.62	9.81
4c	A_3 B_3 Random point A_3 B_3	100.57	95.39	5.15
G ratio				
5c	A_2 B_2 Optimal point A_2 B_2	44.68	39.15	13.01
6c	A_1 B_3 Random point A_1 B_3	39.27	42.66	8.63
Surface roughness R_a (μm)				
7c	A_3 B_3 Optimal point A_3 B_3	0.14	0.15	7.14
8c	A_2 B_2 Random point A_2 B_2	0.18	0.20	11.11
Surface roughness R_z (μm)				
9c	A_3 B_3 Optimal point A_3 B_3	1.11	1.14	2.70
10c	A_1 B_2 Random point A_1 B_2	2.29	2.12	7.42

15.3.4 Microstructure and Morphology Analysis

Surface roughness and surface morphology in surface features are important indexes evaluating surface quality in the grinding performance. Figure 6 shows SEM images of the workpiece obtained under different factors and levels. When different lubricants are used for different workpiece materials, workpiece surface morphologies are obviously different, indicating that they can exert a great influence on cutting and debris formation mechanisms. In terms of workpiece surface morphology, when NMQL is adopted, surface morphology of each workpiece material is better than that obtained under pure palm oil MQL condition, and the surface features clear texture without obvious surface defects like furrows and side flow, and especially, plastic deformation doesn't appear. On the contrary, surface morphology obtained through pure palm oil MQL is complicated, and plastic deformation layer, ploughing and material precipitation are serious on the surface. When abrasive grains at the bottom of grinding wheel surface are embedded into the workpiece surface for cutting, furrows are formed on workpiece surface due to insufficient lubrication. At the time, metal crystals of workpiece experience displacement under high temperature and high pressure, plastic ploughing occurs, metal particles will be re-precipitated on two furrow sides of workpiece by adhering to abrasive grains, thus forming a plastic deformation layer. This clearly shows that NMQL can improve lubricating properties at the grinding wheel/workpiece interface more effectively so that material removal condition is turned into shear, where nanoparticles form a very thin layer of friction oil film after precipitated on workpiece surface so as to exert anti-friction and anti-wear effects very well. According to EDX chemical composition analysis and detection, after nanofluid MQL was used (Figure 5), contents of related nanoparticle elements obviously increase on workpiece surface, indicating that nanoparticles are transported and precipitated on workpiece surface, which further proves existence of friction film. This result is consistent with conclusions drawn in section 15.3.1, namely: optimal grinding parameters for surface roughness R_a and R_z values are 45 steel (level 3) and Al_2O_3 nanofluid (level 3).

However, Figure 6 shows that workpiece surface quality is the optimal under Al_2O_3 nanofluid MQL grinding of nickel base alloy. Meanwhile, Table 6 shows that when force ratio is 0.38, surface roughness R_a value reaches 0.26 μm, which is the minimum value in the same group. This indicates that Al_2O_3 nanofluid obtains the highest efficiency as grinding fluid of nickel base ally. For surface morphology of 45 steel after grinding, plastic deformation is generated due to sliding of some metal crystal particles. Under the same external working conditions, MoS_2 nanofluid obtains higher surface quality than Al_2O_3 nanofluid. This is the case for both surface roughness value and surface morphology. In addition, according to Table 6, 45 steel obtains the lowest force ratio and the highest G ratio under MoS_2 nanofluid condition, possibly because compared with other MQL lubricants, MoS_2 nanofluid is more suitable to be lubricant for 45 steel machining.

Figure 7 shows grinding debris collected under different MQL grinding conditions. Grinding debris of the same material are similar, but grinding debris obtained under different lubricating environments are significantly different. This indicates that grinding performances of different workpiece materials grinded using different lubricants are different. Under pure oil MQL grinding, grinding debris are short and thick, especially serious scratches can be obviously seen on the undersurface of grinding debris. Compared with pure oil MQL grinding, grinding debris generated by all workpiece materials under nanofluid MQL grinding condition are thinned, especially undersurface of grinding debris become smooth, which is ascribed to superior lubricating properties of nanofluids. Nanofluids can be well adsorbed or embedded into texture of debris surface and furrows on workpiece surface to exert "bearing-

Figure 6. SEM micrographs of different ground workpieces in different lubricating environments

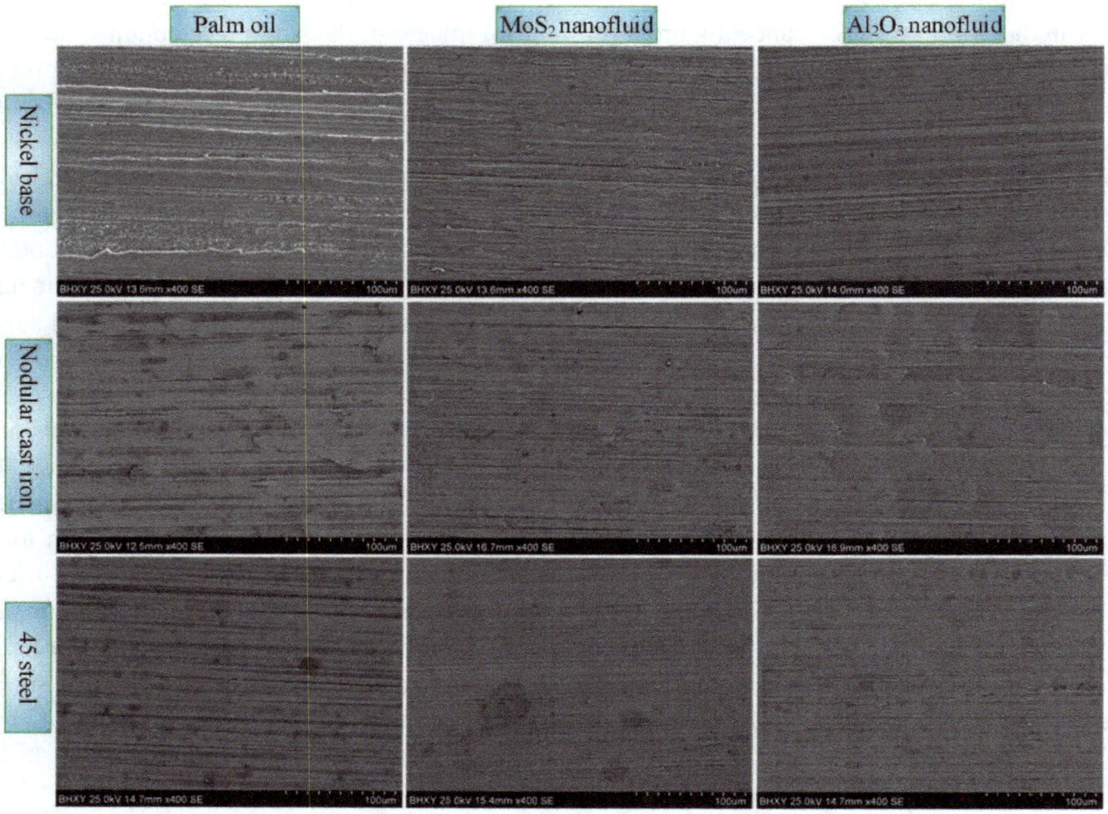

like" effect and act as lubricating media to form lubricating oil film which reduces friction coefficient between abrasive grain and undersurface of grinding debris. Correspondingly, a good workpiece surface morphology is formed (Figure 6).

Grinding machining is realized through cutting of workpiece surface material by a large quantity of single abrasive grains, so observed debris morphology can reflect material removal mechanism to a certain degree. For grinding debris of different workpiece materials, they are greatly different in shape, size and morphological features. This means that different materials have different debris formation mechanisms. In turn, debris formation mechanism of materials will also have a bearing on workpiece surface quality (Figure 6). Therefore, debris formation is surely associated with workpiece surface quality.

In the grinding process of nickel base alloy, grinding debris presents flaky "C" shape structure in Figure 7. As material is of high strength and high wear resistance, mechanical behaviors like shear slip and elastic-plastic deformation will be generated when abrasive grains are cutting the material, and moreover, as nickel base alloy is of low heat conductivity, enormous heat generated within a short time can't be timely diffused, which will cause thermal softening effect of the material in the main shear area. In addition, grinding is typical negative machining and high-speed grinding time is very short, grinding debris of alloy under high temperature is more approximate to adiabatic shear. Morphological features

Figure 7. SEM micrographs of different grinding debris in different lubricating environments

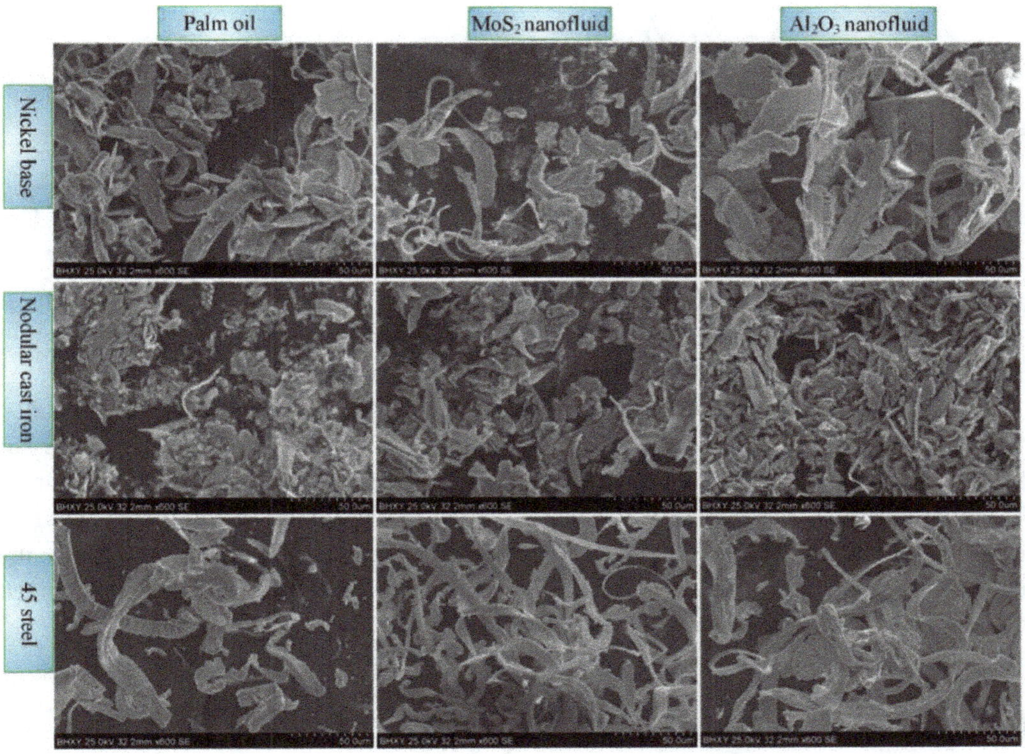

of grinding debris of nickel base alloy indicate that the removal mechanism and debris formation process of this material are different from ordinary materials. Chip formation mechanism is brittle rupture mechanism of brittle hard material as shown in Figure 8a. Workpiece surface quality of this material is poorer than other two materials. Under the grinding condition of nodular cast iron, its grinding debris is short mouldering-shaped debris from micromorphology shown in Figure 7. Nodular cast iron obtains spheroidal granite through balling and inoculation treatment, which effectively improves mechanical properties of cast iron, especially material tenacity is improved, so it obtains higher strength than carbon steel. During the grinding process, nodular cast iron contains much graphite which can exert cutting and lubricating effects. Therefore, cutting resistance of nodular cast iron is smaller than that of steel. When brittle material is machined, metal at the cutting layer experiences rupture without plastic deformation, thus generating irregular mouldering-shaped debris. Debris formation process is shown in Figure 8b. In addition, grinding debris are smaller due to serious crimping and impact. Fine debris are caused by attribution crushing of abrasive grains. Generation of mouldering-shaped debris influences workpiece surface roughness, because while mouldering-shaped debris are generated, a large quantity of heat and force are aggregated nearby abrasive grains. This also has a certain bearing on workpiece surface morphology (Figure 6). When 45 steel is grinded, the formed grinding debris present crimped long strip shape due to its own features and external conditions. Based on a comparison of grinding debris of other two materials, it can be seen from Figure 7 that grinding debris are long and thin under this working condition. 45 steel is medium carbon steel with very good strength and plasticity, but its hardness is

Figure 8. The formation mechanism of grinding debris of different materials

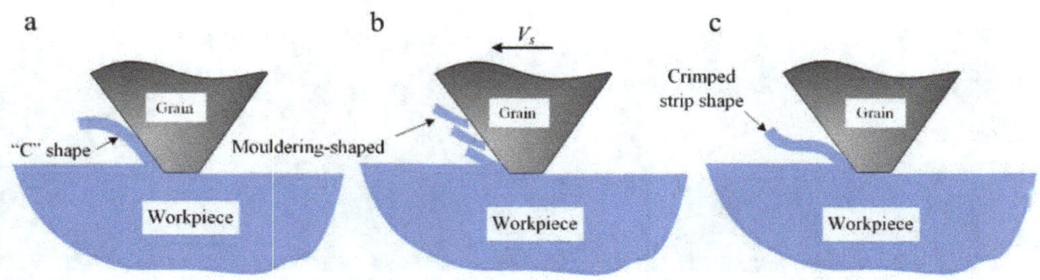

low, so it will easily form cutting machining, and a certain plastic deformation will occur in the grinding formation process. During the machining process, grinding debris present metal spring shape due to temperature different between upper and lower surfaces of grinding debris. If temperature of lower surface is higher than that of upper surface, heat stress will result in debris crimping as shown in Figure 8c. In addition, some long grinding debris are intertwined under high grinding force and temperature. Therefore, a high specific grinding energy is consumed in the material removal process, indicating that energy consumed by plastic removal is higher than that by brittle removal.

REFERENCES

Ali, M. K. A., Hou, X. J., Mai, L. Q., Cai, Q. P., Turkson, R. F., & Chen, B. C. (2016). Improving the tribological characteristics of piston ring assembly in automotive engines using Al_2O_3 and TiO_2 nanomaterials as nano-lubricant additives. *Tribology International*, *103*, 540–554. doi:10.1016/j.triboint.2016.08.011

Bai, X. F., Li, C. H., Dong, L., & Yin, Q. A. (2019). Experimental evaluation of the lubrication performances of different nanofluids for minimum quantity lubrication (MQL) in milling Ti-6Al-4V. *International Journal of Advanced Manufacturing Technology*, *101*(9-12), 2621–2632. doi:10.100700170-018-3100-9

Dambatta, Y. S., Sarhan, A. A. D., Sayuti, M. B., & Bin Abd Shukor, M. H. (2019). Fuzzy logic method to investigate grinding of alumina ceramic using minimum quantity lubrication. *International Journal of Applied Ceramic Technology*, *16*(4), 1668–1683. doi:10.1111/ijac.13219

Dambatta, Y. S., Sayuti, M., Sarhan, A. A. D., Bin Ab Shukor, H., Derahman, N. A. B., & Manladan, S. M. (2019). Prediction of specific grinding forces and surface roughness in machining of AL6061-T6 alloy using ANFIS technique. *Industrial Lubrication and Tribology*, *71*(2), 309–317. doi:10.1108/ILT-03-2018-0098

Dambatta, Y. S., Sayuti, M., Sarhan, A. A. D., Hamdi, M., Manladad, S. M., & Reddy, M. (2019). Tribological performance of SiO_2-based nanofluids in minimum quantity lubrication grinding of Si_3N_4 ceramic. *Journal of Manufacturing Processes*, *41*, 135–147. doi:10.1016/j.jmapro.2019.03.024

Dong, P. Q., Duc, T. M., & Long, T. T. (2019). Performance Evaluation of MQCL Hard Milling of SKD 11 Tool Steel Using MoS_2 Nanofluid. *Metals*, *9*(6), 658. doi:10.3390/met9060658

Emami, M., Sadeghi, M. H., Sarhan, A. A. D., & Hasani, F. (2014). Investigating the Minimum Quantity Lubrication in grinding of Al_2O_3 engineering ceramic. *Journal of Cleaner Production*, *66*, 632–643. doi:10.1016/j.jclepro.2013.11.018

Fernandes, L. D., Lopes, J. C., Volpato, R. S., Diniz, A. E., de Oliveira, R. F. M., de Aguiar, P. R., ... Bianchi, E. C. (2018). Comparative analysis of two CBN grinding wheels performance in nodular cast iron plunge grinding. *International Journal of Advanced Manufacturing Technology*, *98*(1-4), 237–249. doi:10.100700170-018-2133-4

Fox, N. J., & Stachowiak, G. W. (2007). Vegetable oil-based lubricants—A review of oxidation. *Tribology International*, *40*(7), 1035–1046. doi:10.1016/j.triboint.2006.10.001

Ganesan, K., Babu, M. N., Santhanakumar, M., & Muthukrishnan, N. (2018). Experimental investigation of copper nanofluid based minimum quantity lubrication in turning of H 11 steel. *Journal of the Brazilian Society of Mechanical Sciences and Engineering*, *40*(3), 160. doi:10.100740430-018-1093-9

Han, D. B., Guan, D. L., & Song, X. G. (2006). *Effect of Nano-Al_2O_3 in Lubricating Oil on the Characteristics of Wear and Friction of Cylinder Liner/Piston ring Rubbing Pair of Diesel Engine*. Materials for Mechanical Engineering.

Jamil, M., Khan, A. M., Hegab, H., Gong, L., Mia, M., Gupta, M. K., & He, N. (2019). Effects of hybrid Al_2O_3-CNT nanofluids and cryogenic cooling on machining of Ti-6Al-4V. *International Journal of Advanced Manufacturing Technology*, *102*(9-12), 3895–3909. doi:10.100700170-019-03485-9

Jia, D. Z. (2014). *Mechanism and experimental study on the effect of flow field characteristics of nanoparticle jet Micro-lubrication grinding on lubrication*. QingDao University of Technology.

Jia, D. Z., Li, C. H., Zhang, D. K., Zhang, Y. B., & Zhang, X. W. (2014). Experimental verification of nanoparticle jet minimum quantity lubrication effectiveness in grinding. *Journal of Nanoparticle Research*, *16*(12), 2758. doi:10.100711051-014-2758-7

Jia, D. Z., Li, C. H., Zhang, Y. B., Yang, M., Zhang, X. P., Li, R. Z., & Ji, H. J. (2019). Experimental evaluation of surface topographies of NMQL grinding ZrO_2 ceramics combining multiangle ultrasonic vibration. *International Journal of Advanced Manufacturing Technology*, *100*(1-4), 457–473. doi:10.100700170-018-2718-y

Kalita, P., Malshe, A. P., & Rajurkar, K. P. (2012). Study of tribo-chemical lubricant film formation during application of nanolubricants in minimum quantity lubrication (MQL) grinding. *Cirp Annals-Manufacturing Technology*, *61*(1), 327–330. doi:10.1016/j.cirp.2012.03.031

Khan, A. M., Jamil, M., Ul Haq, A., Hussain, S., Meng, L. H., & He, N. (2019). Sustainable machining. Modeling and optimization of temperature and surface roughness in the milling of AISI D2 steel. *Industrial Lubrication and Tribology*, *71*(2), 267–277. doi:10.1108/ILT-11-2017-0322

Kumar, A., Ghosh, S., & Aravindan, S. (2019). Experimental investigations on surface grinding of silicon nitride subjected to mono and hybrid nanofluids. *Ceramics International*, *45*(14), 17447–17466. doi:10.1016/j.ceramint.2019.05.307

Kumar, C. S., & Patel, S. K. (2018). Effect of WEDM surface texturing on Al_2O_3/TiCN composite ceramic tools in dry cutting of hardened steel. *Ceramics International*, *44*(2), 2510–2523. doi:10.1016/j.ceramint.2017.10.236

Lee, K., Hwang, Y., Cheong, S., Choi, Y., Kwon, L., Lee, J., & Kim, S. H. (2009). Understanding the Role of Nanoparticles in Nano-oil Lubrication. *Tribology Letters*, *35*(2), 127–131. doi:10.100711249-009-9441-7

Lee, P. H., Kim, J. W., & Lee, S. W. (2018). Experimental characterization on eco-friendly micro-grinding process of titanium alloy using air flow assisted electrospray lubrication with nanofluid. *Journal of Cleaner Production*, *201*, 452–462. doi:10.1016/j.jclepro.2018.07.307

Li, B. K., Li, C. H., Zhang, Y. B., Wang, Y. G., Jia, D. Z., Yang, M., ... Sun, K. (2017). Heat transfer performance of MQL grinding with different nanofluids for Ni-based alloys using vegetable oil. *Journal of Cleaner Production*, *154*, 1–11. doi:10.1016/j.jclepro.2017.03.213

Li, B. K., Li, C. H., Zhang, Y. B., Wang, Y. G., Yang, M., Jia, D. Z., ... Ding, W. F. (2017). Numerical and experimental research on the grinding temperature of minimum quantity lubrication cooling of different workpiece materials using vegetable oil-based nanofluids. *International Journal of Advanced Manufacturing Technology*, *93*(5-8), 1971–1988. doi:10.100700170-017-0643-0

Li, B. K., Li, C. H., Zhang, Y. B., Wang, Y. G., Yang, M., Jia, D. Z., ... Wu, Q. D. (2017). Effect of the physical properties of different vegetable oil-based nanofluids on MQLC grinding temperature of Ni-based alloy. *International Journal of Advanced Manufacturing Technology*, *89*(9-12), 3459–3474. doi:10.100700170-016-9324-7

Li, C. H., Zhang, D. K., Jia, D. Z., Wang, S., & Hou, Y. L. (2015). Experimental evaluation on tribological properties of nano-particle jet MQL grinding. *International Journal of Surface Science and Engineering*, *9*(2-3), 159–175. doi:10.1504/IJSURFSE.2015.068243

Li, M., Yu, T. B., Zhang, R. C., Yang, L., Li, H. Y., & Wang, W. S. (2018). MQL milling of TC_4 alloy by dispersing graphene into vegetable oil-based cutting fluid. *International Journal of Advanced Manufacturing Technology*, *99*(5-8), 1735–1753. doi:10.100700170-018-2576-7

Liu, G. T., Li, C. H., Zhang, Y. B., Yang, M., Jia, D. Z., Zhang, X. P., ... Zhai, H. (2018). Process parameter optimization and experimental evaluation for nanofluid MQL in grinding Ti-6Al-4V based on grey relational analysis. *Materials and Manufacturing Processes*, *33*(9), 950–963. doi:10.1080/10426914.2017.1388522

Luo, T., Wei, X. W., Huang, X., Huang, L., & Yang, F. (2014). Tribological properties of Al_2O_3 nanoparticles as lubricating oil additives. *Ceramics International*, *40*(5), 7143–7149. doi:10.1016/j.ceramint.2013.12.050

Mao, C., Huang, Y., Zhou, X., Gan, H., Zhang, J., & Zhou, Z. X. (2014). The tribological properties of nanofluid used in minimum quantity lubrication grinding. *International Journal of Advanced Manufacturing Technology*, *71*(5-8), 1221–1228. doi:10.100700170-013-5576-7

Mao, C., Tang, X. J., Zou, H. F., Huang, X. M., & Zhou, Z. X. (2012). Investigation of Grinding Characteristic using Nanofluid Minimum Quantity Lubrication. *International Journal of Precision Engineering and Manufacturing*, *13*(10), 1745–1752. doi:10.100712541-012-0229-6

Mao, C., Zhang, J., Huang, Y., Zou, H. F., Huang, X. M., & Zhou, Z. X. (2013). Investigation on the Effect of Nanofluid Parameters on MQL Grinding. *Materials and Manufacturing Processes*, *28*(4), 436–442. doi:10.1080/10426914.2013.763970

Nam, J., & Lee, S. W. (2018). Machinability of Titanium Alloy (Ti-6Al-4V) in Environmentally-Friendly Micro-Drilling Process with Nanofluid Minimum Quantity Lubrication Using Nanodiamond Particles. *International Journal Of Precision Engineering And Manufacturing-Green Technology*, *5*(1), 29–35. doi:10.100740684-018-0003-z

Nguyen, D., Lee, P. H., Guo, Y., Park, K. H., & Kwon, P. (2019). Wear Performance Evaluation of Minimum Quantity Lubrication With Exfoliated Graphite Nanoplatelets in Turning Titanium Alloy. *Journal Of Manufacturing Science And Engineering-Transactions Of the Asme*, *141*(8), 081006. doi:10.1115/1.4043899

Pashmforoush, F., & Bagherinia, R. D. (2018). Influence of water-based copper nanofluid on wheel loading and surface roughness during grinding of Inconel 738 superalloy. *Journal of Cleaner Production*, *178*, 363–372. doi:10.1016/j.jclepro.2018.01.003

Paul, S., & Ghosh, A. (2019). Suitability of aqueous MoS_2 nanofluid for small quantity cooling lubrication-assisted diamond grinding of WC-Co cermets. *Proceedings of the Institution of Mechanical Engineers. Part B, Journal of Engineering Manufacture*, *233*(2), 426–442. doi:10.1177/0954405417739037

Ping, C., & Gao, Q. (2010). *Tribological Characteristics of MoS_2 Film under Different Lubrication Conditions*. Lubrication Engineering.

Sayuti, M., Sarhan, A. A. D., Tanaka, T., Hamdi, M., & Saito, Y. (2013). Cutting force reduction and surface quality improvement in machining of aerospace duralumin AL-2017-T4 using carbon onion nanolubrication system. *International Journal of Advanced Manufacturing Technology*, *65*(9-12), 1493–1500. doi:10.100700170-012-4273-2

Sen, B., Mia, M., Gupta, M. K., Rahman, M. A., Mandal, U. K., & Mondal, S. P. (2019). Influence of Al_2O_3 and palm oil-mixed nano-fluid on machining performances of Inconel-690: IF-THEN rules-based FIS model in eco-benign milling. *International Journal of Advanced Manufacturing Technology*, *103*(9-12), 3389–3403. doi:10.100700170-019-03814-y

Seyedzavvar, M., Shabgard, M., & Mohammadpourfard, M. (2019). Investigation into the performance of eco-friendly graphite nanofluid as lubricant in MQL grinding. *Machining Science and Technology*, *23*(4), 569–594. doi:10.1080/10910344.2019.1575403

Sharma, A. K., Singh, R. K., Dixit, A. R., & Tiwari, A. K. (2017). Novel uses of alumina-MoS_2 hybrid nanoparticle enriched cutting fluid in hard turning of AISI 304 steel. *Journal of Manufacturing Processes*, *30*, 467–482. doi:10.1016/j.jmapro.2017.10.016

Shen, B., Malshe, A. P., & Kalita, P. (2008). Performance of novel MoS2 nanoparticles based grinding fluids in minimum quantity lubrication grinding. *Transactions of NAMRI/SME, 36*, 357-364.

Shen, B., Shih, A. J., & Tung, S. C. (2008). Application of Nanofluids in Minimum Quantity Lubrication Grinding. *Tribology Transactions*, *51*(6), 730–737. doi:10.1080/10402000802071277

Shi, Y., Wang, Z. H., Xu, S. Z., Yu, T. B., & Sun, Z. L. (2019). Study on the grindability of nano-vitrified bond CBN grinding wheel for nickel-based alloy. *International Journal of Advanced Manufacturing Technology*, *100*(5-8), 1913–1921. doi:10.100700170-018-2807-y

Uysal, A. (2018). Effects of nano graphene particles on surface roughness and cutting temperature during MQL milling of AISI 430 stainless steel. *Materials Testing*, *60*(5), 533–537. doi:10.3139/120.111184

Wang, Y. G., Li, C. G., Zhang, Y. B., Li, B. K., Yang, M., Zhang, X. P., ... Zhai, M. G. (2017). Comparative evaluation of the lubricating properties of vegetable-oil-based nanofluids between frictional test and grinding experiment. *Journal of Manufacturing Processes*, *26*, 94–104. doi:10.1016/j.jmapro.2017.02.001

Wang, Y. G., Li, C. H., Zhang, Y. B., Li, B. K., Yang, M., Zhang, X. P., ... Liu, G. T. (2016). Experimental evaluation of the lubrication properties of the wheel/workpiece interface in MQL grinding with different nanofluids. *Tribology International*, *99*, 198–210. doi:10.1016/j.triboint.2016.03.023

Xu, X. F., Lv, T., Luan, Z. Q., Zhao, Y. Y., Wang, M. H., & Hu, X. D. (2019). Capillary penetration mechanism and oil mist concentration of Al_2O_3 nanoparticle fluids in electrostatic minimum quantity lubrication (EMQL) milling. *International Journal of Advanced Manufacturing Technology*, *104*(5-8), 1937–1951. doi:10.100700170-019-04023-3

Yang, M., Li, C. H., Zhang, Y. B., Wang, Y. G., Li, B. K., & Hou, Y. L. (2017). Experimental research on microscale grinding temperature under different nanoparticle jet minimum quantity cooling. *Materials and Manufacturing Processes*, *32*(6), 589–597. doi:10.1080/10426914.2016.1176198

Yang, W. H., & Tarng, Y. S. (1998). Design optimization of cutting parameters for turning operations based on the Taguchi method. *Journal of Materials Processing Technology*, *84*(1–3), 122–129. doi:10.1016/S0924-0136(98)00079-X

Yildirim, C. V., Sarikaya, M., Kivak, T., & Sirin, S. (2019). The effect of addition of hBN nanoparticles to nanofluid-MQL on tool wear patterns, tool life, roughness and temperature in turning of Ni-based Inconel 625. *Tribology International*, *134*, 443–456. doi:10.1016/j.triboint.2019.02.027

Yin, Q. A., Li, C. H., Dong, L., Bai, X. F., Zhang, Y. B., Yang, M., ... Li, R. Z. (2018). Effects of the physicochemical properties of different nanoparticles on lubrication performance and experimental evaluation in the NMQL milling of Ti-6Al-4V. *International Journal of Advanced Manufacturing Technology*, *99*(9-12), 3091–3109. doi:10.100700170-018-2611-8

Yin, Q. A., Li, C. H., Zhang, Y. B., Yang, M., Jia, D. Z., Hou, Y. L., ... Dong, L. (2018). Spectral analysis and power spectral density evaluation in Al_2O_3 nanofluid minimum quantity lubrication milling of 45 steel. *International Journal of Advanced Manufacturing Technology*, *97*(1-4), 129–145. doi:10.100700170-018-1942-9

Yuan, S. M., Hou, X. B., Wang, L., & Chen, B. C. (2018). Experimental Investigation on the Compatibility of Nanoparticles with Vegetable Oils for Nanofluid Minimum Quantity Lubrication Machining. *Tribology Letters*, *66*(3), 106. doi:10.100711249-018-1059-1

Zhang, D. K., Li, C. H., Zhang, Y. B., Jia, D. Z., & Zhang, X. W. (2015). Experimental research on the energy ratio coefficient and specific grinding energy in nanoparticle jet MQL grinding. *International Journal of Advanced Manufacturing Technology*, *78*(5-8), 1275–1288. doi:10.100700170-014-6722-6

Zhang, J. C., Li, C. H., Zhang, Y. B., Yang, M., Jia, D. Z., Hou, Y. L., & Li, R. Z. (2018). Temperature field model and experimental verification on cryogenic air nanofluid minimum quantity lubrication grinding. *International Journal of Advanced Manufacturing Technology*, *97*(1-4), 209–228. doi:10.100700170-018-1936-7

Zhang, X. P., Li, C. H., Jia, D. Z., Gao, T., Zhang, Y. B., Yang, M., ... Ji, H. J. (2019). Spraying parameter optimization and microtopography evaluation in nanofluid minimum quantity lubrication grinding. *International Journal of Advanced Manufacturing Technology*, *103*(5-8), 2523–2539. doi:10.100700170-019-03642-0

Zhang, X. P., Li, C. H., Zhang, Y. B., Jia, D. Z., Li, B. K., Wang, Y. G., ... Zhang, X. W. (2016). Performances of Al_2O_3/SiC hybrid nanofluids in minimum-quantity lubrication grinding. *International Journal of Advanced Manufacturing Technology*, *86*(9-12), 3427–3441. doi:10.100700170-016-8453-3

Zhang, X. P., Li, C. H., Zhang, Y. B., Wang, Y. G., Li, B. K., Yang, M., ... Zhang, N. Q. (2017). Lubricating property of MQL grinding of Al_2O_3/SiC mixed nanofluid with different particle sizes and microtopography analysis by cross-correlation. *Precision Engineering-Journal Of the International Societies for Precision Engineering And Nanotechnology*, *47*, 532–545. doi:10.1016/j.precisioneng.2016.09.016

Zhang, Y. B., Chang-He, L. I., Jia, D. Z., & Zhang, D. K. (2015). *Experimental Evaluation into Lubricating Property of Nanoparticles Jet MQL Grinding Nickel Base Alloy*. Modular Machine Tool & Automatic Manufacturing Technique.

Zhang, Y. B., Li, C. H., Jia, D. Z., Zhang, D. K., & Zhang, X. W. (2015a). Experimental evaluation of MoS_2 nanoparticles in jet MQL grinding with different types of vegetable oil as base oil. *Journal of Cleaner Production*, *87*, 930–940. doi:10.1016/j.jclepro.2014.10.027

Zhang, Y. B., Li, C. H., Jia, D. Z., Zhang, D. K., & Zhang, X. W. (2015b). Experimental evaluation of the lubrication performance of MoS_2/CNT nanofluid for minimal quantity lubrication in Ni-based alloy grinding. *International Journal of Machine Tools & Manufacture*, *99*, 19–33. doi:10.1016/j.ijmachtools.2015.09.003

Chapter 16
Modeling and Simulation of Surface Topography in Single Abrasive Grain Grinding

ABSTRACT

This study establishes a kinematics model, elastic deformation model, and plastic accumulation model of a single grinding wheel, simulates the grain distribution on the surface of the common grinding wheel by using the grain vibration method, and examines the effect of different grinding parameters on the surface topography of the workpiece. Results show that the peaks and valleys on the profile curve of the workpiece surface increase, and the corresponding Ra and Rz heights decrease, as the peripheral velocity of the grinding wheel increases. The peaks and valleys on the profile curve of the workpiece surface decrease, and the corresponding Ra and Rz heights increase as the feed speed of the workpiece increases. Experiments are conducted to verify the simulation results. The results show that the simulation method can predict the surface roughness of the workpiece, which is a factor in selecting the grinding parameters.

16.1 INTRODUCTION

There are mainly two researches method on formation process of surface morphology: experimental method and modeling and simulation method Ekberg et al. (2019) and Shi et al. (2019). A great number of machining experiments are needed in the experimental method, and this method can only be used to study the finally generated workpiece surface morphology and it's difficult to reflect material removal mechanism in the machining process; modeling and simulation method realizes modeling of the machining process, comprehensively analyzes influences of various parameters on machining process, reveals removal mechanism and shorten experimental time. Surface roughness is an important parameter evaluating workpiece surface morphology. Establishing the relationship between grinding parameter and surface roughness will be of certain guiding effect on optimization of grinding parameters.

Domestic scholars and institutions have conducted certain theoretical and experimental research work for predicting workpiece surface morphology generation and surface roughness.

DOI: 10.4018/978-1-7998-1546-4.ch016

Onishi et al. (2014) from Suhzou University and his team conducted computer simulation of workpiece surface morphology generation in the grinding process, established a related mathematical grinding model and completed simulation of morphology generation process of workpiece surface in the surface grinding using MATLAB software. Material parameters and grinding parameters were considered in the simulation, but randomness of abrasive grains on grinding wheel surface was not fully taken into consideration in the simulation of grinding wheel.

Hong (2000) from Hongkong Polytechnic University established a 3D workpiece surface morphology model related to precision machining simulation. This model considered influence factors of workpiece surface morphology like geometric shape of tool, interference between tool and workpiece and cutting parameters, morphology generation of workpiece surface was simulated and experimental verification was implemented. Results showed that workpiece surface morphology generated by this model was consistent with that obtained through the experiment.

Chen (2010) from Tianjin University generated grinding wheel surface morphology using Johnson transform and linear filter technique, realizes simulation of generation process of workpiece surface morphology through precision grinding under different morphologies of grinding wheel and different machining conditions and verified correctness of surface morphology generation algorithm.

Hao et al. (2013) from Jilin University established relating grinding models based on time response, including grinding force model, grinding heat model and surface roughness model, through which generation process simulation of workpiece surface morphology was simulated and workpiece surface roughness was predicted. The model could be of a certain guiding effect on selection of grinding parameters and prediction of grinding quality.

Gong et al. (2002) from Northeastern University used virtual simulation technology, C++ programming language and OpenGL graphical tool to realize generation process simulation of workpiece surface morphology under surface grinding and built a simulation platform for surface morphology generation. Spherical abrasive grains were used on grinding wheel surface in this method, and influences of grinding wheel parameters and grinding parameters on surface morphology generation were considered. Su et al. 2008) used the same method to realize generation process simulation of surface morphology and investigated the relationship between different grinding parameters and workpiece surface roughness, and the only different was that abrasive grains used on grinding wheel surface in this simulation were cubes with different arrangement directions.

Foreign scholars have carried out a large quantity of theoretical and experimental studies on prediction of morphology generation and roughness of workpiece surface.

Sakakura et al. (2008) measured section profiles of 50 abrasive grains as standard abrasive grains which are randomly distributed on the surface of grinding wheel, and implemented digital simulation of grinding wheel morphology. In consideration of elastic deformation of abrasive grains, this grinding wheel was used to complete simulation of generation process of workpiece surface morphology in the cylindrical grinding process, and surface roughness of sampling area was calculated.

Shiau et al. (2011) from Michigan Technological University sampled grinding wheel surface morphologies using a surface roughometer, reconstructed sampled morphologies and implemented digital simulation of surface morphology of grinding wheel. Reconstructed grinding wheel morphology and related grinding model were used to complete morphology generation simulation of workpiece surface, and workpiece surface morphology was evaluated. However, elastic-plastic deformation of abrasive grains and workpiece was not considered in the simulation, so it's difficult to simulate the actual morphology generation process of workpiece surface.

Chakrabarti et al. (2008) used triangular pyramid abrasive grains, which were under uniform distribution, to generate grinding wheel surface morphology. All abrasive grains on the grinding wheel surface had the same angle and arrangement direction. According to the grinding kinematics, the morphology generation of workpiece surface was simulated, roughness value of workpiece surface was calculated, and then influence laws of grinding parameters on surface roughness and the relationship between surface roughness and maximum undeformed chip thickness were studied. The study indicated that as the surface roughness increased, maximum undeformed debris thickness presented linear growth.

Li and Rong (2011) studied manufacturing process of ordinary grinding wheel. A computer was used to realize simulation of grinding wheel manufacturing processes like abrasive grain-bonding agent mixing, grinding wheel modeling, grinding wheel roasting and grinding wheel dressing and generate surface morphology of ordinary grinding wheel, simulated generation process of workpiece surface morphology using this grinding wheel and analyzed grinding force, grinding heat and surface morphology.

Zhou and Xi (2002) conducted gridding of 3D grinding wheel entity. Abrasive grains of grinding wheel were located at different grid points, and surface morphology of grinding wheel was simulated. According to the theory of grinding kinematics, this grinding wheel was used to simulate grinding simulation, simulated morphology generation process of workpiece surface through the search method and conducted an experimental verification of workpiece surface roughness generated through simulation. Results showed that surface roughness output by simulation was smaller than experimental value. It's noteworthy that due to randomness of surface simulation of grinding wheel, it's difficult to simulate abrasive grains randomly distributed on the surface of ordinary grinding wheel through this method.

Grinding machining, an indispensable machining form, is usually applied to precision machining procedures of parts (Goh et al., 2019; Nguyen & Butler, 2005; Sun et al., 2009; Zhang et al., 2019). However, grinding machining of materials features low removal rate and high energy consumption, and most of this energy is converted into heat. Under high-speed revolution of grinding wheel, contact time between each abrasive grain and workpiece is extremely short, and meanwhile, debris volume generate din the grinding process is very small, so heat carried away by grinding debris is very restricted, most heat will be transferred into grinding wheel base and workpiece, and this will generate an adverse effect on workpiece surface morphology under the coupling effect with grinding force.

Professor Tönshoff from University of Hannover in Germany vividly compared the grinding process to "black art" in the CIRP report (Tönshoff et al., 2002; Zhang et al., 2019; Zhang et al., 2019). Grinding system converts input parameters and system parameters into all kinds of static and dynamic output variables, during which complicated changes may occur. If the grinding process is studied from the whole system, this will cause a certain difficulty to the research work. Meanwhile, abrasive grains are randomly distributed on the grinding wheel surface, which aggravates the research difficulty. Therefore, the current researches related to grinding usually take single abrasive grains as study objects and then transit into the whole grinding wheel body. Given this, models related to single-abrasive-grain grinding process are investigated in this chapter to lay a foundation for the follow-up research work.

16.2 ESTABLISHING OF WORKPIECE SURFACE COORDINATE SYSTEM

Workpiece surface morphology obtained through surface grinding is mainly studied in this thesis (Avery & Overton, 2019; Yin et al., 2019; Zhao et al., 2019). For the convenient of the establishment of follow-up models, the lowest contact point between abrasive grain and workpiece is origin O of coordinates

Figure 1. The coordinate of grinding workpiece surface

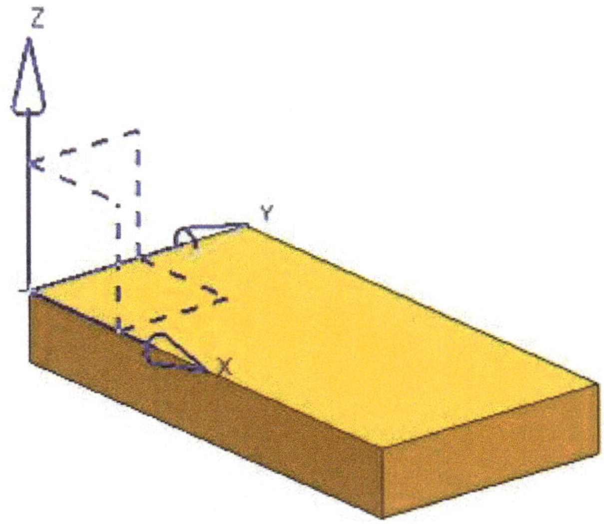

and coordinate system of workpiece surface is established, where axis x and axis y are longitudinal feed direction and transverse feed direction of the workbench respectively and axis z is radial feed direction of grinding wheel as shown in Figure 1.

16.3 KINEMATIC MODEL OF SINGLE ABRASIVE GRAIN

Due to complexity of the grinding process, simplification of the machining process is especially important in the kinematical simulation of grinding process. For the abovementioned reasons, the following hypotheses are made (Gupta et al., 2019; Sakakura et al., 2008; Zhou & Xi, 2002):

(1) Vibration of grinding wheel body is neglected;
(2) When abrasive grains interfere with workpiece surface, the workpiece material contacting abrasive grains will be totally removed;
(3) Generation of side flow and built-up edges in the material removal process is not considered.

Within the *XOZ* plane of coordinate system of workpiece surface, the lowest contact point between abrasive grain and workpiece is origin O of coordinates and rectangular coordinate system is built as shown in Figure 2.

Trajectory equation of abrasive grain G can be written as:

$$x = v_w t + r_s \sin\theta \tag{1}$$

Figure 2. The schematic diagram of single particle trajectory

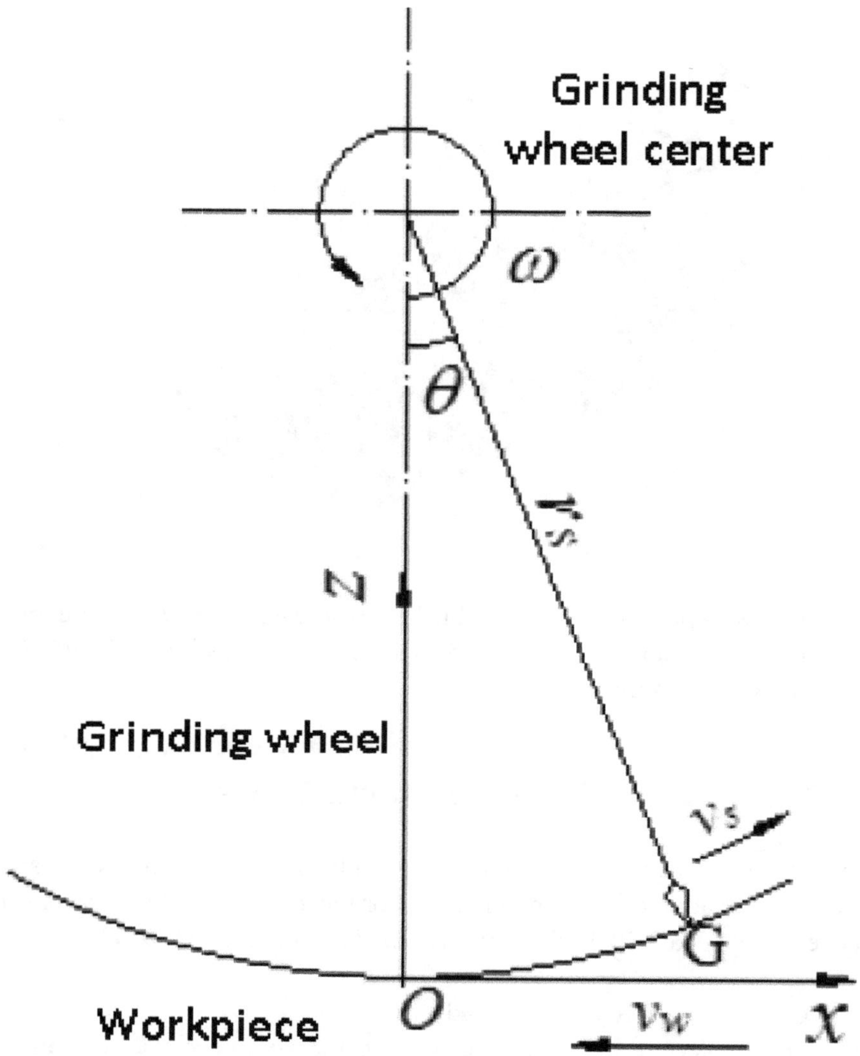

$$z = r_s\left(1 - \cos\theta\right) \tag{2}$$

where v_w——workpiece feed velocity;

r_s——radius of grinding wheel;
t——grinding time;
θ——rotation angle of grinding wheel, as θ is very small, $\sin\theta \approx \theta$ and $\theta^2 = 2(1-\cos\theta)$.

Linear velocity v_s of grinding wheel can be expressed as (Sakakura et al., 2008):

$$v_s = r_s \omega \tag{3}$$

where ——angular velocity of grinding wheel.

Rotation angle θ of grinding wheel can be expressed as (Meng et al., 2019; Xiong, 2019):

$$\theta = \omega t \tag{4}$$

Based on simultaneous equations (3) and (4), ω is eliminated, and grinding time t can be derived as:

$$t = \frac{r_s \theta}{v_s} \tag{5}$$

Equation (5) is substituted into equation (1) to obtain the following:

$$x = r_s \theta \left(\frac{v_w}{v_s} + 1 \right) \tag{6}$$

The following can be obtained by substituting θ^2=2(1-cosθ) into equation (2):

$$z = \frac{1}{2} r_s \theta^2 \tag{7}$$

Based on simultaneous equations (6) and (7), θ is eliminated and trajectory equation of abrasive grain G can be obtained as shown in equation (8).

$$\frac{z}{x^2} = \frac{1}{2r_s \left(\frac{v_w}{v_s} + 1 \right)^2} \tag{8}$$

It can be further written as:

$$z = \frac{x^2}{d_s \left(\frac{v_w}{v_s} + 1 \right)^2} \tag{9}$$

where d_s——diameter of grinding wheel.

Figure 3. The shapes of the abrasive grains

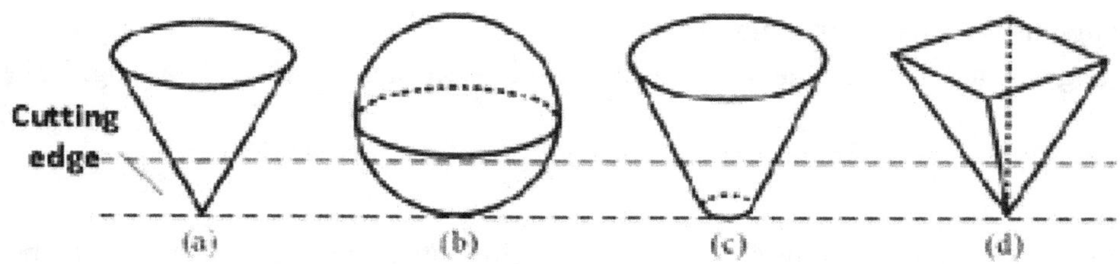

16.4 GEOMETRIC MODEL OF SINGLE ABRASIVE GRAIN

16.4.1 Shape of Abrasive Grain

Abrasive grains, as the main part participating in the cutting process, exert a major influence on workpiece surface quality (Wang et al., 2019). The present abrasive materials used to manufacture grinding wheel are (Dai et al., 2015): natural and man-made abrasive materials. The former is obtained by exploiting natural mines while the latter is obtained through manual smelting. Even though abrasive materials have different sources, they must pass through the same working procedure to obtain abrasive grains needed to manufacture grinding wheel namely crushing. After massy abrasive material is crushed, abrasive grains with different shapes and sizes will be generated (Sakthivel et al., 2019). (Rahmati et al., 2014) have divided shape of abrasive grains into four types: conical, spherical, truncated cone shape and rectangular pyramid shape as shown in Figure 3.

In the present grinding simulation, the most commonly used abrasive grains are conical, spherical and rectangular pyramid shaped. Through a scratching experimental study of single abrasive grains, results show that when cutting depth is large, cutting edge of abrasive grains can be regarded as conical; when cutting depth is small, it can be regarded as spherical. Meanwhile, grinding simulation for conical, spherical and truncated cone shaped abrasive grains indicates that after grinding wheel is dressed, workpiece surface morphology is unrelated to shape of abrasive grains (Dai et al., 2015).

In this paper, grinding depth simulated in the grinding process is small. According to the above conclusions, abrasive grains can be simplified as spherical shape, and meanwhile, minor cutting edges on the surface of abrasive grains are neglected for the convenient of simulation and improving simulation efficiency.

16.4.2 Size of Abrasive Grain

Size of abrasive grains is a very important technical parameter (Hwang et al., 2006). For abrasive materials with maximum size of abrasive grains being greater than 40 μm, mechanical sieving method is usually used to determine grit number of abrasive grains which is number of pores of the sieve which this abrasive grain can pass through on per inch length (Nguyen & Butler, 2005). According to GB/T6 406—2016 (size and composition of abrasive grains), the corresponding relationship between grit number and size of abrasive grains can be obtained as shown in Table 1.

Table 1. The abrasive grit number and grain size

Grid number	Grain size (μm)
20	1180~1000
22	1000~850
24	850~710
30	710~600
36	600~500
40	500~425
46	425~355
54	355~300
60	300~250
70	250~212
80	212~180
90	180~150
100	150~125

When grit number of abrasive grains is given as m, the corresponding scope of abrasive grain size under this grit number can be obtained by looking up Table 1, which is convenient to select proper abrasive grains and manufacture and produce the needed grinding wheel.

16.5 INTERACTION OF ABRASIVE GRAIN WITH WORKPIECE

16.5.1 Elastic Deformation Model

Stress acting upon normal line of a single abrasive grain can be regarded as identical with stress-bearing condition in the test of Brinell hardness (Gong et al., 2002; Wan et al., 2019; White & Lewis, 2019). Deformation process is restricted in an elastic-plastic deformation boundary. When the ball makes horizontal movement, plastic deformation area on the ball surface starts inclining (Sheng et al., 2014). Workpiece materials are stacked forward and torn from workpiece surface to form grinding debris as shown in Figure 4.

According to the stress analysis before motion of single abrasive grains in Figure 4 (a), positive pressure R borne by single abrasive grains can be obtained using the Brinell hardness test method (Wang et al., 2019):

$$R = \pi b^2 H(C' / 3) \tag{10}$$

where C'——ratio of average pressure to axial stress in the contact area, taken as 3 under general circumstances;

b——sectional radius of indentation, equal to 1/2 of grinding width of abrasive grains;

Figure 4. Stress analysis model of single abrasive grains: (a) before the motion (b) after the motion

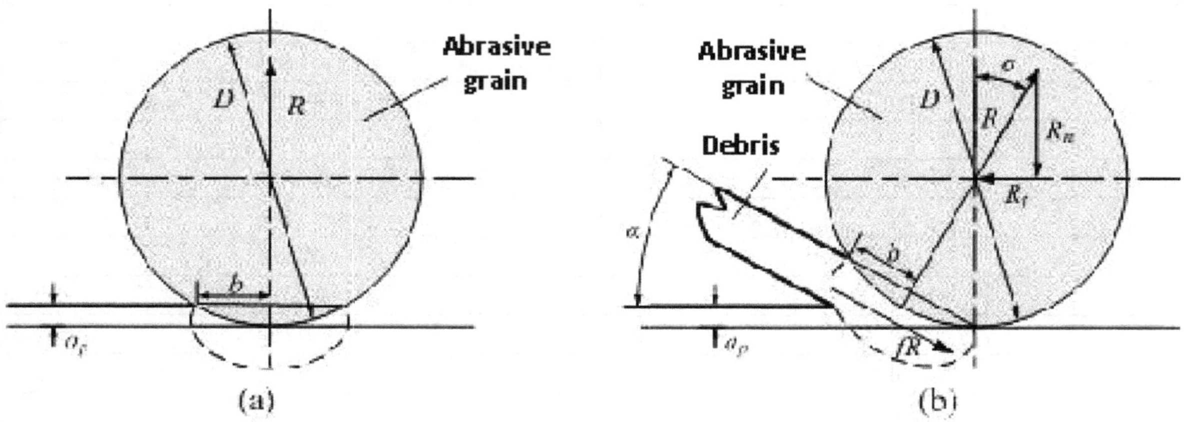

H——Brinell hardness of workpiece material.

According to the stress analysis after motion of single abrasive grains in Figure 4 (b), tangential and normal grinding forces of single abrasive grains can be expressed as:

$$\begin{cases} F_t = R(\sin\varphi + f\cos\varphi) \\ F_n = R(\cos\varphi - f\sin\varphi) \end{cases} \tag{11}$$

Where: *f*——friction coefficient between abrasive grains and workpiece, it is related to lubricating effect at the grinding wheel/workpiece interface, and it is usually taken as 0.4-0.55 under nanofluid MQL conditions;

φ——rotation angle of abrasive grains.

Rotation angle φ can be expressed as:

$$\varphi = \arccos\frac{D - 2a_p}{D} \tag{12}$$

where a_p——grinding depth of abrasive particles;

D——diameter of abrasive grains.

During the grinding process, as abrasive grains are supported by elasticity of bonding agent. Center of abrasive grains rises under the effect of grinding force so that actual contact curve between grinding wheel and workpiece is higher than theoretical contact curve (Liu et al., 2019; Su et al., 2008). In addi-

Figure 5. The interaction curves between abrasive grains and workpiece

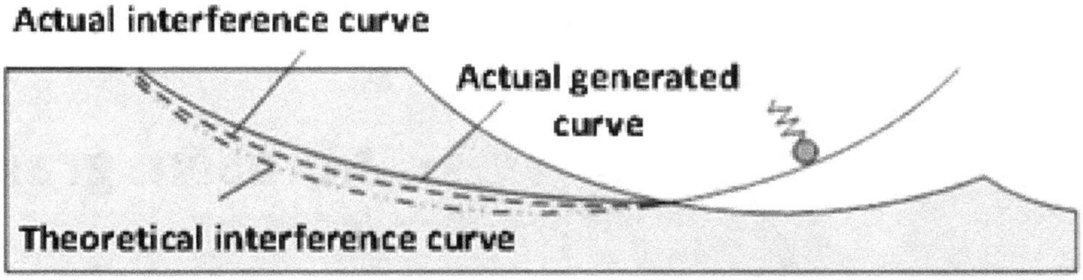

tion, elastic recovery will also occur on workpiece surface after grinding so that finally formed surface generation curve is higher than actual contact curve between grinding wheel and workpiece as shown in Figure 5.

Actual generation curve should be added with concession δ_c of abrasive grain center and elastic recovery δ_w of workpiece based on the theoretical trajectory. The following can be obtained through discretization processing of workpiece surface (2D) (Shipulin et al., 2019; Yaghmaei & Yazdanpanah, 2019):

$$Z_i = z_i + \delta_{ci} + \delta_{wi} \tag{13}$$

where Z_i——actual coordinate value of workpiece surface at point *I* after abrasive grains cut the workpiece;

z_i——theoretical coordinate value at point *i* after abrasive grains cut the workpiece;
δ_{ci}——concession of abrasive grain center at point *i*;
δ_w——elastic recovery of workpiece at point *i*.

Concession δ_c of abrasive grain center and elastic recovery δ_w of workpiece are respectively:

$$\delta_c = CF_n^{2/3} = C\left[R(\cos\theta - f\sin\theta)\right]^{2/3} \tag{14}$$

$$\delta_w = F_n / k = R(\cos\theta - f\sin\theta)/k \tag{15}$$

where *C*——constant, value range is 0.08~0.25 and average value is 0.15;

k——stiffness coefficient of workpiece.

Through the calculation, concession δ_c of abrasive grain center is much smaller than elastic recovery δ_w of workpiece, so elastic recovery of workpiece plays a leading role.

Figure 6. Plastic stacking model

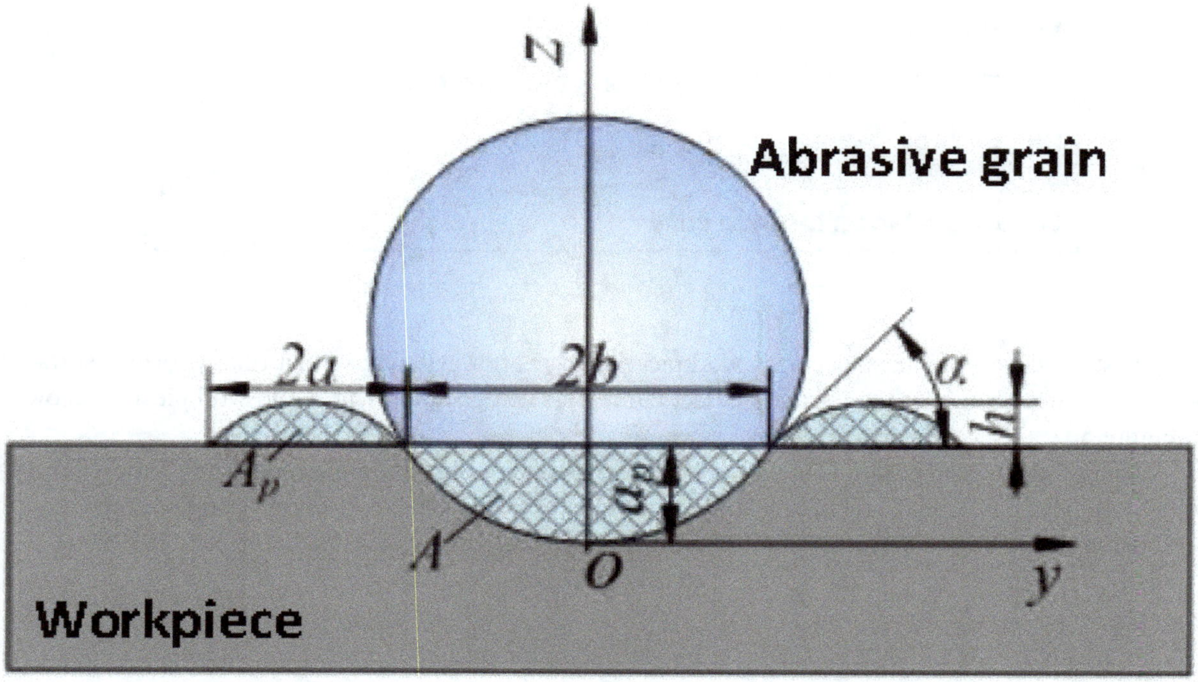

16.5.2 Plastic Accumulation Model

abrasive grains in the grinding process, and unremoved materials go through plastic deformation and are stacked at two sides of abrasive grains. Therefore, a cutting efficiency β can be introduced to quantify material stacking degree (Chakrabarti et al., 2008; You et al., 2019). Its definition is ratio of volume of removed undeformed material to total volume cut by abrasive grains in the cutting area. Schematic diagram of plastic stacking model is shown in Figure 6.

Area A_p of materials experiencing plastic deformation and stacked at two sides of abrasive grains is (Ekberg & Palsson, 2019):

$$A_p = \frac{(1-\beta)A}{2} \tag{16}$$

where A——volume of removed material;

In the YOZ plane of coordinate system of workpiece surface, the lowest contact point between abrasive grains and workpiece is origin O of coordinates, rectangular coordinate system is built, and contour line equation of lower hemisphere is:

$$z = \frac{D}{2} - \sqrt{\left(\frac{D}{2}\right)^2 - y^2} \tag{17}$$

where D——diameter of abrasive grains.

Grinding width of abrasive grains is $2b$, and then b is:

$$b=\sqrt{\left(\frac{D}{2}\right)^2-\left(\frac{D}{2}-a_p\right)^2} \tag{18}$$

Stacking shape of materials at two sides of abrasive grains can be approximately regarded as a parabola, intersection point between axis of this parabola and workpiece surface is taken as origin of coordinates to establish the local coordinate system, and it's set that workpiece materials are stacked at two sides along the direction of angle α, sectional height and width of material stacking on workpiece surface are h and$2a$ respectively, and equation of this parabola in this local coordinate system is:

$$z_1=h-\frac{hy_1^2}{a^2} \tag{19}$$

Profile equation in the global coordinate system is:

$$z=a_p+h-\frac{h(y-a-b)^2}{a^2} \tag{20}$$

Area A of removed material is:

$$A=\frac{\alpha D^2}{2}-\frac{Db}{2}+ba_p \tag{21}$$

Area A_p of stacked material is:

$$A_p=4ah/3 \tag{22}$$

As workpiece materials are stacked at two sides along the direction of angle α, tangent slope k at left intersection point between parabola and axis y_1 in the local coordinate system should be:

$$k=2h/a=\tan\alpha \tag{23}$$

A and h values can be derived as:

$$a=\sqrt{\frac{3A_p}{2\tan\alpha}} \tag{24}$$

$$h = \sqrt{\frac{3A_p \tan\alpha}{8}} \tag{25}$$

where $\tan\alpha = a_p / b$.

16.6 UNDEFORMED CHIP THICKNESS

Undeformed chip thickness, though being an assumed sectional size of chips, fully reflects the influence of grinding parameters on chip layer (Aggestam & Nielsen, 2019; Liao et al., 2019; Sheng et al., 2014). Cylindrical cut-in grinding experiment indicates that maximum undeformed chip thickness presents linear relations with workpiece surface roughness, material removal rate and grinding force. Therefore, it can reveal the relationship between grinding parameters and surface roughness very well (Dai et al., 2015; Zhou & Xi, 2002).

There are three main calculation methods of maximum undeformed chip thickness: average volume method at cutting layer, average chip sectional area method and geometrical relationship method of grinding (Li et al., 2019; Rahmati et al., 2014; Xu et al., 2019).

1. Average Volume Method at Cutting Layer

Average volume method at cutting layer refers to calculate the maximum undeformed chip thickness by calculating average volume at the cutting layer (Li & Rong, 2011; Rowe et al., 1997). Average volume at cutting layer refers to the ratio of volume of material removed within unit time to number of abrasive grains participating in the grinding process, namely:

$$\bar{a}_g \bar{b}_g l_s = \frac{v_w a_p B}{v_s B N_d} \tag{26}$$

where $\bar{a}_g$ —— average cutting thickness grinded by each abrasive grain;

$\bar{b}_g$ ——average cutting width grinded by each abrasive grain;
l_s——contact arc between grinding wheel and workpiece;
N_d——number of abrasive grains participating in the grinding process;
B——grinding width of grinding wheel.

Equation (16-26) is organized, and average cutting thickness $\bar{a}_g$ can be expressed as:

$$\bar{a}_g = \left[\frac{v_w}{v_s} a_p\right]\left[N_d \bar{b}_g l_s\right]^{-1} \tag{27}$$

Maximum undeformed chip thickness can be expressed as:

$$\bar{a}_{g\max} = 2\bar{a}_g = \left[\frac{v_w}{v_s}a_p\right]\left[\frac{N_d \bar{b}_g l_s}{2}\right]^{-1} \tag{28}$$

2. Average Sectional Area Method of Chips

Average sectional area method of chips refers to calculating maximum undeformed chip thickness by calculating average sectional area of chips (Mokarram et al., 2019; Zhang et al., 2015). Sectional area of chips is equal to the ratio of sum of sectional areas of materials removed by grinding wheel within unit width to number of dynamic abrasive grains participating in the grinding process on contact surface within unit width of grinding wheel, namely

$$\bar{a}_g \bar{b}_g = \frac{v_w a_p B / B v_s}{N_{dl} l_s / \bar{b}_g} \tag{29}$$

Equation (29) is organized, and average cutting thickness $\bar{a}_g$ can be expressed as:

$$\bar{a}_g = \left[\frac{v_w a_p}{v_s}\right]\left[\frac{1}{N_{dl} l_s}\right] \tag{30}$$

Maximum undeformed chip thickness can be expressed as:

$$\bar{a}_{g\max} = 2\bar{a}_g = \left[\frac{v_w a_p}{v_s}\right]\left[\frac{1}{N_{dl} l_s}\right] \tag{31}$$

where N_{dl}——number of dynamic abrasive grains participating in the grinding process on unit width of grinding wheel.

Calculation formula of number N_{dl} of dynamic abrasive grains on unit width is:

$$N_{dl} = A_{g1}^{p}\left[2^{p} c_1 k_s\right]^{\frac{1}{p+1}}\left[\frac{v_w}{v_s}\right]^{\frac{1}{p+1}}\left[\frac{a_p}{d_e}\right]^{\frac{p}{2(p+1)}} \tag{32}$$

where A_{g1}——number of dynamic abrasive grains participating in the grinding process on unit width of grinding wheel;

c_1——coefficient related to density of cutting edges on grinding wheel;
k_s——coefficient related to shape of cutting edges on grinding wheel;
p——index;

Figure 7. The interference diagram of wheel and workpiece in cylindrical grinding

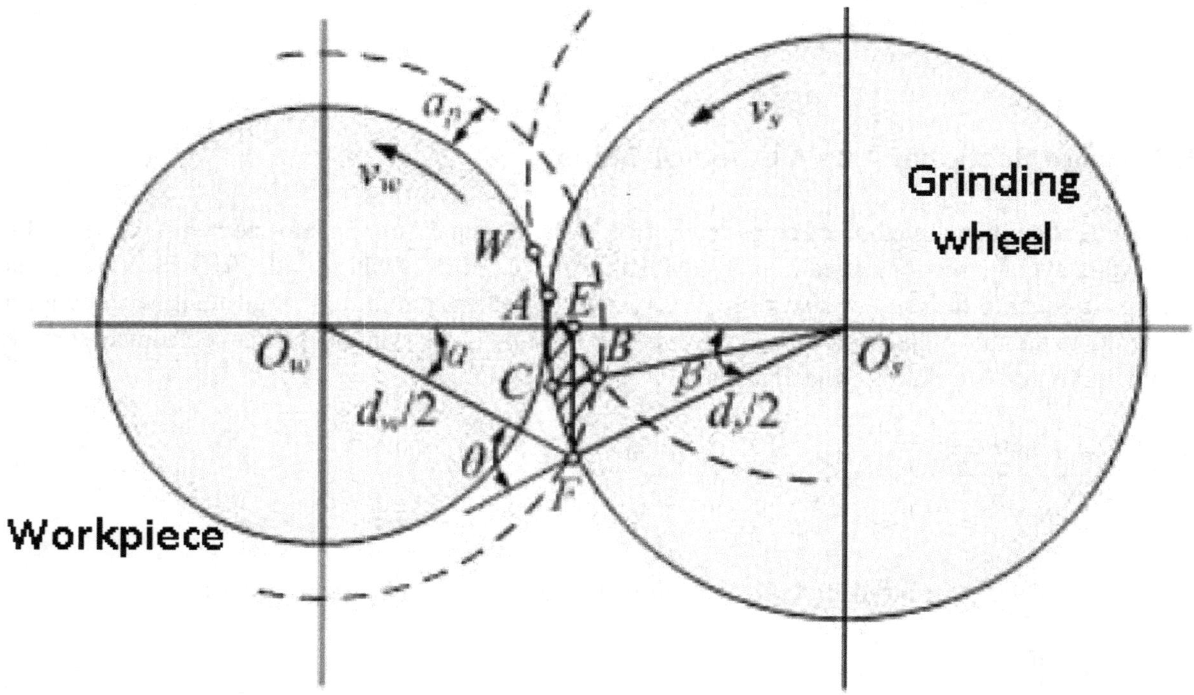

d_e——equivalent diameter of grinding wheel.

The following can be obtained by substituting equation (32) into (31):

$$\bar{a}_{g\max} = \frac{1}{A_{g1}^{p}}\left[\frac{2}{c_1 k_s}\right]^{\frac{1}{p+1}}\left[\frac{v_w}{v_s}\right]^{\frac{1}{p+1}}\left[\frac{a_p}{d_e}\right]^{\frac{1}{2(p+1)}} \tag{33}$$

3.Geometrical Relationship Method of Grinding

Geometrical relationship method of grinding refers to calculating maximum undeformed chip thickness by studying geometrical interference relationship between grinding wheel and workpiece (Ni et al., 2019; Zhang et al., 2015). Figure 7 shows schematic diagram of interference between grinding wheel and workpiece in cylindrical grinding.

In cylindrical grinding, cutting trajectory of cutting edge of one abrasive grain on the grinding wheel surface is $\overline{WB}$, that of cutting edge of the next adjacent abrasive grain is $\overline{AF}$, and then ABF sectional area is a chip section. $\overline{BC}$ is maximum undeformed chip thickness, and then:

$$\overline{BC} = \overline{BF}\sin\theta \tag{34}$$

If spacing of grinding edges under continuous grinding is λ_{sl}, then:

$$\overline{BF} = \lambda_{sl} \frac{v_w}{v_s} \tag{35}$$

As grinding depth is far smaller than diameter of workpiece and grinding wheel, $\angle\alpha$ and $\angle\beta$ are very small, and then:

$$\sin\theta = \sin(\alpha+\beta) \approx \sin\alpha + \sin\beta = 2\overline{EF}\left(\frac{1}{d_s}+\frac{1}{d_w}\right) = 2\sqrt{a_p\left(\frac{d_w+d_s}{d_w d_s}\right)} \tag{36}$$

where d_w——diameter of workpiece.

Equation (2-35) is substituted into (2-36), and then maximum undeformed chip thickness under cylindrical grinding is:

$$a_{g\max} = 2\lambda_{sl}\frac{v_w}{v_s}\sqrt{a_p\left(\frac{d_w+d_s}{d_w d_s}\right)} \tag{37}$$

Correspondingly, maximum undeformed chip thickness under internal grinding is:

$$a_{g\max} = 2\lambda_{sl}\frac{v_w}{v_s}\sqrt{a_p\left(\frac{d_w-d_s}{d_w d_s}\right)} \tag{38}$$

Maximum undeformed chip thickness under surface grinding is:

$$a_{g\max} = 2\lambda_{sl}\frac{v_w}{v_s}\sqrt{\frac{a_p}{d_s}} \tag{39}$$

16.7 SURFACE MORPHOLOGY MATRIX OF WORKPIECE

Height value at any point on workpiece surface is z_{mn}, and then morphology matrix of workpiece surface is:

$$Z(z_{mn}) = \begin{bmatrix} z_{11} & z_{12} & \cdots & z_{1n} \\ z_{21} & z_{22} & \cdots & z_{2n} \\ \vdots & \vdots & \ddots & \vdots \\ z_{m1} & z_{m2} & \cdots & z_{mn} \end{bmatrix} \tag{40}$$

where m and n are numbers of discrete points in width direction and length direction of workpiece respectively.

16.8 SIMULATION OF THE SINGLE GRAIN GRINDING PROCESS

According to the above single-abrasive-grain kinematical model, geometrical model of abrasive grains, elastic deformation model and plastic stacking model, MATLAB programming language is used to program single-abrasive-grain surface grinding process. Simulation analysis of this process is realized through numerical operation of the computer.

16.8.1 Simulation Parameters

Simulation parameters used in the simulation of single-abrasive-grain grinding process include grinding wheel parameters, workpiece parameters and grinding parameters as shown in Table. 2, 3 and 4 respectively.

As mainly corundum grinding wheel with grain size of 80# is simulated in this paper, average grain size of abrasive material with grain size D of 80# in Table. 2 is selected as simulation parameter.

Table 2. The parameters of grinding wheel

Diameter of grinding wheel d_s	Diameter of abrasive grains D
300 mm	196 μm

Table 3. The parameters of workpiece

Material type	Rockwell hardness	Stiffness coefficient k
Quenched 45# steel	HRC=45	300 kg/mm

Table 4. The parameters of grinding process

Linear velocity of grinding wheel v_s	Workbench feed velocity v_w	Grinding depth a_p
30 m/s	0.05 m/s	20 μm

Figure 8. The grinding simulation flow diagram of single grain

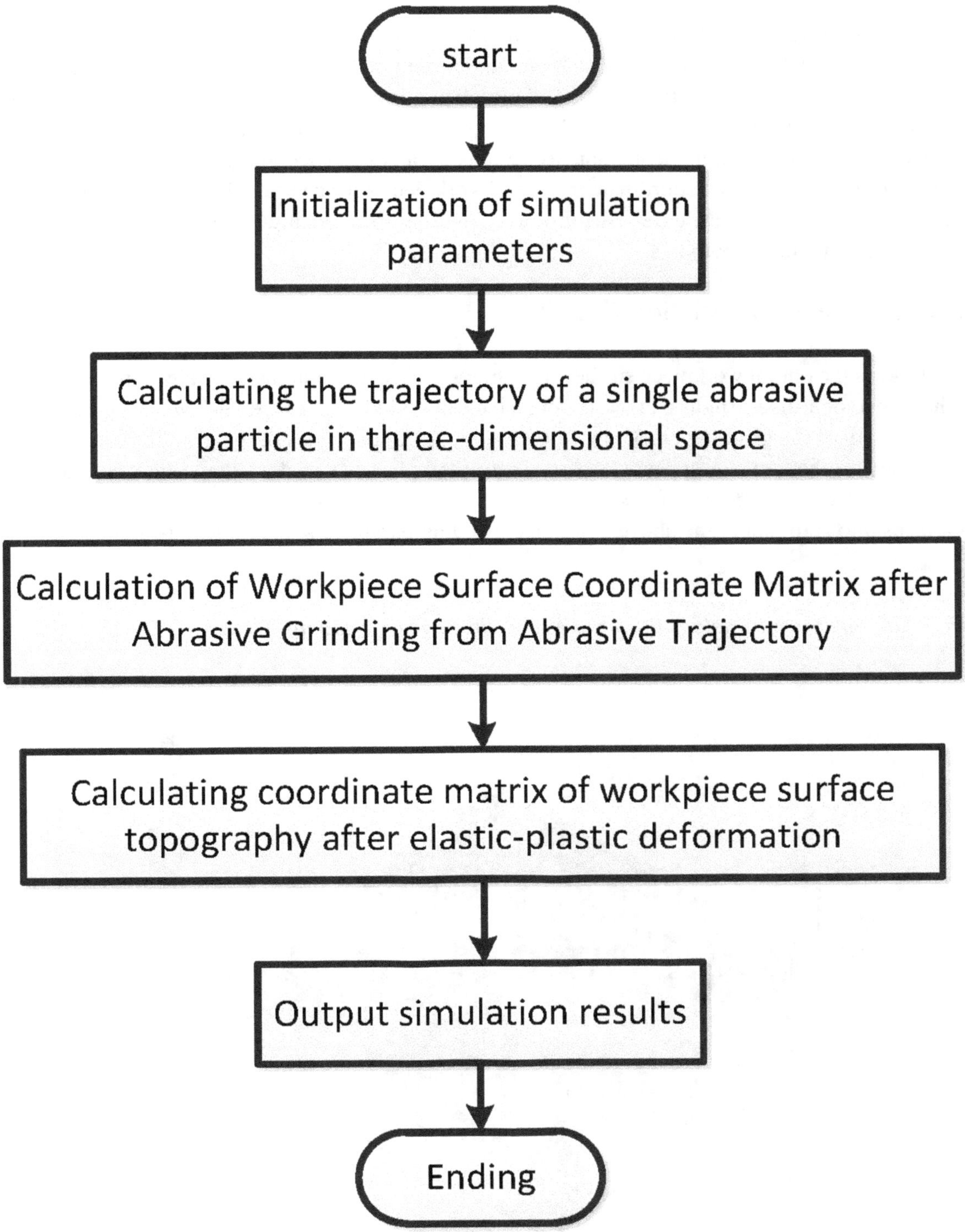

16.8.2 Procedures of Simulation Program

Figure 8 shows simulation flowchart of single-abrasive-grain grinding process. Concrete simulation flow is (Li et al., 2019; Shiau et al., 2011): simulation parameters of single-abrasive-grain grinding are firstly input, and the simulation process is initialized; motion trajectory of single abrasive grain in the 3D space is solved using the single-abrasive-grain kinematical model, and coordinate matrix of workpiece surface after grinding using abrasive grains is calculated according to the motion trajectory calculation of abrasive grains; coordinate matrix of workpiece surface after elastic-plastic deformation is calculated according to elastic deformation model and plastic stacking model is calculated, and the simulation result is finally output.

16.8.3 Analysis of Simulation Results

Figure 9 shows the output workpiece surface morphology after single-abrasive-grain grinding. It can be seen that a parabola-like grinding crack is generated on workpiece surface after single-abrasive-grain grinding, length of the grinding crack is about 2.4 mm, and at two sides of the grinding crack, obvious material stacking phenomenon appears. The maximum workpiece material stacking height output through simulation is 4.82 mm, and it reduces successively with motion of abrasive grains on workpiece surface until it turns into zero; under the abovementioned grinding parameters, theoretical grinding depth is 20

Figure 9 The surface morphology of workpiece by single grain grinding

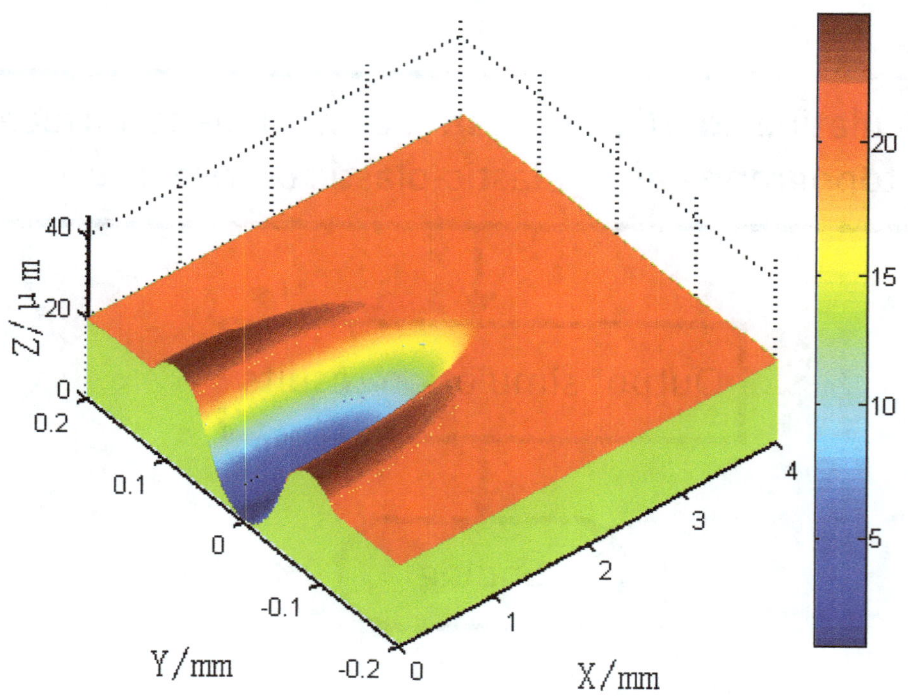

μm, but actual maximum grinding depth output through simulation is 19.12 μm which is smaller than theoretical grinding depth by 0.88 μm, indicating that elastic-plastic deformation on workpiece surface will generate a great influence on surface morphology. Among related studies regarding workpiece surface morphology under surface grinding, elastic-plastic deformation on workpiece surface is nonnegligible.

Single-abrasive-grain kinematical model, geometrical model of abrasive grains, elastic deformation model and plastic stacking model are established for single-abrasive-grain surface grinding, and workpiece surface morphology after single-abrasive-grain grinding is studied, and results show that obvious material stacking phenomenon appears at two sides of grinding crack after single-abrasive-grain grinding, and meanwhile, elastic-plastic deformation occurs on workpiece surface at the grinding crack, and this deformation is nonnegligible in the study of workpiece surface morphology.

REFERENCES

Aggestam, E., & Nielsen, J. C. O. (2019). Multi-objective optimisation of transition zones between slab track and ballasted track using a genetic algorithm. *Journal of Sound and Vibration*, *446*, 91–112. doi:10.1016/j.jsv.2019.01.027

Avery, J. J., & Overton, S. H. (2019). The Wheel in the Lab: Innovative Learning Strategy With Simulation. *Clinical Simulation in Nursing*, *35*, 1–4. doi:10.1016/j.ecns.2019.05.007

Chakrabarti, S., & Paul, S. (2008). Numerical modelling of surface topography in superabrasive grinding. *International Journal of Advanced Manufacturing Technology*, *39*(1-2), 29–38. doi:10.100700170-007-1201-y

Chen, D. (2010). Modeling and Simulation Methodology of the Machined Surface in Ultra-precision Grinding. *Journal of Mechanical Engineering*, *46*(13), 186. doi:10.3901/JME.2010.13.186

Dai, J., Ding, W., Zhang, L., Xu, J., & Su, H. (2015). Understanding the effects of grinding speed and undeformed chip thickness on the chip formation in high-speed grinding. *International Journal of Advanced Manufacturing Technology*, *81*(5-8), 995–1005. doi:10.100700170-015-7265-1

Ekberg, A., & Palsson, B. A. (2019). Multiscale modelling of train-track interaction phenomena with focus on contact mechanics. *Wear*, *430*, 393–400. doi:10.1016/j.wear.2019.05.018

Goh, H. P., Heng, P. W. S., & Liew, C. V. (2019). The Effects of Feed Frame Parameters and Turret Speed on Mini-Tablet Compression. *Journal of Pharmaceutical Sciences*, *108*(3), 1161–1171. doi:10.1016/j.xphs.2018.09.005 PMID:30237030

Gong, Y. D., Wang, B., & Wang, W. S. (2002). The Simulation of Grinding Wheels and Ground Surface Roughness Based on Virtual Reality Technology. *Journal of Materials Processing Technology*, *129*(1), 123–126. doi:10.1016/S0924-0136(02)00589-7

Gupta, P., Wu, Y., He, X. Y., Zhuge, W. L., & Ma, F. H. (2019). Life cycle analysis of HCNG light-duty vehicle demonstration project. *Materia-Rio De Janeiro*, *24*(2). doi:1517-7 07620190002.0696 doi:10.1590

Hao, T., Deng, Z., Wei, L., Wan, L., & Wu, Q. (2013). Present research situation and future perspectives of grinding process simulation system. *Key Engineering Materials*, *531-532*, 759–764.

Hong, K. (2000). Prediction of 3-D Surface Topography in Ultra-precision Machining. *Zhongguo Jixie Gongcheng*.

Hwang, Y., Park, H. S., Lee, J. K., & Jung, W. H. (2006). Thermal conductivity and lubrication characteristics of nanofluids. *Current Applied Physics*, *6*(1), e67–e71. doi:10.1016/j.cap.2006.01.014

Li, C. Y., Ji, C. J., Shi, M., & Sun, Q. (2019). Influence of rotor-stator component partition structure on the aerodynamic performance of centrifugal compressors. *Engineering Applications of Computational Fluid Mechanics*, *13*(1), 1080–1094. doi:10.1080/19942060.2019.1669075

Li, L., Li, Y., Yang, M. S., Xiao, X. D., Cui, L. M., & Cui, F. K. (2019). Effects of process parameters on forming force and accuracy in cold roll-beating forming external tooth groove. *International Journal of Advanced Manufacturing Technology*, *100*(9-12), 2229–2242. doi:10.100700170-018-2844-6

Li, X., & Rong, Y. (2011). *Framework of grinding process modeling and simulation based on microscopic interaction analysis*. Academic Press.

Liao, J. Y., Xie, J., Yang, H., & Sun, J. X. (2019). Modeling of curved diamond wheel errors for improvement of freeform grinding accuracy. *International Journal of Advanced Manufacturing Technology*, *103*(5-8), 1879–1892. doi:10.100700170-019-03679-1

Liu, Z. M., Tang, Q., Liu, N., & Song, J. (2019). A profile error compensation method in precision grinding of screw rotors. *International Journal of Advanced Manufacturing Technology*, *100*(9-12), 2557–2567. doi:10.100700170-018-2841-9

Meng, Q. H., Qian, C. J., & Sun, Z. Y. (2019). Finite-time stability control of an electric vehicle under tyre blowout. *Transactions of the Institute of Measurement and Control*, *41*(5), 1395–1404. doi:10.1177/0142331218780967

Mokarram, M., Khoei, A., & Hadidi, K. (2019). A fuzzy Anti-lock braking system (ABS) controller using CMOS circuits. *Microprocessors and Microsystems*, *70*, 47–52. doi:10.1016/j.micpro.2019.05.001

Nguyen, T. A., & Butler, D. L. (2005). Simulation of precision grinding process, part 1: Generation of the grinding wheel surface. *International Journal of Machine Tools & Manufacture*, *45*(11), 1321–1328. doi:10.1016/j.ijmachtools.2005.01.005

Ni, J., Feng, K., Al-Furjan, M. S. H., Xu, X. J., & Xu, J. (2019). Establishment and Verification of the Cutting Grinding Force Model for the Disc Wheel Based on Piezoelectric Sensors. *Sensors (Basel)*, *19*(3), 725. doi:10.339019030725 PMID:30754669

Onishi, T., Sakakura, M., Nakano, Y., Harada, M., Kawakami, K., Sakane, T., ... Tsukamoto, S. (2014). Simulation Analysis and In-Process Measurement of the Workpiece Temperature Distribution in Large Surface Grinding. *Key Engineering Materials*, *656-657*(24), 353–356.

Rahmati, B., Sarhan, A. A. D., & Sayuti, M. (2014). Investigating the optimum molybdenum disulfide (MoS2) nanolubrication parameters in CNC milling of AL6061-T6 alloy. *International Journal of Advanced Manufacturing Technology*, *70*(5-8), 1143–1155. doi:10.100700170-013-5334-x

Rowe, W. B., Black, S. C. E., Mills, B., Morgan, M. N., & Qi, H. S. (1997). Grinding Temperatures and Energy Partitioning. *Proceedings Mathematical Physical & Engineering Sciences, 453*(1960), 1083-1104.

Sakakura, M., Tsukamoto, S., Fujiwara, T., & Inasaki, I. (2008). Visual Simulation of Grinding Process. *Intelligent Production Machines & Systems*, *222*(10), 1233–1239.

Sakthivel, R., Mohanapriya, S., & Almakhles, D. J. (2019). Robust Tracking and Disturbance Rejection Performance for Vehicle Dynamics. *IEEE Access: Practical Innovations, Open Solutions*, *7*, 118598–118607. doi:10.1109/ACCESS.2019.2936655

Sheng, W., Li, C., Zhang, D., Jia, D., & Zhang, Y. (2014). Modeling the operation of a common grinding wheel with nanoparticle jet flow minimal quantity lubrication. *International Journal of Advanced Manufacturing Technology*, *74*(5-8), 835–850. doi:10.100700170-014-6032-z

Shi, X. L., Zhou, Y. S., Zhang, W. J., & Tang, J. Y. (2019). A new worm grinding method of face gears based on the optimization of dressing wheel profile. *Forschung Im Ingenieurwesen-Engineering Research*, *83*(3), 751–757. doi:10.100710010-019-00353-6

Shiau, T. N., Chen, K. H., Wang, F. C., Chio, C. T., & Hsu, W. C. (2011). The effect of dynamic behavior on surface roughness of ball screw under the grinding force. *International Journal of Advanced Manufacturing Technology*, *52*(5-8), 507–520. doi:10.100700170-010-2731-2

Shipulin, L. V., Ardashev, D. V., & Dyakonov, A. A. (2019). Description of topography of grinded surface during flat grinding with the periphery of a wheel. *Surface Topography-Metrology and Properties*, *7*(2), 025007. doi:10.1088/2051-672X/ab1631

Su, C., Zhu, L. D., Hou, J. M., & Wang, W. S. (2008). Research and Development of Simulation System for Virtual Grinding. *International Symposium on Computer Science & Computational Technology*. 10.1109/ISCSCT.2008.264

Sun, L. (2009). Prediction for Surface Roughness of Profile Grinding and Optimization of Grinding Parameters Based on Least Squares Support Vector Machine. *Journal of Mechanical Engineering*, *45*(10), 254–260. doi:10.3901/JME.2009.10.254

Tönshoff, H. K., Friemuth, T., & Becker, J. C. (2002). Process Monitoring in Grinding. *CIRP Annals*, *51*(2), 551–571. doi:10.1016/S0007-8506(07)61700-4

Wan, N., Xia, C., Zhao, H., & Zhang, S. T. (2019). 5-axis grinding path generation for free-form surface based on plane instantaneous grinding engagements. *International Journal of Advanced Manufacturing Technology*, *103*(5-8), 1861–1877. doi:10.100700170-019-03664-8

Wang, K., Fu, P. H., Peng, L. M., Wang, Y. X., & Ding, W. J. (2019). A Simplified Hot-Tearing Criterion for Shape Castings Based on Temperature-Field Simulation. *Metallurgical and Materials Transactions. A, Physical Metallurgy and Materials Science*, *50A*(11), 5271–5280. doi:10.100711661-019-05408-5

Wang, Z. P., Wu, J. Y., Zhang, L., & Wang, Y. C. (2019). Vehicle sideslip angle estimation for a four-wheel-independent-drive electric vehicle based on a hybrid estimator and a moving polynomial Kalman smoother. *Proceedings of the Institution of Mechanical Engineers Part K-Journal of Multi-Body Dynamics*, *233*(1), 125-140. 10.1177/1464419318770923

White, B., & Lewis, R. (2019). Simulation and understanding the wet-rail phenomenon using twin disc testing. *Tribology International*, *136*, 475–486. doi:10.1016/j.triboint.2019.03.067

Xiong, X. W. (2019). Artificial Intelligence Control Algorithm for the Steering Motion of Wheeled Soccer Robot. *International Journal of Pattern Recognition and Artificial Intelligence*, *33*(10), 1959034. doi:10.1142/S0218001419590341

Xu, D. F., Zhang, J. W., Zhou, B., & Yu, H. S. (2019). Investigation of mode transition coordination for power-split hybrid vehicles using dynamic surface control. *Proceedings of the Institution of Mechanical Engineers Part K-Journal of Multi-Body Dynamics*, *233*(3), 696–713. doi:10.1177/1464419319838931

Yaghmaei, A., & Yazdanpanah, M. J. (2019). Output control design and separation principle for a class of port-Hamiltonian systems. *International Journal of Robust and Nonlinear Control*, *29*(4), 867–881. doi:10.1002/rnc.4407

Yin, H., Chen, Y. H., & Yu, D. J. (2019). Controlling an Underactuated Two-Wheeled Mobile Robot: A Constraint-Following Approach. *Journal of Dynamic Systems Measurement and Control-Transactions of the Asme*, *141*(7), 071002. doi:10.1115/1.4043112

You, B., Li, Z., Ding, L., Gao, H. B., & Xu, J. Z. (2019). A new local path planning approach based on improved dual covariant Hamiltonian optimization for motion planning method. *Advances in Mechanical Engineering*, *11*(5). doi:10.1177/1687814019851007

Zhang, D., Li, C., Zhang, Y., Jia, D., & Zhang, X. (2015). Experimental research on the energy ratio coefficient and specific grinding energy in nanoparticle jet MQL grinding. *International Journal of Advanced Manufacturing Technology*, *78*(5-8), 1275–1288. doi:10.100700170-014-6722-6

Zhang, Q. L., Guo, N., Chen, Y., & To, S. (2019). Effects of binder concentration on the nanometric surface characteristics of WC-Co materials in ultra-precision grinding. *International Journal of Refractory Metals & Hard Materials*, *85*, 105048. doi:10.1016/j.ijrmhm.2019.105048

Zhang, Z. S., Dhanasekar, M., Ling, L., & Thambiratnam, D. P. (2019). Effectiveness of a raised road: Rail crossing for the safety of road vehicle occupants. *Engineering Failure Analysis*, *97*, 258–273. doi:10.1016/j.engfailanal.2019.01.046

Zhao, B., An, X. Z., Zhao, H. Y., Shen, L. L., Sun, X. D., & Zhou, Z. Y. (2019). DEM simulation of the local ordering of tetrahedral granular matter. *Soft Matter*, *15*(10), 2260–2268. doi:10.1039/C8SM02166J PMID:30762865

Zhou, X., & Xi, F. (2002). Modeling and predicting surface roughness of the grinding process. *International Journal of Machine Tools & Manufacture*, *42*(8), 969–977. doi:10.1016/S0890-6955(02)00011-1

Chapter 17
Modeling and Simulation of the Surface Topography Generation With Engineered Grinding Wheel

ABSTRACT

Precision grinding can obtain workpiece with high surface quality and high precision, but random distribution of abrasive grains on the grinding wheel surface poses a certain difficulty to improvement of machining precision and quality. This study established kinematic model of multiple grains, simulated the grain distribution on the surface of the common grinding wheel by using the grain vibration method, and examined the effect of different grinding parameters on the surface topography of the workpiece. Results show that the peaks and valleys on the profile curve of the workpiece surface increase and the corresponding Ra and Rz heights decrease, as the peripheral velocity of the grinding wheel increases. The peaks and valleys on the profile curve of the workpiece surface decrease, and the corresponding Ra and Rz heights increase as the feed speed of the workpiece increases. The number of grinding cracks on the surface of the workpiece decreases, the length of each crack increases, and the bump height on the surface increases slightly as the grinding depth increases.

17.1 INTRODUCTION

The American scholar Xunkun Li (Li, 2010; Li & Rong, 2011) has studied the manufacturing process of the common grinding wheel. He simulated the manufacturing process of the grinding wheel mixture, which involves mixing the grains and binder, grinding wheel modeling, roasting, and dressing to generate the surface topography of the common grinding wheel. He also completed the simulation of the surface topography generation process of the workpiece with the grinding wheel and analyzed the resulting grinding force, grinding heat, and surface topography.

The Canadian scholar X. Zhou (Zhou & Xi, 2002) has performed grid processing of the 3D grinding wheel in which the grains are located at each grid point to simulate the surface topography of the workpiece. Following the grinding kinematical theory, he conducted the grinding simulation using the

DOI: 10.4018/978-1-7998-1546-4.ch017

grinding wheel, completed the surface topography generation process of the workpiece using the search method, and experimentally verified the surface roughness of the workpiece generated by the simulation. The results demonstrated that the surface roughness output by the simulation was less than the experimental value. Because of the random simulation of the surface of the grinding wheel, this method had difficulty simulating the grains randomly distributed on the surface of the common grinding wheel.

The aforementioned studies on surface topography were conducted based on a number of hypotheses without considering the random distribution of the grains on the surface of the grinding wheel. In addition, the previous models assume that all grains would participate in the cutting work, despite not being the case, so the simulation results of the previous models could not reflect the actual machining process. Thus, the present study examines the grinding process of the common grinding wheel using nanoparticle jet flow, fully considering the bump height and random distribution of the grains on the surface of the grinding wheel and predicting the surface topography of the grinding workpiece using the surface generation mechanism.

In order to simulate the surface grinding process of engineered grinding wheel and evaluate the generated workpiece surface morphology, abrasive grain arrangement on the surface of engineered grinding wheel should be firstly determined to realize parameterization of the simulated grinding wheel; the process simulation was realized according to grinding kinematical model, elastic deformation model and plastic stacking model, and then workpiece surface morphology matrix was generated; the optimal arrangement was determined by calculating surface roughness R_a and R_z values.

17.2 ENGINEERED GRINDING WHEEL

17.2.1 Definition of Engineered Grinding Wheel

Precision grinding can obtain workpiece with high surface quality and high precision (Dold et al., 2011; Guo et al., 2017; Jia, Li, Zhang, Zhang, & Zhang, 2014; Jia, Li, Zhang, Zhang, & Zhang, 2016; Xu et al., 2019), it is usually taken as the final part machining procedure, but random distribution of abrasive grains on the grinding wheel surface poses a certain difficulty to improvement of machining precision and quality (Koshy, Iwasald, & Elbestawl, 2003; Ren & Hua, 2011). In order to solve the above problem, Canadian scholar P. Koshy (Chen, Wang, Yushan, Liu, & Wang, 2015; Heinzel, Rickens, & Trumpold, 2009; Koshy et al., 2003) proposed the concept of engineered grinding wheel in the 53rd CIRP annual meeting, namely grinding wheel manufactured according to manually set spatial distribution matrix and protruding height of abrasive grains with the same size.

17.2.2 Characteristics of Engineered Grinding Wheel

As abrasive grains present orderly arrangement on the surface of engineered grinding wheel, protruding heights of abrasive grains are uniform, and this grinding wheel has the following features:

1. Abrasive grains present single-layer orderly arrangement on the surface of grinding wheel body, thus avoiding stacking phenomenon;
2. Gaps exist between abrasive particles, which enlarge chip space on the grinding wheel surface and make it easy for grinding fluid to enter the grinding area;

Figure 1. Diamond grinding wheel manufactured using electroplating method

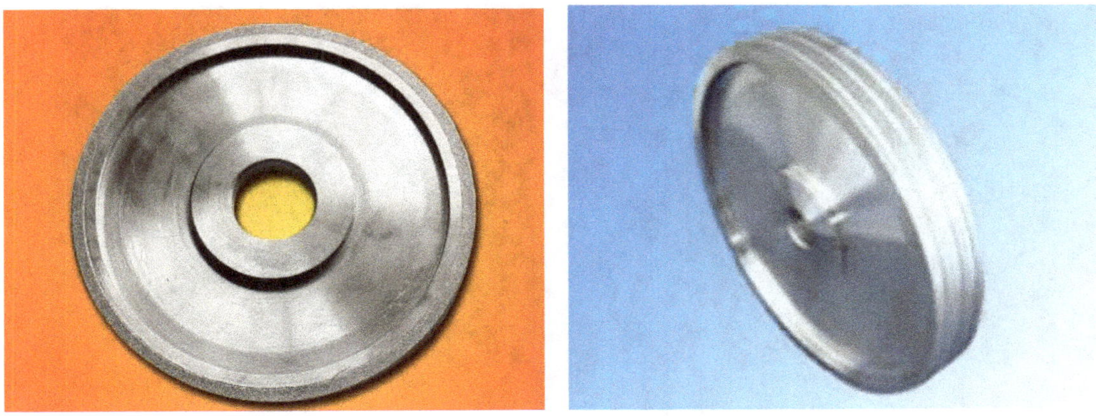

3. Heat conductivity is elevated, temperature in the grinding area is lowered, burning phenomenon on workpiece surface is reduced, and workpiece surface morphology after grinding is improved;
4. Surface properties of grinding wheel have good consistency, which improves grinding imbalance of grinding wheel and machining precision;
5. As abrasive grains are solidly bonded with grinding wheel base, abrasive grains only go through wear but will not fall off in the grinding process, and service life of grinding wheel is lengthened.

17.2.3 Preparation of the Engineered Grinding Wheel

With the development of new technologies and new processes, there are more and more manufacturing method of engineered grinding wheel. Already known manufacturing methods are mainly: electrochemical plating method, high-temperature brazing method and rapid prototyping method.

(1) Electroplating Method

Electroplating method refers to precipitating a layer of metal bonding agent on the grinding wheel base and fixing abrasive grains on the base surface according to working principles of electroplating: metal bonding agent with thickness being 20% of abrasive grain size is precipitated on the surface of grinding wheel base; metal bonding agent with thickness being 2/3 of abrasive grain size is precipitated again to fix abrasive grains. Figure 1 shows cylindrical diamond grinding wheel and forming diamond grinding wheel manufacturing using the electroplating method (Dold et al., 2011; Kitzig, Tawakoli, & Azarhoushang, 2016; Z. Li, Ding, Shen, Xi, & Fu, 2016; Lyu, Yu, Wang, Zhao, & Liu, 2017; Zhou & Xi, 2002).

Electroplating method, which is used for manufacturing of grinding wheel, features simple machining process and easy operation. The whole manufacturing process can be implemented under normal temperature, thus reducing energy consumption and manufacturing cost of grinding wheel as well as residual stress on the grinding wheel surface (Gift & Misiolek, 2004; Yan, Zhang, & Syoji, 2003; Zhao et al., 2016); meanwhile, protruding heights of abrasive grains on manufactured grinding wheel present favorable consistency, so no dressing is needed. Protruding height of abrasive grains is about 30%~40% of abrasive grain size and chip space is large. However, electroplating is a special machining process, so

Figure 2. Brazing agglomerate grinding wheel

it can only be used to manufacture metal bonded grinding wheel, abrasive grains on the grinding wheel surface are only simply mechanically embedded and inlaid into the coated metal, chemical metallurgical bonding between abrasive grains and grinding wheel base can't be reached, so bonding strength is low, and they can easily fall off under high-load condition.. In addition, treatment of electroplate liquid is difficult, which will generate environmental pollution and pose a certain harm to health of operators.

(2) High-Temperature Brazing Method

High-temperature brazing method refers to a grinding wheel manufacturing method which melts solder alloy under high temperature and brazes abrasive grains onto the grinding wheel base according to the set arrangement pattern (Chen, Fu, Li, Gao, & He, 2017; Lankiewicz, Baranowski, Babul, & Kowalski, 2015; Shen, Wang, & Sun, 2018; Shen, Wang, & Sun, 2019). Solder alloy must have good wetting characteristics, and commonly used solder alloys include Ni-Cr alloy, Cu-Sn alloy, Ag-Cu alloy, etc. Figure 2 shows single-layer brazing agglomerate grinding wheel designed by Professor Ding et al. from Nanjing University of Aeronautics and Astronautics (Chen, Xiong, Mao, & Xiao-Hong, 2008; Ding, Xu, Chen, Su, & Fu, 2010).

High-temperature brazing can realize chemical metallurgical bonding between abrasive grains and solder alloy, bonding strength between abrasive grains and grinding wheel base is high, machining task can be completed under high-load condition, grinding wheel failure will not be generated, service life of grinding wheel is lengthened, protruding heights of abrasive grains are favorably consistent so no dressing is needed. In addition, protruding height of abrasive grains is higher than electroplating grinding wheel, being about 70%~80% of abrasive grain size, and chip space on the grinding wheel surface is enlarged. However, high-temperature brazing process is complicated, it can only manufacture single-layer metal bonded superhard abrasive grinding wheel, and melting temperature of solder alloy is generally above 1000°C, which will generate thermal injuries to superhard abrasive material, and after grinding wheel base is cooled, residual stress exists on the base surface and the base experiences deformation, which will impact accuracy of precision machining (Alves et al., 2009; Amamou, Ben Fredj, & Fnaiech, 2008; Aurich, Linke, Hauschild, Carrella, & Kirsch, 2013; Fathallah, Fredj, Sidhom, Braham, & Ichida, 2009; Ding et al., 2017).

(3) Rapid Prototyping Method

Rapid prototyping method refers to generating 3D geometrical information among abrasive material, bonding agent and filler constituting the grinding wheel according to CAD modeling using the principles of rapid prototyping and then sintering and stacking materials layer by layer through laser beams to form a grinding wheel with 3D free profile and optimized arrangement of abrasive grains (Li, Fang, Cui, & Ding, 2010). Manufacturing process of this method is divided into two parts: hierarchical digitalization of grinding wheel body; stacking using rapid prototyping machine for grinding wheel manufacturing. The concrete flow is: CAD grinding wheel entity model is discretized layer by layer along axial direction of grinding wheel, hierarchical data files are then imported into the rapid prototyping machine which controls feed of abrasive material, bonding agent and filler, and layer-by-layer sintering and stacking are implemented, and thus manufacturing of grinding wheel is completed.

Rapid prototyping method can realize manufacturing of grinding wheel with 3D hook face and any shape. The manufactured grinding wheel features high protruding height of abrasive grains, large bonding strength, large chip space and good heat dissipation. Therefore, this method can significantly lengthen service life of grinding wheel, reduce burning on workpiece surface and improve workpiece surface quality so as to improve part performance. Moreover, it can realize high-efficiency forming and grinding of impeller and vane.

17.3 KINEMATIC MODEL OF MULTIPLE GRAINS

In order to establish the multi-abrasive-grain kinematical model, it's necessary to simplify the grinding process and make certain hypotheses which are identical with those of the single-abrasive-grain kinematical condition (Chen & Huang, 2013; Li, Yu, Zhu, & Wang, 2015; Wang, Li, Zhang, Jia, & Zhang, 2014; Zhang & Lin, 2012), namely:

(1) Vibration of grinding wheel is neglected;
(2) When abrasive grains interfere with workpiece surface material, and workpiece material contacting abrasive grains is totally removed;
(3) Side flow and built-up edge generation in the material removal process are not considered.

A local coordinate system is established at the contact point O_1 between the first abrasive grain G_1 and workpiece. This coordinate system overlaps with the global coordinate system as shown in Figure 3.

According to equation (16-9), trajectory equation of abrasive grain G_1 in the local coordinate system is (Wang, Li, & Qiang, 2014):

$$z_1 = \frac{r_1 x_1^2}{2\left(r_1 + \frac{v_w}{w}\right)^2} \tag{1}$$

Figure 3. Schematic diagram of motion trajectories of two adjacent abrasive grains

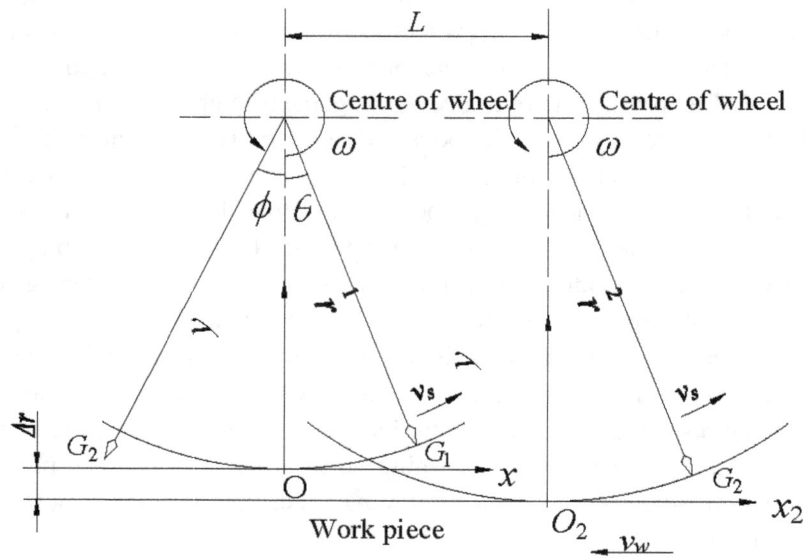

where subscript 1 means the first abrasive grain; r_1 is radius of grinding wheel at this abrasive grain; ω is revolving speed of grinding wheel.

As the local coordinate system of the first abrasive grain overlaps with the global coordinate system, then trajectory equation of the first abrasive grain in the global coordinate system is:

$$z = \frac{r_1 x^2}{2\left(r_1 + \frac{v_w}{w}\right)^2} \tag{2}$$

A local coordinate system is established at the contact point O_2 between the second abrasive grain G_2 and workpiece, and then trajectory equation of the second abrasive grain G_2 in the local coordinate system is:

$$z_2 = \frac{r_2 x_2^{\ 2}}{2\left(r_2 + \frac{v_w}{w}\right)^2} \tag{3}$$

where the subscript 2 means the second abrasive grain and r_2 is radius of grinding wheel at this abrasive grain.

The distance L from origin O of the global coordinate system to origin O_2 of the local coordinate system along the direction of axis X is:

$$L = \frac{v_w}{v_s} r_s \varnothing \tag{4}$$

The distance Δr from origin O of the global coordinate system to origin O_2 of the local coordinate system along the direction of axis Y is:

$$\Delta r = r_2 - r_1 \tag{5}$$

Trajectory equation of abrasive grain G_2 in the global coordinate system is:

$$z + (r_2 - r_1) = \frac{r_2 \left(x - \frac{v_w}{w} \varnothing \right)^2}{2 \left(r_2 + \frac{v_w}{w} \right)^2} \tag{6}$$

Trajectory equation of the i (th) abrasive grain G_i in the global coordinate system is:

$$z + (r_i - r_1) = \frac{r_i \left(x - \frac{v_w}{w} \theta_i \right)^2}{2 \left(r_i + \frac{v_w}{w} \right)^2} \tag{7}$$

where: r_i——radius of grinding wheel at the i (th) abrasive grain;

θ_i——included angle between the i (th) abrasive grain and the position where the first abrasive grain is located.

Through simultaneous equations (2) and (6), the following can be obtained:

$$\frac{r_1 x^2}{2 \left(r_1 + \frac{v_w}{w} \right)^2} + (r_2 - r_1) = \frac{r_2 \left(x - \frac{v_w}{w} \varnothing \right)^2}{2 \left(r_2 + \frac{v_w}{w} \right)^2} \tag{8}$$

Equation of point of intersection of adjacent abrasive grains in the global coordinate system can be written as:

$$\frac{r_i\left(x-\frac{v_x}{w}\theta_i\right)^2}{2\left(r_i+\frac{v_w}{w}\right)^2}+(r_{i+1}-r_i)=\frac{r_{i+1}\left(x-\frac{v_w}{w}\theta_{i+1}\right)^2}{2\left(r_{i+1}+\frac{v_w}{w}\right)^2} \tag{9}$$

where: r_i+1——radius of grinding wheel at the *i*+1 (th) abrasive grain;

θ_i+1——the included angle between *i*+1 (th) abrasive particle and the position where the first abrasive grain is located.

Grinding study of engineered grinding wheel is mainly conducted. As abrasive grains on the surface of this engineered grinding wheel have equal protruding heights, then equation (7) can be simplified as:

$$z=\frac{(x-iL)^2}{d_s\left(1+\frac{v_w}{v_s}\right)^2} \tag{10}$$

Equation (17-9) of point of intersection between adjacent abrasive grains can be simplified as:

$$\begin{cases} z=\dfrac{(x-iL)^2}{d_s(1+v_w/v_s)^2} \\ z=\dfrac{L^2}{4d_s(1+v_w/v_s)^2} \end{cases} \tag{11}$$

Coordinates of point of intersection between adjacent abrasive grains can be solved as:

$$\begin{cases} x=\dfrac{(2i+1)L}{2} \\ z=\dfrac{L^2}{4d_s(1+v_w/v_s)^2} \end{cases} \tag{12}$$

17.4 EVALUATION OF SURFACE ROUGHNESS

17.4.1 Profile Centerline

In order to evaluate workpiece surface roughness, profile centerline on workpiece surface should be determined firstly (Liu & Shi, 2008). There are two common determination methods of profile centerline: least square profile centerline; arithmetic profile centerline. As least square profile centerline is of uniqueness, profile centerline on workpiece surface is determined according to the least square profile centerline method based on data simulated through simulation (Chen & Zhou, 2008; Yu, 2016).

Equation of the solved least square profile centerline is set as:

$$y = a_0 x + a_1 \tag{13}$$

where a_0 and a_1 are arbitrary constants.

Coordinates (x_i, y_i) of datapoint are already known, and the following is set:

$$M = \sum_{i=1}^{m} \left[y_i - (a_0 x_i + a_1) \right]^2 \tag{14}$$

According to the least square principle, in order to solve a_0 and a_1 which satisfy minimum M value, equation (14) is regarded as independent variables of a_0 and a_1, M is multivariate function of dependent variable, and the conditions to obtain extremum of this multivariate function are:

$$\begin{cases} M_{a_0}(a_0, a_1) = 0 \\ M_{a_1}(a_0, a_1) = 0 \end{cases} \tag{15}$$

The following can be obtained by substituting equation (14) into equation (15):

$$\begin{cases} \dfrac{\partial M}{\partial a_0} = -2\sum_{i=1}^{m} \left[y_i - (a_0 x_i + a_1) \right] x_i = 0 \\ \dfrac{\partial M}{\partial a_1} = -2\sum_{i=1}^{m} \left[y_i - (a_0 x_i + a_1) \right] = 0 \end{cases} \tag{16}$$

Through the organization, the following can be obtained:

$$\begin{cases} a_0 \sum_{i=1}^{m} x_i^2 + a_1 \sum_{i=1}^{m} x_i = \sum_{i=1}^{n} x_i y_i \\ a_0 \sum_{i=1}^{m} x_i + m a_1 = \sum_{i=1}^{m} y_i \end{cases} \tag{17}$$

Solution of equations in (17) can be solved as:

$$\begin{cases} a_0 = \left(m \cdot \sum_{i=1}^{m} x_i y_i - \sum_{i=1}^{m} x_i \cdot \sum_{i=1}^{m} y_i \right) / \left[m \cdot \sum_{i=1}^{m} x_i^{\,2} - \left(\sum_{i=1}^{m} x_i \right)^2 \right] \\ a_1 = \frac{1}{m} \cdot \sum_{i=1}^{m} y_i - a_0 \cdot \frac{1}{m} \cdot \sum_{i=1}^{m} x_i \end{cases} \tag{18}$$

Solution of equation (18) is substituted into equation (13) to solve the least square profile centerline equation.

Workpiece surface morphology matrix obtained through simulation is set as Z_{mn}, coordinates of arbitrary point on workpiece surface are (x_{ij}, y_{ij}, z_{ij}), and then equation of least square profile centerline on workpiece surface within the section *YOZ* is:

$$z = a_0 y + a_1 \tag{19}$$

Corresponding coefficients a_0 and a_1 are respectively:

$$\begin{cases} a_0 = \left(m \cdot \sum_{i=1}^{m} y_{ij} z_{ij} - \sum_{i=1}^{m} y_{ij} \cdot \sum_{i=1}^{m} z_{ij} \right) / \left[m \cdot \sum_{i=1}^{m} y_{ij}^{\,2} - \left(\sum_{i=1}^{m} y_{ij} \right)^2 \right] \\ a_1 = \frac{1}{m} \cdot \sum_{i=1}^{m} z_{ij} - a_0 \cdot \frac{1}{m} \cdot \sum_{i=1}^{m} y_{ij} \end{cases} \tag{20}$$

17.4.2 Evaluation Parameters

According to stipulations specified in national standard GB/T 3505-2009, evaluation parameters of surface roughness include amplitude parameter, spacing parameter, hybrid parameter, curve and related parameters (Mao, 1991). Common evaluation parameters include arithmetic mean profile deviation R_a, maximum profile height R_z, average width RS_m of profile units and bearing length rate R_{mr}, etc. In consideration that arithmetic mean profile deviation R_a and maximum profile height R_z can objectively and comprehensively reflect micro-geometrical features of workpiece surface, the two parameters are selected to evaluate workpiece surface roughness (Yang & Jiang, 2002).

(1) Arithmetic Mean Profile Deviation R_a

Arithmetic mean profile deviation refers to arithmetic mean value of absolute values of vertical coordinates *z(x)* of all points on the profile curve within a sampling length, and it can be expressed as:

$$R_a = \frac{1}{l_r} \int_0^{l_r} \left| z(x) \right| dx \tag{21}$$

where: l_r—— sampling length.

(2) Maximum Profile Height R_z

Arithmetic maximum profile height value refers to the sum of maximum profile peak height and maximum profile valley depth within a sampling length, and it can be expressed as:

$$R_z = R_p + R_v \tag{22}$$

where: R_p——maximum profile peak height;

R_v——maximum profile valley depth.

17.5 GRAIN ARRANGEMENT ON THE SURFACE OF GRINDING WHEEL

17.5.1 Grain Arrangement Parameters

Abrasive grains on the surface on the engineered grinding wheel are under human intervened orderly arrangement (Yu, Lyu, & Wang, 2019). In order to standardize arrangement scheme of abrasive grains, arrangement parameters of abrasive grains should be firstly determined. Commonly used arrangement parameters of abrasive grains mainly include (Aurich, Herzenstiel, Sudermann, & Magg, 2008; Zhang, 2001; Zhang & Xu, 2019): circumferential pitch ΔX, axial pitch ΔY and offset distance ΔY_t of abrasive grains as shown in Figure 4.

Axes x, y and z in Fig. 17-4 are circumferential, axial and radial directions of grinding wheel. Circumferential pitch ΔY of abrasive grains is the distance between two adjacent abrasive grains on the same straight line along axial direction of grinding wheel; offset distance ΔY_t of abrasive grains is offset of abrasive grains on two adjacent straight lines parallel to axial direction of grinding wheel.

17.5.2 Grain Arrangement Scheme

When protruding heights of abrasive grains on the surface of grinding wheel are consistent, the workpiece surface quality obtained through grinding is the best (Chen, Huang, & Xu, 2009; Ding et al., 2013; Exner & Muller, 2009; Herman & Krzos, 2009; Pal, Chattopadhyay, & Chattopadhyay, 2012; Sanchez, Pombo, Cabanes, Ortiz, & de Lacalle, 2008). Given this, it's set that abrasive grains on the surface of engineered grinding wheel have the same protruding height, abrasive grains are corundum abrasive materials with the same diameter, abrasive grains are the mostly densely arranged in axial direction of grinding wheel, offset ΔY_t of abrasive grains on two circumferential adjacent straight lines parallel to grinding wheel

Figure 4. The abrasive arrangement parameters on the surface of grinding wheel

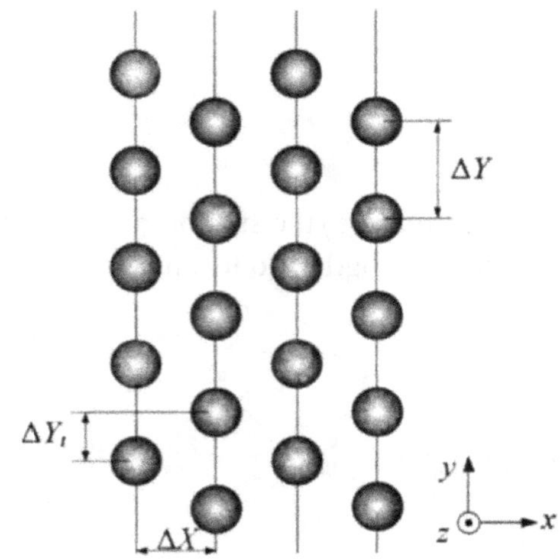

Table 1. The arrangement parameters under different arrangement schemes of abrasive grains

Scheme	Circumferential pitch of abrasive grains ΔX	Axial pitch of abrasive grains ΔY	Offset distance of abrasive grains ΔY_t
1	5D	D	0% D
2	5D	D	25% D
3	5D	D	50% D

Note: D is diameter of abrasive grains.

Figure 5. Arrangement schemes of abrasive grains

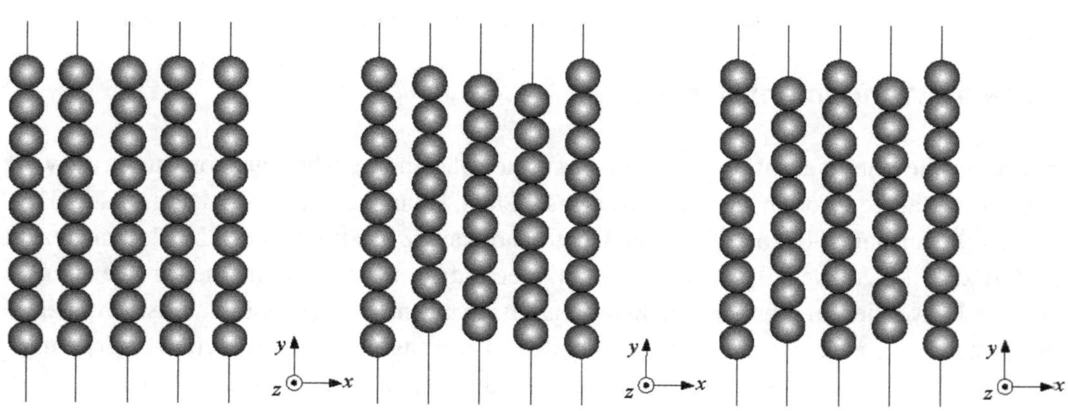

is 0%, 25% and 50% of diameter of abrasive grains, and concrete arrangement parameters of abrasive grains are shown in Table 1.

Arrangement schematic of the above three schemes is shown in Figure 5, where (a), (b) and (c) are corresponding to Scheme I, Scheme II and Scheme III respectively.

(a) (b) (c)

Figure 6. Simulation flow chart of engineered grinding wheel

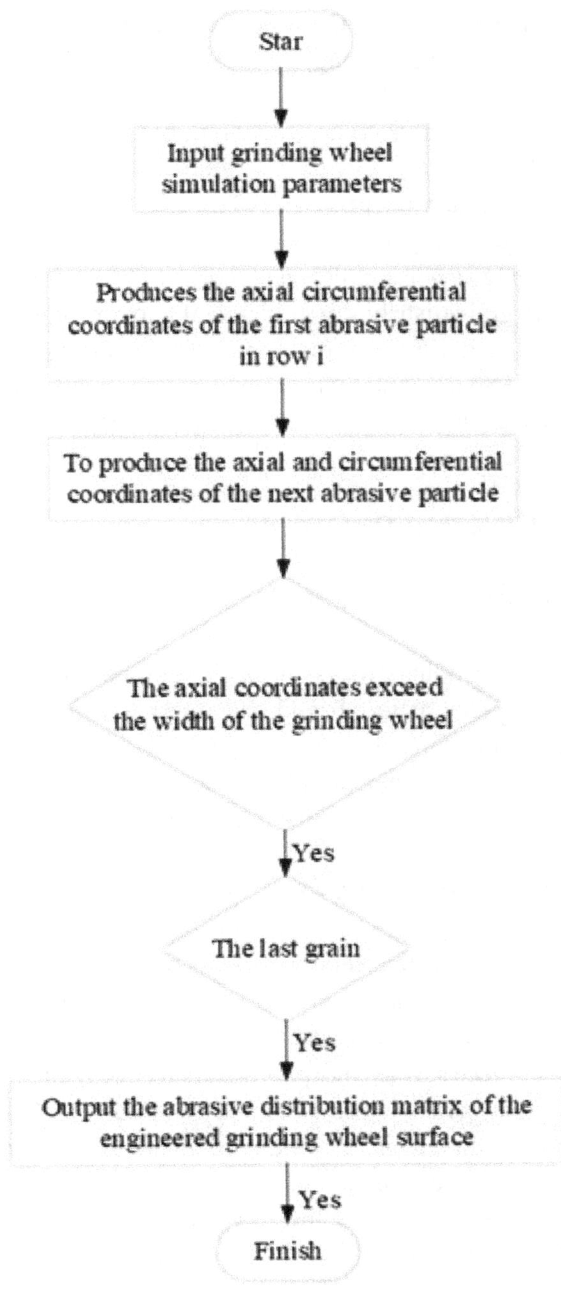

17.5.3 Simulation of the Engineered Grinding Wheel Morphology

Figure 6 shows flowchart of morphology simulation of engineered grinding wheel. The concrete flow is: simulation parameters are firstly input, and axial circumferential coordinates of the first abrasive particle in the *i* (th) row are generated; axial circumferential coordinates of the next abrasive grain are generated, and whether axial coordinates exceed width of grinding wheel will be judged, if not, axial circumferential coordinates of the next abrasive grain will be continuously generated, and if yes, whether it is the last row will be judged, and if it is not the last row, continue to execute the above steps, and if it is, abrasive grain distribution matrix on the surface of engineered grinding wheel is output (Li, Hou, Li, Han, & Ding, 2011; Nie, Wang, Lin, Su, & Rong, 2016; Palmer, Ghadbeigi, Novovic, & Curtis, 2018; S. Wang, Li, Zhang, Jia, & Zhang, 2014; Zhang et al., 2017; Zhu, Yan, & Li, 2014).

Surface morphologies of engineered grinding wheel output through simulation for the above three schemes are shown in Figure 7.

(a), (b) and (c) in Fig. 17-7 are surface morphologies of the engineered grinding wheel generated using Scheme I, Scheme II and Scheme III respectively. Being the same as single-abrasive-grain simulation, abrasive grain size is still the average grain size corresponding to80# abrasive materials, namely 196 μm. According to Fig. 17-7, abrasive grain distribution is identical with the preset arrangement scheme, axial pitches of abrasive grains are all 196 μm, circumferential pitches of abrasive grains are all 5 times of abrasive grain diameter, namely 980 μm, and their offset distances are 0, 49 and 98 98 μm respectively. Centers of sphere of all abrasive grains are in the same plane, so they have the same protruding height.

17.6 SIMULATION OF GRINDING PROCESS

17.6.1 Setting of Simulation Parameters

Simulation parameters used in the grinding process simulation of the engineered grinding wheel include grinding wheel parameters, workpiece materials and grinding parameters. Grinding wheel parameters derive from abrasive grain arrangement matrix on grinding wheel surface generated through morphology simulation of the engineered grinding wheel, simulation object is quenched 45# steel, and workpiece and grinding parameters are identical with those in single-abrasive-grain simulation.

17.6.2 Procedures of Simulation Program

Fig. 17-8 shows simulation flowchart of the grinding process of the engineered grinding wheel, and the concrete simulation flow is: grinding simulation parameters of the engineered grinding wheel are firstly input, and the simulation process is initialized; abrasive grain distribution matrix is loaded on the surface of ordinary grinding wheel, workpiece surface morphology matrix generated by the *i* (th) row of abrasive grains is solved according to single-abrasive-grain kinematical model, multi-abrasive-grain kinematical model, elastic deformation model and plastic stacking model, and the workpiece morphology matrix is updated; the above steps are repeatedly executed until the corresponding workpiece surface morphology matrix is generated for the abrasive grains in the last row, and simulation result is finally output(X. Li, Wolf, Geng, & Rong, 2014; S. Wang, C. H. Li, X. W. Zhang, et al., 2014).

Figure 7. Surface morphologies of the engineered grinding wheel

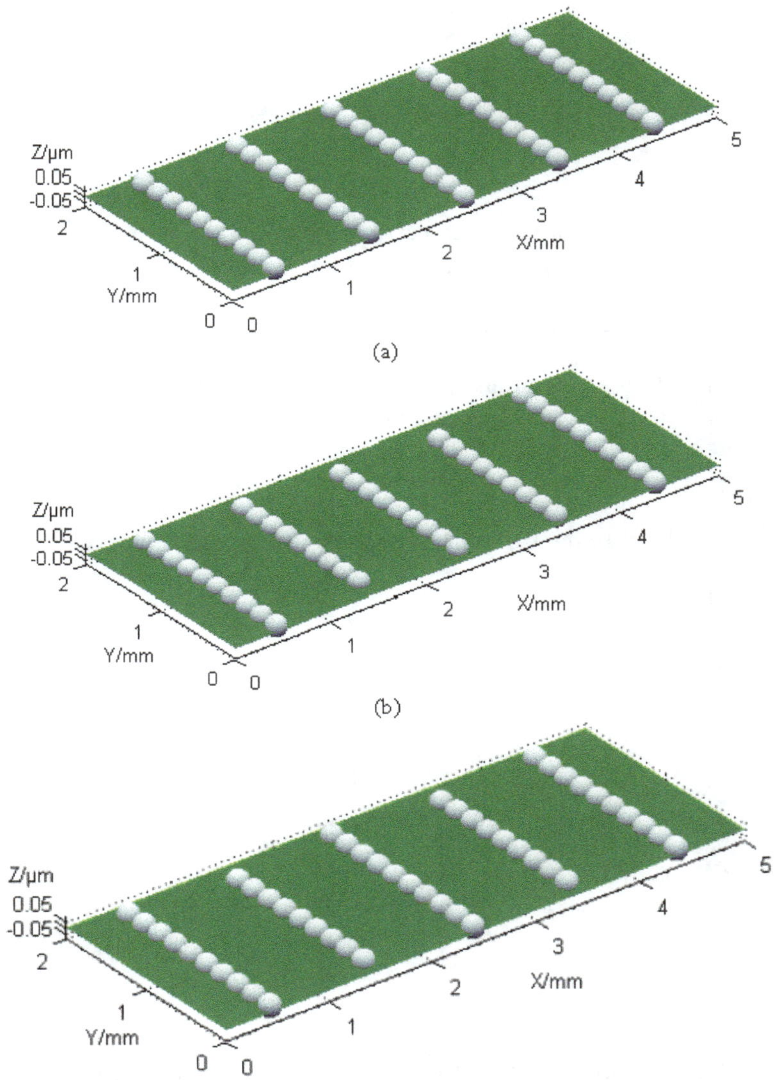

17.6.3 Analysis of Simulation Results

Workpiece surface morphology generated through simulation is shown in Figure 9.

(a), (b) and (c) in Figure 9 are workpiece surface morphologies generated using Scheme I, Scheme II and Scheme III respectively. It can be seen that workpiece surfaces show the same texture, number of workpiece textures obtained through Scheme II is the greatest with the minimum protruding height, and the surface is the poorest in Scheme I. It's noteworthy that actual grinding depth in Scheme I is

Figure 8. Grinding simulation flowchart of the engineered grinding wheel

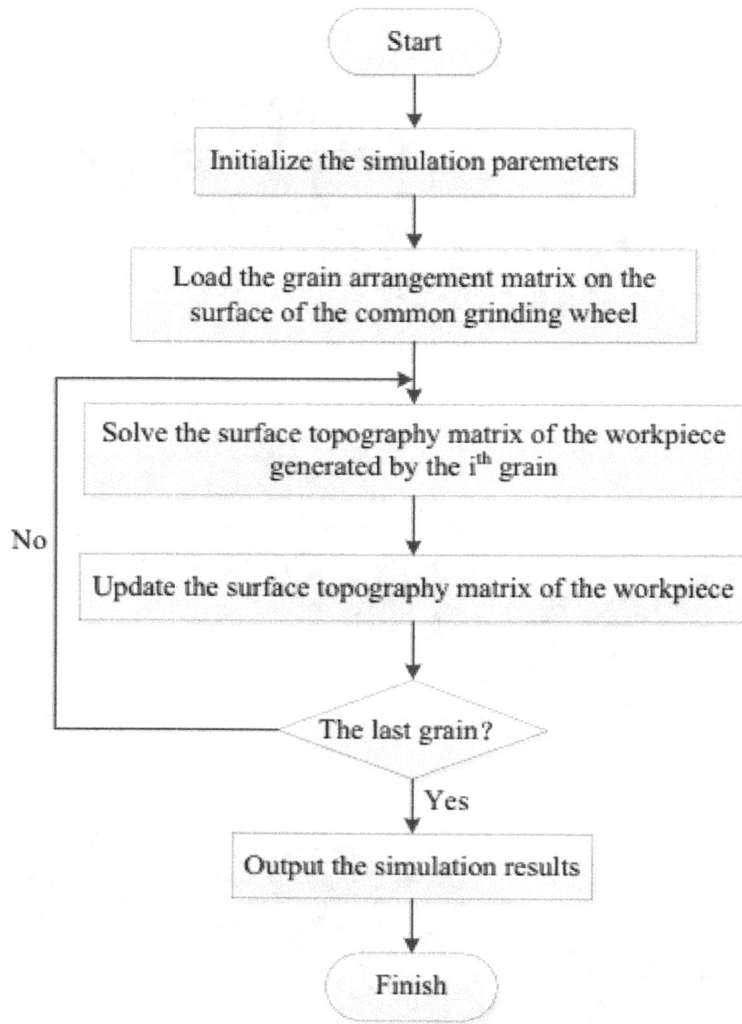

smaller than 20 μm because elastic-plastic deformation occurs on workpiece surface (Wang, Li, Jia, & Hou, 2014; S. Wang, Li, Jia, Zhang, & Zhang, 2015).

Cross section vertical to grinding texture direction on workpiece surfaces in Figure 9 is taken. Workpiece surface profile curves obtained through intersection of this section with workpiece surface are shown in Figure 10.

(a), (b) and (c) are workpiece surface profile curves corresponding to Scheme I, II and III respectively in Figure 10. The height of profile curve obtained through Scheme II is the minimum with most profile peaks and valleys, and the profile curve obtained through Scheme I is the poorest. By calculating surface roughness values corresponding to the profile curves, workpiece surface roughness histogram after grinding of the engineered grinding wheel corresponding to the above three schemes can be obtained as shown in Figure 11.

Figure 9. Workpiece surface morphology after grinding of the engineered grinding wheel

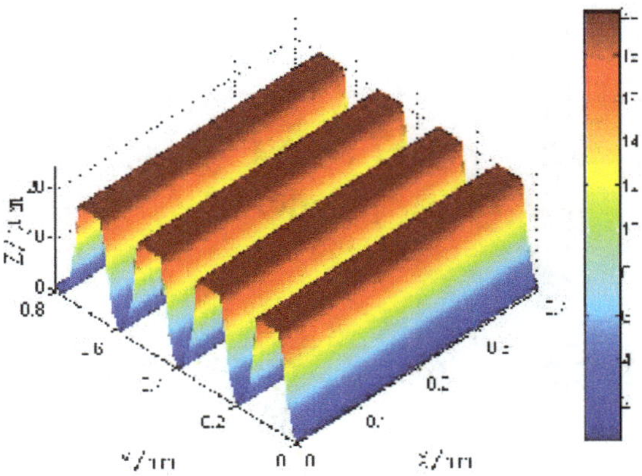

(a)

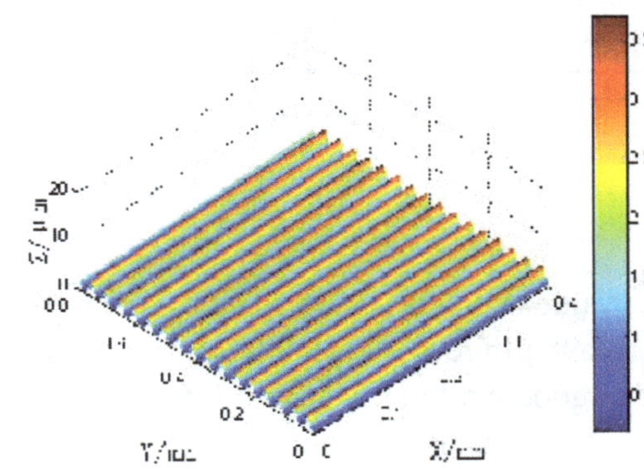

(b)

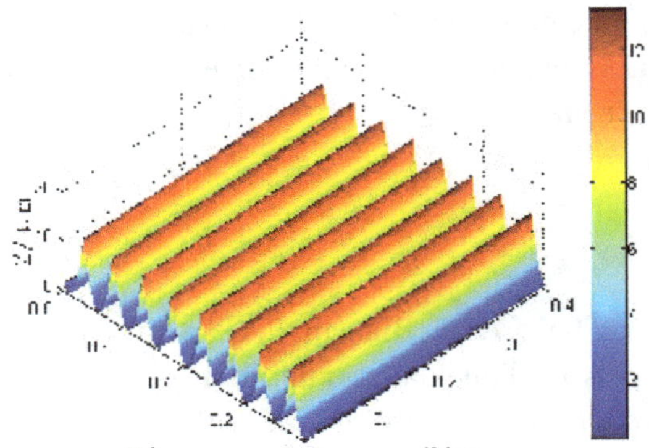

Figure 10. Workpiece surface profile curves after grinding of the engineered grinding wheel

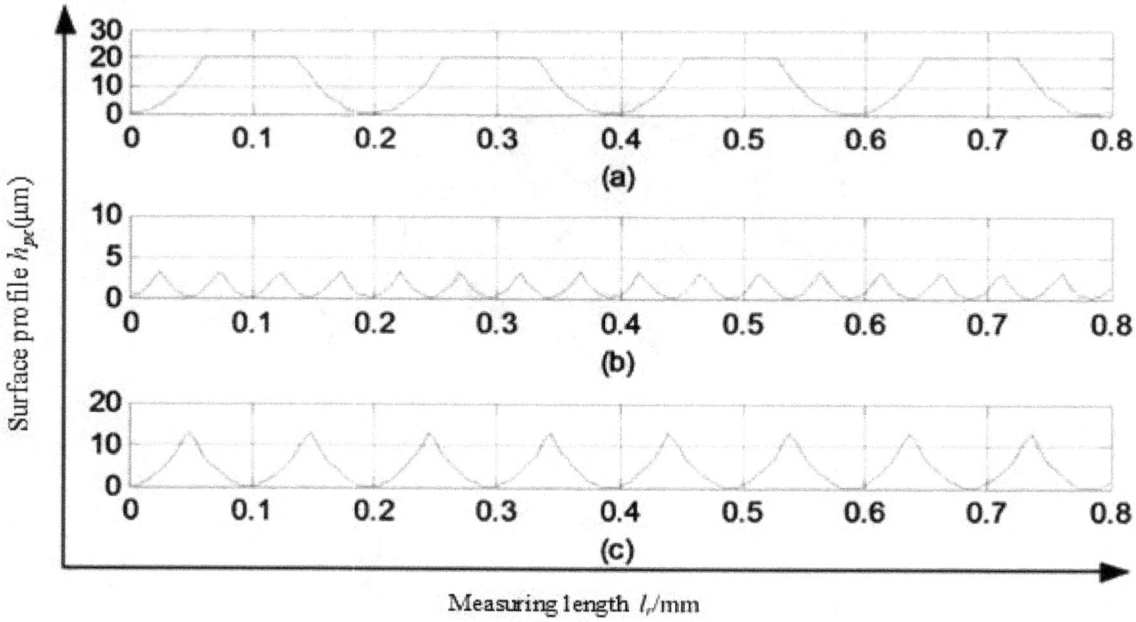

In Figure 11, R_a value and R_z value are 7.51 μm and 19.69 μm respectively under Scheme I; those under Scheme II are 0.79 μm and 2.98 μm respectively; those under Scheme III are 3.31 μm and 12.56 μm respectively. It can be seen that good workpiece surface morphology can be obtained using Scheme II (namely offset distance of abrasive grains is 25% of abrasive grain size).

When offset distance of abrasive grains continues to decline, the obtained surface roughness will be smaller, and meanwhile, the workpiece surface roughness generated when offset distance of abrasive grains is 25% of average grain size is already smaller than ordinary wheel with the same grain size. Therefore, when abrasive grains are most densely arranged in axial direction of grinding wheel, in order to obtain a good surface roughness, offset distance of abrasive grains should be greater than 0, but should be small than 25 of average grain size.

REFERENCES

Alves, J. A. C., Fernandes, U. D., Diniz, A. E., Bianchi, E. C., de Aguiar, P. R., & Canarim, R. C. (2009). Analysis of the Influence of Sparkout Time on Grinding Using Several Lubrication/Cooling Methods. *Journal of the Brazilian Society of Mechanical Sciences and Engineering*, *31*(1), 47–51. doi:10.1590/S1678-58782009000100007

Amamou, R., Ben Fredj, N., & Fnaiech, F. (2008). Improved method for grinding force prediction based on neural network. *International Journal of Advanced Manufacturing Technology*, *39*(7-8), 656–668. doi:10.100700170-007-1264-9

Figure 11. Workpiece surface roughness after grinding of the engineered grinding wheel

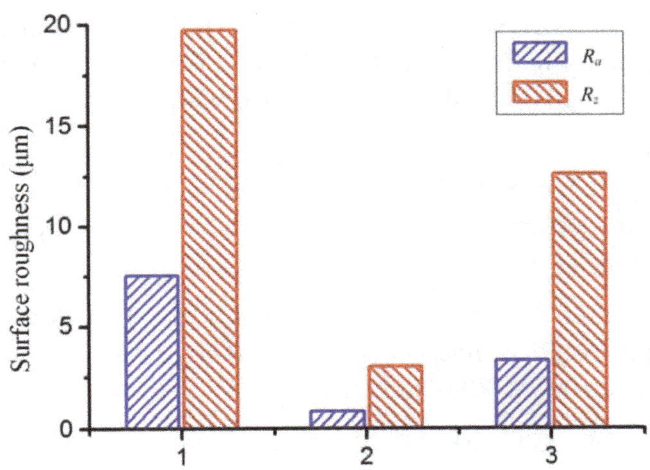

Aurich, J. C., Herzenstiel, P., Sudermann, H., & Magg, T. (2008). High-performance dry grinding using a grinding wheel with a defined grain pattern. *Cirp Annals-Manufacturing Technology*, *57*(1), 357–362. doi:10.1016/j.cirp.2008.03.093

Aurich, J. C., Linke, B., Hauschild, M., Carrella, M., & Kirsch, B. (2013). Sustainability of abrasive processes. *Cirp Annals-Manufacturing Technology*, *62*(2), 653–672. doi:10.1016/j.cirp.2013.05.010

Ben Fathallah, B., Ben Fredj, N., Sidhom, H., Braham, C., & Ichida, Y. (2009). Effects of abrasive type cooling mode and peripheral grinding wheel speed on the AISI D2 steel ground surface integrity. *International Journal of Machine Tools & Manufacture*, *49*(3-4), 261–272. doi:10.1016/j.ijmachtools.2008.10.005

Chen, B., Xiong, H. P., Mao, W., & Xiao-Hong, L. I. (2008). Wettability of Several High-temperature Brazing Fillers on C/C Composites. *Journal of Materials Engineering*.

Chen, C., Wang, J., Yushan, L. V., Liu, Z., & Wang, X. (2015). *Experimental investigation of grinding temperature when grinding TC_4 with engineered grinding wheels*. Diamond & Abrasives Engineering. doi:10.13394/j.cnki.jgszz.2015.5.0006

Chen, J. J., Fu, Y. C., Li, Q. L., Gao, J. J., & He, Q. S. (2017). Investigation on induction brazing of revolving heat pipe grinding wheel. *Materials & Design*, *116*, 21–30. doi:10.1016/j.matdes.2016.11.057

Chen, J. Y., Huang, H., & Xu, X. P. (2009). An experimental study on the grinding of alumina with a monolayer brazed diamond wheel. *International Journal of Advanced Manufacturing Technology*, *41*(1-2), 16–23. doi:10.100700170-008-1459-8

Chen, Y., & Huang, G. Q. (2013). Development and Research on the Dynamics Simulation System for Surface Grinding Process with Diamond Wheel. *Key Engineering Materials, 589-590*, 658-664.

Chen, Y. P., & Zhou, Z. Y. (2008). Basics of interchangeability and measurement technology. Beijing: China Machine Press.

Ding, W. F., Xu, J. H., Chen, Z. Z., Su, H. H., & Fu, Y. C. (2010). Grindability and Surface Integrity of Cast Nickel-based Superalloy in Creep Feed Grinding with Brazed CBN Abrasive Wheels. *Chinese Journal of Aeronautics*, *23*(4), 501–510. doi:10.1016/S1000-9361(09)60247-8

Ding, W. F., Xu, J. H., Chen, Z. Z., Yang, C. Y., Song, C. J., & Fu, Y. C. (2013). Fabrication and performance of porous metal-bonded CBN grinding wheels using alumina bubble particles as pore-forming agents. *International Journal of Advanced Manufacturing Technology*, *67*(5-8), 1309–1315. doi:10.100700170-012-4567-4

Ding, W. F., Zhang, L. C., Li, Z., Zhu, Y. J., Su, H. H., & Xu, J. H. (2017). Review on grinding-induced residual stresses in metallic materials. *International Journal of Advanced Manufacturing Technology*, *88*(9-12), 2939–2968. doi:10.100700170-016-8998-1

Dold, C., Transchel, R., Rabiey, M., Langenstein, P., Jaeger, C., Pude, F., ... van Griethuysen, J. P. (2011). A study on laser touch dressing of electroplated diamond wheels using pulsed picosecond laser sources. *Cirp Annals-Manufacturing Technology*, *60*(1), 363–366. doi:10.1016/j.cirp.2011.03.117

Exner, H. E., & Muller, C. (2009). Particle Rearrangement and Pore Space Coarsening During Solid-State Sintering. *Journal of the American Ceramic Society*, *92*(7), 1384–1390. doi:10.1111/j.1551-2916.2009.02978.x

Gift, F. C., & Misiolek, W. Z. (2004). Fluid Performance Study for Groove Grinding a Nickel-Based Superalloy Using Electroplated Cubic Boron Nitride (CBN) Grinding Wheels. *Journal of Manufacturing Science and Engineering*, *126*(3), 451–458. doi:10.1115/1.1763181

Guo, S. M., Li, C. H., Zhang, Y. B., Wang, Y. G., Li, B. K., Yang, M., ... Liu, G. T. (2017). Experimental evaluation of the lubrication performance of mixtures of castor oil with other vegetable oils in MQL grinding of nickel-based alloy. *Journal of Cleaner Production*, *140*, 1060–1076. doi:10.1016/j.jclepro.2016.10.073

Heinzel, C., Rickens, K., & Trumpold, H. (2009). Engineered wheels for grinding of optical glass. *Cirp Annals-Manufacturing Technology*, *58*(1), 315–318. doi:10.1016/j.cirp.2009.03.096

Herman, D., & Krzos, J. (2009). Influence of vitrified bond structure on radial wear of cBN grinding wheels. *Journal of Materials Processing Technology*, *209*(14), 5377–5386. doi:10.1016/j.jmatprotec.2009.03.013

Jia, D. Z., Li, C. H., Zhang, D. K., Zhang, Y. B., & Zhang, X. W. (2014). Experimental verification of nanoparticle jet minimum quantity lubrication effectiveness in grinding. *Journal of Nanoparticle Research*, *16*(12). doi:10.100711051-014-2758-7

Jia, D. Z., Li, C. H., Zhang, Y. B., Zhang, D. K., & Zhang, X. W. (2016). Experimental research on the influence of the jet parameters of minimum quantity lubrication on the lubricating property of Ni-based alloy grinding. *International Journal of Advanced Manufacturing Technology*, *82*(1-4), 617–630. doi:10.100700170-015-7381-y

Kitzig, H., Tawakoli, T., & Azarhoushang, B. (2016). A novel ultrasonic-assisted dressing method of electroplated grinding wheels via stationary diamond dresser. *International Journal of Advanced Manufacturing Technology*, *86*(1-4), 487–494. doi:10.100700170-015-8156-1

Koshy, P., Iwasald, A., & Elbestawl, M. A. (2003). Surface Generation with Engineered Diamond Grinding Wheels: Insights from Simulation. *CIRP Annals - Manufacturing Technology, 52*(1), 271-274. doi:10.1016/S0007-8506(07)60582-4

Lankiewicz, K., Baranowski, M., Babul, T., & Kowalski, S. (2015). The study of the impact of surface preparation methods of inconel 625 and 718 nickel-base alloys on wettability by BNi-2 and BNi-3 brazing filler metals. *Archives of Metallurgy and Materials*, *60*(2), 739–745. doi:10.1515/amm-2015-0200

Li, C. H., Fang, Z., Cui, Y., & Ding, Y. C. (2010). Layered Manufacturing by Reverse Engineering-Principle and Application. *Advanced Materials Research, 102-104*, 436-440.

Li, C. H., Hou, Y. L., Li, J. Y., Han, Z. L., & Ding, Y. C. (2011). Mathematical Modeling and Simulation of Fluid Velocity Field in Grinding Zone with Smooth Grinding Wheel. *Advanced Science Letters*, *4*(6-7), 2468–2473. doi:10.1166/asl.2011.1344

Li, H., Yu, T., Zhu, L., & Wang, W. (2015). Analysis of loads on grinding wheel binder in grinding process: Insights from discontinuum-hypothesis-based grinding simulation. *International Journal of Advanced Manufacturing Technology*, *78*(9-12), 1943–1960. doi:10.100700170-014-6767-6

Li, X. (2010). *Modeling and simulation of grinding processes based on a virtual wheel model and microscopic interaction analysis*. Doctoral Dissertations.

Li, X., Wolf, S., Geng, Z., & Rong, Y. (2014). The modelling and experimental verification of the grinding wheel topographical properties based on the 'through-the-process' method. *International Journal of Advanced Manufacturing Technology*, *70*(1-4), 649–659. doi:10.100700170-013-5301-6

Li, X. K., & Rong, Y. M. (2011). Framework of grinding process modeling and simulation based on microscopic interaction analysis. *Robotics and Computer-integrated Manufacturing*, *27*(2), 471–478. doi:10.1016/j.rcim.2010.06.029

Li, Z., Ding, W. F., Shen, L., Xi, X. X., & Fu, Y. C. (2016). Comparative investigation on high-speed grinding of TiCp/Ti-6Al-4V particulate reinforced titanium matrix composites with single-layer electroplated and brazed CBN wheels. *Chinese Journal of Aeronautics*, *29*(5), 1414–1424. doi:10.1016/j.cja.2016.01.005

Liu, Z. P., & Shi, L. Y. (2008). Least square principles and its MATLAB implementation. *Science and Technology of West China*, *7*(17), 33–34.

Lyu, Y. S., Yu, H. Y., Wang, J., Zhao, G. W., & Liu, Z. Z. (2017). Improved performance of electroplated grinding wheels using a new method of controlled grain size sorting. *Journal of Manufacturing Processes*, *30*, 336–342. doi:10.1016/j.jmapro.2017.10.004

Mao, Q. G. (1991). *Evaluation and measurement of surface roughness*. China Machine Press.

Nie, Z. G., Wang, G., Lin, Y. L., Su, X., & Rong, Y. M. (2016). Modeling and Simulation of Single Abrasive-Grain Cutting Process in Creep Feed Grinding. *Key Engineering Materials, 693*, 1241-1246.

Pal, B., Chattopadhyay, A. K., & Chattopadhyay, A. B. (2012). Performance Study of Brazed Type cBN Grinding Wheel on Hardened Bearing Steel and High Speed Steel. *International Journal of Precision Engineering and Manufacturing*, *13*(5), 649–654. doi:10.100712541-012-0084-5

Palmer, J., Ghadbeigi, H., Novovic, D., & Curtis, D. (2018). An experimental study of the effects of dressing parameters on the topography of grinding wheels during roller dressing. *Journal of Manufacturing Processes*, *31*, 348–355. doi:10.1016/j.jmapro.2017.11.025

Ren, J. X., & Hua, A. D. (2011). *Grinding principles* (Vol. 1). Academic Press.

Sanchez, J. A., Pombo, I., Cabanes, I., Ortiz, R., & de Lacalle, L. N. L. (2008). Electrical discharge truing of metal-bonded CBN wheels using single-point electrode. *International Journal of Machine Tools & Manufacture*, *48*(3-4), 362–370. doi:10.1016/j.ijmachtools.2007.10.002

Shen, X. T., Wang, C. C., & Sun, F. H. (2018). The effects of deposition parameters on the grain morphology and wear mechanism of monolayer diamond grinding tools fabricated by hot filament CVD method. *Diamond and Related Materials*, *89*, 312–321. doi:10.1016/j.diamond.2018.09.021

Shen, X. T., Wang, X. C., & Sun, F. H. (2019). Fabrication and evaluation of monolayer diamond grinding tools by hot filament chemical vapor deposition method. *Journal of Materials Processing Technology*, *265*, 1–11. doi:10.1016/j.jmatprotec.2018.10.001

Wang, S., Li, C., Jia, D., & Hou, Y. (2014). Investigation on the grinding temperature field of nano-ZrO~2 dental ceramics with a nanoparticle jet of MQL. *International Journal of Computational Materialsence & Surface Engineering*, *9*(6), 11–15. doi:10.1504/IJCMSSE.2014.063757

Wang, S., Li, C., Jia, D., Zhang, Y., & Zhang, Q. (2015). Modelling and simulation of the surface topography generation with engineered grinding wheel. *International Journal of Computational Materials Science & Surface Engineering*, *6*(2), 111–129. doi:10.1504/ijcmsse.2015.071647

Wang, S., Li, C., & Qiang, Z. (2014). *Prediction and experiment on surface roughness of grinding process in minimum quantity lubrication with nanoparticles*. Modern Manufacturing Engineering.

Wang, S., Li, C. H., Zhang, D. K., Jia, D. Z., & Zhang, Y. B. (2014). Modeling the operation of a common grinding wheel with nanoparticle jet flow minimal quantity lubrication. *International Journal of Advanced Manufacturing Technology*, *74*(5-8), 835–850. doi:10.100700170-014-6032-z

Wang, S., Li, C. H., Zhang, X. W., Jia, D. Z., & Zhang, Q. (2014). Modeling and Simulation of the Single Grain Grinding Process of the Nano-Particle Jet Flow of Minimal Quantity Lubrication. *Open Materials Science Journal*, *8*(1), 55–62. doi:10.2174/1874088X01408010055

Xu, X. F., Feng, B. H., Huang, S. Q., Luan, Z. Q., Niu, C. C., Lin, J. B., & Hu, X. D. (2019). Capillary penetration mechanism and machining characteristics of lubricant droplets in electrostatic minimum quantity lubrication (EMQL) grinding. *Journal of Manufacturing Processes*, *45*, 571–578. doi:10.1016/j.jmapro.2019.07.036

Yan, Q. S., Zhang, Z. Q., & Syoji, K. (2003). Machining precision in the face grinding of a groove bottom using a cup CBN quill. *Journal of Materials Processing Technology*, *138*(1), 527–530. doi:10.10160924-0136(03)00140-7

Yang, P. Z., & Jiang, S. W. (2002). Study of 3D evaluation of surface roughness. *Machine Design and Research*, *18*(2), 64–67.

Yu, H. (2016). Simulation of grinding surface roughness using the grinding wheel with an abrasive phyllotactic pattern. *International Journal of Advanced Manufacturing Technology*, *84*(5-8), 861–871. doi:10.100700170-015-7722-x

Yu, H. Y., Lyu, Y. S., & Wang, J. (2019). Experimental Investigation on Grinding Temperature of Ti-6Al-4V Using Biomimetic Engineered Grinding Wheel. *International Journal Of Precision Engineering And Manufacturing-Green Technology*, *6*(2), 163–173. doi:10.100740684-019-00050-3

Zhang, H. M. (2001). Ordinary abrasive grain manufacturing. Beijing: China Standards Press.

Zhang, X., & Lin, B. (2012). Research on the analytical thermal model in surface grinding by cup wheel. *International Journal of Advanced Manufacturing Technology*, *66*(1-4), 1–13. doi:10.100700170-012-4294-x

Zhang, Y. B., Li, C. H., Ji, H. J., Yang, X. H., Yang, M., Jia, D. Z., ... Wang, J. (2017). Analysis of grinding mechanics and improved predictive force model based on material-removal and plastic-stacking mechanisms. *International Journal of Machine Tools & Manufacture*, *122*, 67–83. doi:10.1016/j.ijmachtools.2017.06.002

Zhang, Y. Z., & Xu, X. P. (2019). Influence of surface topography evolution of grinding wheel on the optimal material removal rate in grinding process of cemented carbide. *International Journal of Refractory Metals & Hard Materials*, *80*, 130–143. doi:10.1016/j.ijrmhm.2019.01.009

Zhao, Z., Fu, Y., Xu, J., Zhang, Z., Liu, Z., & Jian, H. (2016). An investigation on high-efficiency profile grinding of directional solidified nickel-based superalloys DZ125 with electroplated CBN wheel. *International Journal of Advanced Manufacturing Technology*, *83*(1-4), 1–11. doi:10.100700170-015-7550-z

Zhou, X., & Xi, F. (2002). Modeling and predicting surface roughness of the grinding process. *International Journal of Machine Tools & Manufacture*, *42*(8), 969–977.

Zhu, D. H., Yan, S. J., & Li, B. Z. (2014). Single-grit modeling and simulation of crack initiation and propagation in SiC grinding using maximum undeformed chip thickness. *Computational Materials Science*, *92*, 13–21. doi:10.1016/j.commatsci.2014.05.019

Chapter 18
Modeling and Simulation of the Surface Topography Generation With Ordinary Grinding Wheel

ABSTRACT

In order to simulate the surface grinding process of the common grinding wheel and evaluate the surface topography of the workpiece, the mathematical model of the common grinding wheel is established by using the abrasive vibration method to realize the parameterization of the simulated grinding wheel. The grinding wheel body with random arrangement of abrasive grains is generated, and the simulation of common grinding wheel morphology is completed. Then, according to the grinding kinematics model, elastic deformation model and plastic accumulation model, the simulation of the process is realized, and the workpiece surface morphology matrix is generated. Finally, the influence of grinding parameters and grinding wheel parameters on the workpiece surface morphology and surface roughness is studied by calculating the surface roughness value.

18.1 INTRODUCTION

At present, there are two main methods to study the process of surface topography creation: one is experimental method, the other is modeling and simulation method (Chen, Cheung, Zhang, & Zhao, 2018). The experimental method needs a lot of processing experiments, and can only study the final workpiece surface morphology, which is difficult to reflect the removal mechanism of materials in the process of processing; while the modeling and simulation method models the processing process, fully analyzes the influence of various parameters on the processing process, reveals the removal mechanism, and reduces the experimental time (Arizmendi et al., 2009; C. Zhang, Shi, Ehmann, & Li, 2016). Surface roughness is an important parameter to evaluate the surface topography of workpiece (Cheng, Gong, & Wang, 2013). By establishing the relationship between grinding parameters and surface roughness, it will play a guiding role in the optimization of grinding parameters. Li et al. (2018) have done theoretical and experimental research on the generation of surface morphology in slow tool servo machining of

DOI: 10.4018/978-1-7998-1546-4.ch018

free-form surfaces. The effectiveness of slow tool servo process in manufacturing free-form surfaces with nano morphology is proved. Zhang et al. (2013) proposed a milling surface morphology prediction model considering tool wear. Based on the model of surface topography, an algorithm for simulating the generation of milling surface profile is proposed. The experimental model of stainless steel and cemented carbide milling cutter on five axis milling center was established and verified. The cutting test results of plane and cylinder are in good agreement with the model prediction. Biondani and Bissacco (2019) introduced how to build the model of this interaction, at the same time, considering the cutting edge shape, material deformation and cutting edge trajectory error, to achieve a reliable prediction of the formation of ball end milling cutter surface shape. Zhang et al. (2016) used ultrasonic elliptical vibration to assist the micro groove turning on the cylinder surface, and studied the micro groove shape model on the cylinder surface. Chen et al. (2018; 2017) studied the modeling of real surface topography of grinding wheel and the method of digital generation of grinding workpiece surface based on the real surface of grinding wheel. Zou et al. (2019) used a three-dimensional shape optical scanner to detect the surface morphology of the abrasive belt, and carried out surface fitting and noise filtering. Based on Johnson transform system and filter pulse function, the surface morphology of non-Gaussian sand belt is generated. The simulation results are in good agreement with the measured topography. Considering the effect of vibration in actual grinding, Wang et al. (2016) studied the surface morphology of workpiece under forced vibration. The results show that the grinding speed has a significant effect on the surface topography, mainly on the amplitude and frequency. Finally, the simulation results are verified by grinding experiments. Considering the relative vibration of grinding wheel and workpiece, Cao et al. (2013) established the analysis and simulation model of grinding process surface morphology, and established the equation of abrasive trajectory. Chakrabarti & Paul (2008) put forward a numerical simulation technology to generate the grinding wheel shape by using the square and cone abrasive particles, to generate the grinding workpiece surface, to simulate the movement track of all abrasive particles, and to calculate the average roughness of the workpiece surface. Finally, the variation of surface roughness with the maximum chip thickness is studied. Chi et al. (2016) put forward a simulation model of the workpiece surface topography which is different from the existing surface grinding. Considering the displacement and micro interaction between grinding wheel and workpiece, the simulation model of grinding process is established according to the particle trajectory mapping between grinding wheel and workpiece. Zhou et al. (2019) established a comprehensive simulation model of uvag considering Poisson effect, predicted the three-dimensional surface morphology, and calculated the parameters (undeformed chip thickness, dynamic effective grain number) that can reflect the characteristics of the corresponding material removal process. Liao et al. (2018) proposed a moving average model based on time series method to generate grinding wheel surface topography. Ding et al. (2017) carried out grinding experiments by using the measured wheel shape. The surface morphology of single-layer CBN wheel was reconstructed by Johnson transform and its inverse transform. The influence of topological evolution of grinding wheel on the thickness inhomogeneity of undeformed chip is determined by using the improved model. Liu et al. (2019) put forward the basic shape parameters of the particles and two new indexes to characterize the surface shape of the grinding wheel, and established a three-dimensional model of the surface shape of the grinding wheel. Wang et al. (2019) proposed a new method to control and predict the machined surface morphology of engineering ceramics. According to the kinematic analysis and fracture mechanics of brittle materials, the surface forming model is gradually established. The influences of grinding force and material removal rate on carpet technology are discussed. The dynamic relationship among ultrasonic energy, grinding parameters and surface microstructure is revealed by establishing mathematical model. The randomness of grains has a great influence on the performance of grinding process, which is also a difficulty in grinding simulation. In order to solve this problem, Chen et al. (2019) proposed a new method to generate CBN grain model with arbitrary shape. Based on the geometry of CBN grains, the method uses the statistics of geometric characteristics to assist the modeling of CBN grains. The next research considers the randomness of CBN particles in attitude and position. In order to predict the grinding process, two-dimensional convex shell is used to obtain the projection profile of random grain shape. The continuous cutting track of cutting points is analyzed, and the surface topography is simulated. Considering ploughing and friction, Chen et al. (2018) proposed a new modeling method of grinding surface topography. Firstly, the depth of abrasive participation is defined as the real-time cutting depth varying with time, and the relationship between the undeformed chip cross-sectional area and the real-time cutting depth is established. Secondly, the three-dimensional cutting path of abrasive particles is defined on the premise of plowing. Chen et al. (2017) put forward a new method to predict the three-dimensional profile of the gear surface in the process of generating grinding, and derived the parameter equation to describe the trajectory of each ideal spherical particle on the gear surface. In order to determine the final three-dimensional gear surface morphology produced by thousands of single abrasive particles with randomly distributed positions and protruding heights, a geometric analysis algorithm is proposed to systematically treat the gear surface profile. Du (2009) have done some research on the computer simulation of workpiece surface topography generation in grinding process. The mathematical model of grinding process is established, and the surface topography of the workpiece is simulated by MATLAB software. The material parameters and grinding parameters are considered in the simulation, but the randomness of the grinding wheel surface is not fully considered in the simulation of the grinding wheel body. Chen (2010) used Johnson transform and linear filtering technology to generate the surface morphology of grinding wheel. According to the related theory of grinding kinematics, the simulation of the surface topography generation process of precision grinding workpiece under different grinding wheel morphology and different processing conditions is realized, and the correctness of the surface topography generation algorithm is verified. Wang (2002) established a grinding process model based on time response. It includes grinding force model, grinding thermal model and surface roughness model. With the help of these models, the generation process simulation of workpiece surface morphology and the prediction of workpiece surface roughness are realized. The model can be used to guide the selection of grinding parameters and the prediction of grinding quality. Gong (2002) realized the simulation of the creation process of the surface topography of the surface grinding workpiece. A simulation platform for surface topography generation is established. In this method, spherical abrasive particles are used on the surface of grinding wheel, and the influence of grinding wheel parameters and grinding parameters on the surface topography is considered. Su (2008) studied the relationship between different grinding parameters and workpiece surface roughness. The difference is that in the simulation, the abrasive grains on the surface of the grinding wheel adopt the cube with different arrangement directions. Sakakura (2008) realizes the digital simulation of the wheel shape by measuring the cross-section profile of 50 abrasive particles as standard abrasive particles and randomly distributing them on the surface of the wheel. Under the premise of considering the elastic deformation of abrasive grains, the surface topography of the workpiece in cylindrical grinding is simulated with the help of the grinding wheel, and the surface roughness of the sampling area is calculated. Salisbury (2001a, 2001b) sampled and reconstructed the surface topography of the grinding wheel by the surface roughness instrument, and realized the digital simulation of the surface topography of the grinding wheel. Based on the reconstruction of the grinding wheel and the related grinding models, the process of creating the workpiece surface morphology is simulated, and the workpiece surface morphology is evaluated. However, it is difficult to simulate the actual process of workpiece surface formation without considering the elastic-plastic deformation of abrasive particles and workpiece. Chakrabarti and Paul (2008) produced the surface morphology of the grinding wheel by using the regular triangular cone abrasive. All the abrasive grains on the surface of the grinding wheel have the same angle and arrangement direction. According to the grinding kinematics, the simulation of the surface topography of the workpiece is realized, and the surface roughness value of the workpiece is calculated. The influence of grinding parameters on the surface roughness and the relationship between the surface roughness value and the maximum undeformed chip thickness are studied. The research shows that with the increase of the surface roughness value, the surface roughness value increases. The maximum undeformed chip thickness increases linearly. Li and Rong (2011) studied the manufacturing process of ordinary grinding wheel. The computer is used to simulate the manufacturing process of grinding wheel, such as abrasive, binder mixing, grinding wheel modeling, grinding wheel baking and grinding wheel dressing. The surface morphology of the common grinding wheel is generated. The process of creating the surface morphology of the workpiece is simulated by using the grinding wheel. The grinding force, grinding heat and surface morphology are analyzed. Zhou and Xi (2002) gridded the three-dimensional solid of the grinding wheel, and the grinding wheel was on every gridding point. The simulation of the surface morphology of the grinding wheel was completed. According to the theory of grinding kinematics, the grinding simulation is realized by using the grinding wheel, and the simulation of workpiece surface topography generation process is realized by searching method, and the simulated workpiece surface roughness is verified by experiments. Cui et al. (2017) used the homogeneous transformation, through the establishment of three-dimensional simulation model, give the geometry and motion of the workpiece and wheel. The model and related simulation can be used to study the kinematics and rounding process of workpiece under different geometry and grinding parameters, and further analyze the influence of these parameters on the roundness generation of workpiece.

18.2 ORDINARY GRINDING WHEEL

18.2.1 Definition of Ordinary Grinding Wheel

Grinding wheel, as a commonly used machining tool in the grinding process, consists of abrasive material, bonding agent and pore (Ren J, 2011). Abrasive materials used to manufacture grinding wheel are usually divided into two types: ordinary abrasive material and ultrahard abrasive material; there are mainly four bonding agents: resinoid bond, ceramic bond, metallic bond and rubber bond. Just as its name implies, ordinary grinding wheel is manufactured through high-temperature sintering after mixing of ordinary abrasive material, bonding agent and other auxiliary materials at the designated proportion. Bonding agent mainly exerts connection and fixation effects on abrasive material in the grinding wheel; abrasive materials in the grinding wheel body mainly exert the effect of cutting workpiece surface material (Wenfeng, Jiuhua, Zhenzhen, Honghua, & Yucan, 2010).

18.2.2 Characteristics of Ordinary Grinding Wheel

Ordinary grinding wheel has been mostly extensively applied to grinding process. Compared with the engineered grinding wheel, it doesn't need complicated manufacturing equipment or consume too many energy sources with simple manufacturing technology and low cost, and all kinds of universal grinding wheel and forming grinding wheel can be manufactured. High-porosity grinding wheel can be used to coarse grinding with large removal rate, and it can significantly improve machining efficiency, mitigate the insufficient heat exchange phenomenon in the grinding process, reduce burning on machined workpiece surface and improve workpiece surface quality (Yang, 2016). The most significant feature of ordinary grinding wheel is of self-sharpening property. As abrasive grains are connected together through bonding agent, blunted abrasive gains will automatically fall off under the effect of grinding force in the grinding process. New abrasive grains with cutting capability will participate in the machining process, thus reducing dressing times and replacement time of grinding wheel. It's worth mentioning that ordinary grinding wheel also has deficiencies: as abrasive grains are under random arrangement on the surface of ordinary grinding wheel and their protruding heights are uncertain, grinding wheel surface bears nonuniform force in the grinding process, which will degrade workpiece machining precision, reduce service life of grinding wheel and generate an adverse effect on spindle of grinding wheel (Y. Zhang, Li, C., Yang, M., Jia, D., Zhang, D., & Zhang, X., 2015).

18.2.3 Preparation of Ordinary Grinding Wheel

The main manufacturing method of ordinary grinding wheel is sintering method. So-called sintering method refers to grinding wheel manufacturing method through sinter molding under high temperature (Deng Zhaohui, 2010; Li B, 2003). Figure 1 shows ordinary abrasive grinding wheel manufactured through sintering method.

Sintering method for manufacturing of ordinary grinding wheel is mainly divided into five steps: raw material weighing, raw material mixing, compaction and molding, high-temperature sintering and dressing as shown in Figure 2.

Figure 1. Ordinary abrasive grinding wheels

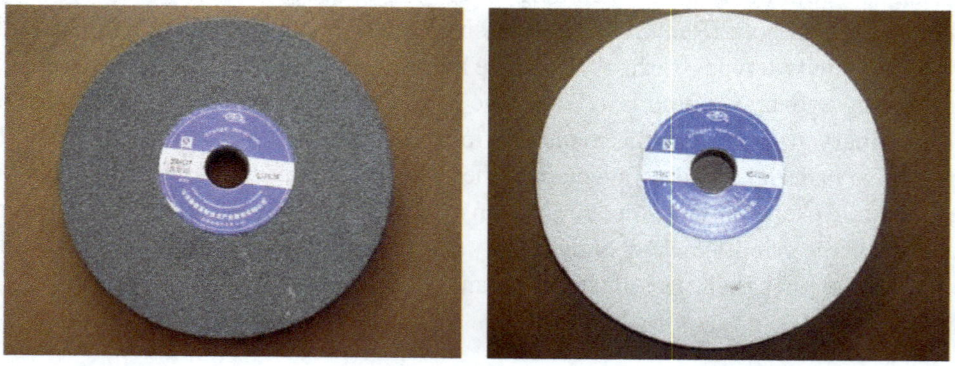

Figure 2. Onventional grinding wheel fabrication steps

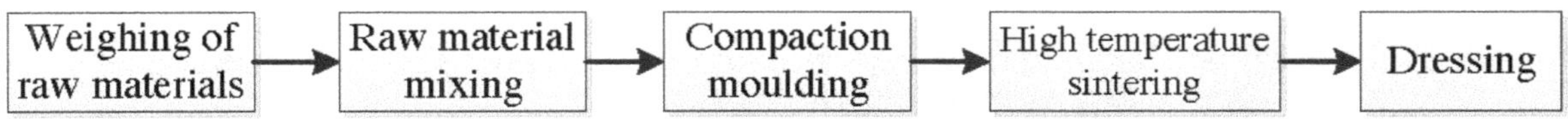

Concrete steps are as follows:

1. Raw materials for grinding wheel manufacturing mainly include abrasive materials, bonding agent and other auxiliary materials. Masses of needed raw materials are weighed according to the preset matching ratio, and auxiliary materials generally account for 1% of total mass of raw materials;
2. Weighed raw materials—abrasive material, bonding agent and auxiliary material—are blended so the manufactured grinding wheel has good structural uniformity;
3. Blended raw materials are stuffed in the designated grinding wheel die and compacted using a press machine, grinding wheel workblank with the preset shape is obtained, pressure acted by press machine is generally 100~3000 N/cm^2, and action time is 10~30s;
4. Compacted forming grinding wheel workblank is placed in the sintering furnace for high-temperature sintering treatment. Sintering temperature of grinding wheel depends on type of bonding agent. For resin bonded grinding wheel, sintering temperature is generally 300~400 °C, but that of ceramic bonded grinding wheel is 1700~2300 °C;
5. Sintered grinding wheel is dressed so that grinding wheel satisfies requirements for dimensional accuracy and shape accuracy, and necessary balance adjustment should be conducted.

18.3 ESTABLISHMENT OF ORDINARY GRINDING WHEEL MODEL

18.3.1 Grain Size

The corresponding relationship between grit number and grain size is already introduced in Chapter 17. According to the above relationship, abrasive materials with grain size satisfying requirements can be selected. For the given grit number, range of abrasive grain size corresponding to this grit number can be determined. Figure 3 shows schematic diagram of grain size distribution (Ali & Zhang, 1999; Doman, 2006).

Average abrasive grain size *D*ave of abrasive grains contained in the grinding wheel with grit number of *m* can be expressed as:

$$D_{ave} = \frac{D_{\max} + D_{\min}}{2} \tag{1}$$

where: $D_{\max}$ is maximum abrasive grain size of abrasive materials with given grit number, *mm*; and $D_{\min}$ is minimum abrasive grain size of abrasive materials with given grit number, *mm*.

Figure 3. The schematic diagram of abrasive grain size distribution

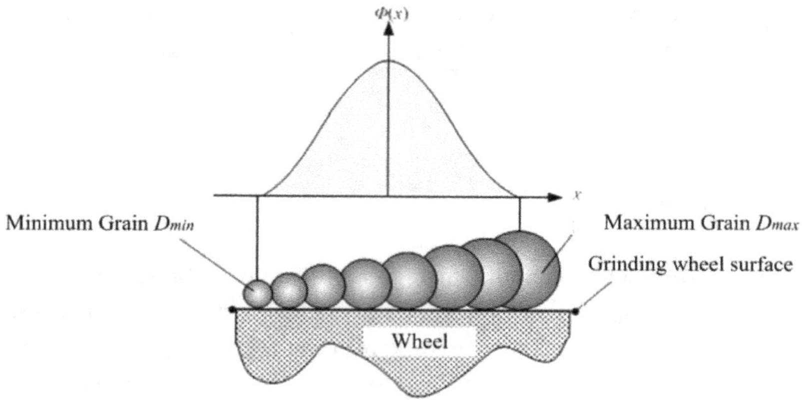

According to 6σ principles of normal distribution, standard deviation σ of abrasive grain size distribution can be determined as:

$$\sigma = \frac{D_{\max} - D_{\min}}{6} \tag{2}$$

Grain size of the above abrasive grains contained in the above grinding wheel is D, and then probability density function of this grain size can be expressed as:

$$\Phi(D) = \frac{1}{\sqrt{2\pi}\sigma} \exp\left[-\frac{(x - D_{ave})^2}{2\sigma^2} \right] \tag{3}$$

According to the above derivation, gran size D_i of arbitrary abrasive grain on the surface of grinding wheel body can be expressed as:

$$D_i = D_{ave} + rand\ n \cdot \sigma \tag{4}$$

where: *randn* is random number of standard normal distribution.

18.3.2 Grain Distribution

It's well known that grinding wheel consists of abrasive grain, pore and bonding agent. For grinding wheel with fixed model number, volume content V_g of abrasive grains in this grinding wheel body can be calculated through the organization number S as:

$$V_g = 2(31 - S) \tag{5}$$

Width of sampled grinding wheel body is set as *W* and length along circumferential direction of the grinding wheel is *L*. In order to calculate number of abrasive grains on the surface of sampled grinding wheel body, abrasive grains are uniformly distributed on the surface, and then initial distance of abrasive particles is:

$$L_r = D_{ave}\left(\sqrt{\frac{\pi}{4V_g}} - 1\right) \tag{6}$$

According to distribution distance and width *W* and circumferential length *L* of the sampled grinding wheel body, numbers N_x and N_y of abrasive grains in circumferential and axial directions of grinding wheel are respectively:

$$N_x = \frac{L}{D_{ave} + L_r} \tag{7}$$

$$N_y = \frac{W}{D_{ave} + L_r} \tag{8}$$

Total number *N* of abrasive grains of the sampled grinding wheel body is:

$$N = N_x N_y \tag{9}$$

Related studies indicate that abrasive grains follow uniform distribution in the plane of the grinding wheel body. In order to simulate random distribution of abrasive grains on the surface of grinding wheel body, abrasive grain vibration method is adopted in this paper (Liu, Warkentin,, Bauer, & Gong, 2013).

So-called abrasive grain vibration method refers to that all abrasive grains in the grinding wheel body freely move in the grinding wheel at their initial positions and preset vibration times with random vibration quantity so as to realize random arrangement of abrasive grains in the grinding wheel. Iterative equation of abrasive grain vibration is:

$$\begin{bmatrix} x_{gkn} \\ y_{gkn} \end{bmatrix} = \begin{bmatrix} x_{gk0} \\ y_{gk0} \end{bmatrix} + \begin{bmatrix} \delta x_{gk1} + \delta x_{gk2} + \ldots + \delta x_{gkn} \\ \delta y_{gk1} + \delta y_{gk2} + \ldots + \delta y_{gkn} \end{bmatrix} \tag{10}$$

where: (x_{gk0}, y_{gk0}) - initial abrasive grain coordinates;

(x_{gkn}, y_{gkn}) - final abrasive grain coordinates;

δx_{gkl}, δy_{gkl} - random number following uniform distribution within the interval $[-L_r, L_r]$.

Subscripts *k* and *n* - number and vibration times of abrasive grains.

In order to prevent abrasive grains from going beyond the sampling area of grinding wheel in the vibration process, center coordinates (x_{gi}, y_{gi}) of each abrasive grain participating in the vibration should satisfy:

$$\begin{cases} \frac{D_i}{2} \le x_{gi} \le L - \frac{D_i}{2} \\ \frac{D_i}{2} \le y_{gi} \le W - \frac{D_i}{2} \end{cases} \tag{11}$$

Meanwhile, multiple abrasive grains may experience overlapping phenomenon in the vibration process. In order to prevent this phenomenon, center coordinates of any two abrasive grains participating in the vibration process should satisfy:

$$\sqrt{(x_{gi} - x_{gj})^2 + (y_{gi} - y_{gj})^2} \ge \frac{D_i + D_j}{2}, i \ne j \tag{12}$$

Abrasive grain distribution on the grinding wheel surface is shown in Figure 4.

The concrete distribution process in Figure 4 is: abrasive grains are at equally spaced positions on the grinding wheel surface (Figure 4(a)); abrasive grain size on the grinding wheel surface is adjusted according to normal distribution of their grain size (Figure 4(b)); each abrasive grain vibrates according to the preset vibration times along width direction and circumferential direction of grinding wheel.

The above abrasive grain distribution process realizes random distribution of abrasive grains in the plane XOY. In order to realize random distribution of abrasive grains in the whole 3D space, abrasive grains will vibrate along the direction vertical to grinding wheel (namely direction of axis Z). According to equation (10), the corresponding iterative equation is:

$$z_{gkn} = z_{gk0} + \delta z_{gk0} + \delta z_{gk2} + \ldots + \delta z_{gkn} \tag{13}$$

Figure 4. Grain distribution process: (a) Initial configuration;(b) Adjust grain size;(c) Grain vibration

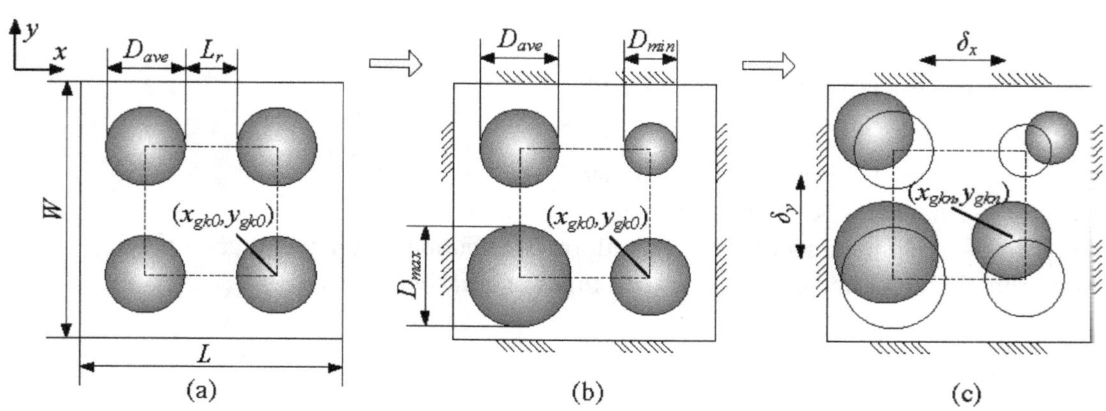

where: z_{gk0} - initial coordinate value of abrasive grains in direction Z;
z_{gkn} - final coordinate value of abrasive grain in direction Z;
δz_{gkn} - random number satisfying uniform distribution within the interval $[-L_r, L_r]$.

18.3.3 Mathematical Model of Ordinary Grinding Wheel

Center coordinates and radius of abrasive grains on the surface of ordinary grinding wheel are (x_g, y_g, z_g) and R_g respectively, and then mathematical model of abrasive grain distribution on the surface of ordinary grinding wheel can be expressed using a $N\times 4$ matrix, specifically being:

$$G_N(x_g, y_g, z_g, R_g) = \begin{bmatrix} x_{g1} & y_{g1} & z_{g1} & R_{g1} \\ x_{g2} & y_{g2} & z_{g2} & R_{g2} \\ \vdots & \vdots & \vdots & \vdots \\ x_{gN} & y_{gN} & z_{gN} & R_{gN} \end{bmatrix} \quad (14)$$

where N is total number of abrasive grains on the surface of ordinary grinding wheel mentioned in section 18.2.2.

18.4 SIMULATION AND ANALYSIS OF ORDINARY GRINDING WHEEL

According to the above established mathematical model of ordinary grinding wheel and the proposed abrasive grain vibration method, programming is conducted using MATLAB language. Morphology simulation of ordinary grinding wheel surface is realized through the program operation, and corresponding abrasive grain distribution matrix on the surface is generated for the follow-up grinding process simulation.

18.4.1 Simulation Parameters

In the morphology simulation of workpiece surface under plane grinding, the used ordinary abrasive grinding wheel is white corundum bonded grinding wheel, its model number is WA80MV12P and concrete grinding wheel parameters are shown in Table 1.

Parameters use in morphology simulation of grinding wheel mainly include abrasive grain grit number, organization number and geometrical parameters of grinding wheel.

Table 1. The grinding wheel parameters of WA80MV12P

Abrasive material	Grain size	Hardness	Bonding agent	Organization number	Shape	Outer diameter mm	Thickness mm	Inner diameter mm
White corundum	80#	Medium soft	Ceramic	12	Parallel grinding wheel	300	20	76.2

18.4.2 Simulation Scheme

Figure 5 shows simulation flow of ordinary grinding wheel. Concrete simulation flow is: simulation parameters of ordinary grinding wheel are firstly input, and initial abrasive grains spacing is calculated according to simulation parameters; numbers N_x and N_y of abrasive grains along circumferential and axial directions are calculated; according to initial abrasive grains spacing and numbers of abrasive grains along two directions of grinding wheel, abrasive grain positions are initialized on the grinding wheel surface; abrasive grain sizes on the grinding wheel surface are adjusted according to normal distribu-

Figure 5. The simulation flow of ordinary grinding wheel

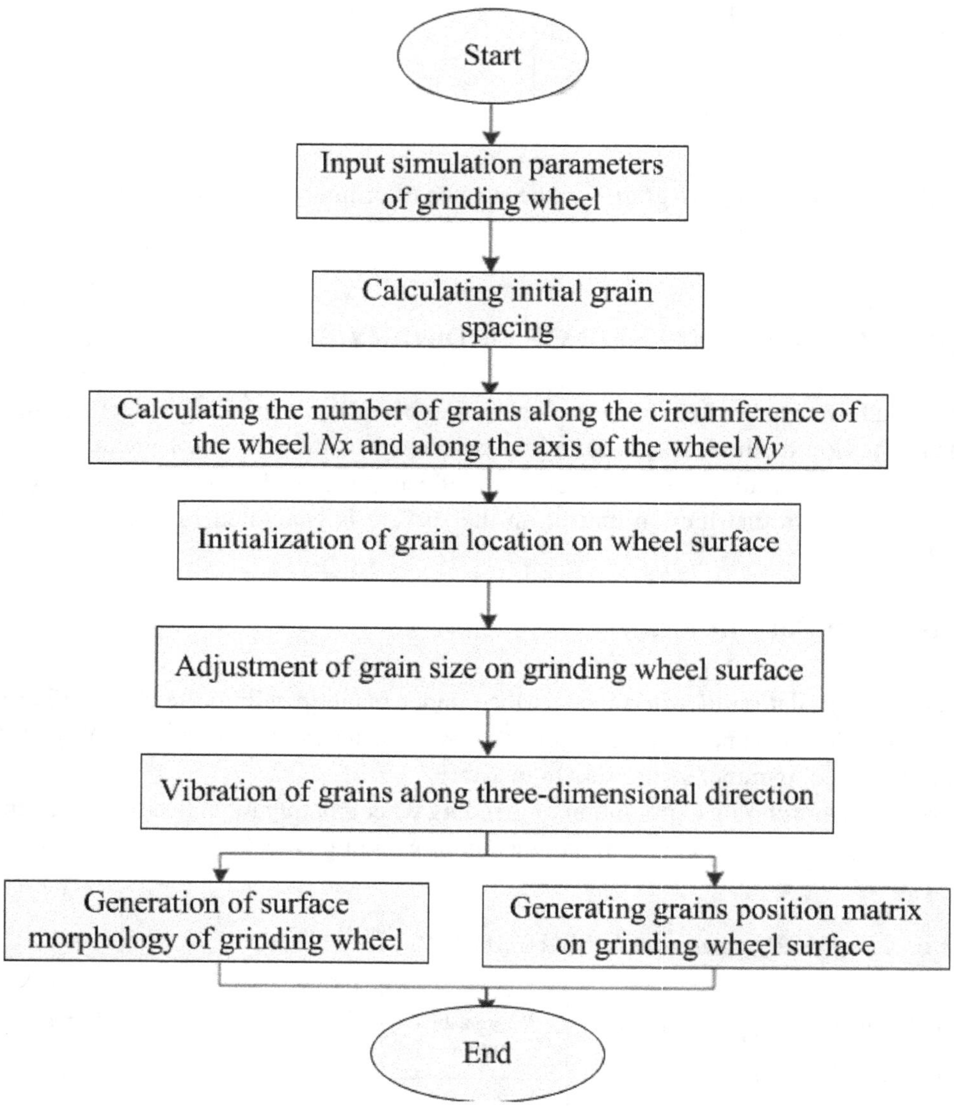

tion of abrasive grain sizes; abrasive grains on the grinding wheel surface vibrate along 3D direction, and surface morphology graph of ordinary grinding wheel and abrasive grain position matrix on the grinding wheel surface are generated.

Figure 6 shows flowchart of vibration of all abrasive grains once along circumferential and axial directions on the grinding wheel surface. The concrete flow is: variable quantities of X-coordinate and Y-coordinate of center of sphere of abrasive grains are initialized; variable quantity δx of X-coordinate of the i (th) abrasive grain is randomly generated; according to judging criterion (11) for abrasive grains boundary, whether abrasive grains with X-coordinate changed are within the boundary is judged. If not, return to the last step to generate new variable quantity δx of X-coordinate for judgement, but if yes, the program will be continued; according to judging criterion (12) for abrasive grains overlapping, whether overlapping phenomenon happens to abrasive grains is judged. If yes, return to the last step to generate new variable quantity δx of X-coordinate, judgment is conducted again according to the two judging criteria. If not, the program will be continued; X-coordinate of center of sphere of this abrasive grain is updated; continue to execute the program step similar to updating of X-coordinate and realize updating of Y-coordinate of center of sphere of this abrasive grain, and at the time, vibration of one abrasive grain along two coordinate directions is completed, the above program step is repeatedly executed, and abrasive grain position matrix after all abrasive grains on grinding wheel surface vibrate along two coordinate directions is obtained (Zhenxuan, 2012).

Vibration times of abrasive grains influence surface morphology simulation accuracy of ordinary grinding wheel. When vibration times reaches one fixed value, random distribution of abrasive grains will be stable. In order to seek for this fixed value, a section vertical to grinding wheel surface is made along axial direction of grinding wheel, as this cross-sectional area is the sum of intercepted areas of abrasive grains, standard deviation of the cross-sectional area can reflect abrasive grain distribution. Figure 7 shows the relation curves between vibration times and standard deviation of cross-sectional area.

According to Figure 7, as the vibration times of abrasive grains increases, standard deviation of cross-sectional area reduces, and meanwhile, change gradient of curve becomes smaller and smaller. For 60# grinding wheel, when vibration times of abrasive grains changes from 1 to 1500, standard deviation of cross-sectional area changes from 0.35 to 0.112. When vibration times is 1000, relation curves basically level with each other; for 80# grinding wheel, when vibration times of abrasive grains increases from 1 to 1500, standard deviation of cross-sectional area changes from 0.27 to 0.073, and when vibration times is 800, relation curves basically level with each other. Therefore, in the morphology simulation of 80# grinding wheel, vibration times of abrasive grains is selected as 800, and this can not only accurately complete morphology simulation of grinding wheel but also can improve operating speed of simulation program,

18.4.3 Analysis of Simulation Results

Through the calculation of simulation program, surface morphology graph and abrasive grain distribution matrix of ordinary grinding wheel are acquired. In order to intuitively display abrasive grain distribution and surface morphology on the surface of ordinary grinding wheel, the grinding wheel surface morphology generated through simulation is sampled, and sampling area is grinding wheel body with 5mm length and 2mm width. Figure 8, Figure 9 and Figure 10 show morphology under initial abrasive grain distribution on the grinding wheel surface, morphology with adjusted abrasive grain distribution and the final morphology of the grinding wheel respectively (S. Wang, Li, Zhang, Jia, & Zhang, 2014).

Figure 6. Flowchart of abrasive grain vibration process

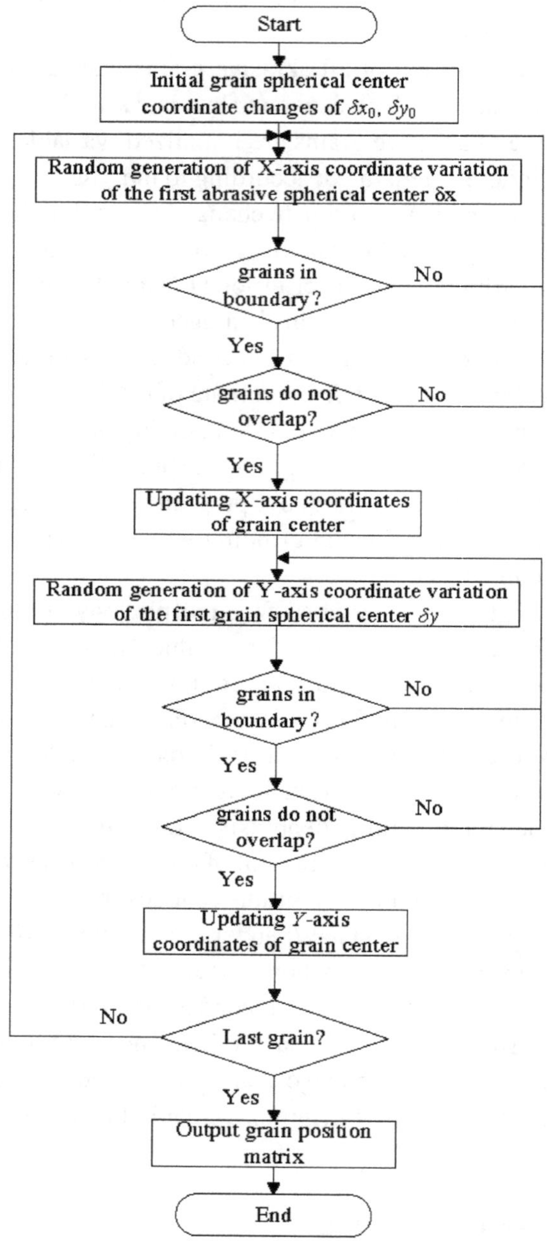

Figure 8 shows that abrasive grains generated through simulation are uniformly distributed on the grinding wheel surface according to the same abrasive grains spacing, all abrasive grans have the same grain size (196 μm), and meanwhile, centers of spheres of all abrasive grains are located in the same plane, so they have the same protruding height. In Figure 9, abrasive grain sizes on the grinding wheel surface are not equal, but ranging from 180 μm to 212 μm, and they follow normal distribution. Figure

Figure 7. Relationship between vibration times and standard deviation of cross-sectional area

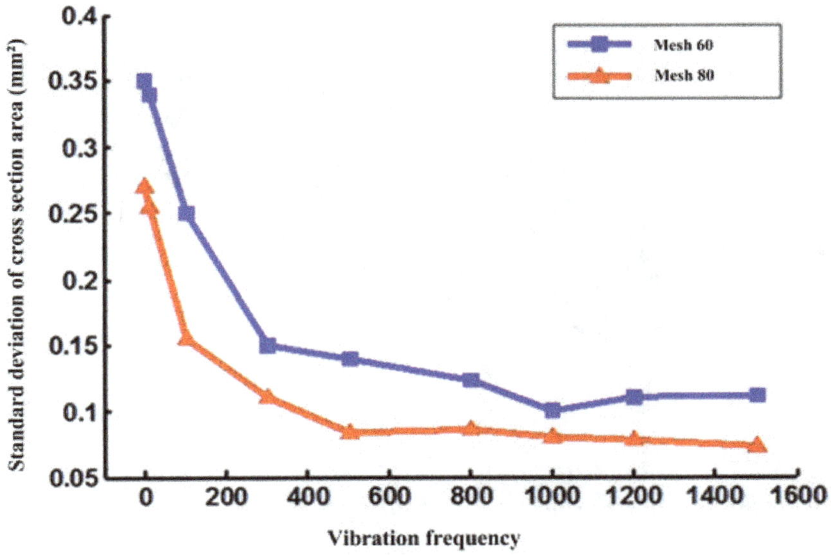

Figure 8. Morphology under initial grain distribution

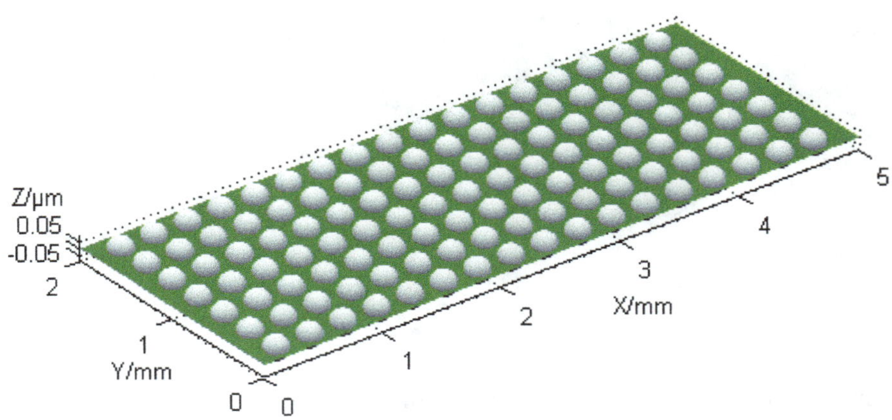

10 shows the final grinding wheel surface morphology when abrasive grains vibration along 3D direction, abrasive grains are disorderly arranged on the grinding wheel surface and their protruding heights are different, which is similar to actual surface morphology of ordinary grinding wheel.

In order to verify whether abrasive grain sizes generated through simulation follow normal distribution, all abrasive grains on the grinding wheel body generated through simulation are calculated, and abrasive grain size is divided into 11 intervals, and ranges of abrasive grain sizes corresponding to different intervals are shown in Table 2.

Figure 9. Morphology under adjusted grain distribution

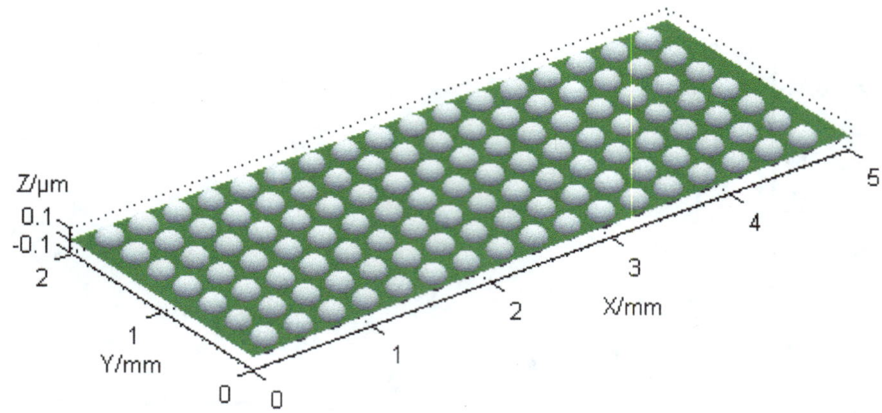

Figure 10. Final morphology of the grinding wheel surface

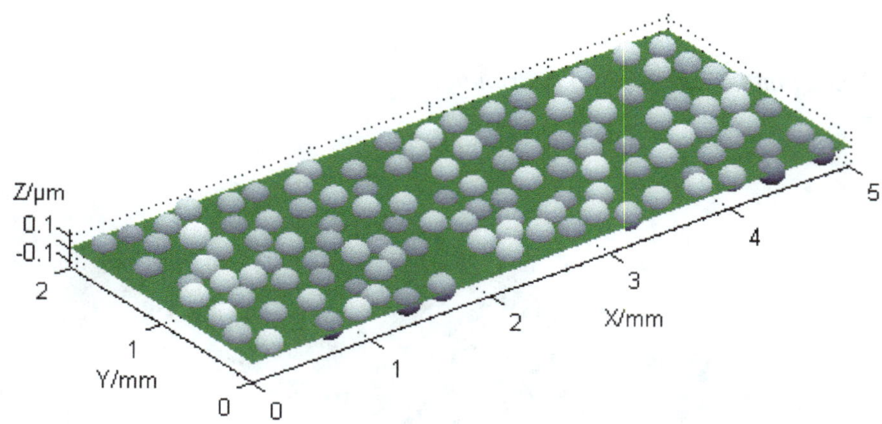

Figure 11 shows grain number distribution histogram and probability density curves corresponding to different intervals of grain sizes, where left coordinate is grain number and right coordinate is probability density. It can be seen that number of abrasive grains distributed in each interval fits in well with probability density value. Therefore, it can be said that abrasive grain sizes generated through simulation satisfy normal distribution and grinding wheel morphology generated through simulation can simulate real ordinary grinding wheel.

Table 2. The distribution interval of grain size

Interval number	Range of grain size(μm)	Interval number	Range of grain size (μm)	Interval number	Range of grain size (μm)	Interval number	Range of grain size (μm)
1	<181.6	4	188~191.2	7	197.6~200.8	10	207.2~210.4
2	181.6~184.8	5	191.2~194.4	8	200.8~204	11	>210.4
3	184.8~188	6	194.4~197.6	9	204~207.2		

Figure 11. The grain number distribution histogram and the probability density curve

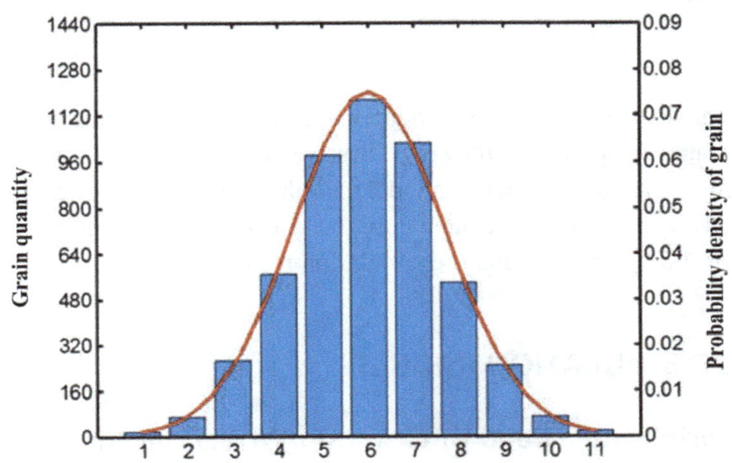

18.5 SIMULATION OF GRINDING PROCESS

18.5.1 Setting of Simulation Parameters

Simulation parameters used in the grinding process simulation of ordinary grinding wheel include grinding wheel parameters, workpiece parameters and grinding parameters. Grinding wheel parameters derive from abrasive grain distribution matrix on grinding wheel surface generated through morphology simulation of grinding wheel, simulation object is quenched 45# steel, and concrete grinding wheel parameters, workpiece parameters and grinding parameters are shown in Table 3,4 and 5.

As influence laws of grinding parameters and grinding wheel parameters on workpiece surface morphology in the grinding process of ordinary grinding wheel are mainly investigated in this chapter and univariate method is used, grinding parameters given in Table 5 are range values, and concrete grinding parameters will be given in the simulation result analysis.

18.5.2 Procedures of Simulation Program

Figure 12 shows simulation flowchart of grinding process of ordinary grinding wheel. Concrete simulation flow is: grinding simulation parameters of ordinary grinding wheel are firstly input, and the simulation

Table 3. The parameters of grinding wheel

Diameter of grinding wheel d_s	Diameter of abrasive grains D
300 mm	196 μm~212 μm

Table 4. The parameters of workpiece

Material type	Workpiece size	Rockwell hardness	Stiffness coefficient k
45#	0.8 mm×0.8 mm	HRC=45	300 kg/mm

Table 5. The parameters of grinding process

Linear velocity of grinding wheel v_s	Workbench feed velocity v_w	Grinding depth a_p
5 m/s ~60 m/s	0.01 m/s~0.1 m/s	10 μm ~45 μm

process is initialized; abrasive grain distribution matrix is loaded on the grinding wheel surface; workpiece surface morphology matrix generated by the i (th) abrasive grain is solved according to single-abrasive-grain kinematical model, elastic deformation model and plastic stacking model, and then it is updated; the above program steps are repeatedly executed until the last abrasive grain generates the corresponding workpiece surface morphology matrix and the simulation result is output.

18.6 ANALYSIS OF SIMULATION RESULT

18.6.1 Effect of Peripheral Speed of Grinding Wheel on Surface Morphology

Under diameter d_s of grinding wheel of 300 mm, workpiece feed velocity v_w of 0.05 m/s and grinding depth of 20 μm and 40 μm respectively, influence laws of linear velocity of grinding wheel on surface morphology are investigated by changing linear velocity of grinding wheel, which is taken as 5 m/s, 10 m/s, 15 m/s, 20 m/s, 25 m/s, 30 m/s, 35 m/s, 40 m/s, 45 m/s, 50 m/s, 55 m/s and 60 m/s respectively.

Figure 13 shows workpiece surface morphologies under different linear velocities of grinding wheel. (a)~(d) show workpiece surface morphologies under grinding depth of 20 μm and linear velocity of 5 m/s, 20 m/s, 35 m/s and 50 m/s respectively; (e)~(h) show workpiece surface morphologies under grinding depth of 40 μm and linear velocity of 5 m/s, 20 m/s, 35 m/s and 50 m/s respectively. It can be seen that: as linear velocity of grinding wheel increases, number of grinding cracks on workpiece surface increases, surface texture tends to be regular, and protruding height reduces; this is because elevating linear velocity of grinding wheel increases the number of abrasive grains contacting the workpiece within unit time and decreases grinding force borne by each abrasive grain, elastic-plastic deformation of workpiece surface is reduced and workpiece surface morphology turns good.

A section vertical to grinding texture direction in the surface morphologies in Figure 13 is taken, and the workpiece surface profile curve obtained through intersection between this section and workpiece surface is shown in Figure 14.

(a)~(d) in Figure 14 show workpiece surface profile curves under grinding depth of 20 μm and linear velocity of 5 m/s, 20 m/s, 35 m/s and 50 m/s respectively; (e) ~(h) show workpiece surface profile curves under grinding depth of 40 μm and linear velocity of 5 m/s, 20 m/s, 35 m/s and 50 m/s respectively. It can be seen that: as linear velocity of grinding wheel increases, peak and valley on the workpiece surface

Figure 12. The grinding simulation flowchart of ordinary grinding wheel

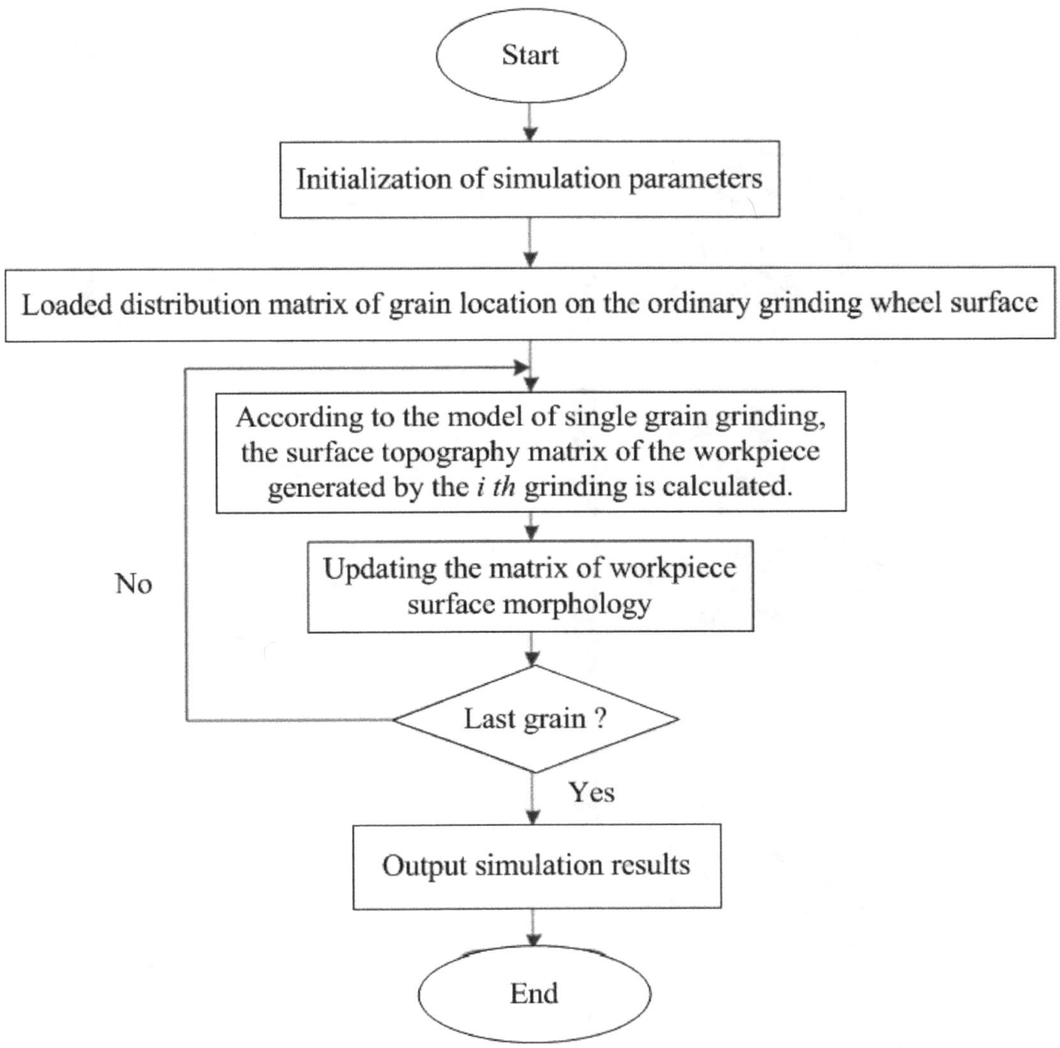

profile curve increase in number, and peak height and valley depth reduce; this is because elevating linear velocity of grinding wheel increases the number of abrasive grains participating in the cutting process within unit time and decreases the grinding force, and generated grinding heat decreases and workpiece surface quality turns good (Wang, Li, Jia, Zhang, & Zhang, 2015).

By calculating surface roughness values corresponding to different profile curves, the relation curves between linear velocity of grinding wheel and workpiece surface roughness can be obtained as shown in Figure 15.

In Figure 15, (a) shows the relation curve between linear velocity of grinding wheel and R_a value and (b) is the relation curve between linear velocity of grinding wheel and R_z value. It can be seen that as linear velocity of grinding wheel increases, R_a value and R_z value on workpiece surface are reduced, so does change gradient of curve. When grinding depth is 20 μm and linear velocity of grinding wheel

Figure 13. Workpiece surface morphologies under different linear velocities of grinding wheel

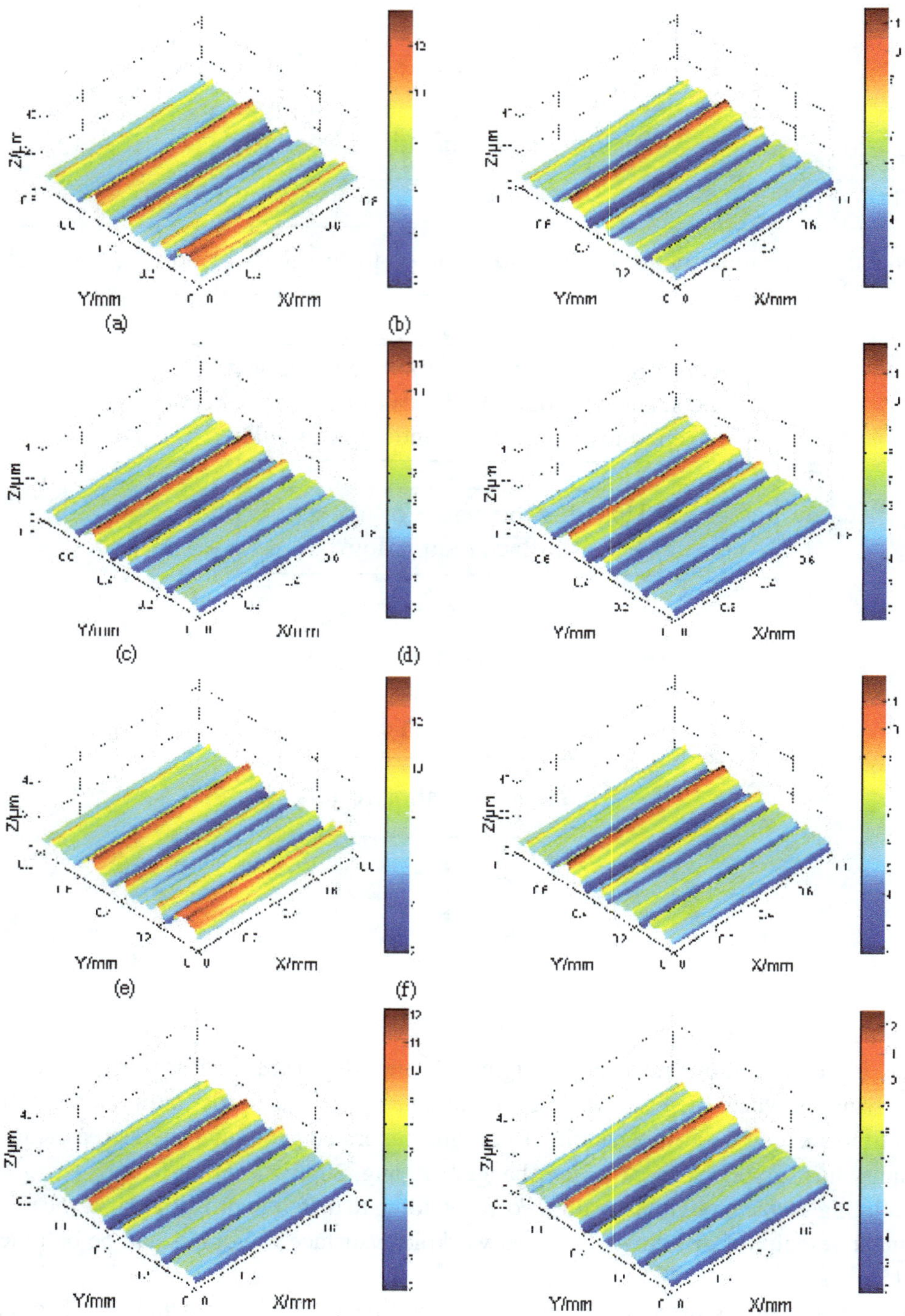

Figure 14. Surface profile curves under different linear velocities of grinding wheel

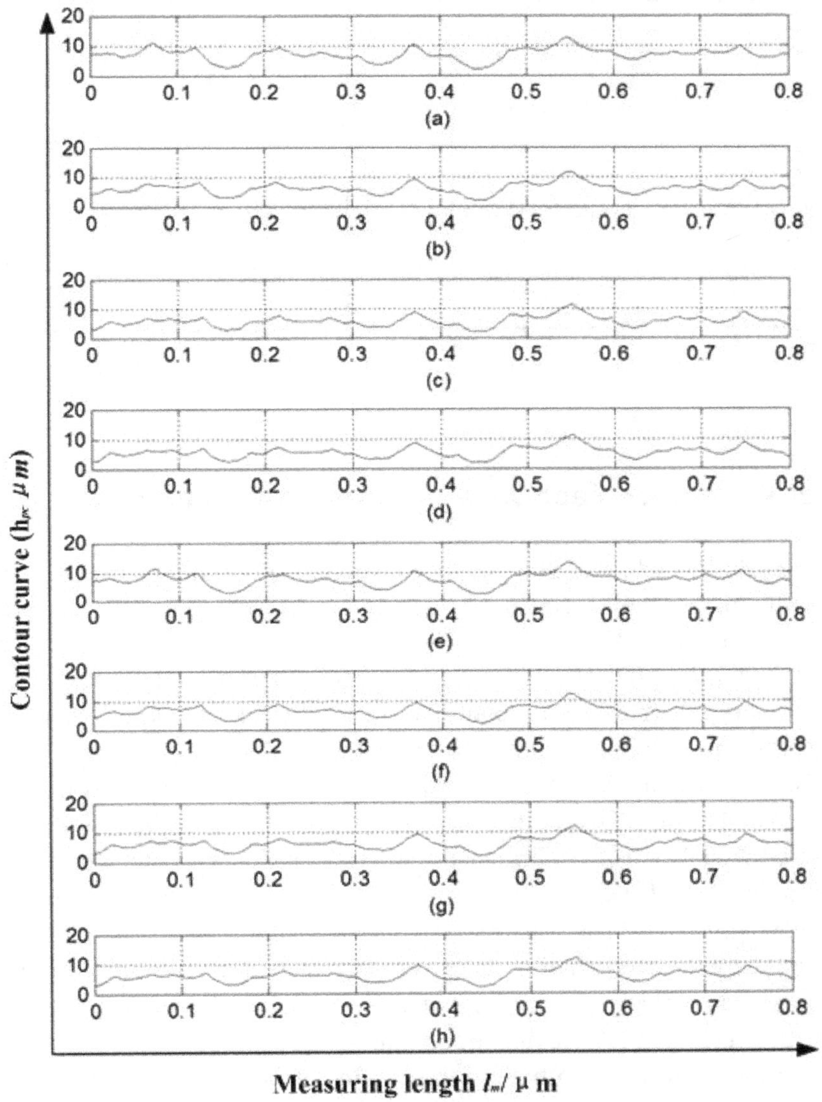

changes from 5 m/s to 60 m/s, R_a value changes from 2.05 μm to 1.31μm and R_z value changes from 10.62 μm to 9.55 μm; when grinding depth is 40 μm and linear velocity of grinding wheel changes from 5 m/s to 60 m/s, R_a value changes from 2.09 μm to 1.33 μm and R_z value from 10.79 μm to 9.71 μm. This is because elevating linear velocity of grinding wheel increases number of abrasive grains participating in the grinding process within unit time and decreases grinding force each abrasive grain bears, and the influence of elastic-plastic deformation on surface roughness is reduced; meanwhile, maximum undeformed chip thickness of single abrasive grain is reduced, so is the volume of materials removed by single abrasive grain, and then surface roughness values become small (Wang, Zhao, Zhao, & Lu, 2013).

Figure 15. The relationship between linear velocity of grinding wheel and surface roughness

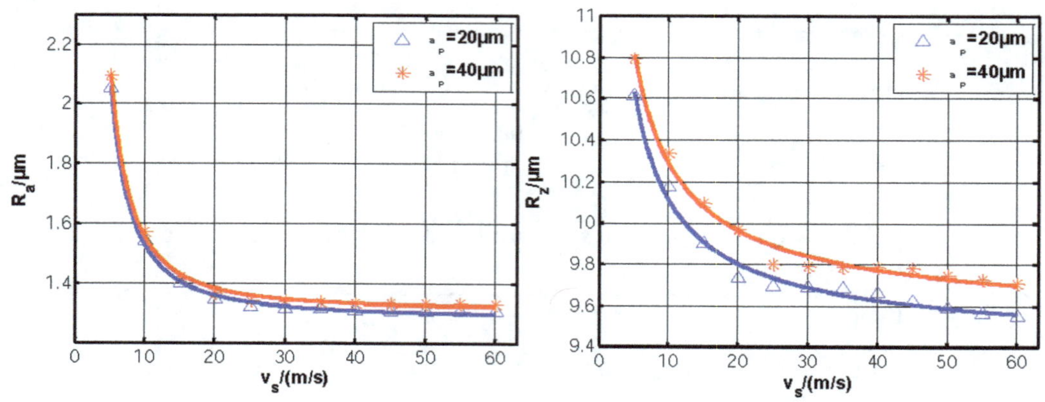

18.6.2 Effect of Workpiece Feed Rate on Surface Morphology

Under diameter d_s of grinding wheel of 300 mm, linear velocity of grinding wheel of 30 m/s and grinding depth of 20 μm and 40 μm respectively, influence laws of workpiece feed velocity on surface morphology are investigated by changing workpiece feed velocity, which is taken as 0.01 m/s, 0.02 m/s, 0.03 m/s, 0.04 m/s, 0.05 m/s, 0.06 m/s, 0.07 m/s, 0.08 m/s, 0.09 m/s and 0.1 m/s respectively.

Figure 16 shows workpiece surface morphologies under different workpiece feed velocities. (a)~(d) show workpiece surface morphologies under grinding depth of 20 μm and workpiece feed velocity of 0.03 m/s, 0.05 m/s, 0.07 m/s and 0.09 m/s respectively; (e)~(h) show workpiece surface morphologies under grinding depth of 40 μm and workpiece feed velocity of 0.03 m/s, 0.05 m/s, 0.07 m/s and 0.09 m/s respectively. It can be seen that: as the workpiece feed velocity increases, number of grinding cracks on workpiece surface reduces while length of each grinding crack increases, so does protruding height; this is because elevating workpiece feed velocity reduces number of abrasive grains contacting the workpiece within unit time and increases the grinding force borne by each abrasive grain, and then elastic-plastic deformation on workpiece surface increases and workpiece surface morphology becomes poor.

A section vertical to grinding texture direction in the surface morphologies in Figure 18-16 is taken, and the workpiece surface profile curve obtained through intersection between this section and workpiece surface is shown in Figure 17.

(a)~(d) in Figure 17 show workpiece surface profile curves under grinding depth of 20 μm and workpiece feed velocity of 0.03 m/s, 0.05 m/s, 0.07 m/s and 0.09 m/s respectively; (e) ~(h) show workpiece surface profile curves under grinding depth of 40 μm and workpiece feed velocity of 0.03 m/s, 0.05 m/s, 0.07 m/s and 0.09 m/s respectively. It can be seen that: as workpiece feed velocity increases, peak and valley on the workpiece surface profile curve are reduced in number, but peak height and valley depth increase; this is because elevating workpiece feed velocity reduces the number of abrasive grains participating in the cutting process within unit time and increases the grinding force, and generated grinding heat also increases and workpiece surface quality is degraded (Wang, 2013).

Figure 16. The workpiece surface morphologies under different workpiece feed velocities

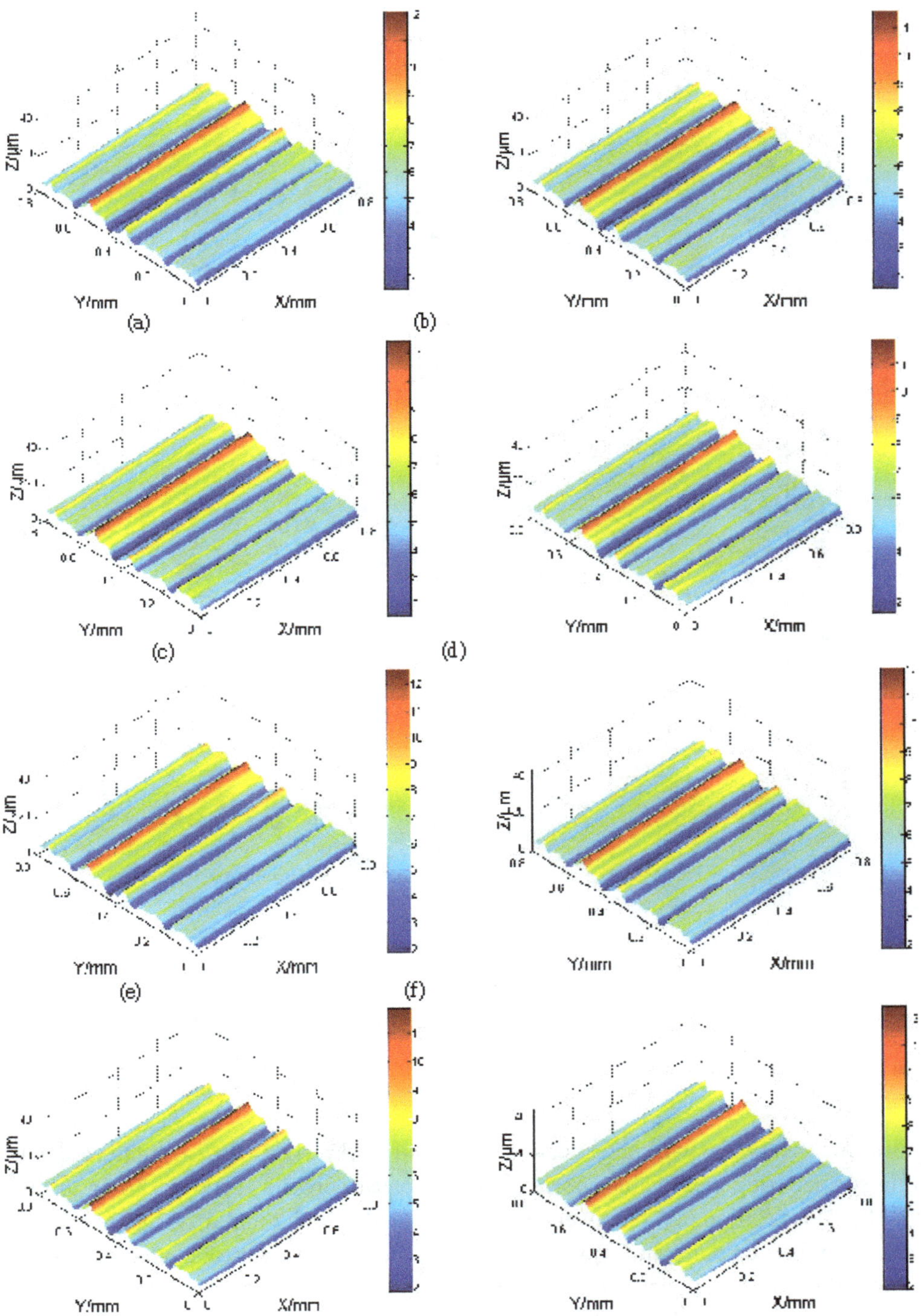

Figure 17. Surface profile curves under different workpiece feed velocities

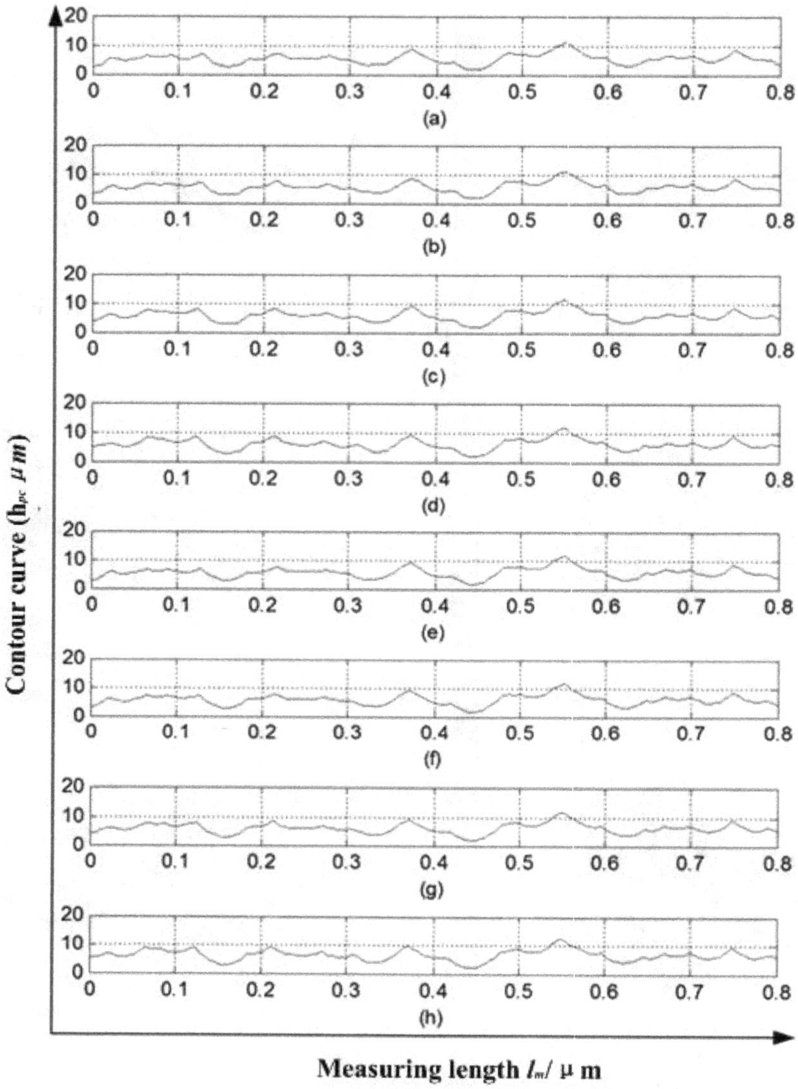

The relation curve between workpiece feed velocity and workpiece surface roughness can be obtained by calculating surface roughness values corresponding to each profile curve as shown in Figure 18-18. In Figure 18-18, (a) is relation curve between workpiece feed velocity and R_a value; (b) is relation curve between workpiece feed velocity and R_z value. It can be seen that as workpiece feed velocity increases, R_a value and R_z value on workpiece surface increase, and change gradient of curve also increases. When grinding depth is 20 μm and workpiece feed velocity changes from 0.01 m/s to 0.1 m/s, R_a value changes from 1.31 μm to 1.40 μm while R_z value changes from 9.47 μm to 9.69 μm; when grinding depth is 40 μm and workpiece feed velocity changes from 0.01 m/s to 0.1 m/s, R_a value changes from 1.33 μm to

Figure 18. The relationship between workpiece feed velocity and surface roughness

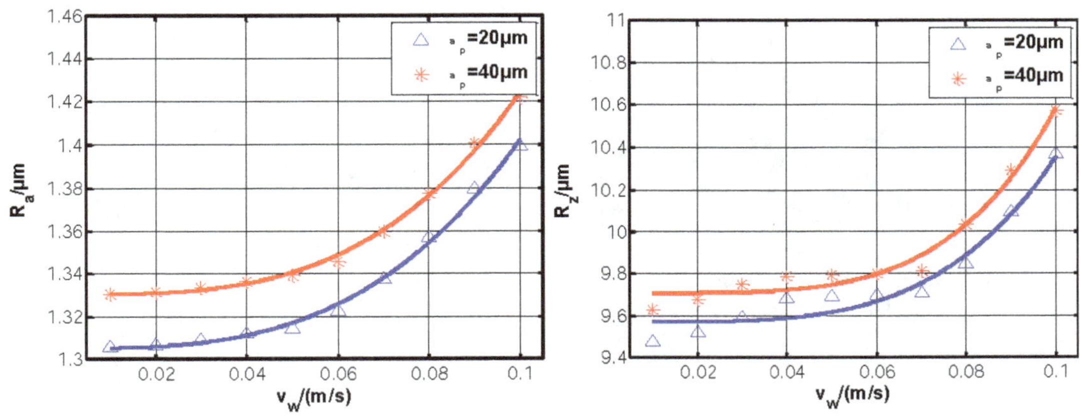

1.42 μm while R_z value changes from 9.63 μm to 10.57 μm. This is because elevating workpiece feed velocity reduces number of abrasive grains participating in the grinding process within unit time and increases grinding force borne by each abrasive grain, and the influence of elastic-plastic deformation on surface roughness is enhanced; meanwhile, maximum undeformed chip thickness of single abrasive grains increases, so does the volume of materials removed by single abrasive grain, so surface roughness values increase.

18.6.3 Effect of Grinding Depth on Surface Morphology

Under diameter d_s of grinding wheel of 300 mm, linear velocity 30 m/s of grinding wheel, and workpiece feed velocity v_w of 0.05 m/s and 0.1 m/s respectively, influence laws of grinding depth on surface morphology are studied by changing grinding depth. Grinding depth in the simulation is selected as 10 μm, 15 μm, 20 μm, 25 μm, 30 μm, 35 μm, 40 μm and 45 μm.

Figure 19 shows workpiece surface morphologies under different grinding depths. (a)~(d) show workpiece surface morphologies under workpiece feed velocity of 0.05 m/s and grinding depth of 10 μm, 20 μm, 30 μm and 40 μm respectively; (e)~(h) show workpiece surface morphologies under workpiece feed velocity of 0.1 m/s and grinding depth of 10 μm, 20 μm, 30 μm and 40 μm respectively.

According to Figure 19: as the grinding depth increases, number of grinding cracks on workpiece surface is reduced, length of each grinding crack increases, and protruding height also slightly increases, because enlarging grinding depth increases undeformed chip thickness of single abrasive grain and grinding force borne by each grain, and as a result, elastic-plastic deformation on workpiece surface increases and workpiece surface morphology becomes poor.

A section vertical to grinding texture direction in the surface morphologies given in Figure 19 is taken, and workpiece surface profile curve obtained through intersection between this section and workpiece surface is shown in Figure 20.

Figure 19. The workpiece surface morphologies under different grinding depths

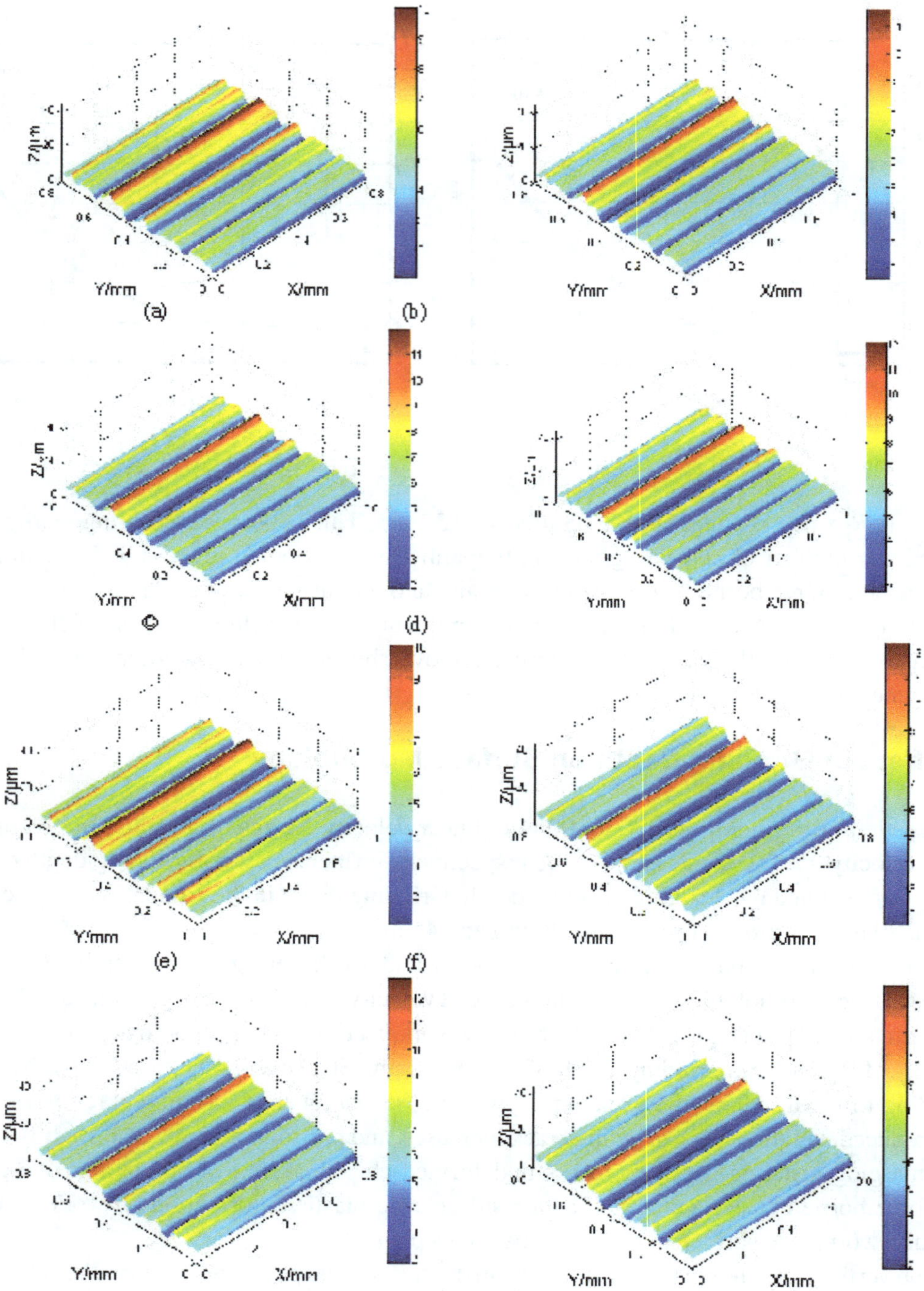

Figure 20. Surface profile curves under different grinding depths

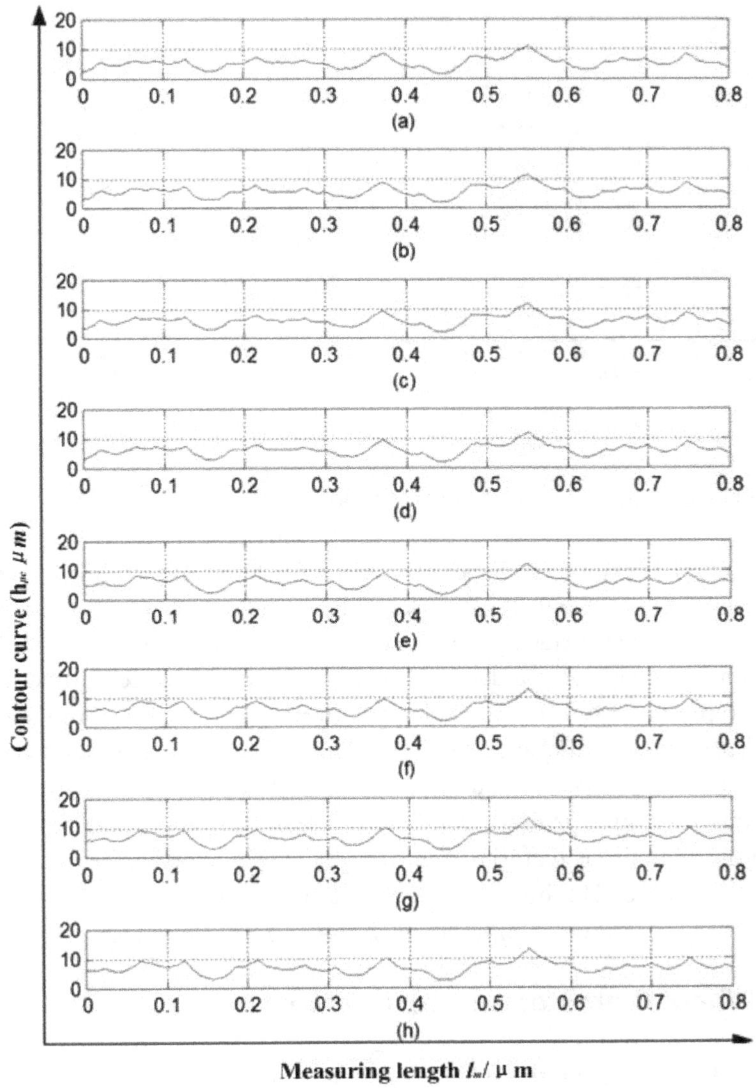

(a)~(d) in Figure 20 show workpiece surface profile curves under workpiece feed velocity of 0.05 m/s and grinding depth of 10 μm, 20 μm, 30 μm and 40 μm respectively; (e) ~(h) show workpiece surface profile curves under workpiece feed velocity of 0.1 m/s and grinding depth of 10 μm, 20 μm, 30 μm and 40 μm respectively. It can be seen that: as grinding depth increases, peak and valley on the workpiece surface profile curve both decrease to different degrees, and peak height and valley depth increase, because enlarging grinding depth increases undeformed chip thickness of single abrasive grain and the grinding force it bears, and moreover, the generated grinding heat also increases and workpiece surface quality becomes poor.

Figure 21. The relationship between grinding depth and surface roughness

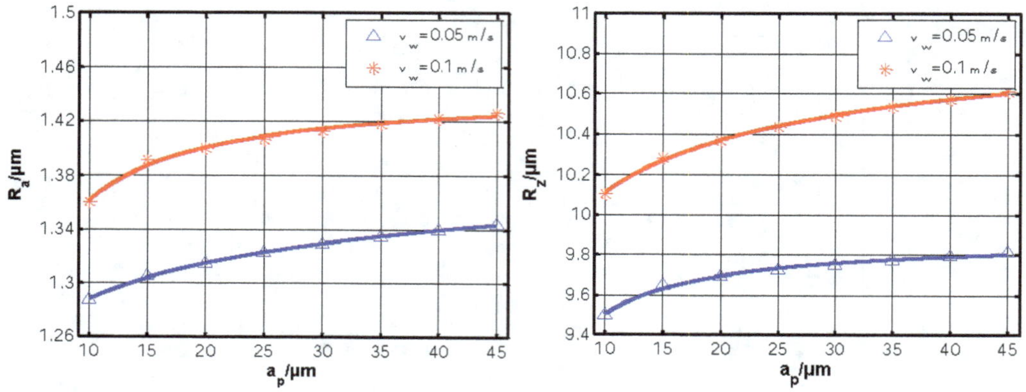

By calculating surface roughness values corresponding to each profile curve, the relation curves between grinding depth and workpiece surface roughness can be obtained as shown in Figure 21.

(a) in Figure 21 shows the relation curve between grinding depth and R_a value and (b) shows the relation curve between grinding depth and R_z value. It can be seen that: as the grinding depth increases, R_a value and R_z value on workpiece surface correspondingly increase but increase amplitude is not large. Meanwhile, curve change gradient also reduces. When workpiece feed velocity is 0.05m/s and grinding depth changes from 10 μm to 45 μm, R_a value changes from 1.28 μm to 1.34 μm and R_z value changes from 9.49 μm to 9.81 μm; when workpiece feed velocity is 0.1m/s and grinding depth changes from 10 μm to 45 μm, R_a value changes from 1.35 μmμm to 1.42 μm and R_z value changes from 10.10 μm to 10.61 μm. This is because enlarging grinding depth will increase maximum undeformed cutting thickness of single abrasive grain, volume of materials removed by single abrasive grains will correspondingly increase, so will grinding force. The influence of elastic-plastic deformation on workpiece surface on surface roughness is elevated in the grinding process, and surface roughness values increase.

18.6.4 Effect of Wheel Diameter on Surface Morphology

Under linear velocity 30 m/s of grinding wheel, workpiece feed velocity v_w of 0.05 m/s and grinding depth a_p of 20 μm and 40 μm respectively, influence laws of diameter of grinding wheel on surface morphology are studied by changing diameter of grinding wheel. Diameter d_s of grinding wheel used in the simulation is 50 mm, 100 mm, 150 mm, 200 mm, 250 mm, 300 mm, 350 mm and 400 mm respectively.

Figure 22 shows workpiece surface morphologies under different diameters of grinding wheel.

(a) ~(d) in Figure 22 are workpiece surface morphologies under grinding depth of 20 μm and diameters of grinding wheel of 50 mm, 150 mm, 250 mm and 350 mm respectively; (e)~(h) are workpiece surface morphologies under grinding depth of 40 μm and diameters of grinding wheel of 50 mm, 150 mm, 250 mm and 350 mm respectively. It can be seen that: as diameter of grinding wheel increases, number of grinding cracks on workpiece surface increases, surface texture tends to be orderly, and protruding height on the surface correspondingly reduces; this is because enlarging diameter of grinding wheel reduces

Figure 22. The workpiece surface morphologies under different diameters of grinding wheel

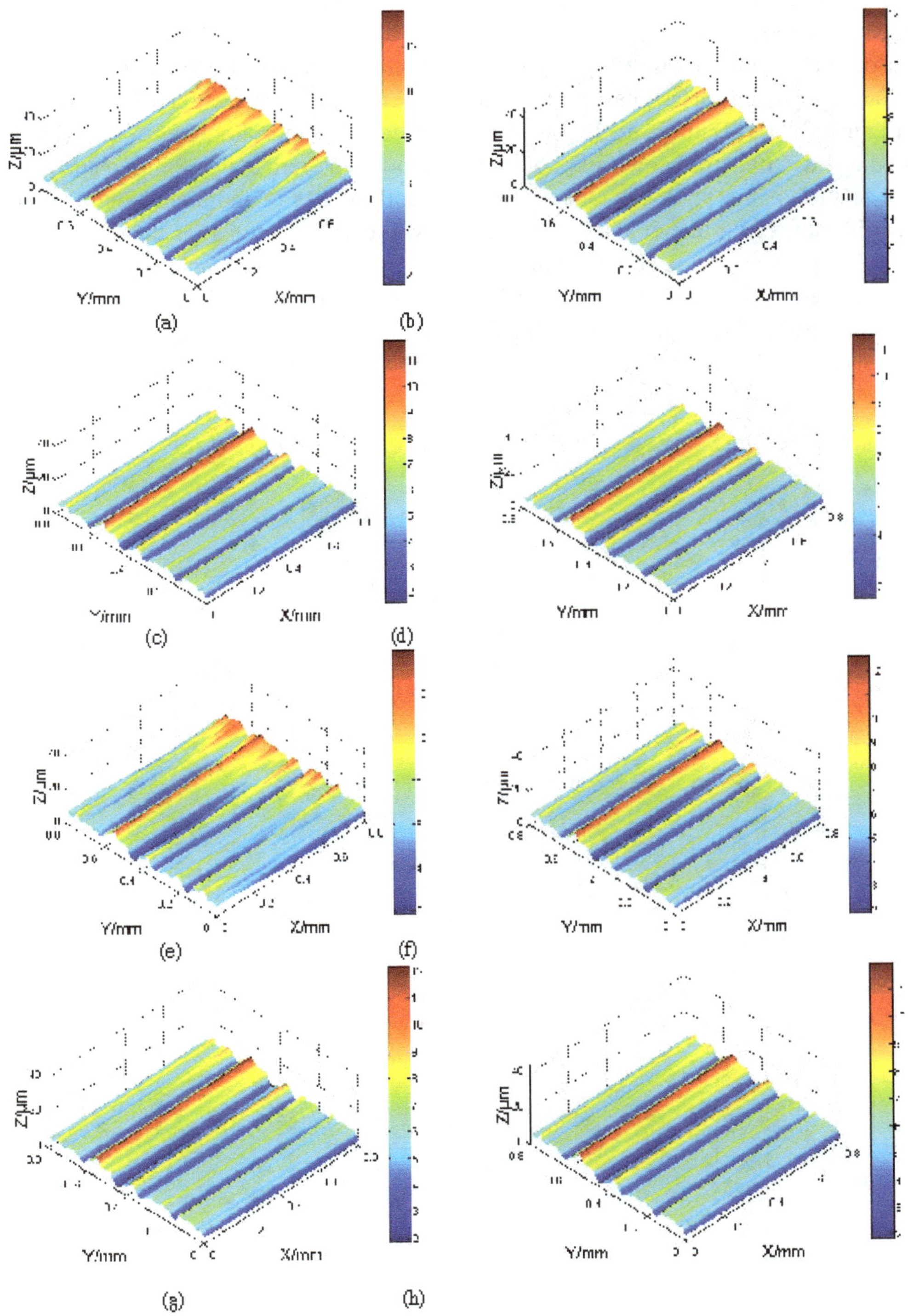

undeformed chip thickness of single abrasive grains and grinding force borne by each abrasive grain is also reduced so that elastic-plastic deformation on workpiece surface is reduced and workpiece surface morphology turns good.

Section of surface morphologies in Figure 22 is taken vertical to grinding texture direction, and workpiece surface profile curve obtained through intersection between section and workpiece surface is shown in Figure 23. (a)~(d) in Figure 23 shows workpiece surface morphologies under grinding depth of 20 μm and diameters of grinding wheel of 50 mm, 150 mm, 250 mm and 350 mm respectively; (e)~(h) are workpiece surface morphologies under grinding depth of 40 μm and diameters of grinding wheel of 50 mm, 150 mm, 250 mm and 350 mm respectively. It can be seen hat: as the diameter of grinding wheel increases, peak and valley on the workpiece surface profile curve increase, and peak height and valley depth are reduced; this is because enlarging diameter of grinding wheel reduces undeformed chip thickness of single abrasive grain and grinding force so that generated grinding heat is also reduced and workpiece surface quality is improved (Wang, Li, Jia, & Hou, 2014).

Figure 23. Surface profile curves under different diameters of grinding wheel

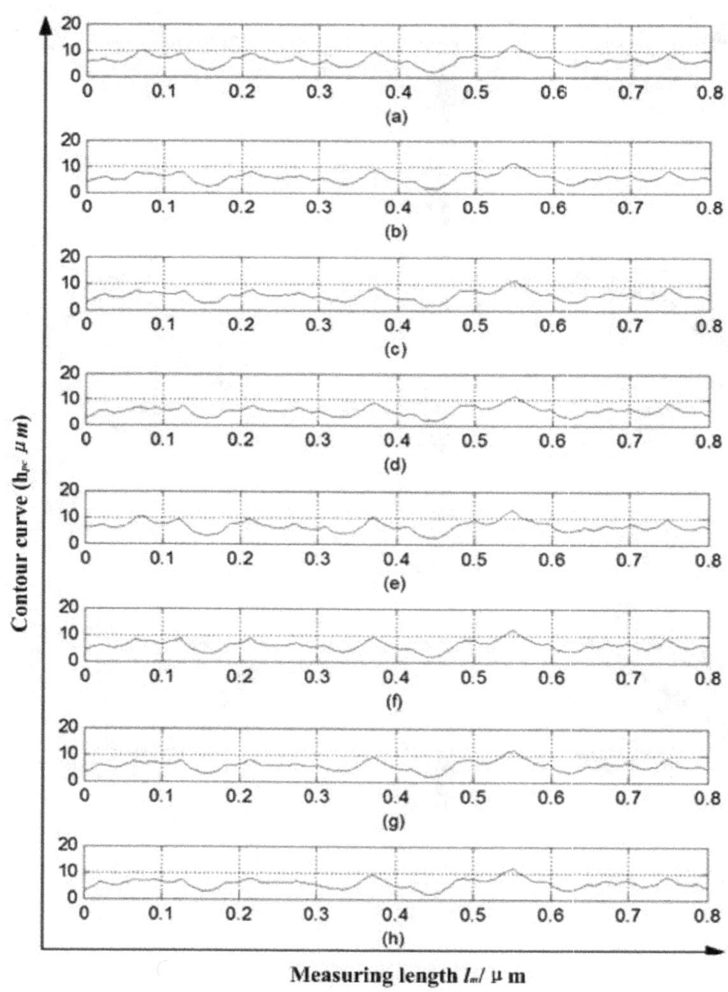

The relation curve between diameter of grinding wheel and workpiece surface roughness can be obtained by calculating surface roughness values corresponding to each profile curve as shown in Figure 24. In Figure 24, (a) is relation curve between diameter of grinding wheel and R_a value; (b) is relation curve between diameter of grinding wheel and R_z value. It can be seen that as the diameter of grinding wheel increases, R_a value and R_z value on workpiece surface reduce, but the reduction amplitude is small. Meanwhile, curve change gradient correspondingly reduces. When grinding depth is 20 μm and diameter of grinding wheel changes from 50mm to 40mm, R_a value changes from 1.46 μm to 1.31 μm while R_z value changes from 10.36 μm to 9.69 μm; when grinding depth is 40 μm and diameter of grinding wheel changes from 50mm to 400mm, R_a value changes from 1.48 μm to 1.34 μm while R_z value changes from 10.53 μm to 9.79 μm. This is because enlarging diameter of grinding wheel adds to the number of abrasive grains contacting workpiece surface within unit time, grinding force borne by each abrasive grain is reduced, and the influence of elastic-plastic deformation on surface roughness is reduced; meanwhile, maximum undeformed chip thickness of single abrasive grains is reduced, so is the volume of materials removed by single abrasive grain, so surface roughness values become small.

Mathematical model of ordinary grinding wheel is built using abrasive grain vibration method, programming is conducted using MATLAB language, and morphology simulation of ordinary grinding wheel is completed through program operation; abrasive grains of grinding wheel generated through simulation present random distribution in 3D direction of grinding wheel, that abrasive grain sizes follow normal distribution is verified by calculating abrasive grain sizes on the grinding wheel surface, and this is identical with theoretical analysis results (Wang, Li, Zhang, Jia, Zhang, & Zhang, 2014).

Plane grinding process simulation of quenched 45# steel according to grinding wheel morphology generated through simulation and set grinding parameters, and simulation results show that:

1. Number of grinding cracks on workpiece surface increases as the linear velocity of grinding wheel rises, surface texture tends to be orderly and protruding heights on the surface reduce; peak and valley on the workpiece surface profile curve increase, while peak height and valley depth are reduced; R_a value and R_z value on workpiece surface are reduced with large reduction amplitude.

Figure 24. The relationship between grinding wheel diameter and surface roughness

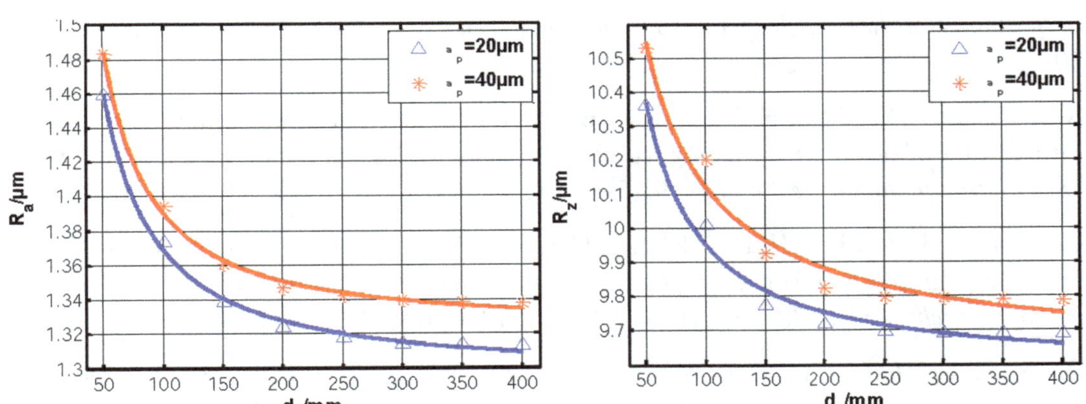

2. As the workpiece feed velocity is elevated, number of grinding cracks on workpiece surface is reduced, length of each grinding crack increases, so does protruding height on surface; peak and valley on the workpiece surface profile curve are reduced, and peak height and valley depth increase; R_a value and R_z value on workpiece surface increase.
3. As the grinding depth increases, number of grinding cracks on workpiece surface increases, length of each grinding crack increases, so does protruding height on surface; peak and valley on the workpiece surface profile curve are reduced to different degrees, and peak height and valley depth increase; R_a value and R_z value on workpiece surface increase, but increase amplitude is not large.
4. As the diameter of grinding wheel increases, number of grinding cracks on workpiece surface increases, surface texture tends to be orderly, and protruding height on surface slightly declines; peak and valley on the workpiece surface profile curve increase, and peak height and valley depth decrease; R_a value and R_z value on workpiece surface decrease, but decrease amplitude is small.
5. Linear velocity of grinding wheel has the greatest influence on workpiece surface morphology, followed by workpiece feed velocity, grinding depth and diameter of grinding wheel successively.

REFERENCES

Ali, Y. M., & Zhang, L. C. (1999). Surface roughness prediction of ground components using a fuzzy logic approach. *Journal of Materials Processing Technology, 89-90*, 561-568.

Arizmendi, M., Campa, F. J., Fernandez, J., de Lacalle, L. N. L., Gil, A., Bilbao, E., ... Lamikiz, A. (2009). Model for surface topography prediction in peripheral milling considering tool vibration. *Cirp Annals-Manufacturing Technology*, *58*(1), 93–96. doi:10.1016/j.cirp.2009.03.084

Biondani, F. G., & Bissacco, G. (2019). Effect of cutting edge micro geometry on surface generation in ball end milling. *Cirp Annals-Manufacturing Technology*, *68*(1), 571–574. doi:10.1016/j.cirp.2019.04.017

Cao, Y. L., Guan, J. Y., Li, B., Chen, X. L., Yang, J. X., & Gan, C. B. (2013). Modeling and simulation of grinding surface topography considering wheel vibration. *International Journal of Advanced Manufacturing Technology*, *66*(5-8), 937–945. doi:10.100700170-012-4378-7

Chakrabarti, S., & Paul, S. (2008). Numerical modelling of surface topography in superabrasive grinding. *International Journal of Advanced Manufacturing Technology*, *39*(1-2), 29–38. doi:10.100700170-007-1201-y

Chen, C. S., Tang, J. Y., Chen, H. F., & Zhao, B. (2018). An active manufacturing method of surface micro structure based on ordered grinding wheel and ultrasonic-assisted grinding. *International Journal of Advanced Manufacturing Technology*, *97*(5-8), 1627–1635. doi:10.100700170-018-2044-4

Chen, C. S., Tang, J. Y., Chen, H. F., & Zhu, C. C. (2017). Research about modeling of grinding workpiece surface topography based on real topography of grinding wheel. *International Journal of Advanced Manufacturing Technology*, *93*(5-8), 2411–2421. doi:10.100700170-017-0668-4

Chen, H., Yu, T. B., Dong, J. L., Zhao, Y., & Zhao, J. (2019). Kinematic simulation of surface grinding process with random cBN grain model. *International Journal of Advanced Manufacturing Technology*, *100*(9-12), 2725–2739. doi:10.100700170-018-2840-x

Chen, H. F., Tang, J. Y., Zhou, W. H., & Chen, C. S. (2017). The equal theoretical surface roughness grinding method for gear generating grinding. *International Journal of Advanced Manufacturing Technology*, *90*(9-12), 3137–3146. doi:10.100700170-016-9651-8

Chen, H. F., Tang, J. Y., & Zhu, C. C. (2018). A new approach to modeling the surface topography in grinding considering ploughing action. *Machining Science and Technology*, *22*(4), 604–620. doi:10.1080/10910344.2017.1382514

Chen, S. S., Cheung, C. F., Zhang, F. H., & Zhao, C. Y. (2018). Three-dimensional modelling and simulation of vibration marks on surface generation in ultra-precision grinding. *Precision Engineering-Journal of the International Societies for Precision Engineering and Nanotechnology*, *53*, 221–235. doi:10.1016/j.precisioneng.2018.04.006

Chen Dongxiang, T. Y. (2010). Modeling and Simulation Methodology of the Machined Surface in Ultra-precision Grinding. *Journal Of Mechanical Engineering*, *46*(13), 186–191. doi:10.3901/JME.2010.13.186

Cheng, J., Gong, Y. D., & Wang, J. S. (2013). Modeling and Evaluating of Surface Roughness Prediction in Micro-grinding on Soda-lime Glass Considering Tool Characterization. *Chinese Journal of Mechanical Engineering*, *26*(6), 1091–1100. doi:10.3901/CJME.2013.06.1091

Chi, J., Guo, J. L., & Chen, L. Q. (2016). The study on a simulation model of workpiece surface topography in external cylindrical grinding. *International Journal of Advanced Manufacturing Technology*, *82*(5-8), 939–950. doi:10.100700170-015-7406-6

Cui, Q., Cheng, K., Chen, S. J., & Ding, H. (2017). An innovative investigation on the workpiece kinematics and its roundness generation in through-feed centreless grinding. *Proceedings of the Institution of Mechanical Engineers. Part B, Journal of Engineering Manufacture*, *231*(7), 1131–1143. doi:10.1177/0954405415585378

Deng, Zhang, & Zhang. (2010). Recent Advances and Future Perspectives in New Type Grinding Wheels. *Zhongguo Jixie Gongcheng*, *21*(21), 2632–2638.

Ding, W. F., Dai, C. W., Yu, T. Y., Xu, J. H., & Fu, Y. C. (2017). Grinding performance of textured monolayer CBN wheels: Undeformed chip thickness nonuniformity modeling and ground surface topography prediction. *International Journal of Machine Tools & Manufacture*, *122*, 52–66. doi:10.1016/j.ijmachtools.2017.05.006

Doman, D. A., Warkentin, A., & Bauer, R. (2006). A survey of recent grinding wheel topography models. *International Journal of Machine Tools & Manufacture*, *46*(3-4), 343–352. doi:10.1016/j.ijmachtools.2005.05.013

Du Haijun, R. Y. (2009). MATLAB-based simulation method of surface's three-dimensional model in grinding process. *Modern Manufacturing Engineering*, *2*, 48–51.

Gong, Y. D., Wang, B., & Wang, W. S. (2002). The simulation of grinding wheels and ground surface roughness based on virtual reality technology. *Journal of Materials Processing Technology*, *129*(1-3), 123–126. doi:10.1016/S0924-0136(02)00589-7

Li, B. Z. B. (2003). *Modern Grinding Technology*. China Machine Press.

Li, D., Qiao, Z., Walton, K., Liu, Y. T., Xue, J. D., Wang, B., & Jiang, X. Q. (2018). Theoretical and Experimental Investigation of Surface Topography Generation in Slow Tool Servo Ultra- Precision Machining of Freeform Surfaces. *Materials (Basel)*, *11*(12), 2566. doi:10.3390/ma11122566 PMID:30562973

Li, X., & Rong, Y. (2011). Framework of grinding process modeling and simulation based on microscopic interaction analysis. *Robotics and Computer-integrated Manufacturing*, *27*(2), 471–478. doi:10.1016/j.rcim.2010.06.029

Liao, D. R., Shao, W., Tang, J. Y., Li, J. P., & Tao, X. (2018). Numerical generation of grinding wheel surfaces based on time series method. *International Journal of Advanced Manufacturing Technology*, *94*(1-4), 561–569. doi:10.100700170-017-0868-y

Liu, W., Deng, Z. H., Shang, Y. Y., & Wan, L. L. (2019). Parametric evaluation and three-dimensional modelling for surface topography of grinding wheel. *International Journal of Mechanical Sciences*, *155*, 334–342. doi:10.1016/j.ijmecsci.2019.03.006

Liu, Y., Warkentin, A., Bauer, R., & Gong, Y. (2013). Investigation of different grain shapes and dressing to predict surface roughness in grinding using kinematic simulations. *Precision Engineering*, *37*(3), 758–764. doi:10.1016/j.precisioneng.2013.02.009

Wang, L. G. (2002). Modelling and Computer Simulation for Grinding Process. *Zhongguo Jixie Gongcheng*, *13*(1), 1–4.

Ren, J. H. D. (2011). Grinding Theory. *Electronic Industry Press.*, *1*, 376–377.

Sakakura, M., Tsukamoto, S., Fujiwara, T., & Inasaki, I. (2008). Visual simulation of the grinding process. *Proceedings of the Institution of Mechanical Engineers. Part B, Journal of Engineering Manufacture*, *222*(10), 1233–1239. doi:10.1243/09544054JEM1032

Salisbury, E. J., Domala, K. V., Moon, K. S., Miller, M. H., & Sutherland, J. W. (2001a). A three-dimensional model for the surface texture in surface grinding, part 1: Surface generation model. *Journal of Manufacturing Science and Engineering*, *123*(4), 576–581. doi:10.1115/1.1391427

Salisbury, E. J., Domala, K. V., Moon, K. S., Miller, M. H., & Sutherland, J. W. (2001b). A three-dimensional model for the surface texture in surface grinding, part 2: Grinding wheel surface texture model. *Journal of Manufacturing Science and Engineering*, *123*(4), 582–590. doi:10.1115/1.1391428

Su, C., Hou, J. M., & Wang, W. S. (2008). Research and development of simulation system for virtual grinding. *2008 International Symposium on Computer Science and Computational Technology*, *1*. 409-412. 10.1109/ISCSCT.2008.264

Wang, J., Zhao, L., Zhao, C., & Lu, Y. (2013). Research on Mill-grinding Experiment of Grinding Wheel with Phyllotactic Pattern. *Tool Engineering*, *47*(3), 26–29.

Wang, S., Li, C., Jia, D., & Hou, Y. (2014). Investigation on the grinding temperature field of nano-ZrO_2 dental ceramics with a nanoparticle jet of MQL. *International Journal of Computational Materials Science and Surface Engineering*, *6*(1), 13–28. doi:10.1504/IJCMSSE.2014.063757

Wang, S., Li, C., Jia, D., Zhang, Y., & Zhang, Q. (2015). Modelling and simulation of the surface topography generation with engineered grinding wheel. *International Journal of Computational Materials Science and Surface Engineering*, *6*(2), 111–129. doi:10.1504/IJCMSSE.2015.071647

Wang, S., Li, C., Zhang, D., Jia, D., & Zhang, Y. (2014). Modeling the operation of a common grinding wheel with nanoparticle jet flow minimal quantity lubrication. *International Journal of Advanced Manufacturing Technology*, *74*(5-8), 835–850. doi:10.100700170-014-6032-z

Wang, S., Li, C., Zhang, X., Jia, D., Zhang, D., & Zhang, Q. (2014). Modeling and simulation of the single grain grinding process of the nano-particle jet flow of minimal quantity lubrication. *Open Mater. Sci. J*, *8*(1), 55–62. doi:10.2174/1874088X01408010055

Wang, S. Z. Q., & Li, C. (2013). Finite element modeling approaches in grinding. *Precise Manufacturing & Automation*, *1*, 40–45.

Wang, X. Z., Yu, T. B., Dai, Y. X., Shi, Y., & Wang, W. S. (2016). Kinematics modeling and simulating of grinding surface topography considering machining parameters and vibration characteristics. *International Journal of Advanced Manufacturing Technology*, *87*(9-12), 2459–2470. doi:10.100700170-016-8660-y

Wang, Y., Guangheng, D. Y., Zhao, J. N., Dong, Y. H., Zhang, X. F., Jiang, X. M., & Lin, B. (2019). Study on key factors influencing the surface generation in rotary ultrasonic grinding for hard and brittle materials. *Journal of Manufacturing Processes*, *38*, 549–555. doi:10.1016/j.jmapro.2019.01.046

Wenfeng, D., Jiuhua, X., Zhenzhen, C., Honghua, S., & Yucan, F. (2010). Grindability and surface integrity of cast nickel-based superalloy in creep feed grinding with brazed CBN abrasive wheels. *Chinese Journal of Aeronautics*, *23*(4), 501–510. doi:10.1016/S1000-9361(09)60247-8

Yang, Z., Zhang, Z., Yang, R., & Liu, A. (2016). Study on the grain damage characteristics of brazed diamond grinding wheel using a laser in face grinding. *International Journal of Advanced Manufacturing Technology*, *87*(1-4), 853–858. doi:10.100700170-016-8454-2

Zhang, C., Guo, S., Zhang, H. Y., & Zhou, L. S. (2013). Modeling and predicting for surface topography considering tool wear in milling process. *International Journal of Advanced Manufacturing Technology*, *68*(9-12), 2849–2860. doi:10.100700170-013-4989-7

Zhang, C., Shi, G. L., Ehmann, K. F., & Li, Y. G. (2016). Modeling and simulation of micro-groove topography on cylindrical surface by elliptical vibration-assisted turning. *International Journal of Advanced Manufacturing Technology*, *86*(5-8), 1407–1424. doi:10.100700170-015-8250-4

Zhang, Y., Li, C., Yang, M., Jia, D., Zhang, D., & Zhang, X. (2015). Modeling and Simulation of Nanofluid Minimum Quantity Lubrication Surface Grinding Thermal Stress. *The Open Mechanical Engineering Journal, 9*(1).

Zhenxuan, L. (2012). Analysis of Numerical Grinding Wheel Topology and Experimental Measurement. *Journal Of Mechanical Engineering*, *48*(23), 184–190. doi:10.3901/JME.2012.23.184

Zhou, W. H., Tang, J. Y., Chen, H. F., & Shao, W. (2019). A comprehensive investigation of surface generation and material removal characteristics in ultrasonic vibration assisted grinding. *International Journal of Mechanical Sciences*, *156*, 14–30. doi:10.1016/j.ijmecsci.2019.03.026

Zhou, X., & Xi, F. (2002). Modeling and predicting surface roughness of the grinding process. *International Journal of Machine Tools & Manufacture*, *42*(8), 969–977. doi:10.1016/S0890-6955(02)00011-1

Zou, L., Liu, X., Huang, Y., & Fei, Y. (2019). A numerical approach to predict the machined surface topography of abrasive belt flexible grinding. *International Journal of Advanced Manufacturing Technology*, *104*(5-8), 2961–2970. doi:10.100700170-019-04032-2

Index

P

S

T

V

W

Z

www.ingramcontent.com/pod-product-compliance
Lightning Source LLC
LaVergne TN
LVHW080533300926
849162LV00007B/1599